Practical Handbook of
Ground-Water Monitoring

Edited by

David M. Nielsen

LEWIS PUBLISHERS

Library of Congress Cataloging-in-Publication Data

Practical handbook of ground-water monitoring/David M. Nielsen, editor.

   p. cm.
Includes bibliographical references and index.
1. Water, Underground—Quality—Measurement. I. Nielsen, David.

TD426P73   1991   628.1'61—dc20   90-36848
ISBN 0-87371-124-6

LEWIS PUBLISHERS, INC.
121 South Main Street, Chelsea, Michigan 48118

PRINTED IN THE UNITED STATES OF AMERICA

3 4 5 6 7 8 9 10

PREFACE

Ground-water monitoring is by no means a new field because hydrogeologists, engineers, and other professionals have been monitoring ground water for decades. Yet, much of the technology used today to perform such basic tasks as water-level measurement, water-quality characterization, and aquifer tests is less than a decade old. Indeed, the decade of the 1980s was a period of explosive growth for the field of ground-water monitoring, and a time of great achievement for those involved in conducting ground-water investigations. During this time, we developed a basic understanding of how our ground-water resources were being impacted by many of our everyday activities, what we needed to do to prevent further degradation, and how we could clean up ground water that had already been detrimentally impacted.

Passage of the Resource Conservation and Recovery Act (RCRA) by Congress in 1976 and subsequent promulgation of the first of the regulations authorized under RCRA by the U.S. Environmental Protection Agency (EPA) in May 1980 provided the primary impetus for the growth of the field of ground-water monitoring. RCRA, which is EPA's main tool for managing hazardous waste from generation through disposal, included provisions for establishing ground-water or vadose-zone monitoring systems at all of this country's hazardous waste treatment, storage, and disposal facilities, which include hundreds of thousands of sites. More recent provisions of RCRA specify similar monitoring systems for each of the country's solid waste landfill facilities (sanitary landfills), which number in the thousands. Still other provisions of recent amendments to RCRA (the Hazardous and Solid Waste Amendments of 1986) call for the installation of ground-water or vadose-zone monitoring systems at many underground storage tank locations, which number in the hundreds of thousands across the country.

Passage of the Comprehensive Environmental Response, Compensation and Liability Act (CERCLA), better known as "Superfund," by Congress in December 1980 addressed the national threat caused by so-called "uncontrolled" hazardous waste sites, which probably number in the tens of thousands. Cleanup of these sites requires the installation of monitoring devices to investigate the extent of environmental contamination and to monitor the progress of the cleanup. Ground-water monitoring is also done under other environmental regulatory programs, and for a variety of nonregulatory purposes, creating a tremendous demand for knowledge in this field.

Like most fields that experience such tremendous surges in growth, the ground-water monitoring field, if it can truly be called that, has seen periods of disorder and disorganization. In the early 1980s persons involved in ground-water monitoring were cautioned that they had to learn from the mistakes that scientists conducting surface-water monitoring programs had made in the 1970s. However, the cautions went largely unheeded. Many ground-water quality monitoring investigations were conducted strictly to meet the letter of the law, and many poor-quality data were produced. No real procedural guidelines or standards were available for those conducting ground-water monitoring investigations to follow. At the same time, the technology for monitoring ground water was evolving at such a rapid rate that it was difficult for practitioners to keep up. Clearly there was a need to step back and take a long, hard look at the direction in which the field was headed.

Many questions had to be answered. Could we effectively address the hundreds of thousands of sites that now fell under govenment regulation? Were enough trained and experienced specialists available to do the work and to evaluate the work that would be done? Could we do the work that we were being asked to do with existing methods and technologies? Could we do the work without the benefit of guidance or standards to follow?

Not much time passed before we learned that the answer to each of these questions was a resounding "NO." The enormous number of regulated sites swamped the state and federal regulatory agencies, not nearly enough environmentally oriented consulting firms existed to handle the workload created, the nation's colleges and universities could not supply enough properly trained professionals to keep up with the demand, many "classic" field methods and much of the equipment used in ground-water investigations required modification or complete redesign, and confusion reigned when we found that our usual practice of "winging it" without following prescribed practices no longer provided the results we needed to stand up to the scrutiny of a courtroom battle.

But we are catching up. Although we still can't answer a single of the aforementioned questions with a strong affirmative, we are beginning to get a handle on things. We have identified most of the real "problem" sites and have begun to prioritize them for appropriate action. We have developed college curricula and continuing education programs to educate future ground-water scientists and engineers. We have developed new technologies, equipment and methods to monitor and clean up ground-water contamination, and we have begun the laborious process of developing appropriate and relevant standards and guidance to follow in conducting ground-water investigations.

With the decade of the 1980s behind us, we can safely say that we learned a great deal in the space of 10 short years—perhaps more in this single decade than in all of the time that has passed since Henry Darcy conducted his sand-tank studies more than a century ago. Much of what we have learned about ground-water monitoring in that short period of time is included in this book. Much more remains to be learned.

The decade of the 1990s promises to be an exciting time for ground-water professionals, for in developing the answers to the questions raised in the preceding decade, we generated an entirely new set of questions. Will field practices be able to keep pace with advances in laboratory detection capabilities? Will remote sensing technology replace direct observation? Will regulators, consultants and industry professionals be able to agree on a significant number of important, yet unresolved, technical issues? The real question is: Are we capable of meeting the challenge of answering these and other new questions? To be sure, this will mean conducting additional research, refining existing methodologies, and developing new and innovative technology.

We have the momentum of the 1980s and a generation of experienced professionals in our favor.

David M. Nielsen is a hydrogeologist and President of Nielsen Ground-Water Science, Inc., a geoscientific consulting firm specializing in ground-water and environmental training. He holds BA and MS degrees in geology from Miami University and Bowling Green University, respectively. In his 13 years of professional experience, he has managed regional offices for two geoscientific and engineering consulting firms, and served as Director of Research and Education for the National Water Well Association (NWWA) and Geologist for both the Ohio Department of Natural Resources Division of Water, and the West Virginia Geological Survey.

Nielsen is Editor of *Ground Water Monitoring Review,* Chairman of the American Society for Testing and Materials (ASTM) Subcommittee D-18.21 on Ground Water and Vadose Zone Investigations, and a member of the Board of Counselors of the Wright State University Department of Geological Sciences. He is a Certified Professional Geologist with the American Institute of Professional Geologists, a Certified Ground Water Professional with the Association of Ground Water Scientists and Engineers, and holds licenses or registrations as a professional geologist in Indiana, Delaware, South Carolina, Florida, Arkansas, and Alaska. Nielsen has implemented and managed ground-water contamination investigations, environmental site assessments, ground-water monitoring and sampling programs, petroleum hydrocarbon spill investigations and abatement projects and remedial action programs at hazardous and nonhazardous sites across the United States. He has published papers on a variety of subjects related to ground water, and he teaches short courses on ground water and vadose zone monitoring and sampling for NWWA, ASTM, and other organizations.

EDITOR'S NOTE

When I first accepted the challenge of assembling a book on ground-water monitoring, I was certain that it was a worthy but hopeless task. After all, four years ago the field was in a state of flux—regulations were constantly changing, field practices were evolving, analytical methods were being developed to detect lower and lower levels of chemical constituents, equipment technology was advancing at light speed, and the effort to develop standards to apply to ground-water and vadose zone investigations was just beginning. Still, nobody had attempted to establish the current state-of-the-art/science—there was no real "ground zero" to serve as a yardstick to measure progress so that the field could advance in an orderly fashion. Thus the decision to proceed with this book.

Early on, I recognized that, given the broad range of disciplines and subjects that comprise the "field" of ground-water monitoring, it would be impossible for one person or even half a dozen people to write this book. To cover the subjects adequately would require a group of writers with a mélange of backgrounds and experiences. So I recruited (some might say coerced) a number of the most talented professionals practicing ground-water monitoring today to provide the multidisciplinary approach necessary to do justice to the subject. Included herein are contributions from representatives of industry, environmental and engineering consulting firms, state regulatory and nonregulatory agencies, research organizations, and academia with backgrounds in geology, geophysics, geochemistry, hydrogeology, hydrology, drilling technology, engineering, soil science, analytical chemistry, statistics, industrial hygiene and other fields. I owe a tremendous debt of gratitude to each one of the contributing authors and to the many other people involved in the production of this book.

As you can imagine, organizing and managing the effort required to produce this book was not accomplished without some difficulty. After all, each of the contributing authors (and the editor) is gainfully employed and few are in a position to devote large blocks of time to writing. Many evening hours and weekends were spent assembling material contained in this book. I strongly believe that the effort, which spanned nearly four years, was well worth the occasional headaches, eye strain and writer's cramp. I am confident that you will find the book a worthwhile investment and a valuable reference.

Is this the definitive work on ground-water monitoring? Well, yes and no. For the present, it may be. In the coming years, regulations will change, technology will advance, standards will be written for many of the practices and procedures described herein, some of the contents of the book will become outdated and some of it will have to be rewritten. But that happens in every rapidly evolving field. That's what second (and third? and fourth?) editions are for.

David M. Nielsen
Galena, Ohio

CONTENTS

1

Regulatory Mandates for Controls on Ground-Water Monitoring

Kathryn S. Makeig

THE NEED FOR GROUND-WATER MONITORING: PROTECTION OF A RESOURCE AT RISK

The need for the regulation of activities that pose a threat to the quality of ground water has become apparent over the years. This once pristine, widely available resource is in a delicate balance between supply and demand. The quantity of useable ground water is closely linked to the quality of the water that is available for various uses, but the apparent ignorance of humans to the finite nature of this resource has led to its exploitation and abuse, particularly as a dumping ground for unwanted waste materials. Since the mid-1970s, there have been increased efforts to protect this resource from further degradation and there are now regulatory mandates in place to both protect useable ground water and to clean up ground water that has suffered the effects of short-sighted waste management practices.

Approximately 50% of U.S. households use ground water as a primary source of drinking water. Yet the demands on this resource are varied and numerous, as industry and agriculture also have many uses for ground water. Along with numerous uses come limitations and constraints placed on the appropriation and quality of this resource. Both federal and state legislators have attempted to address the evolving requirements for ground water that, in some cases, must be clean enough to drink and, in other cases, must only be relatively free of chemicals that could affect the performance of an industrial process. Legislation also has addressed the problem of the potential for ground-water contamination from the use of ground water both as a resource and as a means of disposal. Finally, with the advent of "Superfund," legislation has been passed to clean up ground water that has already been contaminated.

This chapter discusses the role of ground-water monitoring within the framework of existing environmental regulations, focusing on the protection of the resource from overdevelopment and contamination. It places the discussion in the context of the level of protection that is required for various uses.

FEDERAL REGULATORY MANDATES FOR GROUND-WATER MONITORING

There exist a variety of federal agencies whose purposes include the protection of ground water. Among them are the Nuclear Regulatory Commission, the Office of Surface Mining, and the U.S. Environmental Protection Agency. By far the largest body of environmental regulations requiring ground-water monitoring has been promulgated by the U.S. Environmental Protection Agency (EPA). Copies of the regulations are readily available from a number of sources, but the primary source is the Government Printing Office. The primary emphasis of the ensuing discussion focuses on the mandates for ground-water monitoring.

The federal regulatory programs that involve the implementation of ground-water monitoring include the following:

- Resource Conservation and Recovery Act (RCRA), including the Hazardous and Solid Waste Amendments (HSWA), which include the Federal Underground Storage Tank Program
- Comprehensive Environmental Response, Compensation, and Liability Act (CERCLA or "Superfund"), including the Superfund Amendments and Reauthorization Act (SARA)
- Toxic Substances Control Act (TSCA)
- Clean Water Act (CWA)
- Safe Drinking Water Act (SDWA), including the Underground Injection Control Program (UIC), and the Well Head Protection Program
- Surface Mining Control and Reclamation Act (SMCRA)

Each of these pieces of legislation or programs is described briefly, and the ground-water monitoring provisions of each program are summarized in the following sections.

Resource Conservation and Recovery Act (RCRA)

The Resource Conservation and Recovery Act (Public Law 94-580) was passed by Congress in October 1976 as an amendment to the 1965 Solid Waste Disposal

Act because of the need to address the problem of how to safely dispose of the huge volumes of solid and hazardous waste generated nationwide each year (U.S. EPA, 1986). RCRA has evolved from a relatively limited program dealing with nonhazardous solid waste to a far-reaching program that focuses primarily on hazardous waste. Hazardous waste generators, transporters, and owners/operators of treatment, storage, and disposal facilities (TSDFs) comprise the RCRA-regulated community. On November 8, 1984, Congress passed the Hazardous and Solid Waste Amendments (HSWA) to RCRA, thereby greatly expanding the nature and complexity of activities covered under RCRA.

The goals of RCRA, as set forth by Congress, are:

- to protect human health and the environment
- to reduce waste and conserve energy and natural resources
- to reduce or eliminate the generation of hazardous waste as expeditiously as possible

To achieve these goals, three programs were established. The first program, termed Subtitle D, encourages states to formulate comprehensive solid-waste management plans, primarily for nonhazardous waste. The second program, Subtitle C, establishes a program to control hazardous waste from the time it is generated until its ultimate disposal, the so-called "cradle-to-grave" concept. The third program, Subtitle I, regulates certain underground storage systems.

Subtitle D. Subtitle D establishes a voluntary program under which participating states receive federal financial and technical support to develop and implement solid-waste management plans. This program is primarily a planning tool used to clarify state, local, and regional roles in the management of solid waste. One of the objectives of this portion of the act is to identify those facilities that are "open dumps." Although originally there were no specific regulations within this program requiring the monitoring of ground water, HSWA now contains roles governing land disposal units. The proposed version has ground water monitoring requirements.

Subtitle C. Subtitle C is the backbone of RCRA. It calls for the management of hazardous waste from the time it is generated until its ultimate disposal. Subtitle C clearly defines what is considered a hazardous waste and what is not. It also defines the types of facilities that fall under these regulations. These are commonly referred to as TSDFs (treatment, storage, and disposal facilities). The purpose of these regulations is to protect human health and the environment, with the emphasis on the protection of ground water.

There are a number of Parts to Subtitle C. Sections containing requirements for ground-water monitoring include:

- Part 264—Regulations for Owners and Operators of Permitted Hazardous Waste Facilities

- Part 265—Interim Status Standards for Owners and Operators of Hazardous Waste Facilities
- Part 267—Interim Standards for Owners and Operators of New Hazardous Waste Management Facilities*
- Part 270—Regulations for Federally Administered Hazardous Waste Permit Programs (Part B Permits)
- Part 271—Requirements for Authorization of State Hazardous Waste Programs

The regulatory scheme established under RCRA is to grant permits to all TSDFs that are in compliance with RCRA requirements. The standards set forth in Part 264 apply to these permitted facilities. Because there are thousands of facilities currently awaiting permits, RCRA provides the means to regulate unpermitted facilities prior to their final permitting. New facilities waiting to be built or in the process of being built fall under Part 264. Established facilities operating without a final permit, but under the regulatory framework, fall under Part 265. The information needed to submit an application for status as a permitted facility is detailed under Part 270.

Not every TSDF requires ground-water monitoring. In most cases, only surface impoundments, landfills, and land treatment facilities require ground-water monitoring in order to comply with RCRA. Monitoring is required during the active life of the facility, during its closure period, and during any post-closure period that is applicable. Most ground-water monitoring applies to the water quality in the "uppermost aquifer," although in some cases with known contamination, monitoring other hydrogeologic units may be required to characterize the extent of the contaminant plume.

Part 264. For facilities operating under a RCRA permit, there are generally three types of ground-water monitoring that may be required. The monitoring scheme is based on a phased approach, so that facilities that have not released contaminants into the ground water have different requirements than those that have a release. The most rudimentary monitoring scheme is the Detection Monitoring Program (40 CFR 264.98). This program must consist of ". . . a sufficient number of wells, installed at appropriate locations and depths to yield ground-water samples from the uppermost aquifer that:

1) represent the quality of ground water that has not been affected by leakage from the regulated unit, and
2) represent the quality of ground water passing the point of compliance (roughly the boundary of the management unit or units, such as individual or adjacent groups of impoundments or landfills)." (40 CFR 264.97)

---

*This part was only temporary until Part 264 was finalized, but has never been removed from Subtitle C. Specific requirements under Part 267 are no longer applicable.

This program has set forth a vast array of criteria that must be met and a list of chemical parameters characterizing ground-water quality that must be monitored in order for the TSDF to meet the performance criteria of the Detection Monitoring Program. Should there be a statistically significant change in the concentrations of the monitored chemical parameters that could indicate a release from the regulated unit, the owner or operator must implement the next level of ground-water monitoring, the Compliance Monitoring Program (40 CFR 264.99).

The Compliance Monitoring Program applies to units in which there is reason to believe that concentrations of certain chemicals in the ground water exceed established ground-water protection standards (40 CFR 264.92). The Regional Administrator of the program has a certain amount of discretion in identifying the parameters to be monitored, as set forth in the permit. Once it has been established that there is a release of the type and magnitude of concern at the compliance point of the facility, then a Corrective Action Program must be implemented (40 CFR 264.100).

The Corrective Action program requires that the owner or operator remove or treat the wastes that are causing the release, so that the ground-water quality complies with the ground-water protection standards. In this program, the primary purpose of the ground-water monitoring network is to monitor the effectiveness of the corrective action. Ground-water cleanup criteria are usually determined either by the individual states or within a state on a case-by-case basis. In all cases, the cleanup criteria must be as stringent as, or more stringent than, various standards set by the federal government.

After the TSDF ceases operation, the ground-water monitoring network may still be required to monitor the facility during the closure and post-closure periods. The closure period usually runs from the time the facility receives the final volume of waste until all activities cease (40 CFR 264.112 and 264.113). Post-closure monitoring, usually a period of 30 years after closure, is required at facilities in which all of the waste or waste constituents are not removed from the facility at closure. This applies primarily to landfills and land treatment facilities, but can also apply to surface impoundments that are closed with waste constituents remaining in the ground (40 CFR 264.117). Certain demonstrations can be made to reduce the duration of the post-closure monitoring. Table 1.1 lists the citations associated with Part 264 ground-water monitoring requirements.

Table 1.1   Ground Water Monitoring Citations for RCRA Part 264.

Citation	Description
40 CFR 264.97	General groundwater monitoring requirements
40 CFR 264.98	Detection Monitoring Program
40 CFR 264.99	Compliance Monitoring Program
40 CFR 264.100	Corrective Action Program
40 CFR 264.112	TSD facility closure
40 CFR 264.117	TSD facility post-closure
40 CFR 264.221	Design and operation of surface impoundments
40 CFR 264.228	Closure and post-closure of surface impoundments
40 CFR 264.310	Closure and post-closure of landfills

Part 265. Part 265 of RCRA addresses facilities that are under interim status. This applies to existing TSDFs that are waiting to obtain a final permit. The ground-water monitoring requirements under this part are much narrower in scope than those under Part 264, and are explained under 40 CFR 265.91 through 265.93. For interim status, a facility needs only to perform one type of ground-water monitoring, similar in some respects to the detection monitoring of Part 264. However, unlike Part 264 requirements, there is no phased approach and, if a release from the facility is detected by the monitoring system, a Ground-Water Quality Assessment Program is implemented (40 CFR 265.93). There are no provisions that clearly spell out the procedures, once in the Ground Water Quality Assessment Program, to determine whether ground-water remediation is required. Table 1.2 lists the citations associated with Part 265 ground-water monitoring requirements.

Table 1.2.   Ground Water Monitoring Requirements for RCRA Part 265.

Citation	Description
40 CFR 265.90 through .94	Groundwater monitoring program
40 CFR 265.112	Closure of Interim Status TSD
40 CFR 265.117 and .118	Post-closure of Interim Status TSD
40 CFR 265.221	Interim Status surface impoundments
40 CFR 265.301	Interim Status landfill design
40 CFR 265.310	Interim Status landfill closure and post-closure

Part 270. Owners and operators of hazardous waste management facilities are required to file a Part A and Part B permit application to receive their facility permit to operate. A Part A notification serves to notify the Regional Administrator of the existence of the facility and the wastes that are associated with it. A Part B permit application requires the generation of a substantial amount of information about the facility and the activities that take place there. As part of a Part B application for owners of surface impoundments, waste piles, land treatments units, and landfills, information regarding the protection of the ground water is necessary (40 CFR 270.14 and 270.97).

Part 271. Part 271 of RCRA deals with the authorization of state programs. The regulations require that states seeking authority for their programs have regulations similar to those promulgated under RCRA for TSDFs (40 CFR 271.12 and 271.128).

Subtitle I. The U.S. EPA has estimated that as many as 100,000 to 300,000 underground storage tanks (USTs) could be leaking their contents to the environment and polluting ground water (U.S. EPA, 1986). To address this problem, Congress created a program under HSWA, entitled Subtitle I, to control and prevent the leakage of stored products from underground storage tanks. These amendments to RCRA are significant in that they mark the first time that RCRA regulations have applied to raw product as well as to waste. Subtitle I is limited to regulating the storage of petroleum or hazardous chemicals, while Subtitle C regulates the storage of hazardous wastes.

Although there is no specific language in Subtitle I that requires the monitoring of ground water, there are references to a tank owner having the ability to detect releases. Subtitle I also authorizes federal and state personnel to ". . . monitor the surrounding soils, air, surface water, and ground water" (U.S. EPA, 1985a). There is specific language in a number of state UST programs that a ground-water monitoring well shall be installed adjacent to each new tank or tank field.

Final rules covering technical standards and requirements for new and existing underground storage tanks containing petroleum and hazardous chemicals took effect in December of 1988. The purpose of these rules is to regulate the vast numbers of underground tanks and to minimize the environmental impact of leakage from these tanks by implementing early detection techniques, ground-water monitoring, and physical protection of the tanks themselves. After an initial 10-year period, the use of ground-water monitoring wells is one of the specified methods that can achieve the required monthly monitoring for releases from these tanks.

The schedule for technically upgrading and monitoring requirements for existing tanks is dependent on the tank age. However, after 1993, all existing tanks are required to perform monthly leak detection monitoring, either by means of in-tank gauging, vapor monitoring, interstitial monitoring, or ground-water monitoring.

Tanks that are confirmed to be leaking must initiate corrective action. The rules do not specify the types of measurements or site assessment techniques that must be employed. However, it is implied that soil and ground-water samples should be obtained. If there has been a confirmed release that requires corrective action, a corrective action plan must be submitted that will address the remediation of soil and ground water, as required, and the means to verify the success of these actions. It is important to note that many states and local municipalities have additional requirements that may regulate the monitoring or remediation of a petroleum hydrocarbon release.

Comprehensive Environmental Response, Compensation, and Liability Act (CERCLA)

The Comprehensive Environmental Response, Compensation, and Liability Act (CERCLA), more popularly known as "Superfund," was passed by Congress in December 1980 to deal with threats posed to the public by abandoned waste sites. With the Superfund Amendments and Reauthorization Act (SARA) of 1986, CERCLA has assumed a larger role in the cleanup of hazardous waste sites. The main objectives of CERCLA, as established by Congress, are:

- to develop a comprehensive program to set priorities for cleaning up the worst existing hazardous waste sites

- to make responsible parties pay for those cleanups wherever possible
- to set up a Hazardous Waste Trust Fund for the twofold purpose of performing remedial cleanups in cases where responsible parties could not be held accountable, and responding to emergency situations involving hazardous substances
- to advance scientific and technical capabilities in all aspects of hazardous waste management, treatment, and disposal (U.S. EPA, 1987)

There are several steps involved in completing a Superfund cleanup. The initial report of the existence of a site may come from an individual or a facilities manager, either to EPA's National Response Center or to a local or state official. After EPA learns of the site, it collects all available background information to perform a preliminary assessment of the potential hazards posed by the site. In this step, EPA not only tries to identify the size of the problem and the types of wastes at the site, but it also attempts to identify any and all potentially responsible parties (PRPs) associated with the wastes. If the preliminary study reveals evidence that the site may pose a significant threat to human health or the environment, a site inspection is performed and the site is ranked, using the EPA Hazard Ranking System. A high enough ranking will place the site on the National Priorities List (NPL). As of January 1990, 1,219 abandoned waste sites had been either listed (1,010) or proposed for listing (209) by EPA. However, in 1980 there were an estimated 9,000 ''problem'' hazardous waste sites. In 1989, more than 30,000 sites had been entered into EPA's computerized database (CER-CLIS). Current projections are that as many as 2,500 of these sites will require cleanup under the federal Superfund program. (8)

The ultimate objective of placing sites on the NPL is their permanent cleanup. To identify the cleanup strategy that best suits a particular situation, the sites undergo a Remedial Investigation/Feasibility Study (RI/FS). Ground-water monitoring is a critical element of the RI, as it is necessary to establish whether or not ground water at the site serves as a pathway for waste constituents to migrate away from the site. The Feasibility Study (FS) often is heavily dependent on the data gathered during the RI, so that the optimal remedial technology(ies) may be implemented at the site. Ground-water monitoring is also a critical factor in evaluating whether the remedial activities implemented at the site are successful in abating ground-water contamination.

Guidance documents available from EPA set forth the procedures that should be followed to conduct a Remedial Investigation in support of a Feasibility Study (U.S. EPA March 1988). The focus of the RI effort depends on the quality of the existing data, key site problems, the need to provide sufficient technical data to support the FS, and enforcement needs. These factors dictate the study parameters and the type(s) and amount of sampling that will be sufficient to meet the needs of the study. Therefore, unlike RCRA, CERCLA does not set up any specific ground-water monitoring program requirements. The investigator must address each site individually. Although the purpose of the RI is to characterize the hydrogeologic setting and any contamination present at the site, there are several other important aspects to the ground-water monitoring program that are required for the FS.

The collection of data that will help in the evaluation of remedial technology alternatives is essential during the RI. These data may not directly aid in the definition of the problem, but could predict interactions between water quality and certain alternatives. For instance, although the level of iron present in the ground water is not an essential piece of information to establish the presence or extent of ground-water contamination, it may be useful in the FS portion of the project. If, for example, air stripping is proposed in the FS as a candidate remedial technology, the concentrations of iron must be known to devise methods of preventing scale buildup on the air-stripping unit, which would reduce its effectiveness.

Ground-water monitoring also is essential during the cleanup of a contaminated site. The effectiveness of ground-water cleanup can be monitored in a network of wells. This will aid in determinations of when remedial activities can cease and when the site can be declared clean.

Toxic Substances Control Act (TSCA)

The Toxic Substances Control Act (Public Law 94-469), enacted by Congress in 1976, brought significant changes in the day-to-day operation of the U.S. chemical industry. With TSCA, EPA was given the authority to identify and control chemical products that pose an unreasonable risk to human health or the environment through their manufacture, chemical distribution, processing, use, or disposal. To enable EPA to monitor the marketing of new chemicals, TSCA requires manufacturers to submit pre-manufacture notices on new chemical substances. EPA is authorized to take a variety of steps to protect against threats to human health or the environment by the introduction or unrestricted use of new chemicals. Such steps include publication of the chemical inventory, information gathering authority, and permitting access to manufacturing data which could assist in the development of source inventories for ground-water protection planning and investigation. For example, any RCRA facility that handles hazardous wastes that contain more than 50 parts per million of polychlorinated bipenyls (PCBs) is regulated under both RCRA and TSCA; initial ground-water monitoring for background data at PCB disposal sites also is required.

Clean Water Act (CWA)

In the Clean Water Act (Public Law 92-500) of 1972 and in the CWA Amendments of 1977 (Public Law 95-217), Congress provided for the regulation of discharges into all navigable waters of the United States. At the time, the CWA was one of the most far-reaching federal laws ever enacted. It has an application to ground water in several ways. To the extent that surface water and ground water are hydraulically connected, protection of surface-water quality also will protect ground-water quality, and vice-versa. The federal government provides funding to states to set up and implement water-quality management planning programs, including programs related specifically to ground water. In addition, where CWA funds are used to construct municipal sewage treatment plants that use land application techniques, the municipalities are required to design the plants to ensure protection of ground water (40 CFR 35, Appendix A).

A trust fund that was the precursor to Superfund was set up to deal with the problems stemming from such discharges. However, no provision was made to deal with damage to land resources resulting from contamination by hazardous wastes.

The formation of the National Contingency Plan for dealing with emergencies from hazardous waste was an important offshoot of the Clean Water Act. This plan remains the guiding principle behind the implementation of Superfund.

Safe Drinking Water Act (SDWA)

The Safe Drinking Water Act (Public Act 93-523) was passed by Congress in 1974 to respond to accumulating evidence that unsafe levels of contaminants in public drinking water supplies, including ground water, were posing a threat to the public health. There are several major provisions to the Act that impact ground-water quality. The Act provides protection to ground water through the establishment of drinking water standards (Fed. Reg. v.43, no. 243), sole source aquifer designation (42 U.S.C. 300f, sec. 1424), and the establishment of the Well Head Protection Program and the Underground Injection Control Program (40 CFR 144). Standards known as "Maximum Contaminant Levels" (MCLs) were developed under the SDWA and also may be used for enforcement in ground-water monitoring programs conducted both at RCRA interim status and RCRA permitted facilities.

Underground Injection Control Program (UIC)

The Underground Injection Control Program was developed under SDWA to regulate the underground injection of fluids. Injection wells are classified as follows:

- Class I: wells used by generators or owners of hazardous waste facilities to inject hazardous substances beneath the lowermost formation containing a source of drinking water
- Class II: wells which inject fluids in the process of oil or natural gas recovery
- Class III: wells which inject for the purpose of extracting minerals
- Class IV: wells used by generators of hazardous or radioactive waste to dispose of this waste into a formation which is within one-quarter of a mile of an underground drinking water source
- Class V: injection wells not included in the above classifications (40 CFR 144.6)

Under these rules, the EPA Regional Administrator may require ground-water monitoring to evaluate whether an underground source of drinking water may be endangered by injection of fluids into Class II enhanced recovery wells, Class IV wells, and some Class V wells. In addition, the owner or operator of a Class I, II, or III well can be required to "... install and use monitoring wells within the area of review if required by the Director (of the USEPA), to monitor any migration of fluids into and pressure in the underground sources of drinking water. The type, number and location of the wells, the parameters to be measured, and the frequency of monitoring must be approved by the Director" (40 CFR 144.28).

Surface Mining Control and Reclamation Act (SMCRA)

The preparation of a ground-water monitoring plan is required under 30 CFR 780, which deals with the application for a surface mining permit. As part of the minimum requirements for the required Reclamation and Operations Plan, hydrogeologic information must be supplied concerning the quality of the surface water and ground water in the permit and adjacent areas.

Federal Ground-Water Protection Strategy

When EPA established a Ground-Water Protection Strategy in August 1984, it concluded that state governments have the primary responsibility for ground-water protection policies and implementation, yet it set national goals and management strategies for implementing existing federal laws.

The Strategy sets a policy framework to guide EPA's programs affecting ground water. This framework involves developing a system for classification of the nation's ground water. The agency uses this classification system to evaluate the siting of RCRA facilities and will continue to use the immediacy of a threat to ground water as a factor in selecting sites for Superfund cleanup.

Specifically the policy calls for EPA to:

- provide financial support to states for program development and institution building
- assess the problems that may exist from unaddressed sources of ground-water contamination, in particular underground storage tanks, surface impoundments, and landfills
- issue guidelines for agency decisions affecting ground-water protection and cleanup
- establish an Office of Ground-Water Protection within EPA to coordinate agency policies (Bird, 1985)

The classification of ground water is the backbone of a policy to provide consistency in agency decisions.

Ground-Water Classification

The Environmental Protection Agency released a draft document of guidelines for ground-water classification as part of its National Ground-Water Protection Strategy. The document established three classifications for ground water. Class I, special ground water, is ecologically vital or irreplaceable as a source of drinking water. It is rated as highly vulnerable to contamination because of hydrogeologic factors. With its authority under RCRA, EPA will ban the siting of disposal facilities above Class I ground water. Ecologically vital ground water supports habitats for species listed or proposed for listing under the Endangered Species Act. EPA is considering developing special permit conditions for the Underground Injection Control Program to protect these waters.

Class II ground water includes current or potential sources of drinking water and water having other beneficial uses. Class III ground water includes water not considered to be a potential source of drinking water and water that may be contaminated naturally or by human activity which cannot be cleaned up by reasonable efforts.

Essentially, Class I ground water will receive a high level of protection, Class II ground water will receive less protection, and Class III ground water will receive the least protection under the proposed classification system. There is a provision for variances to lower the protection levels.

Discussion of Ground-Water Quality Standards

Ground-water monitoring only becomes meaningful when the results of the analyses for water quality are compared to some useful reference point. In many cases, ground-water quality standards are applied to water used for consumptive purposes as it leaves the tap. At other times, standards are applied to ground water after it has been cleaned up, or as it discharges to a surface water body, or in terms of the risk posed by a specified exposure. Further complicating the issue is the fact that many states have different or more restrictive standards than the federal government.

To achieve a better understanding of the standards that can be applied, a definition of some of the basic terms is appropriate. The Safe Drinking Water Act states:

> The term "Primary Drinking Water Regulation" means a regulation which (1) applies to public water systems, (2) specifies contaminants which may have an adverse effect on the health of persons, (3) specifies for each contaminant either a maximum contaminant level or a reduced level based on treatment, and (4) contains criteria and procedures to assure a supply of drinking water that will comply with the maximum contaminant levels (MCLs) and the requirements for the minimum quality of water that can be taken into the system.

The Secondary Drinking Water Regulations, also defined in the Safe Drinking Water Act, are described as follows:

> The term "Secondary Drinking Water Regulation" means a regulation which applies to public water systems and which specifies the maximum contaminant levels which are requisite to protect the public welfare. This applies to any contaminant in drinking water which may adversely affect the odor or appearance of the water so that a significant number of users discontinue its use (SDWA).

The term "maximum contaminant level" (MCL) refers to the maximum permissible level of a contaminant in water which is delivered to any user of a public water system (SDWA). These are enforceable standards that are set as close to MCLGs (maximum contaminant limit goals) as feasible. These standards are often applied to ground water that is used for drinking water purposes, regardless of whether it is supplied by a public system or a private well. These standards also consider the best technology that is available, treatment technologies that can be applied, and associated costs.

A maximum contaminant level goal (MCLG), previously called recommended maximum contaminant level (RMCL), is the maximum level of a contaminant in drinking water at which no known or anticipated adverse effect on the health of persons would occur, which allows for an adequate margin of safety. They are nonenforceable health goals (40 CFR 141.2, July 1987).

The Clean Water Act also has established water quality criteria which are not limited to ground water. They are known as the 304(a)(1) criteria. They are not rules and have no enforcement authority. Rather, these criteria present specific data and guidance on the environmental effects of pollutants, which can be useful in deriving regulatory requirements based on considerations of water-quality impacts. They are, therefore, comparable to the MCLGs, as they are not based on technology or cost, but are health-based goals. These standards can be used when protection of a drinking water source is not the sole objective and they can be applied to water-quality-based effluent limitations and toxic pollutant effluent standards (*Federal Register,* 1980). Although these standards were derived for surface water, they have application to ground water, particularly where other standards for certain chemicals have not yet been set.

Lists of various national standards and criteria are available. Extreme caution should be exercised in applying these criteria and standards to specific site conditions. These criteria, for the most part, do not take into account some other important factors that should be considered when applying standards to ground water used for consumptive purposes. Some of these considerations are: the population that will be using the water; the exposure from other sources that could contribute to the risk; and some of the other risks of exposure, other than the carcinogenic effects. It can be seen that the application of water-quality standards and criteria is neither simple nor straightforward and requires expertise in other fields, particularly toxicology.

MECHANISMS FOR A WORKABLE
FEDERAL GROUND-WATER PROGRAM

Even the best-laid plans have elements that make implementation more difficult than it first appears — environmental regulations are no exception. Now that major federal waste management regulations have been in place for a decade, there are functional goals that must be kept in mind if the mandated protection of ground water is going to succeed.

The first element is one of communication. The regulations will do no good if the regulated community will not follow them, even under the threat of civil or criminal penalty. Enforcement bodies have limited resources for informing the regulated community of their obligations under the law. As a result, most cases of noncompliance are the result of ignorance rather than malice or the profit motive. It is essential for the regulations to be communicated to those parties that they directly affect. Industrial facilities must be made aware of the limitations placed on their practices, such as management of waste, discharge of process wastewater, standards that treatment works must meet, and the permits that must be obtained. Even individual homeowners must be made aware that they are responsible for protecting their small portion of the ground-water resource. Federal regulations are published daily for public consumption, but state and local regulations are not as easy to access. Every effort must be made to disseminate this information to the people who need it.

The second element is the establishment of standardized evaluation and protection practices and the application of the same basic standards to similar situations. It is widely known that ground water is a dynamic resource and the hydrogeologic settings in which it occurs vary widely. However, there are sound scientific and engineering practices that can be applied to the evaluation of ground water to ensure that suitable and appropriate conclusions and recommendations concerning its potential use or abuse can be drawn. Similarly, the standards that are applied to the protection and cleanup of an aquifer should be made clear and not left to the whimsy of an individual regulator. There is a broad spectrum of standards and policies, ranging from nondegradation to inaction, routinely being applied by regulators across the country who lack direction. Standards that are health-based, technology-sensitive, and use-directed are in the process of being developed and will do much to bring this element into focus. However, a standard baseline for protection and cleanup would do much to minimize the uncertainty associated with the site evaluation.

Finally, there should be a mechanism at all levels for changing and amending the regulations as conditions change. This is found to some degree at federal and state levels with the owner's ability to request a waiver to a portion of a regulation or to apply alternate standards in particular cases. This ability should be expanded and streamlined as much as possible to address the changing nature of the resource. If new practices pose new health risks, or if the chemical or physical nature of an aquifer changes substantially over time, the mechanisms must be in place to revise regulations and applicable standards.

REFERENCES

Bird, J. C., 1985. "Groundwater Protection: Emerging Issues and Policy Challenges." Environmental and Energy Study Institute, Washington, D.C.

(CERCLA) Comprehensive Environmental Response, Compensation, and Liability Act of 1980. (Pub. L. 96-510, as amended by Pub. L. 99–499).

(CWA) Clean Water Act. (Pub. L. 92-500, as amended by Pub. L. 92-217 and Pub. L. 95-576, 33 U.S.C. 1251 et seq.)

Federal Register, Vol. 45, No. 231. November 28, 1980.

(SDWA) Safe Drinking Water Act. (Pub. L. 95-523, as amended by Pub. L. 95-1900, 42 U.S.C. 3001 et seq.)

(SWDA) Solid Waste Disposal Act as amended by the Resource Conservation and Recovery Act of 1976. (Pub. L. 94-580, as amended by Pub. L. 96-609 and Pub. L. 96-482, 42 U.S.C. 6901 et seq.)

U.S. Environmental Protection Agency, 1985a. "Underground Storage Tanks (UST), the New Federal Law." Washington, D.C.

U.S. Environmental Protection Agency, March 1988. "Guidance on Conducting Remedial Investigations and Feasibility Studies under CERCLA, 9355.3-01." Superfund Docket, Washington, D.C. (available from ORD (513)569-7562 Order No. 540-689-004).

U.S. Environmental Protection Agency, 1986. "RCRA Orientation Manual." USEPA Office of Solid Waste, Washington, D.C.

U.S. Environmental Protection Agency, 1987. "The New Superfund; Protecting the People and Their Environment." Office of Public Affairs, Vol. 13, No. 1. Washington, D.C.

2

Ground-Water Monitoring System Design

Martin N. Sara

INTRODUCTION

Selection of the proper locations for monitoring wells should be based on a holistic approach to the evaluation of a specific site. The placement of the wells in this process must weigh and balance data collected in the field, laboratory, and office.

The question "how much monitoring is enough?" when answered in the context of the number of monitoring wells required at a site, will be entirely site-specific. In general, the monitoring system designer should ensure that convincing evidence is established for each assumption, and for demonstrating the basic capability of the system to produce ground-water samples representative of both upgradient (background) and downgradient conditions. General rules of thumb are provided in this chapter, but the reader should bear in mind that "enough" is a subjective determination, both for the questions of how much monitoring is necessary to provide a monitoring system capable of detecting ground-water contamination, and how much demonstration is required to convince a regulatory agency of that capability.

The key word in most regulatory programs that require ground-water quality monitoring is "capable." The owner or operator of a facility required to monitor ground water must install and implement a monitoring system capable of determining the facility's impact on ground water; it must be capable of yielding representative ground-water samples for chemical analysis. The number, locations, and depths of the monitoring wells must be such that the system is capable of the prompt detection of any statistically significant differences in indicator parameters.

The monitoring system designer must base decisions on sufficient numbers and locations of monitoring wells on performance-oriented criteria. It will be a very unusual monitoring situation in which as few as several downgradient wells would ensure system capability. Some very simple geologic environments can be effectively monitored with the USEPA specified (U.S. EPA, 1986) minimum system of one upgradient and three downgradient wells. However, this level of monitoring may be representative of very few sites. It is not uncommon for monitoring systems to employ many more sampling points. This is especially true for sites located in heavily regulated states. This is also true for facilities in operation over long periods of time or facilities that consist of multiple units or expansions.

REGULATORY CONCEPTS IN GROUND-WATER MONITORING

Many of the regulatory concepts surrounding monitoring of all types of facilities, but specifically hazardous waste management facilities, evolve around compliance with U.S. EPA regulations on ground-water monitoring, specifically RCRA regulations (40 CFR 264.97) related to ground-water monitoring at hazardous waste and solid waste sites. For example, the owner or operator of a hazardous waste management facility must comply with the following requirements for any ground-water monitoring program developed to satisfy §264.98, §264.99, or §264.100:

    (a) The ground-water monitoring system must consist of a sufficient number of wells, installed at appropriate locations and depths, to yield ground-water samples from the uppermost aquifer that:

        (1) Represent the quality of background (upgradient) ground water that has not been affected by possible leakage from a facility; and

        (2) Represent the quality of ground water downgradient of the facility, at the compliance point.

    (b) If a facility contains more than one regulated unit, separate ground-water monitoring systems are not required for each regulated unit provided that provisions for sampling the ground water in the uppermost aquifer will enable detection and measurement at the compliance point of hazardous constituents from the regulated units that may have entered the ground water in the uppermost aquifer.

Many millions of dollars, millions of words in reports and millions of minutes of meetings have been spent on defining exactly what these relatively few lines of text really mean in the context of actual monitoring of hazardous waste sites.

Both RCRA Subtitle C (hazardous waste) and Subtitle D (solid municipal waste) facilities are required to meet these basic points of detection monitoring programs.

This federal rule can be depicted in a single figure that illustrates the concept of detection monitoring. Figure 2.1 shows a conceptual presentation of the §264.97 guidance on placement of detection monitoring wells.

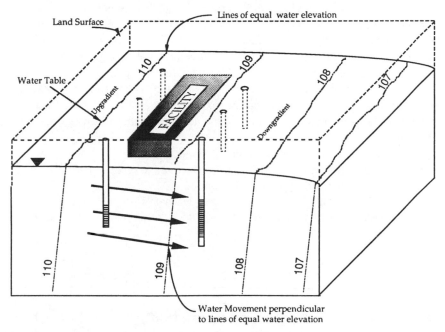

Figure 2.1.   Regulatory concept of detection monitoring.

The RCRA Ground-Water Monitoring Technical Enforcement Guidance Document (TEGD), (U.S. EPA, 1986) provides additional guidance on placement and number of upgradient or background wells by recommending that these wells be:

- located beyond the upgradient extent of possible contamination from the hazardous waste management unit so that they reflect background water quality;
- screened at the same stratigraphic horizon(s) as downgradient wells to ensure comparability of data;
- of sufficient number to account for heterogeneity in background ground-water quality.

The conceptual homogeneous unconfined uppermost aquifer, Figure 2.1, would still meet the above three TEGD requirements. However, this conceptual hydrogeologic condition is seldom observed in the field in such a simple form.

Many waste disposal facilities are located in complex geologic environments in which preliminary site assessment investigations are required to properly locate the wells for detection ground-water monitoring systems. Layering of geologic units of significantly different hydraulic conductivity complicates the simple conceptual picture described by the federal rule. Figures 2.2a and 2.2b show a two-layer system with the uppermost aquifer consisting of homogeneous isotropic sand below a near surface silt/clay unit of lower hydraulic conductivity.

In Figure 2.2a, the uppermost aquifer is unconfined (i.e., potentiometric surface), and in Figure 2.2b, the aquifer is confined by the lower hydraulic conductivity unit. Downgradient well positions are shown as point A in both figures. Both upgradient and background wells are shown in these figures. The concept of background representing not hydraulically upgradient locations but reflecting general water quality of the uppermost aquifer are represented by point B. In each case, the sand unit should be considered as the uppermost aquifer for the following reasons:

- the sand unit has regional extent;
- the sand unit has sufficient hydraulic conductivity to produce usable quantities of water to springs or wells; and
- the sand unit would be the zone in which leachate from the facility could migrate horizontally away from the site to affect human health and the environment.

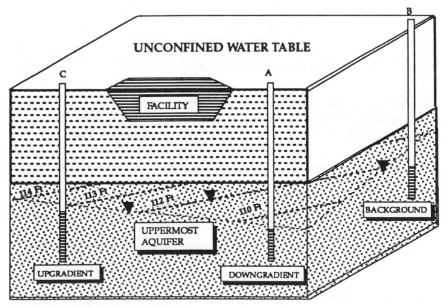

Figure 2.2a.   Unconfined water table.

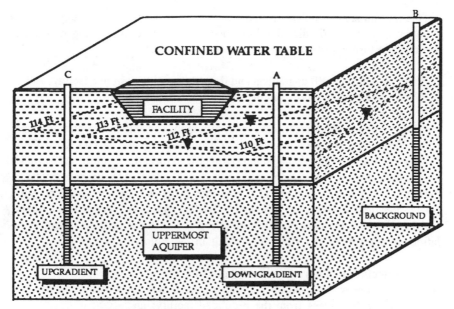

Figure 2.2b. Confined water table.

Much of the concern of regulatory agencies with respect to review of detection monitoring programs is with meeting federal regulations in 40 CFR 265.91, which describe ground-water monitoring system requirements for interim status hazardous waste disposal facilities. These regulations state:

"A ground-water monitoring system must be capable of yielding ground-water samples for analysis and must consist of:
  (1) Monitoring wells (at least one) installed hydraulically upgradient (i.e., in the direction of increasing static head) from the limit of the waste management area. Their number, locations, and depths must be sufficient to yield ground-water samples that are:
     (i) Representative of background ground-water quality in the uppermost aquifer near the facility; and
     (ii) Not affected by the facility; and
  (2) Monitoring wells (at least three) installed hydraulically downgradient (i.e., in the direction of decreasing static head) at the limit of the waste management area. Their number, locations, and depths must ensure that they immediately detect any statistically significant amounts of hazardous waste or hazardous waste constituents that migrate from the waste management area to the uppermost aquifer."

This interim status rule on ground-water monitoring has several key features different from §264.97 rules that have been widely used in defining what a detection monitoring system should consist of, specifically:

- one monitoring well upgradient and three downgradient from a facility; and
- the system must have "immediate" detection capabilities.

While "immediate" detection is open to widely variable interpretation, especially considering the slow movement of ground water, the TEGD (U.S. EPA, 1986) provides some additional guidance on how to meet the "immediate" criteria by placing detection monitoring wells immediately adjacent to the waste management unit. The federal Subtitle D regulations proposed for nonhazardous solid waste sites (early 1989) set 150-meter buffer zones (or property boundary, whichever is less) for placement of monitoring wells. Reducing these federal regulations to a series of criteria would result in the following points:

The detection monitoring system should have:
- sufficent wells, both upgradient (background) and downgradient, to detect discharges from the regulated facility; and
- wells located within a flow path from the regulated facility in an uppermost aquifer.

Furthermore, the uppermost aquifer should have sufficient hydraulic conductivity and extent so that sampling could be conducted within the property buffer zone (not to exceed 150 meters) for nonhazardous solid waste sites and at the waste unit boundary for hazardous waste (Subtitle C) facilities.

An adequate detection monitoring program can be designed for any geologic/hydrogeologic environment using the above criteria. The following sections present conceptual models for detection monitoring programs that provide guidance for monitoring system design for a variety of hydrogeologic environments.

Prior to selecting the locations and depths for screened intervals for ground-water monitoring wells, the ground-water monitoring designer must have, at a minimum, accomplished the following:

- performed a complete site characterization;
- established a conceptual hydrogeologic model for the site;
- constructed a ground-water flow net; and
- located facility boundaries and waste disposal areas.

Each of these tasks provide data that will be used to select the target monitoring zones for the monitoring system. The remaining sections describe the monitoring system design process summarized in Figure 2.3. Examples of the process are included to assist in the design conceptualization process.

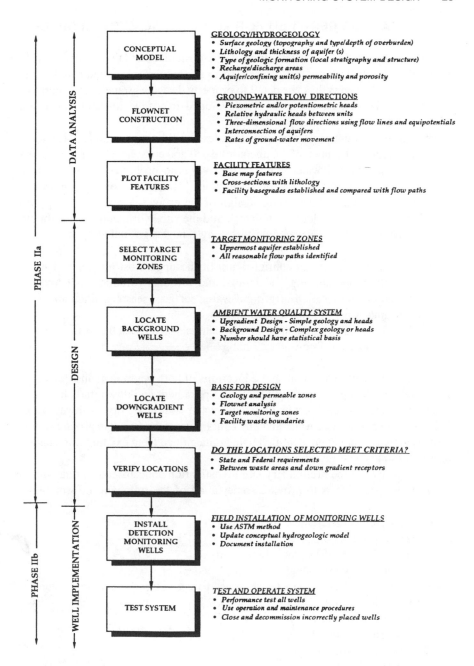

Figure 2.3. Summary of ground-water monitoring system design process.

DATA ANALYSIS REQUIRED FOR MONITORING SYSTEM DESIGN

Geologic factors (related chiefly to geologic formations and their water-bearing properties) and hydrologic factors (related to the movement of water in the formations) must be known in some detail to properly design a ground-water monitoring system. These data are normally developed in a field investigation using methods described elsewhere in this book.

The geologic framework of a site includes the lithology, texture, structure, mineralogy, and distribution of the unconsolidated and consolidated earth materials through which ground water flows. The hydraulic properties of these earth materials depend upon the geologic framework. Thus, the geologic framework of the facility heavily influences the design of the ground-water monitoring system. Elements of the geologic framework and the site hydrogeology that should be considered in ground-water monitoring system design include:

- the spatial location and configuration of the uppermost aquifer and its hydraulic properties (e.g., horizontal and vertical hydraulic conductivities, depth and location of ground-water surface, seasonal fluctuations of ground-water surface elevation);
- hydraulic gradient within the geologic materials underlying the facility; and
- facility operational considerations.

These data are used to establish the locations of both upgradient and downgradient wells in the uppermost aquifer. Both upgradient and downgradients wells should be located in the direction of ground-water flow along flow pathways most likely to transport ground water and any contaminants contained in ground water. These pathways should be identified using data gained from existing information and preliminary site investigations.

The objective of the preliminary site investigation and subsequent data analysis and interpretation is to provide some or all of the following information:

- lithologic characteristics, including:
  —established stratigraphic names
  —classification of hydrogeologic units
  —extent of hydrogeologic units
- key hydrogeologic characteristics used to develop the site conceptual model, including:
  —hydraulic conductivity
  —porosity
  —hydraulic gradient
  —specific yield
- aquifer characteristics, including:
  —boundaries
  —type of aquifer
  —saturated/unsaturated conditions

Each piece of data is an important building block in establishing the conceptual hydrogeologic model and targeting zones to be monitored. These data are used in combination to define the uppermost aquifer and ground-water flow characteristics to allow the identification of aquifer flow pathways so that target monitoring zones can be selected.

SELECTING TARGET MONITORING ZONES

The first task in the design of a ground-water quality monitoring system is the selection of the target monitoring zones. The logic used in selection of the target monitoring zones is illustrated in Figure 2.4. A review of features of the facility to be monitored, used in combination with conceptual models and flow nets, provides the system designer with the information to select those zones that will provide a high level of certainty that releases from the facility will be "immediately" detected. The concept of the target monitoring zone was developed as a means of directing the ground-water monitoring system designer toward placement of well screens in the uppermost aquifer at locations and depths that would have the highest likelihood of detecting leakage from a facility. This target zone usually lies in the saturated geologic unit in which ground-water flow rates are the highest because it possesses the highest hydraulic conductivity of those geologic materials adjacent to or underlying the facility of interest. Figure 2.4 illustrates the process of selection of a target monitoring zone using information on facility features, geologic characteristics, and hydraulic characteristics gathered during the preliminary field investigation. This selection process can be described as a series of steps:

1. Locate Site Features on a Topographic Base Map Format: Site features should be compared to information on geologic and soils maps to define the location of important facility components in relation to the distribution of surficial materials. Any likely recharge/discharge areas (streams, wetlands, or other surface-water bodies) should be located.
2. Cross Section Construction and Conceptual Model Development: Cross sections should be constructed, based on boring logs and/or geophysical traverses. These sections should be compared against the location of site features and facility components. The base grades of the facility should be plotted on cross-sections to establish if any excavations at the facility intersect sensitive geologic units or ground-water flow pathways. A conceptual model should be constructed to establish the site geological framework and to illustrate the distribution of geologic materials of differing hydraulic conductivity.
3. Use Flow Net to Define Likely Direction of Ground-Water Flow: Construction of flow nets will assist in defining the gradient and direction of ground-water flow in the uppermost aquifer. The rates of flow along flow paths can be calculated from the information provided by the

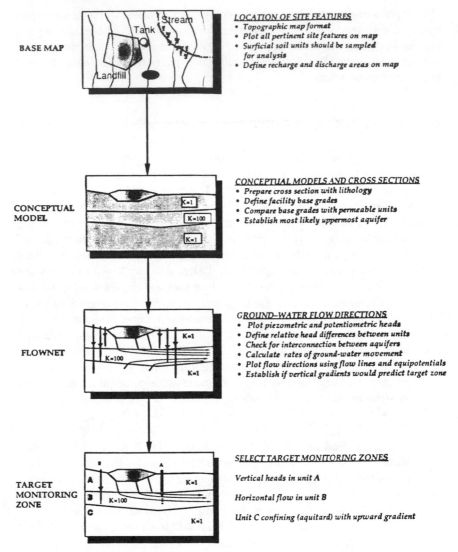

Figure 2.4. Select target monitoring zones.

flow net. Vertical gradients can be used to predict target zones by comparison of relative heads between units. Interconnections between aquifers can be predicted from relationships relating hydraulic conductivity to hydraulic heads for the units defined in the conceptual models.

4. Select Target Monitoring Zones: The zone meeting the regulatory definition of the uppermost aquifer, which also shows primarily horizontal ground-water movement under or adjacent to the facility, would

therefore represent the target monitoring zone. This zone would probably represent a permeable unit that is discharging to other permeable units or to local discharge areas. The system designer should be aware of the flow paths within the uppermost aquifer that would represent the most likely zones of ground-water movement away from the facility. These zones, typically those with the highest hydraulic conductivity, would be the focus of the monitoring system. If interconnected aquifers are present, these units should be monitored as necessary to provide safeguards for downgradient ground-water users.

This four-step procedure for selecting the target monitoring zone must be flexible enough to accommodate seasonal changes in gradient or plans to expand or alter the configuration of the facility. The target zone might include only a portion of a very thick aquifer (for example, the top 30 feet), or span several geologic units (as in the case of a thin, permeable, unconsolidated unit overlying weathered and/or fractured bedrock). These target zones represent the proper location for placement of monitoring wells.

LOCATING BACKGROUND AND DOWNGRADIENT WELLS

Gradients

The basis for developing most monitoring programs is knowledge of the hydraulically upgradient and downgradient direction from the site to be monitored. Figure 2.1 illustrated ground-water movement from higher potentiometric surface elevations (upgradient) to lower elevations (downgradient). This simple conceptual model of a homogeneous aquifer is the basis for much of the regulatory thought on ground-water monitoring.

After the selection of the target monitoring zone(s), the next step in the design of a ground-water monitoring system is the location of upgradient or background monitoring wells. The conceptual geologic model and flow net construction will have defined the uppermost aquifer and the relative direction of ground-water flow, both vertically and horizontally. Selection of upgradient well locations is based not only on this information, but also on other factors mainly relating to the facility. The numbers of upgradient or background wells installed at a site will be based on the size of the facility, the geologic/hydrogeologic, environment, and the ability to satisfy statistical criteria for analysis of water-quality data.

The TEGD (U.S. EPA, 1986) defines upgradient wells as "one or more wells which are placed hydraulically upgradient of the site and are capable of yielding ground-water samples that are representative conditions and not affected by the regulated facility.

This usage of the term "upgradient" is consistent with 40 CFR 265.91, which links background and upgradient for interim RCRA sites. Background wells would meet the 40 CFR 264.97 test to "represent the quality of background water that has not been affected by leakage from a regulated unit and represent the quality of ground water passing the point of compliance." The term "upgradient" can be a difficult concept to demonstrate in ground-water monitoring system design, because field conditions may not match the simple regulatory model. The designer must carefully consider site-specific hydraulic conditions to accurately locate upgradient monitoring wells because ground water does not always flow horizontally from upgradient to downgradient.

Simple single-aquifer flow systems are established by a clear understanding of the directional movements of ground water through definition of the ground-water gradients across a site. Figures 2.5 and 2.6 illustrate, in plan view and cross-section, the flow around a gaining stream, where ground water provides the stream base flow. Figure 2.7 illustrates flow around a losing stream, where surface water supports ground-water levels.

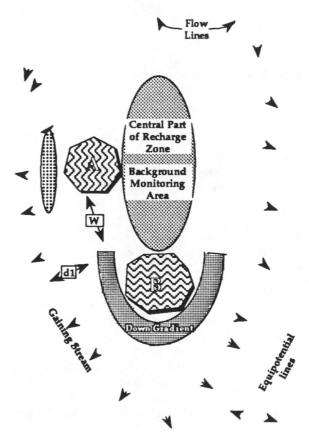

Figure 2.5. Potential target monitoring areas – Plan view.

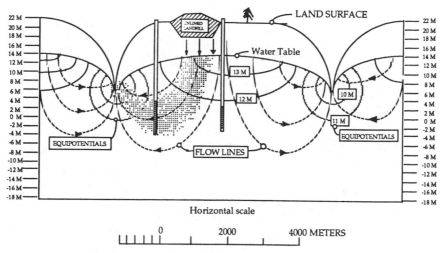

Figure 2.6. Potential target monitoring areas – Cross section.

In each of these cases, this simple system provides directional components to allow the positioning of ground-water monitoring wells. Figures 2.5 and 2.6 illustrate a facility (B) located in a recharge area which discharges to streams on either side of the facility. Ground-water flow lines are shown in plan and cross section. Because the facility is sitting directly atop the recharge area, the down-gradient flow zone is composed of a wide arc around the facility.

Potential target zones for monitoring are shown in Figure 2.5. Upgradient background target zones should be sufficiently above the recharge area so as not to be affected by the facility. Several conclusions can be drawn from Figures 2.5 and 2.6:

- Facility A would have its downgradient monitoring wells located within the ground-water flow lines shown. This facility location would have background monitoring well(s) located in the central recharge area.
- Facility B would have a background well in the area indicated. Because the facility is located directly within the local recharge area, this would not be considered an upgradient well, but rather a background well that represents water quality similar to a well that would be upgradient from the facility.
- Actual flow conditions would result in a water table significantly flatter than that shown in Figure 2.6. Vertical exaggeration (of approximately 125 to 1) makes the flow lines appear to travel deeper than would be represented in real life. The vertical scale indicates that the monitoring wells installed at the site should be screened from 18 to 24 meters below ground surface to intercept the ground-water flow (and any contained contaminants) emanating from beneath the site.

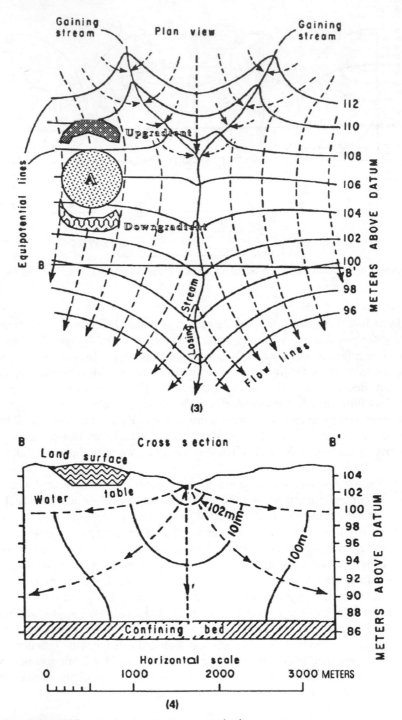

Figures 2.7a and b.   Losing stream target monitoring zones.

Figure 2.7 illustrates a "losing stream" condition and the resultant monitoring target zones for Facility A. Because the stream in this illustration is recharging ground water, and thus represents the highest point of upgradient ground water, target monitoring zones are located along the flow lines shown on Figure 2.7. Depths of screen placement must be based on vertical gradients in the area.

Steep/Flat Gradients

Even simple single-aquifer systems require consideration of local gradients adjacent to the facility of interest. In an area with a flat gradient it is necessary to consider possible ground-water flow in what would normally be an upgradient direction. In an area with a steep gradient on the water-table surface, as shown in Figure 2.8 (typical of lower hydraulic conductivity materials), there is little potential for reversal of flow direction. The target monitoring zone in an area with a steep gradient would normally be narrower than in a flat gradient environment.

Figures 2.9 and 2.10 show an unlined landfill within an area where the hydraulic gradient is low. The target monitoring zone is characteristically thicker than it would be in an area with a high hydraulic gradient (as shown in Figure 2.8).

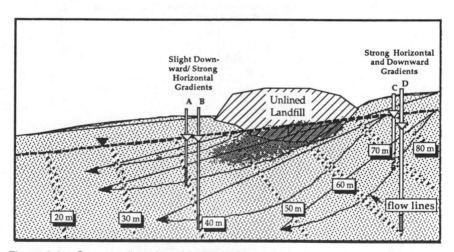

Figure 2.8. Steep gradient detection monitoring.

Procedures for Gradient Controlled Sites

1. Establish lithology and gradients as with single-aquifer systems;
2. Compare natural (baseline) gradients across the site and hydraulic conductivity of the aquifer; and
3. Select the position(s) for upgradient monitoring well(s), as in position A of Figure 2.10.

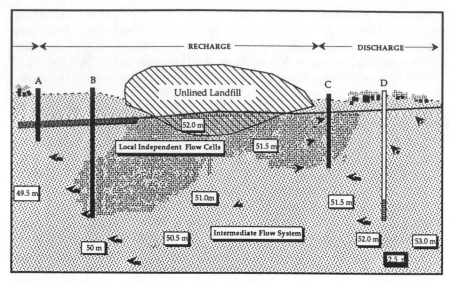

Figure 2.9. Shallow gradient detection monitoring.

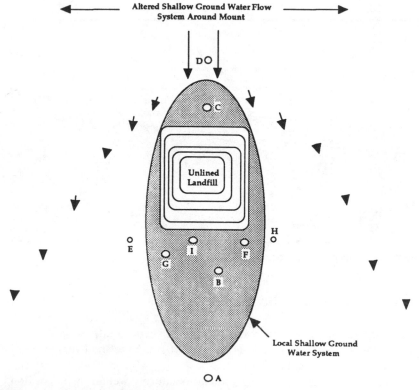

Figure 2.10. Shallow gradient detection monitoring.

Gradient Control/Flow Nets

Figure 2.11 shows an unconfined aquifer separated from a confined aquifer by a low hydraulic conductivity confining bed. Ground-water movement through this system involves flow not only through the aquifers but also across the confining bed.

The hydraulic conductivities of aquifers are several orders of magnitude greater than those of confining beds, the result being that, for a given rate of flow, the head loss per unit of distance along a flow line is several orders of magnitude less in aquifers than it is in confining beds. Consequently, lateral flow in confining beds usually is negligible, and flow in aquifers tends to be parallel to aquifer boundaries, as shown in Figure 2.11.

Example Equipotential Lines in a Geologic Model

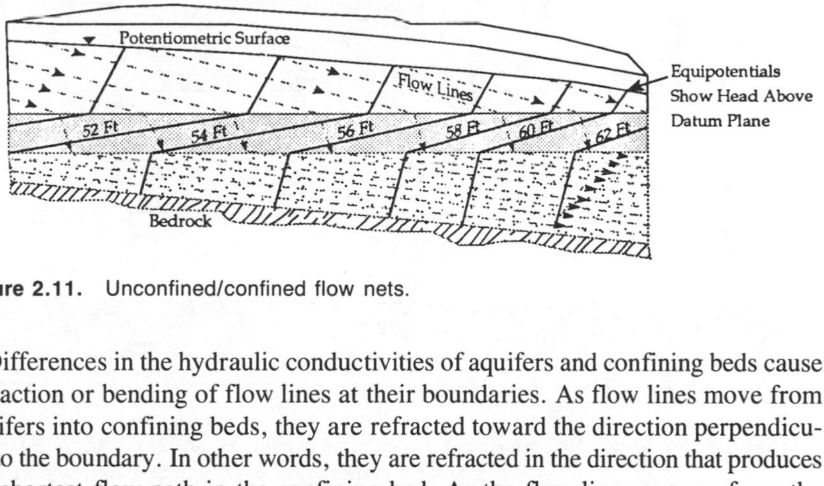

Figure 2.11.   Unconfined/confined flow nets.

Differences in the hydraulic conductivities of aquifers and confining beds cause refraction or bending of flow lines at their boundaries. As flow lines move from aquifers into confining beds, they are refracted toward the direction perpendicular to the boundary. In other words, they are refracted in the direction that produces the shortest flow path in the confining bed. As the flow lines emerge from the confining bed, they are refracted toward the direction parallel to the boundary (Figure 2.11). Hence, ground water tends to move horizontally in aquifers and vertically in confining beds or low hydraulic conductivity materials.

Lateral flow components in aquifers have direct relevance to ground-water monitoring system design. Most monitoring programs concentrate on establishing target monitoring zones in the uppermost aquifer beneath a site. Some programs may involve monitoring the uppermost aquifer, deeper aquifers, and zones between the uppermost and deeper aquifers.

The movement of water through aquifer/confining unit systems is controlled by the vertical and horizontal conductivities, the thicknesses of the aquifers and confining beds, and hydraulic gradients. Because of the relatively large head loss that occurs as water moves across confining beds, the most vigorous circulation of ground water normally occurs through the shallowest aquifers. Movement generally becomes slower as depth increases (Health, 1985). The uppermost aquifer will usually show contamination first (unless a direct conduit for contamination exists in deeper aquifers) and thus must be served by initial monitoring efforts. The concentration of flow lines in aquifers is illustrated further by Figures 2.12a and 2.12b (Freeze and Witherspoon, 1967). Aquifers bounded by a sloping confining layer and a flat-lying confining unit are present, for example, in glaciated regions where low hydraulic conductivity tills overlie higher hydraulic conductivity outwash sand and gravel aquifers. Nearly vertical flow occurs through the generally thick, low hydraulic conductivity materials, while nearly horizontal flow occurs within the underlying aquifer. The aquifer represents the only zone in which ground water moving away from a facility could be properly intercepted and monitored, and thus should be considered the target monitoring zone.

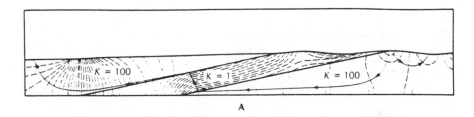

A

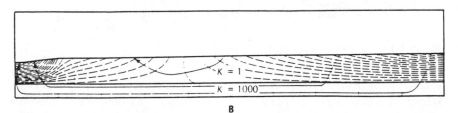

B

Figures 2.12a and b.   Regional ground-water flow in confined aquifers: **A.** Aquifer confined by a sloping confined layer; **B.** aquifer confined by a flat-lying confining layer. *Source:* R. A. Freeze and P. A. Witherspoon, *Water Resources Research,* 3:623⅓34 (1967).

This concept is further illustrated in Figures 2.13 and 2.14. The piezometers installed at different depths in the aquifer and in the confining zone indicate that a strong downward gradient exists in the fine-grained overburden material. The following represents the proper interpretation of flow at this site:

- Monitoring wells located in Figure 2.13 at A-3 and B-3 would represent background and downgradient, respectively. (The entire target zone should be screened in both these locations.)
- Downgradient monitoring wells in the unconfined aquifer, shown in Figure 2.14, should be located in a target zone screened at or below the interval screened by piezometer B-2.

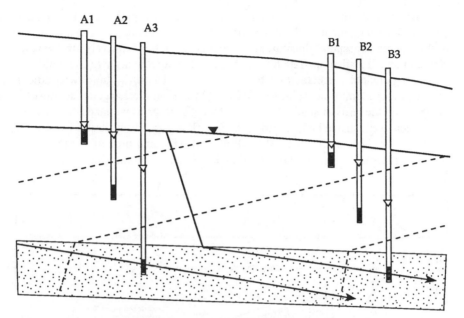

Figure 2.13. Piezometer nests in confining layer and sand aquifer.

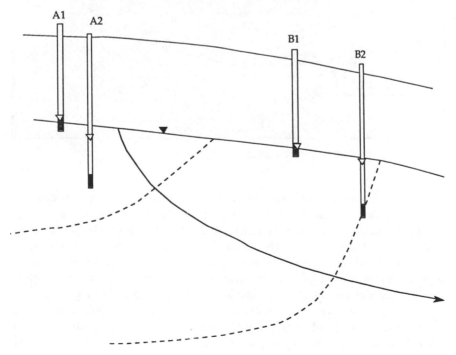

Figure 2.14. Piezometer nests in unconfined aquifer.

Figure 2.15 illustrates the potential ground-water flow paths to a discharging stream. Both upgradient wells (A and B) and a downgradient well (C) are shown in this simple conceptual illustration. However, even this relatively simple conceptual model can demonstrate how a shallow downgradient well (C) would not intercept potential leachate flow from an unlined landfill. The downgradient ground-water monitoring point for facilities located in discharge areas must be designed on the basis of shallow, near-surface discharge to wetlands or streams. Upgradient wells should be screened in shallow flow paths, as illustrated by well B. Deeper upgradient wells (as illustrated by well A) would probably suffice, but may not represent ground water flowing in the target monitoring zone.

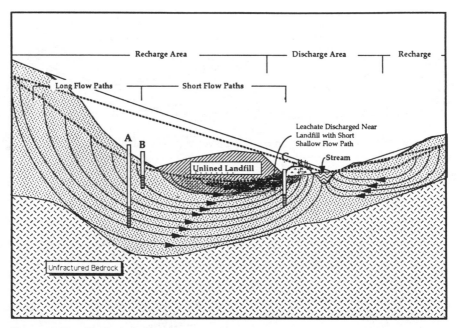

Figure 2.15.   Shallow discharging ground-water system.

Ground-water monitoring in complex alluvial deposits often presents difficult problems with respect to identification of target monitoring zones. These deposits often have shallow sandy zones encapsulated within low hydraulic conductivity sediments. Sand tank experiments have shown that these discontinuous sandy deposits do not affect the downward movement of ground water when strong downward gradients exist. Figure 2.16 shows such a conceptual situation. Shallow permeable zones contained within the low hydraulic conductivity materials do not have significant horizontal gradients; vertical gradients usually predominate in such environments. Monitoring points located adjacent to a facility located in these deposits (such as well A) may not represent a target monitoring point. Only where significant horizontal flow exists, as in the regional (uppermost) aquifer, would a horizontally downgradient target flow path be found. Well B represents a correct downgradient monitoring point. However, upper permeable units may represent uppermost aquifers if they have sufficient hydraulic conductivity and are of sufficient extent to serve as water sources for offsite ground-water users. There are four important criteria for establishing the need to monitor perched-water sand lenses located within lower hydraulic conductivity units. These criteria are:

- differential hydraulic conductivity;
- directional hydraulic heads;
- unit prevalence; and
- unit thickness.

Differential hydraulic conductivity refers to the variation in hydraulic conductivity observed between geologic units. Directional hydraulic head refers to the potential flow directions observed from piezometers located within individual units. Unit prevalence is a qualitative judgment based on the overall site stratigraphic characterization. Unit thickness is defined during the site investigation program and is based on simple thickness of the sandy unit.

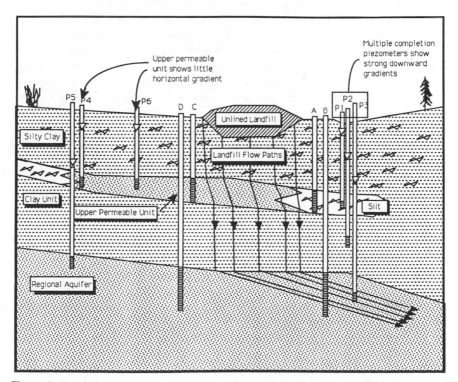

Figure 2.16. Low hydraulic conductivity environments with sandy lenses.

Each of these criteria must be considered in order to decide if a particular permeable unit would require monitoring as a target monitoring zone. Differential hydraulic conductivity represents an order of magnitude comparison of the sandy unit to the adjacent matrix materials. Freeze and Cherry (1979) state that "in aquifer-aquitard systems with permeability contrasts of two orders of magnitude or more, flow lines tend to become almost horizontal in the aquifers and almost vertical in the aquitards." This flow must, however, require that the aquifer discharges into other permeable units, to surface water, or is pumped from the system. Directional heads provide an indication as to the discharge potential of the sandy units. If vertical directional heads are discharging upward and downward into the sandy layers, it is likely that the unit discharges into adjacent areas. The unit prevalance criterion provides an indication as to how continuous the layer is in the field. These data are gathered during the field investigation program to demonstrate the continuity of the unit in the site area. As general guidance, if all soil borings (100%) installed during the site investigation contacted the definable unit at equivalent elevations, it is likely that the geologic stratum is continuous. If this contact percentage falls to 50% or shows an elevation variability, the unit is much less likely to represent a continuous feature that should be monitored. The soil boring program also establishes the relevant thicknesses of these units. If the thicknesses of the saturated permeable units are great (say 100 feet), it would be likely that the unit would require monitoring. As the thickness lessens, the other factors or criteria become important in the overall decision to monitor or not to monitor the unit as the uppermost aquifer. An additional criterion, use of the water contained within the units can outweigh all the other factors, assuming that the unit is hydraulically connected between the facility and the water users. Each of these factors must be weighed in the decision process.

If a discontinuous sand unit with a differential hydraulic conductivity of one order of magnitude shows piezometric heads passing through the unit (i.e., heads continued downward through the sandy unit) and few borings contacted the approximately one-foot thick unit, the unit would not be considered as a target monitoring zone. If that same unit is 10 feet thick and was contacted by only a few borings, it may be necessary to monitor the unit as the uppermost aquifer, or it may not. However, if the unit was 100 feet thick, was penetrated by all borings, and showed piezometric heads discharging into the unit from above and below, the unit would definitely be monitored as the "uppermost aquifer."

Figures 2.17a through 2.17d illustrate the use of this concept with a series of conceptual models showing various levels of discharge from sandy units. The levels of discharge range from almost none in Figure 2.17a to significant discharge between the unconsolidated and bedrock systems. The interpretation of site hydrogeologic conditions and thus the design of the monitoring system in each case would be based on the following key points:

- the lateral extent and thickness of the various geologic units present;
- the hydraulic conductivity of each of the individual units;

- the gradients obtained from piezometers placed in each of the permeable units; and
- the discharge/recharge potentials of the geologic units present on site.

The conceptualization and flow net construction should be based on illustrations with 1 to 1 scales. Figures 2.17a through 2.17d provide some additional keys to interpretation of the appropriate monitoring locations. The conditions depicted in Figures 2.17a and 2.17b would indicate that monitoring need only be conducted in the regional (uppermost) aquifer. Figure 2.17c represents a situation in which a second sandy unit and the regional system should be monitored. This decision to monitor both units should be weighed on the basis of additional site characterization work to determine the regional extent and current/future use of the sandy units. If the second sandy unit represents a likely flow path, and hence a target monitoring zone, it should be included in the monitoring program. The conditions depicted in Figure 2.17d would indicate that the second sandy unit would require monitoring. The first sandy unit in Figure 2.17d would not represent an effective monitoring location because of its discontinuous nature. Ground-water samples obtained from this unit would not be representative of conditions horizontally downgradient of the facility.

Heads established by multiple piezometers can identify the potential flow paths from a facility in homogeneous materials. Figure 2.18 illustrates an upgradient area of recharge and downgradient discharge point as defined by water levels measured in piezometers. The downgradient piezometers show an upward vertical gradient, while upgradient piezometers show a downward vertical gradient. Figure 2.19 illustrates a recharge condition in both background and downgradient piezometers. The heads shown in monitoring wells A and B represent the average of the equipotential heads contacted by the screened area of the wells.

Geologic Control

Geologic controls over ground-water movement represent the most critical factors that should be considered in ground-water monitoring network design. The goal in designing a monitoring network, in most cases, is to define the most likely zone(s) in which ground water moves beneath a facility and, hence, the most likely zone for any possible contaminant movement to occur and be detected. The following discussion first addresses simple geologic systems where design of the monitoring system is relatively straightforward based on the geology and ground-water flow directions. The discussion then moves to more complex systems that require significant site assessment and conceptualization to design an appropriate monitoring system. The discussion also includes design for perched water conditions. Some of the following examples include unlined waste disposal sites where leachate is shown to dramatize the potential flow paths and target monitoring zones.

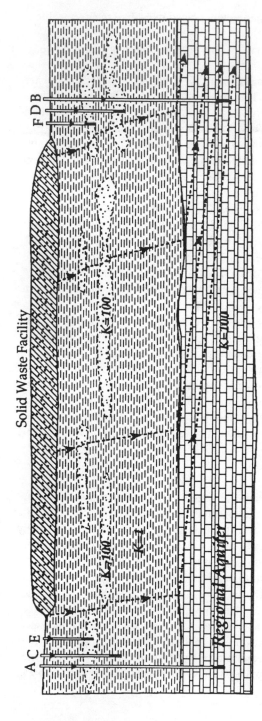

SAND LENSES WITH LITTLE HORIZONTAL DISCHARGE

KEYS IN UNDERSTANDING CONCEPTUAL MODEL
- *Confirm sand lenses in drilling program*
- *Piezometers show vertical gradients in all cases*
- *Regional aquifer known as productive unit*

Figure 2.17a.   Levels of discharge from sandy units.

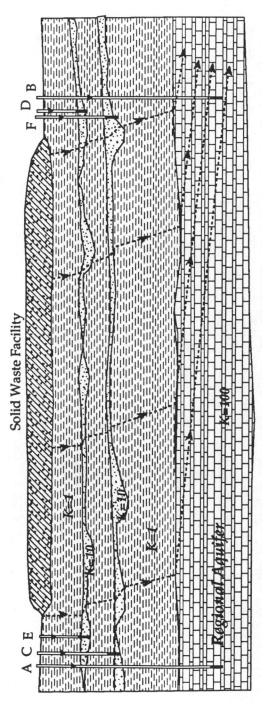

Solid Waste Facility

F D B

F

A C E

K=1

K=10

K=10

K=1

K=1

Regional Aquifer

K=100

CONTINUOUS SAND LAYERS WITH LITTLE HORIZONTAL DISCHARGE

KEYS TO UNDERSTANDING CONCEPTUAL MODEL
- *Well defined sand units present in most boreholes*
- *Sand layers do not have regional extent*
- *Downward gradients present in all cases*
- *Regional aquifer known as productive unit*

- 50 FEET

50 FEET

Scale

Figure 2.17b.   Levels of discharge from sandy units.

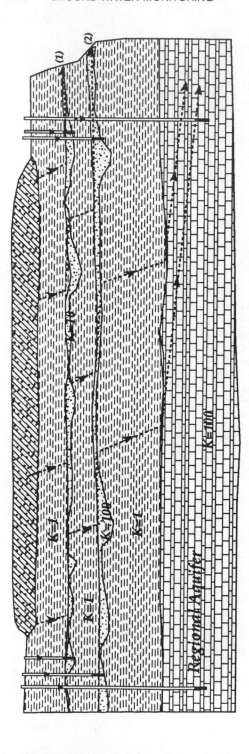

SAND LAYERS WITH SIGNIFICANT HORIZONTAL DISCHARGE

KEYS IN UNDERSTANDING CONCEPTUAL MODEL

- *Confirmed sand layers in drilling program*
- *Vertical gradients in only fine grained units*
- *Regional aquifer known as productive unit*
- *Sand layers used locally or discharge to streams*

Figure 2.17c.   Levels of discharge from sandy units.

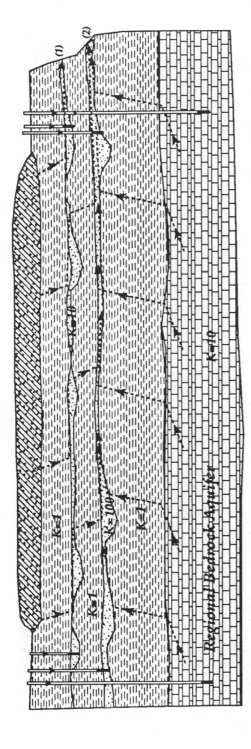

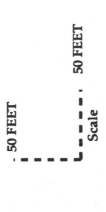

SAND LAYERS DISCHARGING BOTH BEDROCK AND OVERBURDEN

KEYS TO UNDERSTANDING CONCEPTUAL MODEL
- Well defined sand units present in all boreholes
- Sand layers have regional use as water supply
- Gradients discharge into sand layers
- Regional aquifer discharges into sand layer

- 50 FEET

50 FEET

Scale

Figure 2.17d.   Levels of discharge from sandy units.

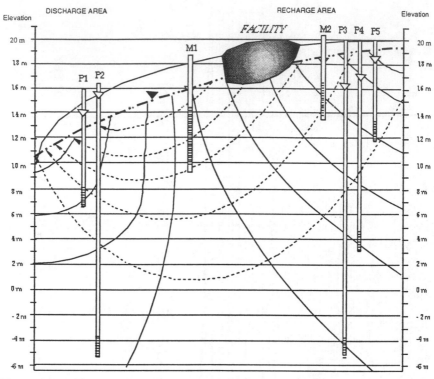

Figure 2.18. Piezometer heads for recharging/discharging system.

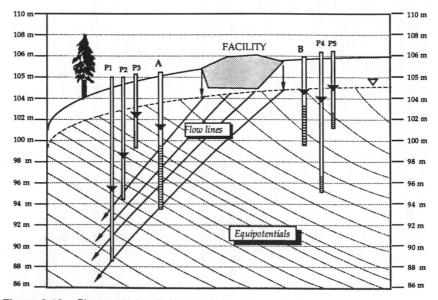

Figure 2.19. Piezometer heads for fully recharging system.

Single Homogeneous Aquifer

The relatively simple single homogeneous aquifer system only requires the following steps to define the target monitoring zone:

1. Evaluate aquifer geometry, thickness, and vertical and horizontal hydraulic conductivity variability through installation of soil borings;
2. Construct flow nets using water level/piezometric head information from piezometers or observation wells;
3. Prepare a conceptual geologic/hydrogeologic model and plot potential target monitoring zones; and
4. Install monitoring points to monitor potential contaminant flow paths.

Figure 2.20 illustrates the subsurface movement of leachate from an unlined solid waste facility in a humid environment. Selection of appropriate screen depths for downgradient wells is relatively simple using the procedure outlined above. Figure 2.21 (Freeze and Cherry, 1979) represents concentrations of chloride adjacent to an unlined solid waste landfill. The contours were based on water quality data obtained from numerous sampling points screened at various vertical intervals. The location of the target monitoring zone here would be the center line

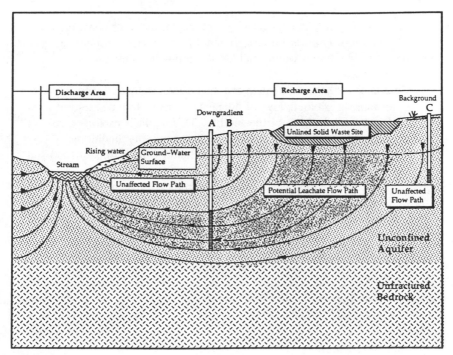

Figure 2.20. Single aquifer/homogeneous geology.

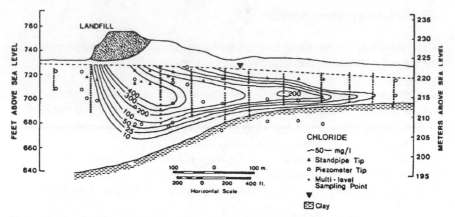

Figure 2.21. Movement of chloride in a homogeneous aquifer.

of the chloride plume. The center point, with the highest chloride concentrations, represents the most direct flow path away from the landfill. Monitoring wells located in this zone would provide the earliest detection of leachate excursion away from the facility.

Single Aquifer of Variable Hydraulic Conductivity

Differences in hydraulic conductivity with depth due to changes in stratigraphy can point the way toward establishment of an effective monitoring system. The procedure for design would include the following steps:

1. Determine the horizontal extent and thickness of individual geologic units by evaluating geologic logs of borings and wells to a depth of at least 25 feet below the base of the facility. This depth is used as a rule of thumb, and actual depths may vary, based on site conditions.
2. Establish hydraulic conductivity for each unit from results of field and laboratory tests;
3. Construct a flow net and a conceptual geologic/hydrogeologic model to select the target monitoring zone(s); and
4. Install monitoring wells based on defined target zones that represent primarily horizontal movement of ground water.

Figure 2.22 depicts a multi-lobed leachate plume from an unlined solid waste facility. Leachate movement in the system is represented by the horizontal excursion in more permeable silty sand and gravel zones. The monitoring system for this facility would consist of wells screened either in the silty sand unit directly adjacent to the facility, in the gravel, or in both. The extent of the geologic units, the potential for offsite migration to occur in the units, and the current or potential use of the water contained in the units are deciding factors in the actual system design. If the silty sand is discontinuous, the gravel would be the primary monitoring target zone. However, if the silty sand extended beyond the site boundaries, both the silty sand and the gravel would be targets for ground-water monitoring. The silty sand would represent the probable first affected unit, and the gravel would most likely represent a water supply for offsite downgradient water users.

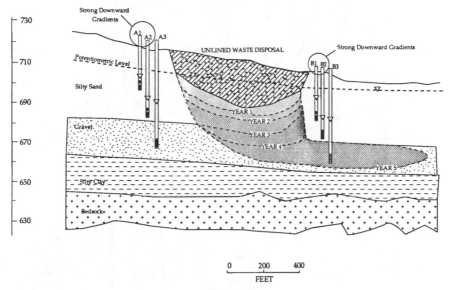

Figure 2.22.   Movement of leachate in a layered aquifer.

Figure 2.23 shows a sand and gravel unit as the uppermost aquifer beneath two tills. Typical of near-surface low hydraulic conductivity units, ground-water flow is nearly vertical in the tills. Ground water then moves horizontally in the much higher hydraulic conductivity sand and gravel aquifer. This unit is the only potential target monitoring zone for a facility located in this type of environment. The dominance of vertical flow in low hydraulic conductivity deposits and horizontal flow in continuous permeable zones is very typical. In glaciated regions, deeper sand and gravel valley fill or outwash deposits are often in direct contact with underlying weathered or highly fractured bedrock. Such systems would represent a composite target monitoring zone. Small lenses of sand within a mass of low conductivity material, however, do not represent adequate targets for monitoring. Thin or discontinuous sand lenses will not provide for horizontal movement of ground water away from a facility. Figures 2.24 and 2.25 represent the idealized cross-section of a facility located in a clay till above a bedrock aquifer.

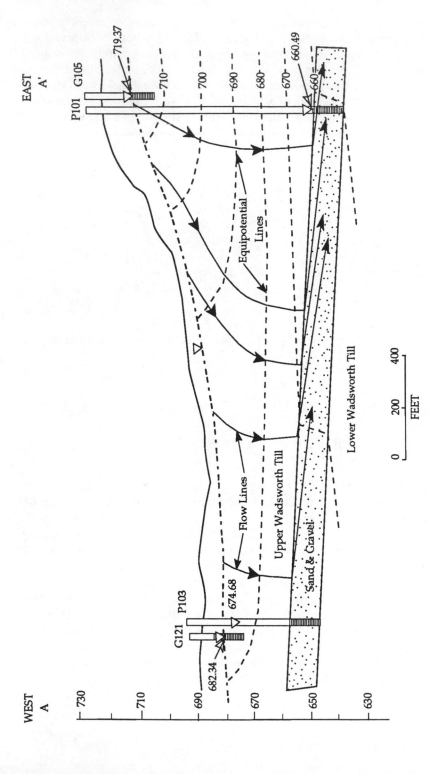

Figure 2.23. Flow net in a layered geologic system.

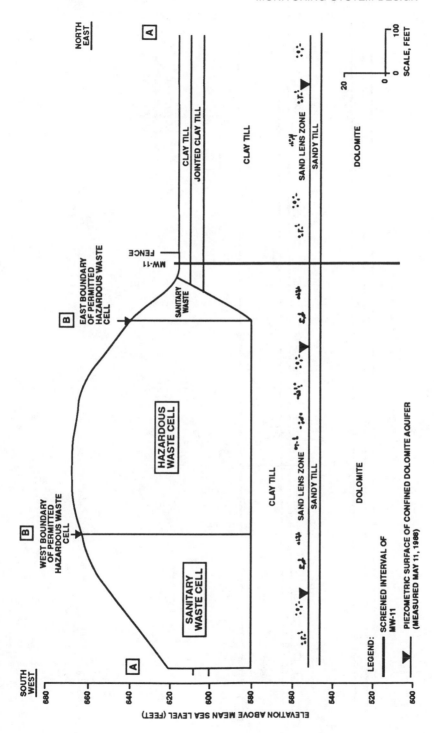

Figure 2.24. Idealized cross section in a layered geologic system.

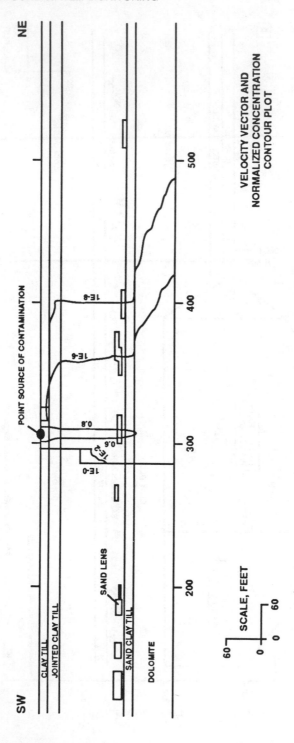

Figure 2.25.   Velocity vector and normalized concentration contour plot.

A series of discontinuous sand seams are present within the clay till. Numerical modeling of the system provided the velocity vector and concentration contour plots shown on Figure 2.25. A point source of contamination was simulated in the modeling exercise. The point source produced a plume that moved horizontally in near-surface material (the jointed till), vertically downward through the clay till and sand lenses, and, finally, horizontally in the underlying dolomite bedrock. The target monitoring zone in this situation is the dolomite, due to the following factors:

- the near-surface jointed till is shallow and does not represent a flow path away from the facility;
- the near-surface till can be influenced by vertical recharge events that are not associated with ground water passing beneath the facility (i.e., not in the flow path);
- the thick clay till and the sand lenses do not represent aquifers;
- the thick clay till and enclosed sand lenses, when considered as a composite unit, has a primarily vertical flow component; and
- the dolomite can yield water to monitoring wells and does represent a horizontal flow path from the facility.

Therefore, the dolomite would represent the target monitoring zone for the facility.

Multiple Aquifers

Multiple aquifers represent a challenge to the ground-water monitoring system designer. Ground water in different aquifers often moves in different directions. Thus, multiple aquifers require more complex three-dimensional hydrogeologic conceptualization to accurately establish a capable monitoring system.

Figure 2.26 shows a two-aquifer system with ground-water flow in opposite directions. Such a geologic environment would require significant geologic and hydrogeologic characterization to establish target flow paths away from the facility. The following procedure should be used to establish a ground-water monitoring system for a two-aquifer system such as that shown in Figure 2.26:

1. Install borings from the surface through all material down to competent bedrock;
2. Complete piezometers in each aquifer so that vertical and horizontal gradients can be established for each of the aquifers;
3. Establish hydraulic conductivity for each unit by conducting in-situ or laboratory hydraulic conductivity tests;
4. Construct flow nets for each aquifer; and
5. Establish the target monitoring zone(s).

If the goal of the monitoring system is to provide immediate detection of any contamination released from a facility (i.e., as in a RCRA detection monitoring

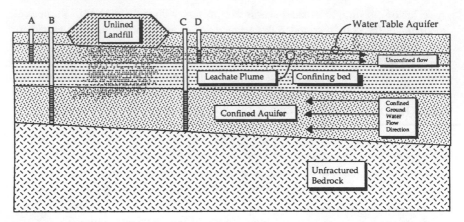

Figure 2.26. Conceptual two aquifer system.

program), the target monitoring zone should be the unconfined uppermost aquifer. If the goal of the system is to assess the extent of contamination emanating from a site (i.e., as in a RCRA assessment monitoring program), defining the rate and extent of contaminant movement would require monitoring in both the upper and lower aquifers. If a nearby surface stream serves as a base flow discharge point for one of the aquifers, the stream would probably also require water-quality monitoring. The monitoring program should also define if there is underflow beneath the stream.

Figure 2.27 shows a three-aquifer system including a deep, interconnected, fractured bedrock aquifer. As with the two-aquifer system, the assessment technique should be as follows:

1. Install borings to take soil samples sufficient to characterize the unconsolidated materials down to competent bedrock. The presence of fractured bedrock indicates that rock core drilling would be required to assess fractures and bedrock hydraulic conductivity;
2. Complete a series of piezometers in each geologic unit to establish vertical and horizontal hydraulic gradients;
3. Establish hydraulic conductivity for each geologic unit, including confining units;
4. Construct flow nets and ground-water contour maps for each aquifer;
5. Develop a geologic/hydrogeologic conceptual model and establish target monitoring zones; and
6. Install monitoring wells. Wells in the deeper units should be double-cased through the overlying units to prevent hydraulic cross-communication between units.

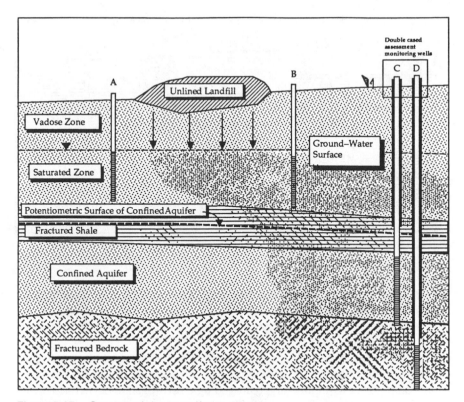

Figure 2.27. Conceptual three aquifer system.

As with the two-aquifer system, a monitoring system installed for the purpose of detecting contamination would focus on the uppermost aquifer to provide immediate detection of any contamination emanating from the facility.

Low Hydraulic Conductivity Environments

Probably the most difficult geologic environment in which to design a ground-water monitoring system is thick, low hydraulic conductivity materials overlying an aquifer at depth. Much of the controversy surrounding ground-water monitoring of hazardous waste sites is based on the difficulties in interpreting ground-water movement in low hydraulic conductivity environments. Figure 2.28 illustrates ground-water flow in a thick low hydraulic conductivity clay overlying a high hydraulic conductivity sand. The sand is confined and the clay contains minor sand lenses.

The shallow, unconfined ground-water surface is affected by the leachate collection system that acts as a subdrain for the adjacent low hydraulic conductivity

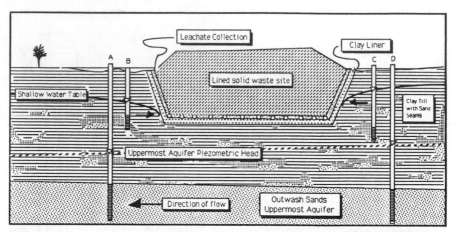

Figure 2.28. Low hydraulic conductivity environment/leachate collection effects.

clay, thus masking the directional components of the shallow ground-water flow. Piezometers should be installed within the clay and the uppermost aquifer (the lower sand), in order to define vertical gradients and to assist in selection of the target monitoring zone.

Geologic environments that consist of primarily low hydraulic conductivity units containing higher hydraulic conductivity deposits of significant lateral extent require comprehensive preliminary hydrogeologic investigations to define the target monitoring zone(s). Figure 2.29 illustrates a conceptual model of a buried channel located in much less permeable shale. One example of this type of system is the Cretaceous Dawson Formation, in the Denver, Colorado area, a formation consisting of materials deposited in a fluvial, deltaic environment. The Dawson stratigraphic sequence consists of depth-uncorrelatable, vertically stacked sandstone channel deposits, each of which is isolated within a fine-grained shale that originated as backswamp delta deposits. Thin, isolated sandstone lenses are present in the sequence that are characteristic of levee splay deposits and minor overbank deposits. These channel deposits can provide discharge pathways to local, recent alluvial materials present in ephemeral stream channels.

Based on an analysis of the depositional environment, the Dawson Formation deposits were laid down in a delta which gradually was uplifted by ancestral Rocky Mountain tectonics, in early Tertiary time, so that the stream gradient increased with time. Different characteristics of each of the sand sequences observed in cored boreholes emphasize that sands were deposited by separate and different stream systems and, therefore, not observed to be vertically interconnected. Minor sand lenses, such as the levee splay deposits or overbank sands which were deposited in backswamps, also have limited areal extent. They are connected horizontally over short distances and are vertically separated from other sandstones in the system by the intervening shales.

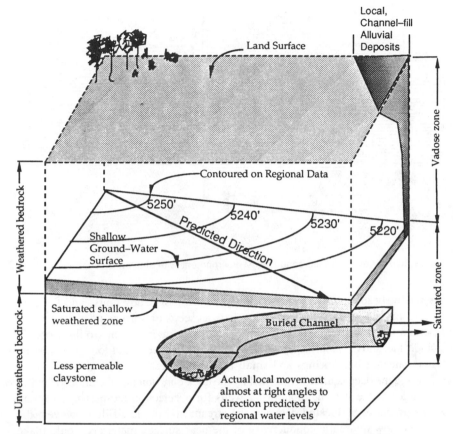

Figure 2.29. Low hydraulic conductivity environment with shallow discharge.

Near-surface Dawson shales are typically weathered and can become seasonally saturated as a perched ground-water system with sufficient hydraulic conductivity to comprise a target monitoring zone. This weathered zone can be easily defined by shallow (< 30 feet) borings and piezometers. The deeper sand channel deposits, however, present a more difficult directional flow analysis problem.

In this environment, each sand channel deposit has its primary component of flow in the stratigraphically down-dip direction. Monitoring such a heterogeneous geologic environment requires detailed field boring log information to define the location of these channels. These channels would serve as secondary target monitoring zones because they serve as subdrain systems for the shallow, unweathered shales.

Low hydraulic conductivity environments also may have a more permeable upper unit discharging locally to a stream or river. Figure 2.30 illustrates a thick clay unit confining a regional uppermost aquifer. The near-surface silty clay unit

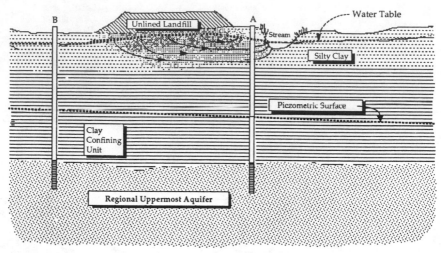

Figure 2.30. Low hydraulic conductivity environment with shallow discharge.

has relatively low hydraulic conductivity with minor sandy units, and the conceptual unlined landfill discharges leachate as seeps or springs near the landfill base and into the stream. Monitoring of such an environment would probably include alternative sampling of the stream. Visual inspection of local streams can provide insight into springs and small discharge areas.

Because the deep aquifer is confined by the thick clay unit and the unlined landfill discharges to the stream, monitoring wells for a detection monitoring program would probably be located between the stream and the landfill. However, the relatively low hydraulic conductivity of the near-surface materials would make monitoring difficult in practice due to long (days to weeks) recovery times for the wells. Additional piezometers should be located across the stream to verify that groundwater discharges along both sides of the stream.

Geologic Structural Control

Geologic structures such as dipping beds, faults, cross-bedding, and facies changes can greatly affect the rate and direction of ground-water movement. The monitoring system designer must consider geologic structural controls throughout the entire preliminary site investigation to ensure adequate site characterization. The development of a conceptual model is the key to successful ground-water monitoring system design in structurally controlled environments. Geologic structures affect ground-water movement in several ways:

- acting as more permeable flow paths, because of higher primary porosity (cross-bedded sands) or through secondary hydraulic conductivity enhancement (natural fractures); and

- acting as either barriers to ground-water flow or as conduits for ground-water flow, as do many fault zones, depending upon the nature of the material in the fault zone.

If a fault zone consists of finely ground rock and clay (gouge), the material may have a very low (i.e., $<1 \times 10^{-6}$cm/s) hydraulic conductivity. Significant differences in ground-water levels can occur across such faults. The monitoring system designer should be alert to large (i.e., $>10\%$), unexplained differences in water levels across a site underlain by faulted strata. These differences may be due to fault gouge retarding ground-water flow across the fault. Impounding faults can occur in unconsolidated materials, as well as in consolidated sedimentary rocks, where interbedded clays or shales can be smeared along the fault. Faultzone barriers are relatively common in the ground-water basins of southern California.

Definition of geologic structures as considerations in ground-water monitoring system design should include the following points:

- identification of major geologic structures specifically and generically, in regional studies conducted during the preliminary site investigation;
- identification of potential fault areas through review of relevant literature and aerial photos;
- identification of springs, vegetation changes, stream alignments, and surface geology through site reconnaissance prior to drilling;
- establishment of an initial conceptual geologic model;
- installation of borings placed to define geologic structure, variable water levels, and gradients in each geologic unit;
- reconciliation of stratigraphic boring logs and piezometer water levels with the conceptual geologic model; and
- construction of structural and ground-water elevation contour maps to develop a conceptual hydrogeologic model with flow nets to identify target monitoring zones.

The installation of the monitoring system should reconcile all available field data to complete the monitoring system.

The effect of fractures on leachate movement is illustrated in Figure 2.31. The steeply dipping, alternating beds of sandstone and shale have significantly different (i.e., a difference of 1×10^{-3} to 1×10^{-4}cm/sec) hydraulic conductivities. The three-dimensional view (Figure 2.31) shows preferential movement of ground water along the strike of the sandstones. The view illustrates the down-dip movement of contaminants. A detection monitoring system located in the shales would not establish the early leakage from the unlined site. One indication of the highly variable hydraulic conductivity of the rock mass is the overland flow of leachate. Because the leachate cannot move rapidly into the sandstone (and less so in the shales), leachate is rejected to the surface over the shale outcrop. A leachate seep

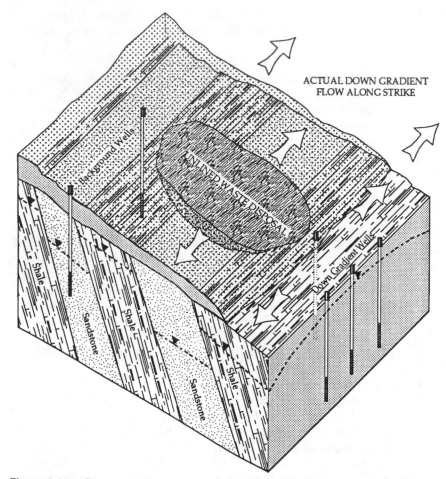

Figure 2.31. Structural effects on ground-water monitoring dipping layered rocks.

occurs at the contact between the sandstone and shale. Surface seeps (and springs) are excellent indicators of changes in formation hydraulic conductivity, and should always be considered in development of conceptual hydrogeologic models.

Perched Ground Water

Monitoring programs for perched ground-water environments present a number of complications to the monitoring system designer. Figure 2.32 presents potential leachate migration pathways away from an unlined waste disposal site. Perched groundwater does not follow regional ground-water gradients, but, rather, flows along an interface of hydraulic conductivity contrast, as shown in Figure 2.32. The approach to designing a monitoring system for this type of condition

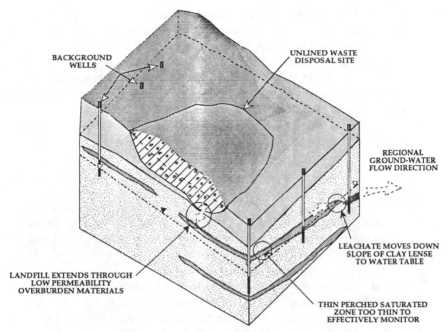

Figure 2.32.   Structural effects on ground-water monitoring perched saturated conditions.

should begin early in the site investigation of the facility. The design procedures would include the following:

1. Evaluate the lateral extent and thickness of various geologic units down to at least 25 feet below the base of the facility through continuous sampling of soil borings. Particular attention must be paid to the presence of saturated zones above low hydraulic conductivity (fine-grained) layers. A rule of thumb is that a potential perched zone may occur at a hydraulic conductivity contrast of two orders of magnitude. A hydraulic conductivity contrast of three orders of magnitude will almost always cause perched ground water. Contrasts between sand and silt or sand and clay will likely show a three-order-of-magnitude variation. Such contrasts in hydraulic conductivity will result in a relatively thin perched water zone that will make interpretation of flow direction very difficult if not properly recognized. Geologic cross sections can help identify potential locations of perched water.
2. Carefully evaluate road cuts in the vicinity of the site for the presence of geologic units that could cause perched conditions, even if the units in the road cut are above the saturated zone.
3. Sufficient borings should be drilled to define the horizontal extent of potential low hydraulic conductivity zones above the regional ground-water surface.

4. Piezometers (at least three) should be completed in each geologic unit, including permeable units above potential perching units, to establish the presence of thin layers of perched water.
5. A contour map of the top surface of the low hydraulic conductivity unit should be constructed to define potential perched water flow directions. The contour map should be combined with cross sections showing water levels, to establish perched water flow paths.
6. If the perched saturated zone is below the base of the facility and sufficiently thick to be characterized as an ''uppermost aquifer,'' or if it is sufficiently thick to allow collection of adequate ground-water samples to serve as an early detection location, it should be considered a target monitoring zone.
7. If the perched saturated zone is too thin to be saturated year-round (i.e., <2 feet), the monitoring system should be installed in the first permanently saturated zone (the uppermost aquifer) beneath the perched zone, as shown in Figure 2.32.

A detection monitoring system would be installed in either the perched water body, if a sufficiently thick (i.e., >10 feet) saturated zone exists, or within the deeper aquifer. A key to determining if there is a potential for perched water bodies is the stratigraphy present at the site. If clay or other fine-grained materials are present at the surface, the potential for occurrence of perched water bodies is greatly reduced, on the basis of limited recharge. If highly permeable (i.e., $>1 \times 10^{-3}$cm/sec) material exists at the surface with less permeable material below, then the presence of perched water bodies is more likely. The lower the amount of recharge, the less likely that the hydraulic conductivity contrast will act as a significant perching mechanism. Monitoring beneath the perched zone for quantification of rate and extent of contaminant movement would concentrate, in this case, on the first aquifer beneath the perched zone rather than on perched zone.

Secondary Hydraulic Conductivity

There are three basic types of ground-water occurrence and movement, as shown in Figures 2.33a through 2.33c. Figure 2.33a shows primary porosity where ground water moves through the interstices (voids) between granular materials. Figure 2.33b shows ground-water movement through fractures, which represent secondary porosity. Figure 2.33c shows ground-water movement through solution channels developed in a carbonate rock, another type of secondary porosity. A particular geologic environment could contain any or all of these media. The preliminary field investigation should determine the dominant flow mechanism beneath the facility to be monitored so that the appropriate locations for monitoring wells can be selected.

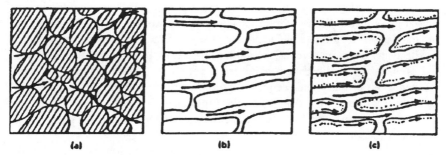

(a) (b) (c)

Figures 2.33a, b, c. Basic types of ground-water occurrence.

Fractured or solution-channeled carbonate rocks provide special problems in ground-water monitoring system design. Often there will be highly directional ground-water movement along discontinuities or solution-widened fractures or joints. The success of any monitoring system in a fractured or solution-channeled environment requires knowledge of the fracture patterns. In some instances, aerial photo interpretation (fracture trace analysis) and special field techniques (i.e., tracer tests) can point toward the target monitoring zone in a secondary porosity environment.

Most consolidated rocks (with the exception of some sandstones) have few primary intergranular openings for ground-water flow, and usually have much lower hydraulic conductivity values than their unconsolidated equivalents. Ground-water flow in bedrock aquifers often takes place through secondary openings such as fractures (joints, bedding planes), and/or solution channels. The investigator designing a monitoring system should fully identify areas in which this factor is important, and should do so at an early stage in the preliminary site investigation. Although regional flow patterns should be well established in the preliminary investigation, it is often very difficult to predict ground-water flow through a set of fractures or solution channels on a site-specific scale. Thus, facilities located directly atop bedrock should employ additional investigative techniques (e.g., fracture-trace analysis, detailed geologic mapping, and pumping tests specifically designed to evaluate anisotropy) to adequately determine likely ground-water flow pathways.

Fractured rock environments require consideration of specific flow paths to define the target monitoring zones. Figure 2.34 conceptually shows the individual ground-water flow paths in a fractured rock environment of a single rock type. Leachate is shown moving down from an unlined landfill toward a series of fracture sets that control local ground-water flow. The monitoring system designer would place wells both upgradient and downgradient of the facility. Individual well screen depths must be based on the results of the boring program and the observed fractures or weathered zones, rather than on only observed heads or ground-water levels. In an assessment monitoring situation, long (i.e., > 15 feet) screened zones should be avoided to reduce the potential for cross-contamination caused by contamination entering an upper zone through the top of the screen and moving downward into uncontaminated zones. Fracture patterns can be highly localized and unpredictable, as shown in Figure 2.34, or more evenly distributed and predictable, as illustrated in Figure 2.35.

Often, both primary and secondary porosity are present in bedrock, as illustrated by Figures 2.34 and 2.35, so the preliminary site investigation must include measurement of the hydraulic and geologic parameters for each of the media present at the site.

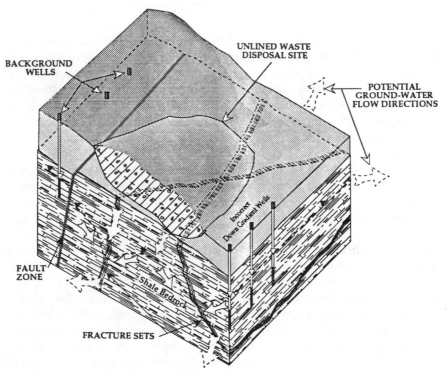

Figure 2.34. Structural effects on ground-water monitoring faulting and permeable fracture sets.

The design of a monitoring system in a fractured geologic environment should follow these procedures:

1. Evaluate fracture patterns using background information, aerial photographs (fracture-trace analysis), and observation of fractures at local exposures;
2. Establish a core drilling program at the site to include rock quality designation (RQD), fracture orientation of cores, and detailed visual

logging. Packer hydraulic conductivity tests should be considered for use in the assessment, and packer intervals should be selected to test specific anomalies observed during drilling. Consideration should also be given to including angle core drilling in those areas where vertical fractures may be present;

3. Implement borehole geophysical surveys, such as caliper logs, flow logs, and temperature surveys;

4. Install multiple piezometers or multilevel monitoring devices to assess hydraulic conditions in individual fracture zones detected in coring and geologic logging;

5. Measure piezometric heads and gradients in relationship to joint patterns or ''sets'' of joints or fractures; and

6. Establish a conceptual geologic model to define the target monitoring zones in plan and cross section.

A detection monitoring system is most effective in a fractured-rock environment if the wells are screened in the highly permeable fractures (those from which water will flow into the borehole) downgradient from the facility. These fractures

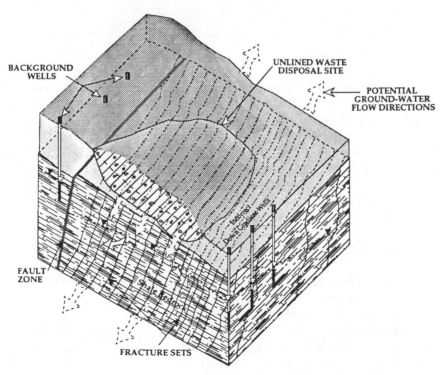

Figure 2.35. Structural effects on ground-water monitoring faulting and homogeneous fracture sets.

can react very quickly to releases from a facility, and thus serve as effective points for early detection of contaminant releases.

Highly solution-channeled limestone bedrock (karst) terrains present additional challenges to the designer of a ground-water monitoring network because monitoring wells can easily miss solution channels and, as a consequence, may end up as dry holes. Figure 2.36 illustrates monitoring of an unlined facility in a karst area, where sinkholes are present beneath and adjacent to the facility. Normally, such a setting is not easy to monitor, but a monitoring system can be developed to determine the facility's impact on the environment.

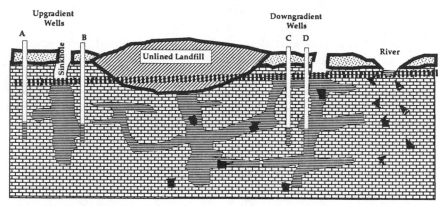

Figure 2.36.   Karst geology monitoring.

The assessment procedure would be similar to that used in a fractured environment with the possible addition of other surface geophysical surveys. Ground-penetrating radar, electromagnetic conductivity surveys, and seismic refraction surveys can help identify zones of solution channeling and deep weathering within the rock mass. Tracer tests are particularly effective in defining solution-channels and specific flow pathways away from a facility.

Definition of target monitoring zones in the plan view and cross section should be prepared before installed of any monitoring wells. In karst systems, gradients are typically very low and require very accurate surveys for definition.

Density Control

In this discussion, various monitoring system designs will be reviewed to illustrate some of the major implications of density and solubility of various chemicals in the design of ground-water monitoring networks.

In the preceding monitoring system designs, it was assumed that the contaminants to be monitored were soluble in water and that the density of the contaminant/water solution was approximately that of water alone. Although most leachates from solid waste disposal facilities fit this profile, there are a number

of monitoring situations in which these assumptions are not appropriate. In particular, wastes such as brines or product spills of halogenated organic solvents
may be soluble in water but have a density significantly greater than that of water.
Petroleum hydrocarbons and some organic solvents may be only slightly soluble
in water and have a density less than that of water. Physical properties of contaminants can result in transport characteristics quite unlike those normally associated with soluble, neutral-density leachates, and can affect the horizontal and
vertical positioning of monitoring wells.

The effects of density on miscible fluid disposal is considered in Bear (1972),
while several references including Schwille (1988), Corey (1977), and Collins
(1961) consider the mechanics of immiscible fluid movement. These are
represented schematically in Figure 2.37.

Figure 2.37a shows a contaminant plume developing as a result of seepage of
a dense miscible fluid into the saturated zone. The dense contaminant tends to
move vertically downward to the bottom of the aquifer. Once on the bottom, the
movement of the dense contaminant is governed largely by the topography of
the confining unit below the aquifer, and thus, the direction of flow will not necessarily be in the direction of ground-water flow. Due to solution and dispersion,
contaminants near the edge of the dense fluid body will be contributed to the local ground-water flow system. Thus, several areas of contamination may be
present, including: (1) the major body of dense miscible fluid; and (2) the adjacent ground-water zone contaminated by diluted levels of dense fluid as a result
of solution and dispersion at the perimeter of the dense fluid body. Establishing
the limits of the area of contaminated ground water would require employing
the typical methods of investigation that are used for any monitoring system design. Locating the major pool of dense miscible fluid would require knowledge
of the location of the upper surface of the first confining zone or low hydraulic
conductivity layer, and the installation of monitoring wells with screened intervals located at or near the bottom of the uppermost aquifer. In the case of high
density, miscible fluids that are leaked from a facility slowly over a period of
time, the density of the fluid as it mixes with ground water would decrease to
the point where the plume would migrate in a manner similar to a fluid with a
density the same as water, or according to the local ground-water flow conditions. In the case of a dense immiscible fluid, the portion of the fluid that is soluble in water would be contributed to the local flow system in the same manner
as with dense, miscible fluids, but much more slowly because of the lower solubility. If a large volume of fluid is leaked, the dense immiscible fluid body would
migrate to the bottom of the aquifer, move vertically according to gravity, and
only slowly solubilize over time. The movement of the dense immiscible fluid
body is controlled by the topography on top of the first confining layer beneath
the uppermost aquifer.

Figure 2.37b represents the migration of a miscible phase having a density less
than that of water. In this case, the major zone of contamination occurs near the
top of the saturated zone and migration is controlled by the slope of the water

VARIABLE TYPES OF PLUME GENERATION

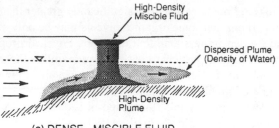

(a) DENSE - MISCIBLE FLUID

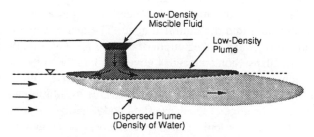

(b) LOW - DENSITY MISCIBLE FLUID

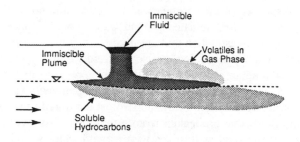

(c) IMMISCIBLE FLUID

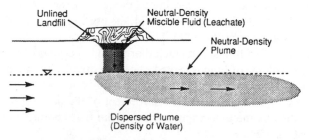

(d) NEUTRAL - DENSITY MISCIBLE FLUID

Figure 2.37a, b, c, and d.   Density control plumes.

table. As a result of solution and dispersion, the contaminant would be contributed to the ground-water flow system and the plume would gradually be dissipated over a potentially long distance. In the case of miscible, less dense than water product spills, monitoring points should be concentrated in the upper part of the aquifer if vertical flow components are not significant.

Figure 2.37c represents the infiltration and migration of an immiscible fluid having a density less than that of water. In this case, the main body of immiscible fluid remains above the top of the saturated zone (i.e., above the capillary fringe) and the soluble portion of the fluid is dispersed in the ground water and moves in the direction of ground water flow (American Petroleum Institute, 1989). As a light, immiscible fluid moves through a porous medium, a residual amount of fluid is retained in the pores of the medium in a relatively immobile state to slowly migrate to ground water as water percolates through the unsaturated zone and slowly solubilizes portions of the fluid. The soluble constituents of the light, immiscible fluid that are leached from the fluid will move to the saturated zone and eventually migrate along with groundwater as part of the regional ground-water flow system. Monitoring wells installed to detect light immiscible fluids should be located so that their screened intervals straddle the water table surface, so that both the soluble portion of the fluid and the insoluble portion can be observed.

Figure 2.37d represents a neutral density (density similar to that of water) miscible fluid moving from an unlined landfill into a local ground-water system. Because the leachate would move in a manner similar to that of the ground water in the aquifer, location of monitoring points would be based on the defined target monitoring zones and three-dimensional ground-water flow components. Ground-water monitoring system design for any of the above types of contaminants should employ both conceptual models and flow net construction, along with consideration of the contaminant density.

REFERENCES

American Petroleum Institute. "A Guide to the Assessment and Remediation of Underground Petroleum Releases," 2nd ed., API Publication #1628, American Petroleum Institute, Washington, D.C., 1989.

Bear, J. *Dynamics of Fluids in Porous Media.* (New York, NY: American Elsevier Publishing Co., 1972).

Collins, R. E. *Flow of Fluids in Porous Materials.* (New York, NY: Van Nostrand/Reinhold Publishing Corporation, 1961).

Corey, A. T. "Mechanics of Heterogeneous Fluids in Porous Media," Water Resources Publications, Fort Collins, CO, 1977.

Freeze, R. A., and P. A. Witherspoon. "Theoretical Analysis of Regional Ground Water Flow," Water Resources Research, Vol. 3, pp. 623–634, 1967.

Schwille, F. *Dense Chlorinated Solvents in Porous and Fractured Media.* Translated by J. F. Pankow. (Chelsea, MI: Lewis Publishers, 1988).

U.S. Environmental Protection Agency. RCRA Ground-Water Monitoring Technical Enforcement Guidance Document. Office of Waste Programs Enforcement and Office of Solid Waste and Emergency Response, OSWER-9950.1, 1986.

3

The Overall Philosophy and Purpose of Site Investigations

Lynne M. Preslo and David W. Stoner

INTRODUCTION

The use, storage, and handling of organic chemical compounds such as petroleum fuels and halogenated solvents, and inorganic materials such as heavy metals, have led to numerous cases of subsurface contamination that has been caused by occurrences such as leaks in underground storage tanks (USTs) or their associated piping; accidental spills during handling of chemicals or during filling of USTs; leaks in industrial sewer systems; or past practices which were acceptable industry standards at one time.

This subsurface contamination becomes increasingly important when one considers that the use of ground water is increasing nationwide, and in fact doubled from 1975 to 1985 (Table 3.1). This increase is due in part to population growth, but as shown on Figure 3.1, the use of ground water within the United States is growing at a rate that is even faster than that of population growth. Along with this increasing usage, environmental regulations have evolved to protect these resources, and it is due to these regulations that the need has surfaced within recent years for site investigations and ensuing engineering feasibility studies. This process then leads to the design and implementation of selected remedial alternatives.

In order to understand the magnitude of the problems presented by subsurface contamination and to devise ways of controlling and cleaning up contamination,

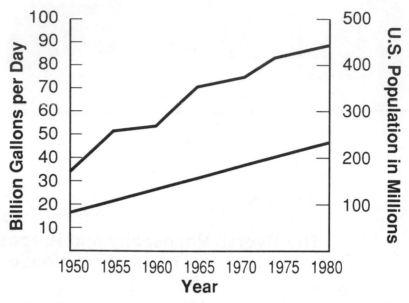

Figure 3.1. Trends in water use and population in the United States.
Source: USGS in U.S. EPA, 1987)

it is first necessary to develop a conceptual model to understand how contaminants behave, both physically and chemically within the subsurface. For example, their chemical properties, such as solubility in water and sorption within the soil organic matrix, dictate how quickly the chemical compounds move within the ground water. Physical properties, such as density and viscosity, determine whether a chemical solution, if present as a separate phase, will sink through or float on the water table within the subsurface. In understanding these concepts and designing mitigation measures, a three-phase approach, as illustrated in Figure 3.2, is typically employed. Briefly, the three phases are summarized below.

Table 3.1.   Ground Water Use Within the United States.

Time		Percentage
1975	•	25% of U.S. water consumption
1985	•	50% of U.S. water consumption
		—96% of all freshwater use
		—95% of rural water supply
		—40% of agricultural irrigation
		—35% of municipal water supplies

Source: Versar, Inc. in U.S. EPA, 1987.

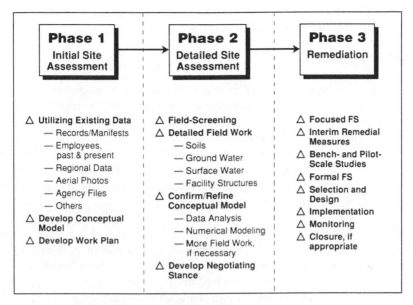

Phase 1

Initial Site
Assessment

Phase 2

Detailed Site
Assessment

Phase 3

Remediation

△ Utilizing Existing Data
— Records/Manifests
— Employees,
    past & present
— Regional Data
— Aerial Photos
— Agency Files
— Others
△ Develop Conceptual
    Model
△ Develop Work Plan

△ Field-Screening
△ Detailed Field Work
— Soils
— Ground Water
— Surface Water
— Facility Structures
△ Confirm/Refine
    Conceptual Model
— Data Analysis
— Numerical Modeling
— More Field Work,
    if necessary
△ Develop Negotiating
    Stance

△ Focused FS
△ Interim Remedial
    Measures
△ Bench- and Pilot-
    Scale Studies
△ Formal FS
△ Selection and
    Design
△ Implementation
△ Monitoring
△ Closure, if
    appropriate

Figure 3.2.   Phased approach to site characterization and cleanup.

Phase 1: Initial Site Assessment

Phase 1 typically consists of utilizing as much existing data as possible to develop a conceptual understanding (or model) of the site and processes. Existing data can include archived records such as manifests or chemical inventories, interviews with past or present employees, as-built drawings, regional information and aerial photographs, among others. In addition to outlining the conceptual model, Phase 1 develops the scope of work to confirm this conceptual understanding and assess whether or not remediation is necessary.

Phase 2: Detailed Site Assessment

Phase 2 consists of the field work portion of the site characterization, and ideally should employ at least three stages: (1) the initial field-screening stage to optimize subsequent, more intrusive, data-gathering points; (2) the more intrusive field program typically consisting of vadose zone monitoring devices, soil borings, and monitoring wells; and (3) the confirmatory stage to refine the conceptual model with data analysis and more field work, if necessary. The end product of this Phase 2 work consists of site characterization data that are sufficient to assess the magnitude and extent of contamination, evaluate whether remediation is necessary, begin the remedial alternative selection process, and set up negotiation stances with the various regulatory agencies.

Phase 3: Remediation

Phase 3 encompasses the remediation portion of the investigation effort and is implemented only if necessary, based on negotiations with the regulatory agencies. If impacts to soils and ground water have exceeded specified action limits or are deemed hazardous due to a risk assessment approach, then remedial measures are selected, designed, and implemented. Aspects of this phase include the selection of the appropriate technologies through a feasibility study process, bench- and pilot-scale testing if needed, interim and long-term remedial measures, subsequent monitoring, and ultimate closure.

The purpose of this chapter is to describe in detail the first two phases, their importance, standard procedures, and some selected case studies. Detailed field procedures will not be included, as other chapters of this book are devoted to this purpose. Phase 3 will be summarized briefly in order to give the reader an appreciation of the final product of site characterization and how it fits into the overall context of site cleanups.

PHASE 1: INITIAL SITE ASSESSMENT

As described in Figure 3.2, Phase 1 consists of the initial site assessment that typically does not involve actual field work. Rather, Phase 1 is designed to maximize the use of existing information and data to develop a preliminary conceptual model of the site and formulate the appropriate remedial alternatives, if necessary.

Utilization of existing information can take various forms. Among these are traditional library literature searches, computer database searches, searches of company records such as purchase requisitions and hazardous waste manifests, interviews with past and present employees, and review of aerial photographs, agency files, and other local or regional information. These sources of information are described within this section. Again, Phase 1 culminates in the generation of the conceptual model or understanding.

Making the Most of Existing Data

One of the most frequently overlooked but most useful tools in conducting site investigation studies is the use of existing data and existing literature as a basis for designing a hydrogeologic and solid investigation program. Such a review may include either the use of traditional literature search techniques or the use of computer literature searches.

Traditional Literature Searches

The easiest and best way to begin a literature search is by locating the most comprehensive and recent references that pertain to the subject that is being addressed. The bibliography contained in such references usually serve as a springboard for further investigation of the literature. The challenge is to find such references without a lengthy search. The least time-consuming method to discover good and timely references is to contact an expert in that particular field who has published on the subject or who is very likely to be familiar with the recent literature.

The next level of effort involves the use of a few select sources of information on the subject matter. In the area of geology and ground water, the U.S. Geological Survey (USGS) Index, state geological surveys, and selected water resources abstracts are three very useful sources of information on geologic and ground water conditions in a particular geographic area. The USGS has an index of publications that is available in most libraries. Additionally, the USGS provides periodic updates on more recent publications. Many state geological surveys have similar indices, most of which also are updated periodically. For example, the New York and Illinois State Geological Surveys have indices which are updated on an annual basis.

When the subject matter being researched is not purely ground water or geology, other sources of information can be tapped. For example, one may be researching characteristics of a particular contaminant and its behavior in the subsurface environment. Pollution Abstracts and Environment Abstracts Indices are good sources of information about the related and recent literature on the subject. With the Pollution Abstracts Index, the subject of interest is indexed alphabetically with references to abstracts contained in larger annually updated volumes. For example, if a researcher were searching for information on chlorobenzene, he would look up chlorobenzene in the index and find references to abstract numbers. He then would refer to the particular abstracts to evaluate whether those articles pertain to the subject of the study. Environment Abstracts Index has an advantage in that the index itself allows you to see the article, title, and subject without referring to the full abstract. Additionally, Pollution Abstracts Index contains mainstream sources of literature which are pragmatically oriented. In contrast, the Environment Abstracts Index tends to contain more esoteric information and contains more pure research than Pollution Abstracts.

Throughout much of the country, very useful sources of information on geology and ground water are unpublished theses at the Master's or PhD level. Most libraries have an index called "Dissertation Abstracts," which is updated periodically and lists the theses completed at accredited universities across the country. Often if one is researching in a library in the same geographic area being investigated, pertinent theses are likely to be in either that library or one in close proximity. If the thesis is located at a university that is far away, it often can be photocopied and sent to the researcher or can be provided by University Microfilms, Intl., of Ann Arbor, Michigan.

Sometimes the challenge is not in finding the literature, but finding a library that is open to use by persons not affiliated with a particular institution. Often universities in the same geographic area will allow persons not affiliated with the university to use library privileges either gratis or for a nominal fee. When such arrangements cannot be worked out, then a local city or county library can be used, and the information, if not available at the municipal library, can be obtained through interlibrary loan.

For information on practically oriented research, the National Technical Information Service (NTIS) is a valuable source of information. NTIS has become the reproduction service for all U.S. Environmental Protection Agency (EPA) documents. In the past, the U.S. EPA published many documents which were provided free to the public. U.S. EPA printing is now done only in limited numbers, but most of these documents are available as reproductions from NTIS after they are out of print at the U.S. EPA. In addition, NTIS carries many other publications from both private noninstitutional sources and other governmental and academic sources. An index is available from NTIS upon request and is entitled "Government Reports, Announcements and Index."

Several other sources of information are available; among them are: Georef Index, Chemical Abstracts Index, and Index to Priority Pollutants. A very comprehensive and long-running source of geological information is the Georef Index. Georef has both a thesaurus to aid in finding the right indexing word and also a guide to indexing. Another source of information is Chemical Abstracts Index, which is updated monthly and recompiled on an annual basis. The monthly index is much more timely, but the annual index takes less work to research a particular topic area. Finally, the Index to Priority Pollutants is a source of literature on particular contaminants.

Computer Literature Searches

Many library systems now have periodicals and books on a computerized database, as opposed to card catalogs. This computerized card catalog is handy, but it is not to be confused with the procedures used in researching a subject using a computerized literature search.

The principal advantage of a computerized literature search is that if done properly, it can scan the appropriate literature and a large number of different databases very quickly. The main disadvantage is that the researcher usually loses some control over the search, because there is usually at least one middleman involved in the process. Perhaps, ironically, even though the computer literature search technique uses computer technology, the success or failure of such a search depends on the human element. The ability of the researcher to communicate with the research librarian is of paramount importance. Normally, most computer literature searches are conducted at a library by a research librarian, although it is possible for a researcher to tie into various databases and indexes by modem from a computer terminal located at the researcher's facility. There are advantages to using a library system, or an established computerized literature search system, because such systems usually tie into a large number of databases, and may tie into a computer search middleman operator whose business it is to broker large numbers of source indices. An example of the large number of computerized databases available is shown in Table 3.2.

Table 3.2. Representative Sampling of Databases for Computerized Literature Searches.

DIALOG BRS ORBIT	DIALOG, BRS, and ORBIT are large commercial systems containing hundreds of computerized databases dealing with a broad scope of disciplines including technical and chemical literature and state and federal regulations.
CELDS	Computer-Aided Environmental Legislative Data System is a collection of abstracted federal and state environmental regulations and standards. CELDS provides quick access to current controls on activities that may affect the environment, as well as data for environmental impact analysis and environmental quality management.
DTIC	Defense Technical Information Center system is the resource for information on Department of Defense Research Development, Test, and Evaluation activities. It provides data on all stages of Defense Research and Development planned work, work in progress, and work completed or terminated.
NLM	The National Library of Medicine system contains a number of computerized databases containing toxicological and chemical information.
HAZARDLINE	HAZARDLINE is a comprehensive databank providing information on over 500 hazardous workplace substances, as defined by OSHA. Also included are OSHA regulations, NIOSH criteria documents, and information necessary for protection of the worker and employer.
CIS	The Chemical Information System is an integrated online system covering a wide variety of subjects related to chemistry.
LEXIS/NEXIS	U.S. federal and state case law, U.S. federal statutes and regulations, tax information, daily news to annual reports, etc.

Source: M. Walker, Librarian, Roy F. Weston, Inc., October 1989.

When a research librarian conducts a computer literature search on behalf of the researcher, it is of critical importance that the research librarian understands the topic of interest to the researcher. Very often the research librarian conducts an interview with the researcher to obtain information about the topic of interest. Based on this interview, the research librarian will choose several key words that will be used in combination to scan the indices for the compatibility of the key words to the literature. The computerized literature indices are indexed by key words. When the information is entered by data entry personnel, they are responsible for selecting the most important key words that pertain to a particular article. As a result, the researcher depends on two levels beyond his own interpretation of a particular article. The person who enters the information into the index interprets the publication and enters key words accordingly, and the research librarian interprets the needs of the researcher and enters key words accordingly. Then the computer matches the research librarian's key words with the key words found in the various indices.

One of the most significant conflicts confronting the research librarian in conducting any computerized literature search is the need to assess whether it is important for a particular search to be more comprehensive or more relevant. The comprehensive search will include a wide range of articles, some of which are not pertinent to the subject being addressed. The focused or "relevant" search may exclude some articles which may be somewhat tangential to the subject, but of interest as well. For example, if a researcher were investigating the literature related to the biodegradation of chlorobenzene in saturated flow systems, the comprehensive literature review might include only one key word, "chlorobenzene" and the result might be any article that related to the characteristics of chlorobenzene. A very relevant or specific search might use the key words chlorobenzene, biodegradation, and ground water. Unfortunately, when a large number of key words are used with a subject such as this one, the result may be that no relevant research is found. A compromise might be the use of two key words. One might try to scan using first, chlorobenzene and ground water, and then chlorobenzene and biodegradation in combination.

For someone to do his or her own computer literature searches, it is necessary for that individual or company to connect a computer terminal via modem and telephone lines to either an intermediate database company, or the database itself. A company called Dialog, a subsidiary of Lockheed, located in California, is a large intermediate computer database source. The National Water Well Association also now maintains a database and computer literature search system which can be accessed either by a research librarian at the National Water Well Association or directly by modem from a computer terminal. Any database system will allow the individual to set up an account number, usually with a minimal annual charge and with a time charge for actual computer use.

For persons desiring to learn more about the art and science of computer literature searches, there are at least two good periodicals published on the subject. One is entitled "Data Base" and the other one is entitled "On Line." "Data Base" tends to dwell more on the usefulness of various databases, while "On Line" tends to focus more on the techniques used to conduct successful literature or data searches. For example, "On Line" may review searching techniques or discuss the difference between the Environmental Index and Pollution Abstracts.

Interpretation of Existing Geologic Information

Frequently, the hydrogeologist must make preliminary hydrogeologic interpretations based on information from the literature. This type of information is used to design the type of ground-water monitoring program; i.e., the construction and location of sampling points. The details of how different monitoring systems are designed are covered elsewhere in this book. The purpose of this brief section is to give examples of geologic interpretations with hydrogeologic implications.

Surficial Geology

There are regions of this country where a knowledge of the surficial geology is critical to understanding ground water and contaminant migration, and hence the proper placement of monitoring wells. The best example of this is the glaciated terrain of the northeastern portion of the country. These deposits can range in permeability from 10^{-8} cm/sec to 10^{-1} cm/sec; that is, from dense lacustrine or marine clays to very permeable outwash gravels. Existing information may give an investigator an idea of the probable permeability of materials to be encountered, but more importantly will offer an interpretation of the deposits and history. The environment of deposition of surficial material can have a profound influence on large-scale flow systems. We may be able to observe from drill cuttings or split spoons that we are in a gravel deposit. We can use geologic interpretation to decide whether we are in a kamic, deltaic outwash, or modern alluvial gravel. The differences between these deposits can be extremely important. While alluvial or even outwash deposits may have linear channel features, deltas and kames tend to have areal changes in permeability; i.e., from coarser grading to fines or vice versa. As described in the previous section, there are numerous sources of geologic information. Sometimes the interpretation work is already done and sometimes it is left to the resources of the investigating hydrogeologist.

Bedrock Geology

Two characteristics of bedrock are most influential with respect to ground water movement and hence the success of a monitoring system: the porosity and the structure. The effect of porosity and structure on flow is treated in great detail in a number of excellent texts, among them Freeze and Cherry (1979), Davis and De Wiest (1966), and Todd (1980).

It is not the intent here to explain the theory of fracture flow, but to demonstrate that a prior knowledge of the bedrock characteristics can be critical to the successful design of a monitoring system. As an example, not all limestones and dolomites are prone to developing solution-enlarged fractures. Does the literature show that a particular formation lies at the diffuse versus the conduit flow end of the spectrum or somewhere in between? One needs to ascertain whether there are broad structural trends, i.e., folding or faulting, which affect the direction of flow. The broad structural features may influence the orientation and development of fractures. While some formations, such as the Ogallala formation of the central plains states, have a strongly developed primary porosity, there are many more which depend upon secondary fracture porosity to conduct flow. If the fracture orientation and density are sufficiently developed, a rock can behave like a relatively uniform porous medium. However, it is much more common for rocks to have fracture systems which create anisotropic, heterogenous conditions.

Very often the information needed by the hydrogeologist to anticipate the nature of bedrock ground-water flow is available in the geologic literature. To be a good hydrogeologist, one must start by being a thoughtful geologist.

Anthropogenic Influences on Ground-Water Systems

Man's activities have profoundly affected the nature and quality of ground water. This is particularly true in urbanized and industrial areas where intense human activity may have occurred over some period of time. Many industrialized and urbanized areas have been constructed near lakes, streams, or rivers. The existing geomorphic features related to these environments often include abandoned or preexisting channels, pools, swamps, and springs. When these lands are developed, many of the preexisting features are filled with debris, creating, in effect, a man-made ground-water flow system within these features, or altering the preexisting flow regime.

In many instances the fill material was not entirely inert. In the past, it was a common practice to use refuse to fill unwanted land. In addition, because many of the nearby land users in these areas have been industrial, pits, swamps or abandoned channels historically have proved to be inexpensive receptacles for industrial waste. Much of that industrial waste is now classified as hazardous.

Additionally, man-made features exist which may function as conduits for ground-water flow. These features include active or inactive storm and sanitary sewers, electrical or telephone conduits, and pipeline facilities. Where these lines or pipes have been constructed below the water table, either the pipe itself or the bedding materials in which the pipe is laid may function as highly permeable conduits for conducting ground water and contaminants.

Man's efforts to keep his environment free of contamination are very recent. Before the Clean Water Act was passed, many liquid industrial waste streams were discharged to surface water bodies. Ironically, once the Clean Water Act was passed and unrestrained discharge of waste to surface water bodies was curtailed, it prompted industry and government to dispose of many wastes in or on the ground instead of into surface water. The types of materials which have been disposed of on and in the ground, either purposefully or via uncontrolled releases, span the spectrum of inorganic and organic chemical compounds. It is important for hydrogeologists investigating a site not to assume that a site is undisturbed and uncontaminated. As more ground-water monitoring programs are being implemented across the country, we are finding that background ground-water quality is far from pristine, even in areas where one might expect the ground-water quality to be quite good. For example, in western Europe, studies have indicated that volatile organics are found at detectable levels and sometimes even at levels causing concern in ground-water flow systems where the only source for such volatiles is precipitation.

Understanding a site prior to initiating any subsurface investigation involves not just a geological research project but a research project focused on man's activities. There are a number of sources of information which can help direct a researcher to provide information about past uses of a particular piece of land. Some of these sources include the following:

Historical Society Records. Local historical societies can provide a wealth of information about past land uses, land owners, and significant occurrences which may give clues about subsurface conditions. For example, historical society records may contain maps showing that a city-owned park was the site 30 years ago of a coal gasification plant. Coal gasification plants were prolific producers of various types of contamination, much of which was disposed of on or adjacent to the plant site.

Aerial Photo Interpretation. Two types of aerial photos are useful for making interpretations about past practices, identifying possible historic source areas for the contaminants of concern, or analyzing ground-water flow pathways. Simple black and white aerial photographs may be useful if a series of these photographs was taken over a period of time. Often these photographs will show the progression of site activity over time, such as landfill filling of some preexisting feature or the operating periods and location of burn pit facility activities which are no longer apparent based on a site reconnaissance. For example, when combined with old topographic maps, aerial photos are useful in defining possible shallow ground-water flow pathways caused by infilling of former drainageways with coarse fill material. Additionally, aerial photographs are used for fracture trace analysis to discern ground-water flow pathways in terrain where the flow is dominated by fracture flow or solution channels.

Color infrared photography is less useful for documenting historical developments, because the history of color infrared aerial photography is relatively short. However, because infrared photography shows vegetation so well, it can be an indicator of the vegetative stress that may occur when plant or tree roots encounter contaminated ground water or contaminated soil. Such photography usually is available through commercial air-photo services, state planning and regulatory agencies, U.S. Department of Agriculture archives, and many times from industries. Photos of historical overflights are compiled and when compared at equivalent scales can provide a wealth of initial data to answer the following questions:

1. *What was the appearance of the site prior to deposition of wastes or before development?* Such information is critical to assess predevelopment drainage, topographic changes, and natural soils and geologic data.
2. *What were the modes and times of deposition?* Initial information on the site size and volume is provided so that the scope of site cleanup can be conceptualized, even before field work begins.

When evaluated by the site investigation teams of scientists and engineers, these photo-generated data can provide a fair degree of quantitation of the problem at hand and sometimes can direct the team to propose initial remedial options to be considered. However, subsurface information is required to quantitatively evaluate these options. For example, if these photographs were available to initially review, along with local ground-water data, the remedial action team might be able to delineate:

- possible source areas
- potential contaminant migration directions
- candidate remedial actions based on the qualitatively estimated volumes at the site

State and Local Regulatory Files. In some instances, state and local governments have attempted to gather information on past site uses which may have included the disposal of hazardous wastes within the area. For example, many states have been keeping files on landfills for at least the period of time during which a health department or environmental agency has been in existence. These files may be of use for indicating whether industrial and hazardous wastes have been disposed of in municipal landfills. Additionally, with the number of site investigations being implemented across the country, many spawned by underground storage tank regulations, agency files often contain records of numerous site investigations within a given area. In addition to waste and landfill records, public agencies also maintain well records and water quality information. These records often are confidential and agencies often require the written permission of the well owner before the agency can release the files to the requesting party (e.g., California). Some states maintain computerized databases that are regularly updated and easily accessed (e.g., Illinois), while others involve more time-consuming manual searches through archival files. Finally, water purveyors in various jurisdictions maintain records of well production, completion information, and water quality.

Industrial Files. For very good reason, large industries are reluctant to open their files and records to anyone, including those who have been retained to help them. Legal counsel often is concerned that information about past practices which is given to professionals may be forced from them during litigation. Although this is a difficult issue, it is one that usually can be resolved if the environmental management and legal representatives of an industry are made to understand the need for their consultant to construct a complete picture of site conditions. If the issue is one of confidentiality, then it usually can be handled by having the hydrogeologist or engineer work directly with the corporate attorney under attorney/client privilege.

The "Conceptual Model"

The science of hydrogeology involves conceptually understanding the site through the generation of models. While some models become very sophisticated and require the assistance of a computer, all work involves the generation of conceptual models or understandings. We develop conceptual models because we are dealing with a "black box" that we can test only remotely from the surface and because we understand only some facets of subsurface materials and processes.

When we get to the point at which we are ready to start drilling at a site, we should already have developed a preliminary model of the subsurface based on existing information, site reconnaissance, and remote sensing information. We then can begin the field work, starting with initial field-screening techniques, followed by confirmatory sampling monitoring, as described within the next subsection.

To illustrate the concept, behavior of various organic fluids in the subsurface will be discussed as examples in the development of a conceptual model. In developing a conceptual model, organic fluids that might be found within the subsurface fall into three categories:

- Light Nonaqueous Phase Liquids (LNAPL) or "Floaters"
- Dense Nonaqueous Phase Liquids (DNAPL) or "Sinkers"
- Solubilized portions of both LNAPL and DNAPL

LNAPL compounds are less dense (or lighter) than water, and if present in a separate fluid phase, will float on top of the water table (Figure 3.3). Petroleum fuel mixtures are the most common LNAPLs. Literally hundreds of individual chemical compounds comprise petroleum fuel mixtures, each mixture containing different proportions of the various classes of compounds. These classes include compounds such as the volatile aromatic fraction (e.g., benzene, toluene, ethylbenzene, and the xylene isomers [BTEX]), short- and long-chain alkanes, as well as additives such as ethylene dibromide (EDB), methyl-tertiary butylether (MTBE), or organo-lead complexes. Some of the more common fuel mixtures include gasoline, diesel, kerosene, No. 2 fuel oil, jet fuel, and No. 6 fuel oil.

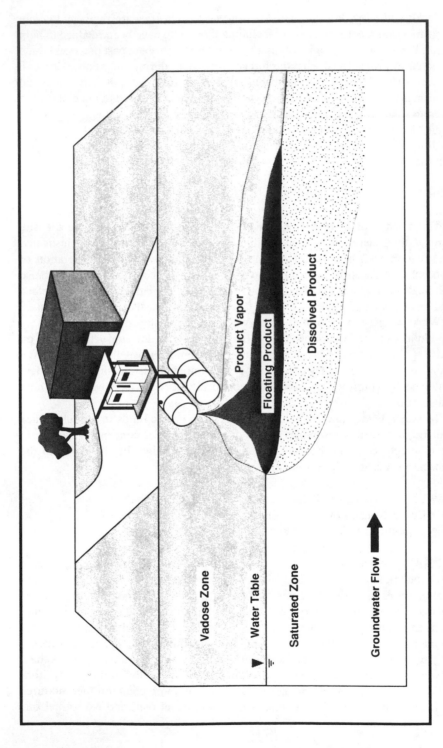

Figure 3.3. Conceptual subsurface distribution of light non-aqueous phase liquid (LNAPL).

To illustrate the different and complex character of these fuel types, gas chromatograms for selected fuel mixtures are presented in Figure 3.4. As demonstrated by these chromatograms, gasoline contains a larger proportion of the lighter and more volatile compounds such as BTEX compounds than does diesel fuel and so on. These differences in composition and behavior are important when designing the proper methods to both characterize and control subsurface contamination by petroleum fuel mixtures.

DNAPL compounds, such as halogenated solvents, are more dense (or heavier) and typically less viscous than water. Several commonly used solvents fall into the DNAPL classification; among these are methylene chloride, chloroform, trichloroethene (TCE), tetrachloroethene or perchloroethene (PCE), 1,1,1-trichloroethane, and various Freons (e.g., Freon-13, Freon-113, Freon-114). The chemical and physical properties of these DNAPL compounds vary, but most are slightly soluble in water and more dense and less viscous than water (Schwille, 1988). These properties cause the DNAPL fluids, when present as a separate phase, to sink rather rapidly through water; hence their common classification as "sinkers."

If DNAPL compounds are present as a separate phase within an aquifer system of porous or fractured media, then the fluid will sink through the water column until it encounters a zone of lower permeability, such as a clay confining layer or a bedrock surface (Figure 3.5). Understanding this behavior is critical when one considers the difficulty inherent in both characterizing as well as cleaning up an aquifer containing DNAPL. In this particular scenario, the DNAPL has leaked from an underground storage tank, and has traveled through the vadose zone to the water table, where it spreads out along the capillary fringe. Assuming enough volume has leaked and the DNAPL reaches a critical entry pressure (or a critical hydraulic head or thickness), it then enters and penetrates through the saturated zone. Once in the saturated zone, the DNAPL sinks to the bedrock surface where it flows with gravity along the bedrock surface, enters fractures, or collects within depressions in the bedrock surface. Again, this behavior is extremely important when characterizing and cleaning up the subsurface environment, because the DNAPL can flow with gravity in a direction *opposite* that of ground-water flow. As a result, both the monitoring and extraction well networks must be designed to accommodate this anomalous occurrence.

Vapor-phase transport within the vadose zone is also important when studying DNAPL behavior. The DNAPL diffuses through the pore spaces of the vadose zone predominantly according to its volatility (vapor pressure) and Henry's law constant for partitioning between the vapor and liquid phases. The vapor densities of DNAPL compounds relative to dry air also tend to be greater than that of air (Schwille, 1988) which allows not only chemical diffusion but also "density flow" of vapors in areas of concentrated DNAPL occurrence.

Another important consideration when studying both LNAPL and DNAPL behavior is residual contamination left within the soil pore spaces after the fluid has passed through either the vadose zone or the saturated zone. These fluids are held within the pore spaces by capillarity, and can be present as disconnected

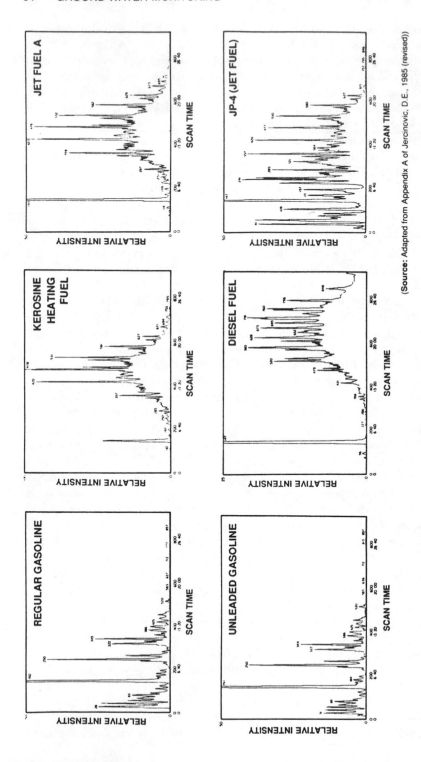

(**Source:** Adapted from Appendix A of Jercinovic, D.E., 1985 (revised))

Figure 3.4.   Gas chromatograms of various petroleum fuel mixtures.

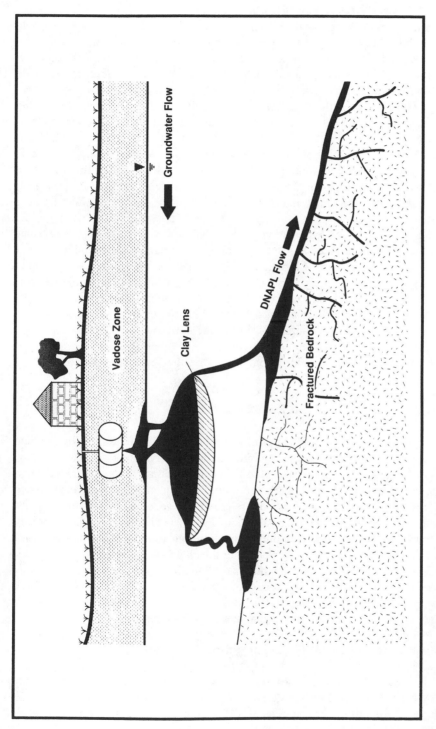

Figure 3.5. Subsurface distribution of dense non-aqueous phase liquid (DNAPL).

or minimally connected globules or tendrils within either the vadose or saturated zones. The more fine-grained a soil is, the more residual contamination it is capable of holding. These residuals then serve as additional sources of contamination for both vapor-phase and ground-water transport within the subsurface. For a more exhaustive discussion of residual contamination, the reader is referred to Schwille (1988), a pioneering work in the field, and Feenstra and Cherry (1988).

The distinctions between LNAPL and DNAPL behavior are important in the development of the conceptual model and in both characterization and cleanup of the sites that have been affected by either or both classes of compounds. When studying LNAPL problems, one must design ground-water monitoring wells so that their screened zones intersect the water table where floating product could be encountered (if separate phases are present). When studying DNAPL problems, the monitoring wells must intersect the bottom of the aquifer or water-bearing zone at the confining layer where a separate phase of DNAPL might be encountered. Similar considerations must be built into the design of ground-water extraction wells designed to extract not only the contaminated gound water, but also the separate phase of LNAPL or DNAPL, if present. It should be noted that while several feet of floating LNAPL have been encountered at numerous sites across the country, it is a rare occurrence to find more than several inches of DNAPL, if it is found at all in a separate phase. In fact, unless significant depressions exist in the confining layer at the bottom of a water-bearing zone, the DNAPL, if present, tends to spread out in a "pancake-like" form without accumulating to any significant thickness for either detection or extraction.

PHASE 2: DETAILED SITE ASSESSMENT

Generation of the conceptual model under Phase 1 often allows a fairly accurate assessment of site problems and an initial indication of possible remedial options. Phase 2 is then designed to collect the field data necessary to support the selection and design of engineering options if required for remediation. Such an approach therefore would include methods required to generate sufficient data for characterization of the site. Data collection ideally would begin first with minimally intrusive techniques, and then utilize the more intrusive techniques such as soil borings, test pits or monitoring wells after the initial site information is compiled.

The principal categories in the minimally intrusive techniques include shallow geophysics and field screening methods using onsite chemical analyses. Both categories contain a variety of techniques that may be applied to large and small sites in a manner that is both efficient and cost-effective.

Shallow Geophysical Field-Screening Techniques

Shallow sounding geophysical techniques include electromagnetic conductivity (EM), shallow and deep resistivity, magnetometric surveys, VLF surveys, IP surveys, ground penetrating radar (GPR) and in certain applications acoustical, gravity, and radiological surveys. Documents such as Pitchford et al. (1988) and Benson et al. (1984) describe these techniques in detail, and these techniques are discussed in detail elsewhere in this book.

These basic techniques often are applied in site investigations and have proved successful for initial site surveys if site conditions are amenable. The costs associated with running single unit or multiconfiguration geophysical surveys are relatively low, and the system responses for a variety of buried wastes are quite acceptable, provided experienced personnel design and interpret the surveys. Further, signal enhancement and filtering techniques using computerized enhancement codes are readily available for microcomputer systems, so that complex data analysis can proceed, even in the field.

A major inhibiting factor to the greater general use of surface geophysical techniques is that in the past, choices of inappropriate techniques were often made because of a general lack of understanding of the limitations and responses of the particular technique. Also, even with appropriate techniques and good data reduction, the vertical resolution of geophysical anomalies is often limited and can be plus or minus 10 feet. However, with selection and implementation of the proper techniques by trained personnel who possess a full understanding of resolution and other limitations, geophysical techniques are very powerful initial survey tools that can refine the preliminary phase of problem definition (often by orders of magnitude from other remote sensing data). Using surface geophysics, the level of understanding of a site and refinement of preliminary remedial options can increase by 10 or 100 fold without incurring massive project costs associated with the more intrusive techniques such as installation of soil borings or monitoring wells.

Onsite Chemical Field-Screening Techniques

In addition to remote sensing and surface geophysical techniques, onsite chemical analytical techniques are being increasingly used for field screening. Chemical screening in the field includes a number of techniques that are minimally intrusive, but still require field efforts costing an order of magnitude or more less than more intrusive techniques. By transferring the analytical laboratory into the field, several techniques are available or can be developed for volatile organics, nonvolatile organics, metallic or organo-metallic compounds and inorganic salts, among others. Field screening techniques provide information which streamlines data collection efforts by optimizing the use of intrusive techniques and the number of samples sent to the laboratory for confirmatory chemical analysis.

Site screening techniques, such as soil-gas screening, are gaining favor with regulatory agencies. For example, care must be exercised in its conduct and interpretation, but soil-gas screening can be used to pinpoint source areas or "hot spots" and delineate the approximately areal extent of volatile organic chemicals (VOCs) in soil and ground water through the monitoring of diffused gases in the overlying vadose zone. The analytical approaches for detection of soil gas reported in the literature vary from qualitative relative indicators such as organic vapor analyzers (OVA) or HNU meters to more quantitative soil gas screening applied in field survey conditions (Preslo, 1987; Lappala and Thompson, 1984). Further, applications of field screening of VOCs in soil vapor routinely are applied to check the integrity of underground storage tanks by both mobile and installed in situ monitors. As described in Table 3.3, the less qualitative meters (OVA, HNU, and IR detectors) are subject to interferences, false positives, and false negatives. The more quantitative methods, gas chromatography with various detectors, and gas chromatography/mass spectrometry (GC/MS), are compound-specific and produce much more reliable data.

Table 3.3.   Examples of Field Screening Techniques.

SOIL GAS METHODS (Volatile Compounds)

Instrument	Quantification	Limitations
OVA or HNU	~1 ppm	—False (+) & (−) —Interferences from water vapor —High detection limits
IR Detectors (e.g., Miran)	~1 ppm	—False (+) & (−) —Interferences —High detection limits

Field-Mounted GC
(with various
detectors)
	Can be ppt,
	depending on
	sample size
	and detector
		—Tentative identification
		—Time-consuming
		—Interferences
		   (Other target compounds or
		   extraneous substances)

Field-Mounted GC/MS	Can be ppt, depending on sample size	—Operator training

NONVOLATILE ORGANIC TECHNIQUES

Compounds	Technique

—PCBs
—PNAs
—Oil and grease
—Petroleum Hydrocarbons
	—Micro-extraction followed by UV,
	   GC/EC, GC/PID, or GC/MS
	
	—Hanby Analytical Services
	   Extraction/Colorimetric Technique

INORGANICS/METALS

Instrument/ Method	Quantification	Limitations
Colorimetric	~1 ppm to %	—Metal group specific —Interferences
Gravimetric	Qualitative	—Interferences
Specific Ion Probes	~1 ppm to %	—Ion-specific —Interferences
X-ray Fluorescence	~1 ppm; maybe lower	—EPA Region 2 exploring this technique
Hach Kits	Qualitative to semiquantitative, depending on analyte	—Interferences

Source: Adapted from Preslo and Leis, 1985.

In addition, field screening has been applied to the detection of less volatile organics on a specific analyte basis (Table 3.3). For example, uptake of polychlorinated biphenyls (PCB) by glabram (sticky sap) plants has resulted in the identification of depressed air plumes of these chemicals at old landfills in the upper Hudson River Valley. Another example is site screening of polynuclear aromatic compounds using ultraviolet light detection of polynuclear aromatics (PNA), techniques which have been applied for years by exploration geologists to detect petroleum hydrocarbons in drill cuttings. Additionally, microextraction techniques followed by conventional gas chromatography have been developed for PCBs and PNAs. Finally, techniques have been developed by Hanley Analytical Services of San Diego, California to semiquantitatively measure the concentration of various petroleum-fuel mixtures in soils or water. The technique shows great promise and growing acceptance by the regulatory community.

Field-screening techniques are not limited to organic chemicals. The technique of trace metal analyses in shallow material and in plants overlying ore bodies has long been an accepted technique to focus on a commercial metal deposit. In field screening for metals, a classic technique has been to use organic chelating agents to create an organo-metallic complex, the analysis of which is accomplished by a color produced by the specific complexing agent and the metallic salt. Techniques that are available include Hach kits for specific metallic or inorganic salts, as well as classical wet chemistry techniques such as colorimetry, gravimetrics, or specific ion probes. These techniques have been used in a semiquantitative fashion to estimate the concentrations of metals in a waste source, soil, or water; caution should be used in interpreting the results as these colorimetric techniques are subject to interferences. Additionally, X-ray fluorescence units that are portable and fieldworthy (Parametrix, HNU), have also been used to measure metals concentrations in soils; these techniques have been used successfully (Spittler, 1989).

Soil Gas Surveying as a Field-Screening Technique

General Usage

Soil-gas surveys have been employed widely using mobile laboratory capabilities as a preliminary step in hazardous waste site investigations throughout the country. By taking the "lab into the field," we have been able to use soil-gas surveys to quickly and cost-effectively screen sites for volatile chemicals.

Soil-gas screening is a tool that optimizes available resources for site investigations without collecting extraneous data. This screening technique produces relative measurements to locate "hot spots" or areas of elevated concentrations of volatile compounds in the soil gas. This information is then used to either locate source areas of volatile compounds within the vadose zone or track these compounds within the ground water if site conditions are amenable. Subsequently, this information is also used to optimize the locations of sampling points for soil borings and/or ground-water monitoring wells. Most importantly, significant overall cost savings results from reducing the number of confirmatory sampling points. In this manner, hazardous waste site investigations can be cost-effectively streamlined to collect only those data necessary to assess what remedial actions, if any, are needed at a particular site.

In the last four to five years, soil-gas sampling has grown from a virtually unknown, seldom used technique to one of the mainstays and essential tools for initially conducting Phase 2 of hazardous waste site investigations. When conducted properly, soil-gas screening is a very effective and comparatively inexpensive technique for the following purposes:

1. *Source-Area Identification.* Source areas of volatile chemicals within the vadose zone are identified using soil-gas techniques.
2. *Delineation of Onsite vs Offsite Sources.* Soil-gas screening assists in the differentiation between on- and off-site sources, including leaking USTs, pipelines, sewer lines, and process areas.
3. *Plume Tracking.* Soil-gas screening is used to track plumes of chemicals in ground water, depending upon site conditions such as depth to ground water and heterogeneity of the geologic materials comprising the vadose zone.
4. *Migration of Landfill Gases.* Soil-gas screening identifies the type of chemicals present and can identify the migration patterns of landfill gases. Care should be taken in trying to use soil-gas results to pinpoint source areas within landfills, as the production of landfill gases such as methane induce ephemeral pressure gradients that can shift apparent source area locations.
5. *Optimize Subsequent Monitoring Points.* Soil-gas screening is used to optimally locate and reduce the total number of more expensive, intrusive monitoring points such as soil borings and ground-water monitoring wells. It also is used to optimize the number of samples sent to the laboratory for confirmatory analyses.

With these purposes in mind, soil-gas screening has been applied to a myriad of site types where volatile chemicals have been historically or are currently used. Manufacturing facilities, fuel storage areas, metals fabricating plants, landfills, pharmaceutical plants, sewer lines, military facilities, aerospace facilities, and degreasing facilities are some of the site types where soil-gas screening has been successfully employed. Typical applications include:

- initial site assessment
- real estate transactions
- location of leaking USTs, leaks in pipelines or sewer lines
- in-field direction of onsite cleanups
- reduction and optimization of the number of soil borings and/or monitoring wells
- reduction and optimization of the number of samples sent to the stationary laboratory for confirmatory analyses

Program Design

Each soil-gas program should be customized to match the requirements of a particular site. All factors such as the size of the affected area, permeability of the subsurface materials, presence of perched water, and depth to ground water should be considered. This information affects the vertical and horizontal spacings and corresponding installation methods of the soil-gas probes across the site and in background areas. The nature of the site cover (asphalt, concrete, or bare soil) affects the design of the actual soil-gas probes. The probes can be easily installed outside or inside buildings and in manufacturing areas without disturbing routine site activities.

The types of compounds at the site, required detection limits, degree of compound specificity, and the objectives of each soil-gas survey define the analytical instruments that must be used. A variety of mobile analytical equipment is available for soil-gas investigations, including mobile gas chromatographs (GC) with appropriate detectors and mobile gas chromatography/mass spectrometers (GC/MS). The more qualitative instruments, such as the HNU and OVA devices, should be avoided as their usage can contribute to false positives and false negatives.

Benefits of Soil-Gas Screening

The use of a mobile laboratory in the field enables field teams to perform soil gas analysis and obtain real-time data that guide concurrent field investigations. As a result, precious time is not lost waiting for offsite laboratory analytical results, and the field investigations are expedited. Typically, results can be available to the field crew on a real-time basis, and a formal report delivered within one or two weeks of completion of the field effort.

In addition to real-time data, another direct benefit is the cost savings achieved by soil-gas surveys when actual costs are compared with those of traditional investigative methods such as soil borings or ground-water monitoring wells. As an example, a soil boring program with comparable coverage could cost two to five times as much as a soil-gas survey, depending on laboratory turnaround time and analyses required for the site.

Again, great care should be exercised in the use of soil-gas surveys and results. These results should be used as a tool to help guide subsequent confirmatory site investigation work and minimize the number of subsequent confirmatory points and samples. Soil-gas values without concentration values from other media should not be used to estimate the actual concentrations of chemicals in other media such as soil or water.

Confirmatory Site Investigation Techniques

While detailed descriptions of the drilling techniques and actual sampling protocols are found elsewhere in this book, several items bear mention and are summarized briefly within this section. These items include design considerations for monitoring both the vadose and saturated zones, and using cost-effective techniques such as onsite mobile laboratories, temporary groundwater sampling probes (i.e., HydroPunch™ or soil-gas probes driven into the water table), and/or cone penetrometer techniques (CPT) to optimize the data collection program. Although temporary sampling points for groundwater have proved quite successful (Edge and Cordry, 1989), care should be exercised with interpreting the results. A selected number of permanent monitoring wells should be installed to confirm actual concentrations. Again, this optimization is important so that data are not ''over-collected,'' yet enough data are gathered to adequately select remedial alternatives.

As the conceptual hydrogeologic model is refined based on the data obtained during Phase 1 and the screening level data obtained during the first part of Phase 2, it then becomes necessary to collect data at a quantitative, defensible level for actual site characterization and the resulting engineering decisions. These data are used to assess the magnitude as well as the lateral and vertical extent (and hence volume) of contaminated material present in the subsurface. These data form the basis for (1) deciding whether remediation is needed, and (2) selecting the appropriate range of remedial options, should they be required.

Two discrete zones are analyzed during this process: the vadose zone with its possible source areas and the saturated zone which could include multiple aquifer systems, as well as localized perched water within the vadose zone. Great care must be exercised, and specialized drilling techniques or configurations, such as conductor casings, often are required to prevent communication and cross-contamination between the various zones.

While soil borings and ground-water monitoring wells enable the hydrogeologist to collect samples and obtain actual quantitative measurements of chemical concentrations, hydraulic parameters, and lithologic data within these media, optimization of the number of sampling points and samples reduces overall program costs and increases efficiency. As stated above, optimization can be achieved by using cost-effective field techniques such as onsite mobile labs, the HydroPunch™ sampling system, and cone penetrometer methods. While each of these systems has several advantages (e.g., ease of operation, comparatively low costs, fewer permitting requirements), disadvantages that depend upon site conditions may render the technique inoperable or yield qualitative data to be used only for screening purposes.

PHASE 3: SELECTION OF REMEDIAL ALTERNATIVES

Once the site has been characterized and chemical concentrations indicate that remediation is required, many technical, economic, institutional, and environmental factors must be considered in the selection and design of remedial alternatives. Many of these factors are highly site-specific; therefore, blanket assumptions are not possible and no single alternative is best for all or even many circumstances.

Selecting and implementing a remedial method depends first of all on an evaluation of cleanup goals, which are the contaminant levels or concentration limits to which the site must be cleaned. These, in turn, are usually based on either regulatory standards or a risk assessment. Remedial options then are chosen by assessing the feasibility of each option to achieve the desired cleanup goal and evaluating the relative cost and site-specific acceptability of the method. It should be noted that the method selected may not always be the most cost-effective.

Four major considerations are important when evaluating the relative feasibility of each remedial alternative:

- technical feasibility
- implementational feasibility
- environmental feasibility
- economic feasibility

Evaluating the technical feasibility of the various alternatives requires applying basic technical principles involved with each method. Various types of information, such as whether a technology is proven or experimental, and results of successful case studies are evaluated with respect to the site at hand and the specific contaminants.

The feasibility of implementing each method then is evaluated in terms of practical considerations that will influence the applicability of a technology to a particular site. Selection criteria for implementational feasibility include:

- design considerations
- equipment requirements
- treatment and disposal needs
- monitoring and permitting requirements
- in the case of non-in situ technologies, onsite versus offsite operation

The environmental feasibility of each method is based primarily on the potential effectiveness of the method to achieve environmental goals for the media of concern—soil, ground water, surface water, and air—as well as exposures to biota in each of those media. Environmental feasibility is based on risk analysis methods that involve two interdependent factors: (1) pathways through which the substances can reach biological receptors, and (2) exposures resulting from the various migration pathways.

Risk analysis then can be performed by combining pathway analyses with potential exposure risk and assessing the resultant health or damage risk. Risk analysis must, by necessity, be based on site-specific considerations.

The economic feasibility of the various methods is a weighted comparison of cost-effectiveness and cleanup effectiveness. This comparison, again, can only be done understanding many site-specific factors. Typically the following cost categories are considered:

- capital costs
- installation costs
- operation and maintenance (O&M) costs
- relative cost-effectiveness

Available remedial technologies for soils and/or ground water are divided into two categories: in situ treatment and non-in situ treatment. In situ treatment refers to treatment of soil or ground water within the subsurface without excavation. In contrast, non-in situ technologies involve excavation and treatment disposal either on- or off site. The remedial technologies can be further subdivided within these categories as follows:

- In situ technologies:
  —vacuum extraction (or volatilization)
  —biodegradation
  —leaching and chemical reaction
  —vitrification
  —passive remediation
  —isolation/containment
- Non-in situ technologies:
  —land treatment
  —thermal treatment
  —asphalt incorporation
  —solidification/stabilization
  —ground-water extraction and treatment
  —chemical extraction
  —excavation and offsite disposal

Several documents are available that describe these technologies in detail; among them are EPRI (1988), Preslo et al. (1989), and U.S. EPA (1987).

REFERENCES

Benson, R. C., R. A. Glaccum, and M. R. Noel, "Geophysical Techniques for Sensing Buried Wastes and Waste Migration," National Water Well Association, Worthington, OH, 1984.

Edge, R. W., and K. Cordry, "The Hydropunch™: An In Situ Sampling Tool for Collecting Ground Water from Unconsolidated Sediments, *Ground Water Monitoring Review,* Summer 1989, pp. 177–183.

Davis, S. N., and R. J. M. De Weist. *Hydrogeology* (New York: John Wiley & Sons, 1966).

Electric Power Research Institute (EPRI), *Remedial Technologies for Leaking Underground Storage Tanks,* by L. M. Preslo, J. Robertson, D. Dworkin, E. Fleischer, P. Kostecki, and E. Calabrese (Chelsea, MI: Lewis Publishers, Inc., 1988).

Feenstra, S., and J. A. Cherry, "Subsurface Contamination by Dense Non-Aqueous Phase Liquid (DNAPL) Chemicals," presented at the International Groundwater Symposium, International Association of Hydrogeologists, Halifax, Nova Scotia, May 1–4, 1988.

Freeze, R. A., and J. A. Cherry. *Groundwater* (Englewood Cliffs, NJ: Prentice-Hall, Inc., 1979).

Lappala, E., and G. Thompson, "Detection of Groundwater Contamination by Shallow Soil Gas Sampling in the Vadose Zone: Theory and Applications," in *National Conference on Management of Uncontrolled Hazardous Waste Sites,* HMCRI, Silver Spring, MD, 1984, pp. 20–28.

Pitchford, A. M., A. T. Mazzella, and K. R. Scarbrough, *Soil-Gas and Geophysical Techniques for Detection of Subsurface Organic Contamination,* U.S. EPA, EPA/600/4-88-019, NTIS, 1988.

Preslo, L., "Soil-Gas Screening as a Tool to Optimize Site Investigations and Cleanup," *Weston Way,* Roy F. Weston, Inc., West Chester, PA, 1987.

Preslo, L., and W. Leis, "Field Screening Techniques for Hazardous Waste Site Investigations," Association of Engineering Geologists Annual Meeting, 1985.

Preslo, L., W. Leis, and R. Pavlick, "Field-Screening Techniques: Quick and Effective Tools for Optimizing Hazardous Waste Site Investigations," *Petroleum Contaminated Soils, Volume 2* (Chelsea, MI: Lewis Publishers, Inc., 1989).

Preslo, L., M. McLearn, M. Miller, P. Kostecki, and W. Suyama, "Available Remedial Techniques for Petroleum Contaminated Soils," *Journal of the Air Pollution Control Association* (APCA), 1988.

Schwille, F., *Dense Chlorinated Solvents in Porous and Fractured Media,* translated into English by James Pankow (Chelsea, MI: Lewis Publishers, Inc., 1988).

Spittler, T., U.S. Environmental Protection Agency, Massachusetts, Personal Communication, 1989.

U.S. EPA, "Managing Underground Storage Tanks," A16404.5500; National Audio Visual Center, Washington, DC, February, 1987.

4

Monitoring and Sampling the Vadose Zone

Thomas Ballestero, Beverly Herzog, O. D. Evans, and Glenn Thompson

INTRODUCTION

This chapter focuses on an aspect of ground water for which there is relatively little knowledge and yet which is increasingly becoming the most important link when considering human-related influences (i.e., artificial recharge, septic systems, landfills, etc.) on the environment: the vadose zone. Definitions and hydraulic theory are presented briefly before the discussion of actual monitoring techniques. The purpose of the chapter is to enlighten the reader concerning vadose zone hydrology and the possibilities for incorporating vadose zone monitoring as an integral part of ground-water investigations.

CHARACTERISTICS OF THE VADOSE ZONE

Definitions and Terminology

The word "vadose" comes from the Latin word "vadosus," meaning "shallow." According to *Webster's Third New International Dictionary,* vadose is defined as, ". . . of, relating to, or resulting from water or solutions in the part of the earth's crust that is above the permanent ground-water level." Thus, the vadose zone is that region of the shallow subsurface bounded on top by the earth's surface and on the bottom by the water table.

The dominant terminology for the porous media and interstitial fluids that exist within these media above the water table is either the vadose zone or the unsaturated zone. These terms are used synonymously. Other descriptors which may be seen intermittently include the tension-dominated zone (saturated and unsaturated) and soil-moisture zone. The following sections identify the basic terms and definitions involved with the vadose zone, the physics of fluid movement and, most importantly, the monitoring and parameter delineation of the vadose zone.

Multiple Phase Components of the Vadose Zone

The two basic categories that can be described for the components of the vadose zone are the solid and nonsolid (or fluid) phases. Due to the hydrologic and agricultural importance of the fluid phase, more descriptive categories exist than for the solid phase, including: water (soil water, soil moisture), vapor (air, soil gas) and immiscible liquids (hydrophobic fluids). Each phase will be described in the following subsections.

Solid Phase

The solid phase of the vadose zone is characterized predominantly by the permanent skeletal structure through which the fluid phases may pass or be retained. This solid skeletal structure is composed of inert particles of fractured rock, cobbles, gravel, sands, silts, and clays as well as organic matter such as roots, leaves, and organism waste products. Small solid particles can be transported by fluids through the interstices of porous media. Due to the relative magnitude of velocities found in the vadose zone, the portion of the solid phase that is mobile is extremely small and for all practical purposes may be neglected.

Fixed and floating microorganisms comprise a second element that can be included with the solid phase. A classic example of the fixed microorganisms is the mat of microorganism buildup that occurs underneath the leach fields of subsurface wastewater disposal systems (i.e., septic systems). An example of the floating microorganisms is found where ground-water samples produce coliform or noncoliform bacteria cultures. Fixed soil bacteria are commonly found at concentrations of 10^6 to 10^{10} cells per gram of soil.

Physical mechanisms used to describe the inert solids in the solid phase include grain-size distribution, porosity, angularity, specific surface, and uniformity. The organic matter is typically separated from the inert solids and just identified as a percentage of the total volume. Many of the physical descriptors, which are interrelated, are also related to the characteristics, content, and mobility of the fluid phases. For example, porosity directly affects the amount of water that can exist in the voids—the lower the porosity, the lower the capacity for water content in the vadose zone. In addition, for a given pore-size distribution, lower porosity means fewer pores and lower capability for transmission of fluids through the vadose zone.

Sedimentary Deposits. The grain-size distribution curve represents the cumulative probability of occurrence of the various grain sizes found in sedimentary deposits. Again, this is manifested in effects on fluid phase content and transmission characteristics. Each individual grain can be physically described by three linear measurements: length, width, and breadth (in descending magnitudes). In trying to fit a grain through a square opening, the smallest sized opening through which the grain may pass is roughly equivalent to the width dimension. By taking a bulk sample of aquifer material and sieving the sample with a sequence of sieves (largest sieve opening first, smallest sieve opening last), each sieve would pass or retain a certain portion of the sample. If the first sieve retained 10% of the sample weight, then 90% would pass this sieve. If the next sieve retained 30% of the sample (by weight), then 60% of the original sample would pass the second sieve and 40% would be retained by the second sieve. (Recall that 10% was on the first sieve which had larger sieve openings; if particles were retained on this sieve, they would also be retained on the second.)

Sieving information is plotted as particle (sieve) diameter on the abscissa axis (usually a log or probability scale) and percent passing on the ordinate axis. Not uncommonly, the percent retained (100% minus the percent passing) is also used for the ordinate axis. For grain size increasing to the right and percent passing increasing upwards (see Figure 4.1), the grain size distribution is characteristically S-shaped. Selecting a particular value from the percent passing scale, say 85%, the grain-size diameter from the grain-size curve is defined as D_{85}.

Two important statistics of the grain-size distribution are the median grain size (D_{50}) and the uniformity coefficient ($C_u = D_{60}/D_{10}$). Figure 4.1 depicts a grain-size distribution for a material comprised predominantly of sand-sized particles. Here, $D_{50} = 0.4$ mm and $C_u = 1.8$ mm. This grain-size distribution was determined by sieve analysis for particles of silt size or larger and by hydrometer analysis for clays (fines). As the grain-size distribution curve becomes more vertical, the uniformity coefficient tends toward unity. For highly diverse particle sizes, the curve is flatter and the C_u is larger.

The importance of D_{50} and C_u with respect to the fluid phases is straightforward. For a constant C_u, average pore size increases with increased D_{50}. For a constant D_{50}, increases in C_u decrease the porosity and average pore size. This latter relationship results from the fact that poorly graded deposits will have smaller grains filling in the interstices between the larger grains.

Particle shape (angularity) can also have a bearing on porosity and fluid transmission. The shape of individual particles can range from spherical to very angular to flat plates. Depending on the packing and mixing of these particles, a wide range of porosities are likely.

The last important descriptor of the solid phase is its specific surface—the ratio of a grain's surface area to its volume (or in some cases, to its mass). Grains with the largest specific surface are flat plates (clays) and those with the smallest specific surface are spheres (weathered silicate grains). Due to the importance of surface chemistry in contaminant transport studies, clay content of the medium is important, partially because of the high specific surface of clays.

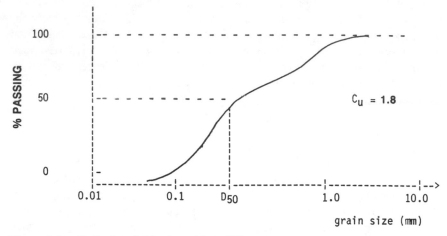

Figure 4.1. Grain size distribution of Lee, NH sand.

Fractured Rock. For fractured rock, a massive structure is decimated by many/few large/small cracks. On a very large scale, the mechanics of such a vadose zone may be no different than a sedimentary deposit vadose zone. The perverse nature of such a medium is that the effect of scale confounds traditional thinking, monitoring, and analysis.

In revisiting what Bear (1979), Corey (1977) and McWhorter and Sunada (1981) describe as the "representative elemental volume" (or REV), the primary difficulty of fractured rock vadose monitoring may be identified. A REV is that volume of material which must be used in order to obtain a valid estimation of a particular parameter. This concept is illustrated for porosity in Figures 4.2 and 4.3. If the sample size is too small, the sample will be heavily influenced by microscopic features, such as individual sand grains or pores. Hence there is much scatter in the data (Figures 4.2 and 4.3). If the sample size is too large, the sample may incorporate more than one type of geologic material (Figure 4.2). Therefore, the REV covers the sample size range between these two extremes. For a sand, the REV is approximate 10^3 mm^3 (Figure 4.3).

Extending this same logic to a fractured rock vadose zone, a sample size on the order of 10^7 m^3 may be required for an accurate estimate of porosity (see Figure 4.3). For a vadose zone of 10 m thick, this would require an area of 1 km by 1 km. When considering that most of our field instruments, at best, investigate between 0.01 m^3 to 5 m^3, it is easy to see why it is difficult to work with fractured rock vadose zone. This fact alone has led us toward two basic avenues of investigating fractured rock vadose zones: (a) disaggregation of the problem into that of a solid mass where no fluid phase occurs and there exists a continuum of interconnected pore spaces, or (b) re-evaluating the REV by either taking many small samples or few large samples. Identification of the REV is paramount in dealing with fractured rock systems.

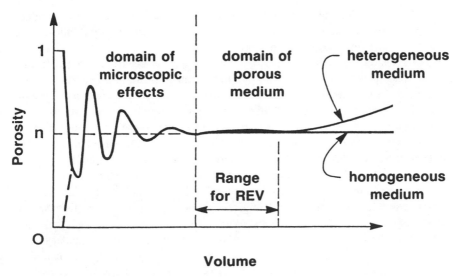

Figure 4.2.   Definition of porosity and representative elementary volume (REV). [From Bear, 1979.]

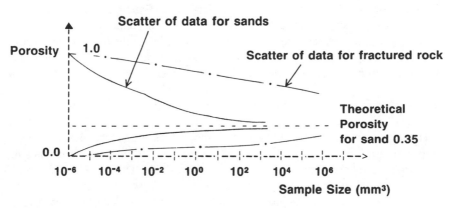

Figure 4.3.   Scatter of porosity data for sands and fractured rock.

Vadose Zone Water

A given molecule of water may reside in the vadose zone from minutes to centuries, depending on the size of a particlar vadose zone and its transport characteristics. Mechanisms by which water may enter the vadose zone from above include precipitation and recharge (i.e., rainfall infiltration, spreading basin, septic system, etc.). From below, water may flow from the saturated zone into the vadose zone. Lastly, and least importantly, water may enter the vadose zone from within due to any of the numerous biological or chemical reactions that have water as an end product. For example, the biodegradation of gasoline leads to the production of carbon dioxide and water. Also, water vapor which enters the vadose zone may be condensed into the liquid phase.

Just as water may enter the vadose zone, it may also exit. At or just below the ground surface, water may exit due to evapotranspiration processes. Below, water may drain into the saturated zone. Finally, vadose zone water may be consumed by certain biological or chemical reactions within the vadose zone.

Fluid properties which are important in describing vadose zone water (as well as other vadose zone fluids) include: density, specific weight, kinematic viscosity, bulk modulus of elasticity, vapor pressure, surface tension, dynamic viscosity and wettability in the presence of air. Detailed descriptions of these fluid properties may be found in most texts on either fluid mechanics or hydrology.

An important characteristic relating any of the liquid phases, in this case soil water, to the solid phase is wettability. Wettability is that property which is characterized by the relative interfacial forces of two fluids at a solid boundary. The two fluids, practically speaking, may be liquid-liquid (i.e., water-oil) or liquid-gas (i.e., water-soil gas). Thus, wettability describes the relative affinity for one fluid over another to a solid surface. With the two fluids in the presence of the solid, one fluid will preferentially coat the solid surface. For example, with water and air on glass, water wets and thus will tend to coat the glass. This explains the shape of the meniscus in a glass capillary tube. Extending this example to the vadose zone, soil water will tend to coat the soil particles and move in very tortuous paths in a porous medium, while air will be left in the larger pore spaces. For petroleum products and soil water, water preferentially wets over the petroleum products, even if the petroleum was there first.

Gas/Vapor Phase In The Vadose Zone

The important practical characteristic of the vadose zone is that the pores contain more than one fluid. Other than soil water, the fluid of most interest is soil gas. There is always a trade-off between the two fluids. During precipitation or recharge events, water in the pores increases and the soil gas must move out. When the vadose zone is draining water to the ground water or when it is drying out due to evapotranspiration, soil gas in the pore space increases at the expense of soil water.

Soil gas has properties similar to those of soil water. In addition, the perfect gas constant is of utility. It is important to recognize that the liquid (soil water) vapor pressure will require that the soil gas and liquid ultimately come into equilibrium. Thus, there will be a certain amount of water vapor found in the soil gas. More importantly, volatile chemicals will also be found in the vapor phase if the solid or liquid phase of the chemical is present in or near the vadose zone. Therefore, the vapor phase can be sampled and analyzed in order to make statements about liquid fluid phases in the vadose zone or below; for example, lying on top of the water table or contained within the saturated zone.

Immiscible Fluids In The Vadose Zone

Fluids other than water or vapor may be found in the vadose zone. Fluids that can easily mix with water (i.e., septic system leachate, etc.) are known as miscible fluids. Miscible fluids in the vadose zone typically should have similar dissolved solids concentrations, temperature, and density as the existing vadose zone water; if not, the fluid may temporarily be considered immiscible. For example, a septic system effluent may act as a fluid unto itself and not readily mix with existing vadose zone water until the leachate temperature moderates with its surroundings.

In other cases, there are fluids in which the primary composition is that of hydrophobic molecules. Here, the fluid may never mix with water and is considered immiscible. Most immiscible fluids exhibit some solubility in water. For practical purposes, it is assumed that the immiscible fluid retains its original volume integrity. For example, gasoline and water do not readily mix, but when keeping the two in a closed container for a few months, traces of some gasoline constituents (i.e., benzene, toluene, xylenes) will be found dissolved in the water phase in the container.

When an immiscible fluid such as gasoline enters the vadose zone, there is increased competition for the void spaces: pore spaces will be filled with a mixture of gasoline, water, and soil gas. In addition, equilibrium thermodynamics will result in volatilization of some gasoline and vaporization of some water, both into the soil gas. Some gasoline products will also dissolve into the water.

The important parameters for immiscible fluids are the same as those for water; unfortunately, it may be more difficult to find them published.

Vadose Zone Moisture and Energy

Hydrostatics

For a constant density, static fluid continuum where gravity is the only acting acceleration field, the law of hydrostatics can be derived as:

$$\frac{dp}{dz} = -\rho g \tag{4.1}$$

$$\frac{dp}{dx} = 0 \tag{4.2}$$

where p is fluid pressure, z is the vertical coordinate axis (positive upwards), ρ is fluid density and g is the acceleration due to gravity. For ground water and vadose zone considerations, by setting $z=0$ at a location where $p=0$ (i.e., at the water table), Equation 4.1 can be integrated to:

$$p = \rho g h \tag{4.3}$$

where p is the pressure at any distance h vertically from the zero pressure datum. This relationship results in a linear pressure distribution above and below the vertical datum (water table) as depicted in Figure 4.4. In this figure, it can be seen that where z=0, p=0. Below this level, z is negative and p is positive, with p increasing linearly as z decreases. When above the level of z=0, z is positive and p is negative. A practical analogy is utilizing a straw in a can of beverage. With no capillarity effects in the straw, the level of the fluid in the straw is at the level of fluid in the can. If you drew the beverage up into the straw and put your thumb over it, the pressure distribution in the fluid in the straw would be described by Equation 4.3 and depicted in Figure 4.4. The zero datum here is the liquid surface of the beverage in the can.

Thus, in the vadose zone, water exists above a zero pressure datum (the water table), the fluid is at negative pressures, and Equation 4.3 accurately predicts the pressure in the vadose zone water as long as saturation is maintained. Due to the breakdown of the fluid continuum (saturation) in the vadose zone, Equation 4.3 does not accurately describe the pressure situation above the capillary fringe.

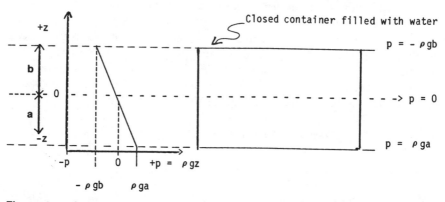

Figure 4.4.   Pressure variation in a static fluid.

Capillarity

When two immiscible fluids exist at an interface, there is a tendency for the molecules of each fluid to move away from the interface and be nearer to like molecules. In order to keep molecules at the interface, energy must be expended on every molecule at the interface. This free surface energy is measured by the fluid property of surface tension. The combination of immiscibility plus surface tension result in the interface acting as a membrane. Given finite fluid masses, the molecular forces which exist at the fluid interface tend to deform the interface into a curved surface. Because the surface is curved, with possibly more than one radius of curvature, there is an imbalance of forces at the surface. This imbalance is offset by a pressure difference across the interface between the fluids.

The difference in pressure across the interface is known as the capillary pressure and can be computed in the vadose zone as:

$$P_c = P_a - P_w \qquad (4.4)$$

where P_c is the capillary pressure, P_a is the soil gas pressure, and P_w is the soil water pressure.

Combining capillarity considerations with wettability (the affinity for a fluid to a solid surface), relationships for static conditions in the vadose zone may be developed. In a very basic analogy, for a single small diameter glass tube (capillary tube) standing vertically and partially submerged in a tank of water, an equilibrium analysis of the forces existing on an element of the curved fluid surface in the tube yields:

$$P_c = \frac{2\gamma}{r} \qquad (4.5)$$

where P_c is again capillary pressure across the interface, γ is the surface tension of the water and r is the radius of the capillary tube. If the upper end of the tube is open to the atmosphere, the pressure on top of the water surface in the tube is also atmospheric. From hydrostatics, Equation 4.3 yields a pressure in the water, right at this interface, of

$$P_w = -\rho gh \qquad (4.6)$$

where h is the height of capillary rise above the surface level of water in the tank. Substituting 4.6 into 4.5, yields:

$$h = \frac{2\gamma}{\rho gr} \qquad (4.7)$$

where it can be seen that the height of capillary rise is inversely proportional to the radius of the capillary tube. Typical values for water yield $h \approx 0.15/r$. For soils, r must be replaced by a measure of the representative pore size d_n. In most practical applications, d_n is a function of the median particle size (d_{50}) where $0.155\ d_{50} < d_n < 0.414\ d_{50}$ (Iwata and Tabuchi, 1988).

Vadose Zone Moisture

The vadose zone pore spaces can be filled with any fluid; the most common fluids are air and water. If the total volume of pores in a sample of the vadose zone is V_p and the volume of pore space occupied by water is V_w, the saturation (S) is calculated as:

$$S = \frac{V_w}{V_p} \qquad (4.8)$$

S can range from 0 to 1.0 and is sometimes reported as a percentage. When evaluating the quantity of water in a sample compared to that of the total sample (of volume V_T), the volumetric water (moisture) content Θ is calculated as:

$$\Theta = \frac{V_w}{V_T} \qquad (4.9)$$

and obviously ranges from 0 to the porosity (ϕ). Thus to relate the two:

$$\Theta = S\phi \qquad (4.10)$$

Under field conditions, gravity alone cannot drain the unsaturated zone because surface tension, osmotic and molecular forces can act against it. Thus, the lower limit for S in the vadose zone is S_r—the residual saturation. At the soil surface, where evaporation may dominate, S_r may approach zero.

From the early part of this century, vadose zone investigators have recognized the relationships between moisture content (or saturation) and the distance above the water table. A typical relationship, for static conditions, is shown in Figure 4.5. At distances above the water table, S approaches S_r. Moving closer to the water table, saturation increases to the field saturation level (S_s). Some residual air will remain at and just below the water table because water-table fluctuations entrap air in this region.

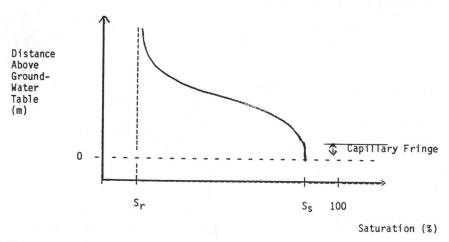

Figure 4.5. Example saturation condition in the vadose zone.

Vadose Zone Suction

As described in the section on capillarity, above the water table, negative pressures (or suction) exist in the liquid phases primarily due to the curvature of the surface of the liquids in this region. A plot of a typical vadose zone (soil) suction relationship is shown in Figure 4.6. Combining this information with that of the last section, as both moisture content and vadose zone suction are functions of the distance from the water table, vadose zone suction can be plotted against soil moisture content for a given soil. Thus, by monitoring vadose zone suction (soil suction or matric potential) the moisture content can be established. Figure 4.7 is an example of this relationship.

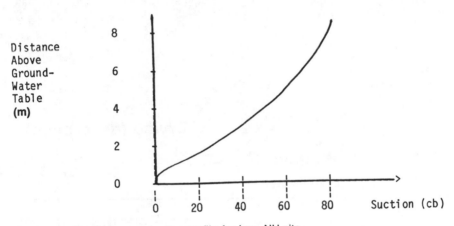

Figure 4.6. Vadose zone suction profile for Lee, NH site.

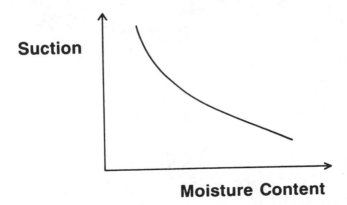

Figure 4.7. Vadose zone suction-moisture content relationship (wetting), Lee, NH.

Hysteresis

The moisture content-suction relationship (and therefore the moisture content-elevation relationship) is not unique for a given vadose zone. Depending on whether or not the vadose zone is undergoing drainage (desorption) or wetting (sorption), the moisture content at a given elevation above the water table will vary. Figure 4.8 depicts this phenomenon, which is known as hysteresis. This process can be explained by analogy of the variation of the pore radii in the vadose zone to that of an ink bottle, depicted in Figure 4.9. In wetting of an initially dry soil, wettability and capillarity will allow water to move vertically into pore spaces. Capillary rise will cease when there is a balance between surface tension and gravitational forces for a given pore size. When the pore size changes from 'r' to a larger value 'R,' although the capillary rise for 'r' could allow water

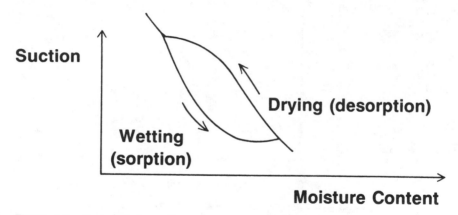

Figure 4.8.   Hysteresis in suction-moisture content relationship, Lee, NH.

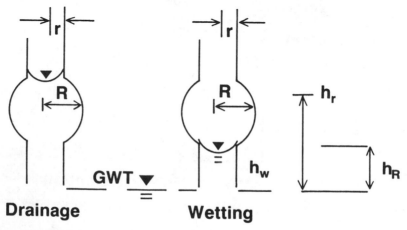

Figure 4.9.   The ink bottle effect, illustrating hysteresis.

to move vertically higher (h_r), the capillary rise for R may be smaller (h_R) than the amount of rise which has already occurred (h_w). Thus, as the pore size increases, capillary rise ceases to maintain the equilibrium in forces. This effect gives rise to the wetting curve in Figure 4.9.

If the vadose zone was initially saturated and then allowed to drain, drainage would occur in more passive heights of capillary rise, while leaving some large pores, below, saturated. In addition, some pores will have their connection to surrounding vadose zone water ruptured by drainage, becoming islands of water in the vadose zone.

Energy Potential in the Vadose Zone

The total status of energy for soil moisture is described by the total moisture potential ψ_T (units of L^2/T^2, i.e., cm^2/sec^2). ψ_T is comprised of the sum of three primary potentials: gravitational (ψ_g), pressure (ψ_p) and osmotic (ψ_o) potential. These potentials are analogous to the Bernoulli sums of gravitational head, pressure head, and velocity head. For vadose zone moisture flow, the velocity potential is negligible. For surface water or piping considerations, osmotic potential is negligible.

ψ_g represents a potential energy due the vertical location of the vadose zone moisture of interest:

$$\psi_g = gz \qquad (4.11)$$

where g is the acceleration due to gravity and z is the distance above ($+$) or below ($-$) some vertical datum.

ψ_p is the hydrostatic pressure existing in the vadose zone water.

$$\psi_p = P/\rho \qquad (4.12)$$

where P is the hydrostatic pressure and ρ is the density of the liquid; here, water.

ψ_o represents the potential resulting from maintenance of concentration gradients of solutes in vadose moisture systems. Normally solute molecules in a zone of high concentration diffuse to zones of lower concentration. If some barrier exists in the vadose zone which prevents the movement of the solute to the zones of lower concentration, yet allows movement of the solvent (water) in any direction, a pressure must exist across this barrier when solvent movement through the membrance equilibrates (solvent flow in one direction is balanced by solvent flow the opposite way). The barrier is commonly referred to as a semipermeable membrane. Figure 4.10 depicts the osmotic pressure at equilibrium. In the figure, h_o is the height of solution yielding a pressure (P_o) on the membrane. In this case:

$$\psi_o = \frac{\psi_o}{\rho} = gh_o = \frac{MRT}{\rho} \qquad (4.13)$$

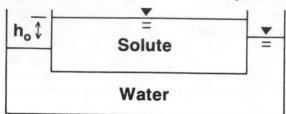

Figure 4.10. Definition sketch for osmotic potential.

where M is the total molar concentration of the solute, T is the temperature (°K) and R is the gas constant. This equation is only good for dilute solutions.

VADOSE ZONE FLOW

Water

Although many of the definitions and characteristics of the vadose zone presented herein may appear different than for aquifer (saturated) systems, the basic equations defining flow for the vadose zone are the same as those for aquifers: Darcy's Law and continuity. In this case, it must be recognized that hydraulic conductivity (or permeability) is a function of moisture content (Θ) in the vadose zone. The total potential (ψ_T) in the vadose zone can be converted to head (as is typically used in Darcy's Law) by dividing by the acceleration due to gravity:

$$h_T = \frac{\psi_T}{g} \tag{4.14}$$

Thus Darcy's Law is now written

$$q = -K(\psi) \, \nabla h_T \tag{4.15}$$

where q is the flux, $K(\psi)$ is the hydraulic conductivity and ∇ is the del operator for spatial vector partial differentiation.

Compared to saturated ground water, flow where the hydraulic conductivity (K) occurs always at saturation, vadose zone mechanics are such that K is a function of saturation or moisture content. Figure 4.11 depicts such a relationship. Quite obviously, water flow becomes more difficult as the degree of saturation decreases. This results from the fact that air takes the most advantageous pore spaces, leaving the most tortuous paths for water.

Combining continuity with Darcy's Law yields the transient vadose zone flow equation:

$$\frac{\partial \Theta}{\partial t} = - \nabla \bullet [K(\psi) \ \nabla h_T]$$ (4.16)

Solutions to this equation have been analytically derived for horizontal and vertical flow cases. One common extension is to reintroduce the moisture content (Θ) for total potential head (h_T) (for example, see Hillel, 1980b, pp 204–207).

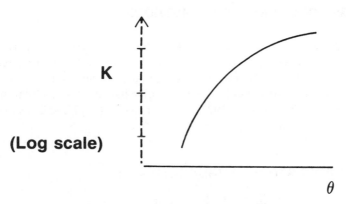

Figure 4.11.   Example of moisture content—hydraulic conductivity relationship.

Vapors (Gases)

Natural vapor flux in the vadose zone is dominated by diffusive (Fickian) transport. As such, the vapor flux (q_v) is computed from a diffusion equation:

$$q_v = -D_v \frac{\partial C}{\partial x}$$ (4.17)

where: D_v is the diffusion coefficient for water vapor in the porous media, C is the concentration of water vapor and x is the spatial coordinate.

In general, if vapor flux is induced, i.e., by soil venting, then vapor flux is dominated by a pressure gradient. That is, instead of diffusion processes accounting for flux, an excessive pressure condition (suction or positive pressure) will drive vapors from regions of high pressure to low pressure. In these instances, velocity and elevation potential are considered insignificant compared to pressure potential. For these types of field conditions, Fick's Law and Darcy's Law are not valid and equations dealing with turbulent fluid transport are necessary, such as a Darcy-Weisbach formulation.

When considering multiple-phase flow, it must be recognized that there is typically a threshold value of fluid content which must be achieved before a particular fluid can move. In the case of three-phase flow (gasoline-water-vapor) there is, for practical purposes, only a small window of saturations at which all three fluids can move. With this in mind, an obvious strategy for immobilization of the gasoline would be to increase water content or, more preferably, vapor (air) content. By such an action (pumping air into the vadose zone) vapor content increases so as to push the system out of the three-phase flow "window." Once immobilization of the gasoline occurs, in situ or other cleanup methodologies can be addressed.

VADOSE ZONE MONITORING METHODS

Three parameters are measured in the vadose zone: storage properties (water content), transmission (flux), and fluid quality. All three are measured in soil water; fluid quality is also measured in soil gas. General sources for methods and applications relating to soil water monitoring include Wilson (1980 and 1981), Everett (1980), Everett et al. (1976), and Fenn et al. (1977). Soil gas monitoring is discussed later in this chapter.

MONITORING STORAGE PROPERTIES

The physical properties of the vadose zone associated with water storage include: bulk density, total thickness, porosity, water content, and soil moisture versus tension relationship. Total potential water storage can be estimated from the first two, easily measured properties. Total porosity can be used in place of bulk density for estimating total potential storage, while pore-size distribution affects fluid transmission. Water content can be measured directly or estimated from the soil moisture characteristic curve. This section discusses measurement of tension and water content, which can be measured using tensiometers, electrical resistance blocks, thermocouple psychrometers, gamma-ray attenuation or nuclear magnetic resonance.

Tensiometers

Tensiometers are used to measure soil matric potential. A tensiometer consists of a porous ceramic cup (or other porous surface) attached to a pressure sensor via a tube filled with water. The pressure sensor is commonly a mercury-filled manometer, a pressure gauge (Figure 4.12) or a pressure transducer. The principle of operation is that water can freely flow into or out of the porous cup at soil tensions which do not exceed the air-entry tension (approximately 1 bar) of the porous cup. As water moves out of the porous cup into the unsaturated soil, a vacuum is formed in the water tube which exerts a corresponding force on the gauge, manometer, or transducer diaphragm. Readings must be corrected by subtracting the height of the water column between the soil surface and the gauge or transducer from the tensiometer reading (Figure 4.12). To assure proper operation, the tensiometer should be installed with as little disturbance to the soil as

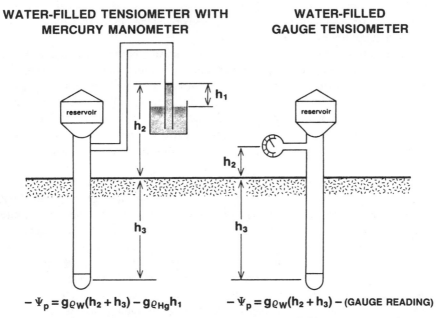

WATER-FILLED TENSIOMETER WITH MERCURY MANOMETER

WATER-FILLED GAUGE TENSIOMETER

$$-\Psi_p = g\varrho_W(h_2 + h_3) - g\varrho_{Hg}h_1$$

$$-\Psi_p = g\varrho_W(h_2 + h_3) - (\text{GAUGE READING})$$

Figure 4.12. Schematic diagram of typical commercially available manometer and gauge tensiometers [from EPRI, 1985].

possible, and the porous cup should be hydraulically connected with the surrounding soil. The latter can be accomplished by forcing the ceramic cup into a snug hole or by placing the cup in a slurry of material removed from the hole.

Tensiometers are inexpensive and can be purchased from any of several manufacturers or custom-designed for special applications. Ethylene glycol solution, or a similar liquid, can be used to make them operational during periods of freezing and thawing. Stannard (1990) presents a number of designs, along with their advantages and disadvantages.

Vacuum gauge tensiometers are durable, and easily operated and maintained. However, they are less accurate and precise than either manometer and transducer tensiometers. Response time with these instruments varies from poor to excellent. Calibration is required before installation and occasionally after installation. Data is collected manually. Gauge tensiometers are not well suited to measuring hydraulic gradients, but can be used for gross measurements of moisture movement. They also cannot measure positive pressure, so they lose their usefulness as the soil becomes saturated.

Manometer tensiometers are also durable and easily operated. Maintenance is dependent on tensiometer design; those with small water lines require frequent purging. Since these devices can be custom made, they are more versatile than gauge tensiometers. Wilson (1990) rates the accuracy of manometer tensiometers as "excellent," precision as "good," and response time as "fair." Data are collected manually, and both positive and negative pressures can be measured. A major advantage of these instruments over the other types of tensiometers is that calibration is never required.

Tensiometers connected to transducers can measure both positive and negative pressure and provide nearly instantaneous in situ readings of soil-water pressure, which can be recorded electronically. As with manometer tensiometers, maintenance and versatility are dependent upon the design of the water conduits. Response time is the most rapid of the three types of tensiometers, making these the best choice for tracking wetting fronts. However, these are the most expensive of the three alternatives and periodic recalibration is required due to instrument drift.

Disadvantages for all types of tensiometers include a bottom tension limit of about 1 bar, with decreased accuracy in readings at tensions greater than about 0.8 bars. Readings are sensitive to temperature changes, atmospheric pressure changes, and air bubbles in the water lines. Tensiometers with small diameter water conduits are especially susceptible to air bubbles and require frequent purging to assure accurate readings. In addition, because the soil-moisture characteristic curve is required for determining soil moisture from tension measurement and the curve is subject to hysteresis, it is necessary to know whether the soil is wetting or drying when the measurement is taken. Other sources of error include operator error in reading the manometer and poor pressure transducer calibration.

Electrical Resistance Blocks

Electrical resistance blocks are inexpensive and can be used to measure either moisture content or soil-water pressure. They consist of two metal plates imbedded in a porous material, usually gypsum. Wires are attached to the plates so that changes in the electrical resistance between the two plates can be measured. As the moisture content (or tension) of the electrical resistance block changes, coincident and in equilibrium with the surrounding soil, the electrical resistance properties of the block are altered. Before use, electrical resistance blocks must be calibrated in the laboratory using soil from the installation site. Calibration produces curves of electrical resistance versus soil moisture or soil-water pressure. Because each block should produce the same curve, calibration allows the user to find faulty blocks before they are installed.

The chief advantages of electrical resistance blocks are: (1) they are suited for general use in the study of soil-water relations; (2) they are inexpensive; (3) they can be used to determine either suction or moisture content; and (4) they require little maintenance.

While electrical resistance blocks present an attractive monitoring alternative, they have problems which may render them unusable in certain situations. Problems include temperature sensitivity, time-consuming calibration, the independent effects of salinity on electrical resistance, and inaccuracy of measurements of high water contents (or low soil-water pressure). They are generally used only for suction in excess of 0.8 atm, which is the upper practical limit on suction for tensiometers. In addition, resistance blocks made of gypsum will eventually dissolve, making them unsuitable for long-term use.

Thermocouple Psychrometers

Thermocouple psychrometers are used to measure in situ soil-water pressure under very dry conditions, where tensiometers cannot be used because of air entry problems. They provide measurements of total water potential. Soil-water pressure is determined based on the relationship between soil-water pressure (potential) and relative humidity in the soil. Psychrometers are composed of a porous bulb to sample the relative humidity of the soil, a thermocouple, a heat sink, a reference electrode, and related circuitry. Calibration is required for each psychrometer unit before field installation.

The major disadvantage with this technology is that psychrometers are very sensitive to temperature fluctuations, so that it is necessary to record and correct for even diurnal temperature changes. However, where very dry conditions prevail, they may be the best monitoring choice. Psychrometers have successfully measured in situ suction values as high as 30 atm (Watson, 1974). Other disadvantages include the expense and complexity of these instruments (Bruce and Luxmore, 1986).

Gamma-Ray Attenuation

Gamma-ray attenuation can be used to indirectly measure moisture content by nondestructively determining soil density. Attentuation of gamma-rays, commonly from a cesium source, passing through a soil column depends on the density of the soil column. If the soil density remains constant (i.e., the soil is nonswelling), changes in attenuation reflect changes in moisture content. This technique requires parallel access holes, one each for the source and the detector. Measurements can be taken as close as 2 cm, either vertically or horizontally, allowing an accurate determination of the location of the wetting front.

Major disadvantages of this technology are that gamma-ray attenuation units are expensive and difficult to use, require special care in the handling of the radioactive source, and changes in bulk density (due to swelling, frost heave, etc.) affect the instrument's calibration. In addition, this technology is unsuitable for applications in which vertical wells cannot be installed.

Nuclear Moisture Logging

A second nuclear method for nondestructively measuring moisture content is nuclear moisture logging. In this method, a probe containing a neutron source (e.g., usually americium or beryllium) and a detector is lowered down an access hole using a cable. The access hole is usually constructed of steel or aluminum. Neutrons emitted from the radioactive source interact with the hydrogen in the water of the surrounding soil. From counts of radioactivity taken at discrete intervals, moisture content can be calculated.

This method had several positive qualities (Schmugge, 1980). Readings are directly related to soil moisture and moisture content can be measured regardless of its physical state. Average moisture contents can be determined with depth. Repeated measurements can be taken at the same site, allowing measurement of rapid changes in moisture content as well as long-term changes. Like several of the other methods, the system can be interfaced with electrical recording equipment.

Nuclear moisture logging equipment also has several disadvantages. Equipment is very expensive. Only moisture content can be measured using this method; no information is provided on soil-water pressure or changes in density. Because of the sphere of influence, accurate measurements cannot be taken near the soil surface. In addition, the accuracy of the method is not high for detecting small changes in water content, especially for dry soils. Like gamma-ray logging, this method requires care in the handling of the radioactive source.

MONITORING VADOSE ZONE TRANSMISSION PROPERTIES

Vadose zone transmission properties are generally of greater interest in ground-water monitoring studies than are storage properties. No direct in situ methods exist for measuring the unsaturated hydraulic conductivity of a geologic material because the hydraulic conductivity is a function of moisture content. However, one can measure flux and use this to calculate fluid transmission rates. The rate of moisture movement is determined indirectly from infiltration rates or measurements of unsaturated flow. When using field data to estimate transmission properties in the vadose zone, it is important to remember that large variations in these parameters can occur because of soil heterogeneities.

Field Measurements of Infiltration Rates

Field measurements of infiltration rates are appropriate for estimating downward fluid transmission during the wetting cycle. Infiltration rates are affected by soil texture and structure (including soil layers), initial moisture content, entrapped air, and water salinity. Waste disposal options for which the principal component of flux is downward include surface spreading or ponding of wastes and installation of landfill liners composed of earthen materials.

Infiltration is determined using infiltrometers; infiltrometers do not directly measure hydraulic conductivity. Infiltration is the process by which water enters a permeable material. When infiltration begins, the infiltration rate is relatively high and is dominated by matric potential gradient. As the matric potential gradient decreases, the infiltration rate asymptotically decreases with time until the gravity-induced infiltration rate, called the steady-state infiltrability, is approached

(Hillel, 1980a). This relationship is shown in Figure 4.13. Steady-state infiltrability is directly proportional to saturated hydraulic conductivity and hydraulic gradient. Therefore, in order to calculate saturated hydraulic conductivity from infiltration data, the hydraulic gradient and the extent of lateral flow must be known. Gradient data can be obtained using many of the instruments described in the previous section. Saturated hydraulic conductivity is of interest even in the vadose zone because it is the upper boundary for unsaturated hydraulic conductivity; use of this value provides a conservative estimate of fluid transmission time.

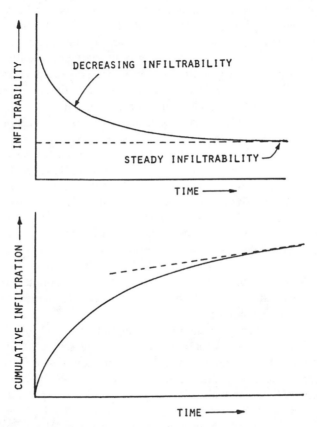

Figure 4.13.   Time dependence of infiltrability and of cumulative infiltration under shallow ponding [from Hillel, 1980b].

While infiltrometers can be designed with either a single ring or a double ring, the double-ring method is preferred because its design minimizes lateral flow, simplifying the calculation of saturated hydraulic conductivity. The principle of operation is based on maintaining a constant head in the inner and outer rings of the infiltrometer. Both rings are sealed in the soil to prevent leakage under the rings. Water is added to the rings to maintain the constant head; if the inner ring is covered to prevent evaporation, the volume of water added to the inner ring is equal to the water infiltrating into the soil. In the design of Daniel and Trautwein (1986) water is added to the inner ring through an IV bag (Figure 4.14). As water from both rings enters the soil, water exits the IV bag and moves into the inner ring to maintain a constant head. The IV bag is weighed periodically to determine weight loss, which is equivalent to the mass of water entering the infiltrometer (and, therefore, the soil). Water is added to the outer ring as needed to maintain a constant head. The IV bag design is well suited to soils with low infiltration rates because the small amount of added water can be measured accurately by weight. For more permeable soils, the water level can be maintained by adding measured volumes of water to the inner ring. Measurements of infiltration are taken until the system reaches steady-state infiltrability. If the test is performed to prove that a soil meets some regulatory requirement, such as the requirement that earthen liners have a saturated hydraulic conductivity of 1×10^{-7} cm/s or less (USEPA, 1988), the test may end when this infiltration rate is achieved because the infiltration rate decreases with time and saturated hydraulic conductivity will be no more than the infiltration rate.

Infiltrometers generally range in size from less than one square foot to about 25 sqaure feet. Large infiltrometers with IV bags were designed for soils with low infiltration rates, generally in the range of 1×10^{-5} to 1×10^{-8} cm/s (Daniel and Trautwein, 1986). The large size is necessary to include macrostructures and to obtain measurable amounts of water loss.

Hydraulic conductivity can be calculated from infiltration rate using either Darcy's Law or the Green-Ampt (1911) approximation. If Darcy's Law is used, the equation is $K = -Q/(AI)$, where Q/A is the measured steady-state infiltrability per unit area and I is the hydraulic gradient determined from tensiometer data. Q/A is negative because flow is downward. This method assumes flow is occurring under saturated conditions.

The Green-Ampt approximation assumes that the wetting front is sharp, the matric potential at the front is constant, and the wetted zone is uniformly wetted and of constant hydraulic conductivity. The assumption of a sharp wetting front may be reasonable for fine-grained soils, as shown by dye studies in an experimental earthen liner (Albrecht et al., 1989). The Green-Ampt approximation differs from the Darcy's Law calculation in that knowledge of the depth of the wetting front is required instead of a measured hydraulic gradient. Under these assumptions, the analytical solution to vertical infiltration produces an equation which resembles the Darcy equation:

$$K = i \left[1 + \frac{h + \psi_f}{L_f} \right]^{-1} \tag{4.18}$$

where i is the steady-state infiltration rate and the bracketed term is the hydraulic gradient. In the bracketed term, h is the height of the water in the infiltrometer,

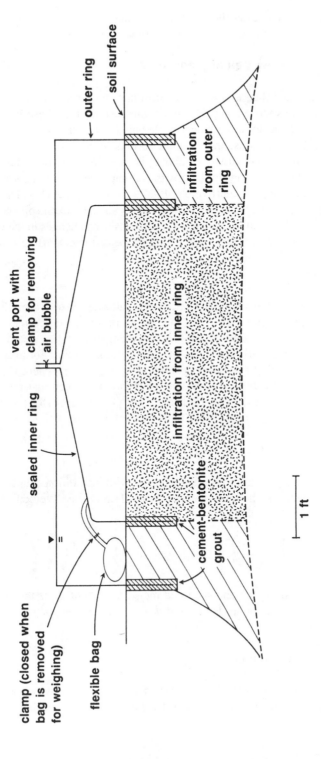

Figure 4.14. Schematic diagram of sealed double-ring infiltrometer.

ψ_f is the matric potential at the wetting front and L_f is the depth of the wetting front. For conservative estimates, ψ_f is estimated to be 0.

Calculation of Flow Velocity and Flux

Flow velocity and flux are two of the most important transmission parameters determined in the vadose zone. However, measurement of these parameters is difficult and most techniques are applicable only to near surface flow. For many waste disposal applications, this is not a serious limitation.

Steady-state infiltration, discussed above, is an appropriate basis for determining flux during the wetting cycle. During the drying cycle, three major approaches to evaluating flux are possible (Everett, 1980). These include: (1) calculating flux from mathematical formulae and empirical relationships between soil-suction, soil-water content, and hydraulic conductivity; (2) measuring changes in the water content of the soil profile over time; and (3) direct measurements using flow meters.

Darcy's Law

The easiest method available for calculating saturated flow from infiltrometer data is the use of Darcy's Law. This method is conservative because it assumes the soil is saturated; it is appropriate for the wetting cycle when steady-state flux is determined from an infiltration test. In simple terms, solving for average linear velocity (V_x), Darcy's Law can be written as $V_x = Q/(n_e A)$, where Q is the discharge, n_e is the effective porosity, and A is the cross-sectional area of flow. Q/A is the measured steady-state infiltration rate per unit area; for use in Darcy's Law Q/A is negative because flow is downward.

Green-Ampt Wetting Front Model

The Green-Ampt wetting front model is used with infiltration data and assumes unsaturated conditions below a wetting front. Travel time (velocity times distance) is predicted from

$$t = \left\{ \frac{\Theta_s - \Theta_i}{K_{sat}} \right\} \left[L_f - (h + \psi_f) \ln \left\{ 1 + \frac{L_f}{h + \psi_f} \right\} \right] \tag{4.19}$$

where Θ_s and Θ_i are initial and saturated moisture content, L_f is the depth to the wetting front, h is the pond depth, and ψ_f is the moisture potential just below the wetting front.

Internal Drainage Method

The internal drainage method can be used to determine the unsaturated hydraulic conductivity in the field by monitoring the transient internal drainage of a near-surface soil profile. The method is described in detail in Hillel (1980b) and was extended to layered profiles by Hillel et al. (1972). It requires simultaneous measurement of moisture content and suction under conditions of internal drainage alone; evapotranspiration must be prevented. The method also assumes that flow is vertical and that the water table is deep enough so that it does not affect the drainage process.

Tensiometers and neutron access tubes or gypsum blocks are installed near the center of the test area. Depth intervals for the instruments should not exceed 30 cm, with a desirable total depth of up to 2 m. The test area is then ponded or irrigated until the soil profile is as wet as practical. Simultaneous measurements of soil suction and moisture content are collected until soil suction exceeds 0.5 bar; at greater suction, the drainage process may be so slow that changes become imperceptible. The measurement period for this test can be several weeks for slowly draining soils. Data are plotted on two graphs (Figures 4.15a and b), moisture content and suction versus time for each measured depth within the soil profile. From these measurements, instantaneous values of potential gradient and flux can be obtained, allowing the calculation of hydraulic conductivity and, hence, flow velocity.

Soil moisture flux is calculated at each time and depth from

$$q_z = Dz(\partial\Theta/\partial t) \tag{4.20}$$

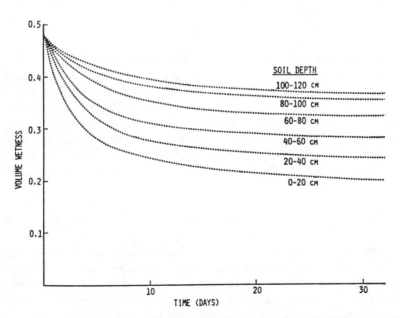

Figure 4.15a.   Graphs used to calculate unsaturated hydraulic conductivity by the internal drainage method [from Hillel, 1980a]. **(a)** Volumetric wetness as a function of time for different dpeth layers in a draining profile.

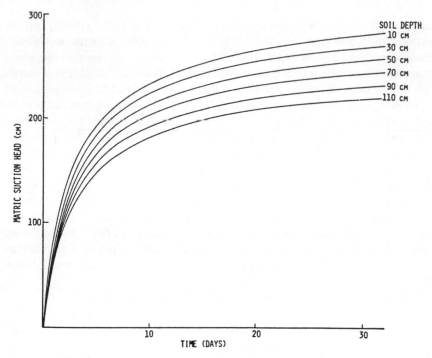

Figure 4.15(b).   Matric suction variation with time for different depth layers in a draining profile.

where $\partial\Theta/\partial t$ is the slope of the wetness curve at the time of interest and Dz is the depth increment. Equation 4.17 is the flux out the bottom of the uppermost increment. Flux through the bottom of each succeeding layer is obtained by summing these incremental fluxes for all layers overlying the depth of interest. Flow velocity can be calculated from Darcy's Law, as discussed above.

Hydraulic head profiles are obtained using the suction versus time data, adding depth to suction to obtain hydraulic head, and graphing hydraulic head versus soil depth for each time (Figure 4.15c). Hydraulic conductivity, K_z, is calculated from

$$K_z = q_z/(\partial H/\partial z) \qquad (4.21)$$

where $\partial H/\partial z$ is the slope of the hydraulic head versus depth curve for the time of interest. K_z is calculated for several depths and times, each of which has a corresponding moisture content. As the final step, moisture content or soil suction is plotted against hydraulic conductivity so that flux and velocity can be calculated using field data at actual monitoring points.

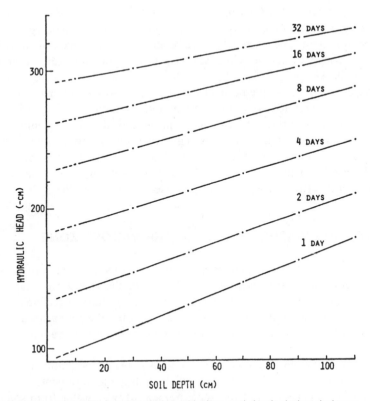

Figure 4.15(c). Hydraulic head variation with time and depth during drainage.

Measurement of Tracer Movement

Tracers are matter or energy carried by ground water which can provide information on the rate and direction of ground water movement. Tracers can be natural, such as heat carried by hot springs; intentionally added, such as dyes; or accidentally introduced, such as oil from an underground storage tank (Davis, et al., 1985). Use of tracers in the saturated zone for determining aquifer parameters is discussed in Chapter 10, so only a brief overview is presented here. The main difference between use of tracers under unsaturated versus saturated conditions is the practical problem of sampling a tracer at increasing depths under unsaturated conditions (Everett, 1980 and 1981).

Davis et al. (1985) present a thorough discussion of tracers. The most important property of any selected tracer is that its behavior in the subsurface should be well understood. Ideally, it should move at the same rate as the ground water, not interact with the soil matrix, and not modify the hydraulic conductivity or other properties of the medium being monitored. Concentration of the tracer should be much greater than the background concentration of the same constituent in the natural system. The tracer should be relatively inexpensive and easily detectable with widely available technology. For most applications, the tracer should also be nontoxic.

A variety of tracers have been successfully used to monitor moisture movement in the unsaturated zone. Fluorescein and rhodamine WT dyes have been successfully used to track the wetting front beneath double-ring infiltrometers and to indicate preferential flow paths in an experimental earthen liner (Albrecht, et al., 1989). Tritium from a low-level radioactive waste disposal site was successfully used to determine the rate of water (and tritium) movement in the unsaturated zone at the waste disposal site (Healy, et al., 1986).

MONITORING WATER QUALITY IN THE VADOSE ZONE

The goal of most vadose zone monitoring programs is to measure the spatial and temporal changes in water quality. Monitoring the vadose zone can provide an early warning system for contaminant migration so that corrective action can begin before any aquifers are contaminated. Wilson (1980) presents a thorough discussion of the chemical reactions affecting contaminant migration in the vadose zone.

Three types of methods are available for monitoring water quality in the vadose zone. These include: (1) indirect methods, including measurements of electrical and thermal properties; (2) direct measurement of pore water from soil cores; and (3) direct soil-water sampling.

Electrical Properties Measurements

Electrical conductivity or its inverse, resistivity, are used extensively to characterize soil salinity and to map shallow contaminant plumes. For shallow soils, electrical conductivity is primarily a function of soil solution (Wilson, 1980). The success of using electrical properties to delineate plumes is dependent upon the contrast between the conductivity of the plume and the natural water, the depth and thickness of the plume, and lateral variations in geology.

Electrical resistivity can be measured using surface geophysical techniques, as discussed in Chapter 5. It can also be measured by using electrical resistance blocks (salinity sensors) to evaluate soil salinity. Electrical resistance blocks were discussed above as a means of measuring in situ moisture content. They can be installed beneath a waste disposal site before the site becomes operational and monitored remotely. Salinity sensors must be calibrated to provide a curve of soil salinity versus electrical conductance. Electrical conductance is highly temperature dependent, so accurate measurement of soil solution temperature is a necessary companion to this device.

Soil Sampling and Water Sampling

Pore Water Extraction

Collection of soil cores is discussed in Chapter 6. Soil cores can provide pore water for water-quality analysis. Because of the difficulty and expense of obtaining soil cores, this is not the most desirable method for obtaining pore-water quality samples. However, when cores are obtained during borehole and monitoring well drilling, the added expense of collecting pore-water samples from cores is significantly reduced.

Certain constituents, including Ph, Eh, and EC, are unstable and must be measured in the field. This requires either extracting the pore water or, more simply, making a saturated paste of the material and taking measurements on the paste. Pore water can be extracted by placing a soil sample in a commercially available filter press, hydraulic ram, or centrifuge and forcing the interstitial water out of the soil sample. This and other methods for pore-water extraction are presented by Fenn et al. (1977). After extraction, standard analytical techniques can be used on the water sample.

Suction Lysimeters

Suction lysimeters allow the collection of in situ soil water. They have a significant advantage over pore water extraction in that repeated samples can be taken at a given location. A typical design, as shown in Figure 4.16, consists of a porous cup attached to a PVC sample accumulation chamber and two access tubes which lead to the soil surface. Porous cups are commonly made of ceramic, alundum (an aluminum oxide), or PTFE; the first two are hydrophilic while the latter is hydrophobic. The sampling radius of a lysimeter is on the order of centimeters, so that many are needed if they are to function as an effective early warning system (Morrison and Lowery, 1990). A variation to standard lysimeter design is the "well-type lysimeter" described by Ball and Coley (1986), which produces a larger sample volume than a standard lysimeter. Thorough discussion of soil water sampling techniques are presented by Litaor (1988) and Wilson (1990).

Lysimeters are installed in a borehole with silica flour packed around the porous cups. Silica flour is necessary to prevent plugging of the cup. Also, without silica flour, lysimeters with PTFE cups will not hold 10 centibars of vacuum (Everett et al., 1988). The sample tube ends at the bottom of the lysimeter, while the air tube ends near the top of the sample accumulation chamber. To collect a sample, the sample tube is clamped and a suction is applied to the lysimeter through the air tube, which is then clamped. This causes an inward gradient gradually drawing water into the sampler. The suction is then released and pressure is applied, forcing the sample to the surface through the sample collection tube. This design has been used to depths of at least 55 feet (Apgar and Langmuir, 1971).

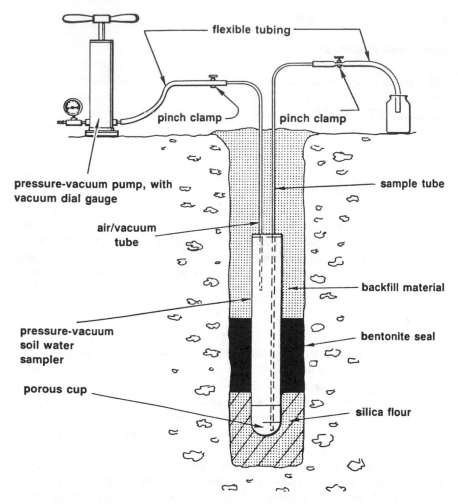

Figure 4.16.   Schematic diagram of installed pressure-vacuum (suction) lysimeter [adapted from Soil Moisture Equipment Corp., undated].

For deep lysimeters in which higher pressures are required to force the sample to the surface, pressure in excess of one atmosphere in the sample will send the sample back through the porous cup into the soil instead of to the surface in a standard lysimeter. Wood (1973) modified the standard lysimeter design to allow sampling from any depth within the vadose zone. This design prevents the pressurization problem by including a check-valve to prevent pressurization of the porous cup. Wood (1978) has been successful in collecting samples from depths in excess of 100 feet.

Questions have been raised as to the validity of samples collected from suction lysimeters. Some studies have indicated that the ceramic cups can alter the chemical composition of samples, making the samples not representative of actual water quality. Wolff (1967) found that new ceramic cups yield several milligrams per liter of Ca, Mg, Na, HCO_3, and SiO_2 even after cleaning with dilute HCl. Grover and Lamborn (1970) and Hansen and Harris (1975) found substantial bias and variability in soil-water samples of NO_3-N, PO_4-P, Na, K, and Ca. Up to a 60% change was noted in sample concentrations caused by sample intake rate, plugging of the ceramic cup and sorption and screening of some ions. Ceramic cups were the source of excessive Ca, Na, and K in samples with low solute concentrations, and served to absorb P. Rinsing the cups with dilute HCl before installation reportedly reduced the problems with Na, K, and P to acceptable levels. Levin and Jackson (1977) found that Ca, Mg, and PO_4 were not altered by lysimeters which were used to collect soil-water samples from intact soil cores.

Lysimeters have also been found to screen certain contaminants. Parizek and Lane (1970) concluded that pressure-vacuum lysimeters are not useful for analysis of soil bacteria, BOD, or suspended solids because of screening. Dazzo and Rothwell (1974) found that screening and adsorption of bacteria by ceramic cups with a pore size of 3 to 8 μm rendered them unusable for fecal coliform. Because the effective pore size of the porous ceramic cup used in most lysimeters is about 1 μm, colloidal particles may pass through. Everett et al. (1988) reported that volatile organics were lost from suction lysimeters, but that the amount of loss was difficult to estimate.

The pre-1980 studies all used solutions with relatively low solute concentrations, which results in high sampling errors and sample variability. Little work had been performed to determine the effects of these samplers on highly contaminated soil solutions. Despite these problems, soil-water samplers are commonly used to monitor highly contaminated soils.

Silkworth and Grigal (1981) studied the effect of porous cup size on sample chemistry. They found less alteration with large cermic cups (4.6-cm diameter) than with small ceramic cups (2.2-cm diameter). Large ceramic samplers compared well with those collected from fritted glass samplers. Large ceramic samplers also produced more sample and had a lower failure rate than either glass or small ceramic samplers.

Creasey and Dreiss (1985) studied the effects of three types of lysimeter cups on trace element and major cation concentrations. Sampler cup materials included ceramics, alundum, and PTFE, all of which are used for commercially available lysimeters. Their study indicated a low potential for significant sample bias by contaminants released from the cleaned and uncleaned samplers when the sample had been buffered to a Ph of 6 to 7. They postulated that differences between their study and eariler studies of ceramic cups may be due to differences in composition of the ceramic cups tested, since many ceramic formulae are available. They concluded that the bias introduced by porous lysimeter cups should only be significant for soil waters with low contaminant concentrations, especially given other sources of error in the collection and analysis of ground-water samples.

These conclusions were supported by Peters and Healy (1988) who found that major ion concentrations collected from pressure-vacuum lysimeters was representative of in situ chemical concentrations where total dissolved solids concentrations were greater than 500 mg/L. However, they found that trace-metal concentrations were significantly affected by sampling with lysimeters.

Based on an extensive program of suction lysimeter testing, Everett et al. (1988) made the following conclusions and recommendations:

1. Prior to field installation, pressure techniques should be used to check all lysimeters for leaks.
2. The approximate bubbling pressure of ceramic low-flow cups is 2.38 atm (35 psi); for ceramic high flow cups 1.224 atm (18 psi); and for PTFE cups 0.068 atm (1 psi).
3. Low-flow ceramic cups are capable of holding their vacuum for several months.
4. PTFE lysimeters must be used with silica flour slurry.
5. The dead space in suction lysimeters must be determined prior to field installation or laboratory tests.
6. Suction lysimeters placed in most types of soil will experience a rapid drop-off of intake rate but will stabilize after about 15 L has been pulled through the porous segments.
7. Use of silica flour around the porous segments negates most plugging associated with finer particles in soils.
8. The effective operating range of ceramic lysimeters is between 0 and 60 centibars of suction regardless of the use of silica flour.
9. The operating range of PTFE lysimeters without silica flour is extremely narrow, but with the use of silica flour is extended to about 7 centibars of suction.
10. Volatile organics can be obtained using a suction lysimeter where equilibrium is established and maintained.
11. Volatile organics are lost from suction lysimeters if the vacuum needs to be intermittently reestablished to draw sufficient sample.

Pan Lysimeters

Pan lysimeters, also called free-drainage samplers, are used to collect water samples by gravity drainage. They are used at waste disposal sites below earthen liners to provide early detection of moisture/solute movement through the liner. The typical pan lysimeter consists of a shallow cone with a drain in the center. The cone is filled with sand and the lysimeter is placed on top of the drainage layer, below a geofabric and the liner material. Less dead space for sample collection exists in the lysimeter than in the drainage layer, so that the breakthrough curve is sharper and sooner than the breakthrough curve from the bottom of the drainage layer. The principle of operation is that under unsaturated conditions sand will have a lower hydraulic conductivity than the surrounding gravel drainage layer. Thus, as the bottom of the earthen liner approaches saturation, water will move preferentially toward the pan lysimeter, instead of into the drainage layer. At saturation, this preferential movement disappears.

Pan lysimeters are relatively inexpensive and can be homemade. For liner monitoring, they should be installed before the soil liner is emplaced. However, they have been installed in tunnels extending from trenches and in buried culverts (Wilson, 1990).

Many pan lysimeters currently in operation have been built with a misunderstanding of the theory of unsaturated zone flow, causing the lysimeters not to provide early information on breakthrough. The error is the belief that under both saturated and unsaturated conditions, water moves faster in coarser material. This has led to lysimeters filled with gravel surrounded by a sand drainage layer. In this case, the preferential flow under unsaturated conditions is away from the lysimeter—the lysimeter will not collect water before the liner becomes nearly saturated.

SOIL GAS TECHNOLOGY

Introduction

During the past several years, sampling and analysis of soil gas for the delineation of subsurface volatile organic compound (VOC) contamination has become very popular. The technology has proven to be effective in a wide range of geologic settings and for many different VOCs. Several methods are currently being used for the collection and analysis of soil gas samples. These methods are generally divided into two types—active and passive techniques. Active sampling is the term applied to those methods which physically remove a gas sample from the vadose zone, usually by pumping (Thompson and Marrin, 1987; Marrin and Thompson, 1987). Passive sampling refers to a technique of burying an absorbent material within the vadose zone and capturing the VOCs present by chemical sorption (Kerfoot and Mayer, 1986; Bisque, 1984). Active sampling techniques have become the more popular because the samples can easily be analyzed in the field and actual concentrations are measured, whereas passive methods measure only relative concentrations across the site. The real-time results made possible by the rapid field analysis using active techniques are very helpful for directing the soil gas investigation. Because the results can be used to direct the investigation, fewer unnecessary samples are collected compared to laboratory-based investigations. This results in both time and cost savings. Due to the greater popularity of active techniques over passive techniques, this discussion concentrates on active techniques.

Presented here are applications and limitations of soil gas technology. Special attention is given to the variables that can impact the effectiveness of a soil gas investigation. These variables include geologic barriers, suitability of the compound to the soil gas application, and the interpretation of soil gas data.

Background on Methodology

Figure 4.17 shows a schematic representation of the driving principles behind soil gas technology.

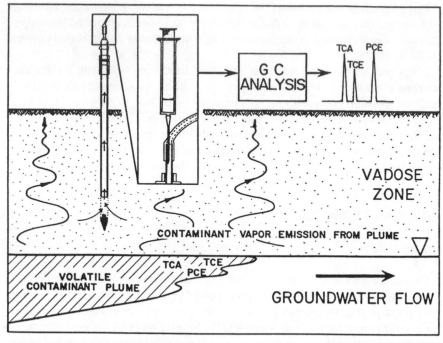

Figure 4.17.   Schematic diagram of the soil gas contaminant investigation technology.

The presence of VOCs in shallow soil gas indicates that the observed compounds may be present either in the vadose zone or in the saturated zone below. Soil gas technology is most effective in mapping low molecular weight halogenated chemical solvents and petroleum hydrocarbons possessing high vapor pressures and low aqueous solubilities. These compounds readily evaporate out of ground water and into the soil gas as a result of the favorable gas/liquid partition coefficients. Once in the soil gas, VOCs diffuse vertically and horizontally through the soil to the ground surface, where they dissipate in the atmosphere. The contamination acts as a source and the above-ground atmosphere acts as a sink, with a concentration gradient typically developing in between. The concentration gradient in soil gas between the source and ground surface may be locally distorted by hydrologic and geologic anomalies (e.g., clays, perched water); however, soil gas mapping generally remains effective because distribution of the contamination is usually broader in areal extent than the local geologic barriers and is defined using a large database. The presence of small-scale geologic obstructions tends to create anomalies in the soil gas-ground water correlations, but generally does not obscure the broader areal picture of contaminant distribution.

Sampling and Analytical Procedures

Soil gas samples can be collected by mechanically advancing a hollow steel probe to a depth generally less than 15 feet into the ground. The aboveground end of the sampling probe is then attached to a vacuum pump. The sampling train is purged by drawing air out of the soil through the probe and an aliquot of the evacuation stream is collected for analysis.

Several methods are currently being used for the analysis of soil gas samples. Field-portable gas chromatographs (GCs), detectors and laboratory-type bench-top GCs are in common usage. The most commonly used detectors are the electron capture detector (ECD), flame ionization detector (FID), photoionization detector (PID), and the Hall electrolytic conductivity detector. The ECD works well for detecting halocarbon compounds and the FID works well with hydrocarbon compounds. These detectors are highly selective to their respective categories of compounds, and thus significantly reduce the problem of misidentification of unknowns. A PID may be used for detecting vinyl chloride, a compound that is not sufficiently detectable using either the ECD or FID. The Hall detector offers reasonable sensitivity to all of the halogenated compounds including vinyl chloride, but is much less sensitive than the ECD to the primary solvents such as trichloroethene (TCE), tetrachloroethane (PCE), and 1,1,1-trichloroethane (TCA).

Quality Assurance/Quality Control Procedures

A very important, yet often overlooked aspect of soil gas investigation is quality assurance/quality control. The following are recommended procedures that have been successful in several applications of soil gas technology.

- Steel probes and sampling train parts are used only once. Before being used, they are washed with a high-pressure soap and hot water spray or steam-cleaned to eliminate the possibility of cross-contamination.
- Prior to sampling each day, system blanks are run to check the sampling train for contamination by drawing ambient air from aboveground through the system and comparing the soil gas analysis to the concurrently sampled ambient air analysis.
- Sample containers and subsampling equipment are used for only one sample before being washed and baked to remove any residual VOC contamination.
- Sample containers and subsampling equipment are checked for contamination by running carrier gas blanks.
- Septa through which soil gas samples are injected into the chromatograph are replaced on a daily basis to prevent possible gas leaks from the chromatographic column.

- Analytical instruments are calibrated each day. Calibration checks are also run after approximately every five soil gas sampling locations or a minimum of three times per day.
- Soil gas pumping is monitored by a vacuum gauge to ensure that an adequate gas flow from the vadose zone is maintained. A negative pressure (vacuum) usually indicates that a reliable gas sample cannot be obtained because the soil has a very low air permeability.

Applications

Case Study

Soil gas investigations are most often applied either for the purpose of defining the areal extent of contamination migrating from a known source or identifying potential sources of ground-water contamination problems. Soil gas data are typically used as a basis for more efficiently locating soil borings and monitoring wells, which are required to confirm the presence and distribution of subsurface contamination. The following case study gives a typical example of the plume mapping and source identification applications of soil gas technology.

Figure 4.18 shows an example of the use of soil gas technology to locate a contamination source. The depth to water was 120 feet, and the geologic materials

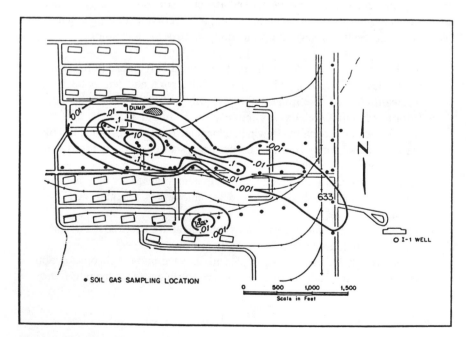

Figure 4.18.   Representative application of the soil gas contaminant investigation technology.

were silty clays. Soil gas samples were collected from a depth of five feet. Well I-1, in the southeast corner of the study area, was contaminated with TCA. A large industrial complex existed on the west side of the road extending more than a mile north and south of the well. Soil gas sampling was initiated along a transect extending several feet along a north-south road between the well and the complex. One soil gas sample on this transect detected TCA slightly above background (Point 633, Figure 4.18). A second east-west transect was initiated along a convenient road into the complex a short distance north of Point 633. The samples along the second transect detected increasingly higher TCA concentrations. Because the soil gas analysis were performed in the field, the sampling plan could be easily directed to "zero in" on the source area. In this case the source was a business with a TCA tank. The long axis of the detectable TCA soil gas plume extended more than 3000 feet from the source toward the contaminated well, which was about one mile away. The investigation left very little doubt about the source of TCA contamination in the I-1 well.

This investigation represents an optimum usage of the soil gas technology. The general distribution of the contaminant can be defined relatively quickly using a probe spacing of between 100 and 300 feet. After the soil gas investigation, verification drilling and soil sampling can proceed very efficiently.

Halocarbon Solvents versus Petroleum Hydrocarbons

The compounds most suited to detection by current soil gas technology are the primary halocarbon solvents. The most common compounds in this group are TCA, TCE, PCE, and 1,1,2-trichlorotrifluorethane or Freon (F113). These compounds readily volatilize out of ground water and into soil vapor as a result of their high gas/liquid partition coefficients. Good detection of these solvent vapors can be expected in most geologic settings. The exceptions are situations where there are geologic barriers to the migration of contaminant vapors. These barriers are discussed further in the following section.

There is no specific depth limitation for remote detection of the primary halocarbon solvents. These vapors tend to resist degradation, and in the absence of the geologic barriers, will migrate through thick unsaturated zones to escape into the atmosphere. Remote detection of certain halocarbons from depths greater than 300 feet has been performed.

The application of soil gas technology to hydrocarbons is more limited than for halocarbons. Good detection of hydrocarbon vapors is common in settings with shallow ground water (<25 feet) and fairly permeable soils. A principal limitation to the application of soil gas technology to hydrocarbon contamination is the relatively rapid degradation of hydrocarbons in well-oxygenated shallow soil. Due to degradation, significant concentrations of the hydrocarbon vapors tend to appear and disappear abruptly in the soil gas profile (Evans and Thompson, 1986). Table 4.1 shows the abrupt change in hydrocarbon (benzene, toluene and total hydrocarbon) concentrations compared to the smooth concentration observed for PCE, a common halocarbon solvent.

Table 4.1.   Comparison of PCE and Hydrocarbon Concentrations in Soil Gas Above a Contaminated Aquifer. (Values are given in μg/L.)

Depth (ft) Below ground Surface	PCE	Benzene	Toluene	Total Petroleum Hydrocarbons
5	0.006	<0.1	<0.1	<0.1
10	0.01	<0.1	<0.1	<0.1
15	0.03	220	31	600

Problems

The most common problems associated with soil gas investigations are geologic barriers, unsuitable target compounds and the tendency to over-interpret soil gas data. An awareness of the limitations of the technology is very important when planning and directing a soil gas investigation.

Geologic Barriers

The most common geologic barrier to the migration of VOC vapors is water saturation of sediments in the vadose zone. A soil gas investigation can be successful in low-permeability clay soils, but if the sediments are completely water saturated, soil gas technology is not effective. Saturated sediments form a nearly impermeable barrier to the migration of contaminant vapors by molecular diffusion, thus preventing remote detection via shallow soil gas samples.

Recharge of significant amounts of clean water over contaminated water commonly limits the area of effective soil gas sampling. Clean recharge acts as a complete barrier only at sites where the recharge is significantly greater than the seasonal fluctuations in the water table. Fluctuations in the water table will allow the contaminated water to be dispersed through the clean water and capillary zone, thus maintaining vapor transport through the capillary fringe. This mechanism also explains the observation that contaminant vapors are commonly easiest to detect near water supply wells where pumping causes variations in the water level, enhancing transport through the capillary fringe.

Figure 4.19 shows the effect of increased soil moisture, or recharge, on the vapors emanating from a TCE contaminated aquifer. The horseshoe indentations correspond to drainage or topographically low areas where a large amount of surface water and runoff is collected or channeled. These areas selectively received greater amounts of recharge which reduced the concentration of the contaminant vapors detected in the shallow soil gas. Contrary to the appearance of the soil gas map, the ground-water contamination does not necessarily diverge in a corresponding manner.

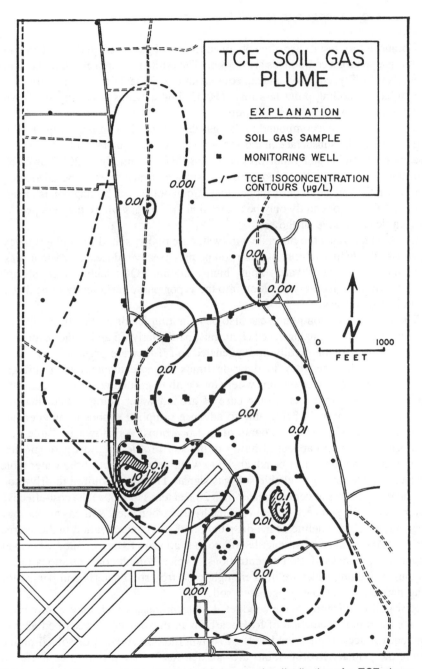

Figure 4.19. Example of the effect of recharge on the distribution of a TCE plume.

Suitability of Compounds to Soil Gas Technology

Unsatisfactory results are often obtained from soil gas investigations when poor or unsuitable target compounds are chosen. The limitations are related to the compound's volatility, stability, and aqueous solubility. Suitable compounds are those which have a boiling point less than 150°C, low aqueous solubility, and relatively good resistance to degradation.

The suitability of a compound to soil gas detection relies on the compound being present in the subsurface in the vapor phase. Compounds with boiling points greater than 150°C and vapor pressures less than 10 mm Hg at 20°C probably will not be present in the vapor phase in sufficient quantities to be adequately detected in the soil gas in most applications. Compounds with boiling points greater than 150°C can commonly only be detected in the soil gas where they are present as significant residue in the soil.

Compounds which are miscible with water are poorly suited for soil gas investigations. The high solubility of these compounds greatly reduces their vapor pressure in the presence of water. Thus, highly soluble VOCs such as alcohols and ketones will not favorably partition into the vapor phase sufficiently to be detectable in the soil gas.

The stability of a compound can also be a limiting factor to the utility of a soil gas investigation. Nonhalogenated chemicals, particularly C5 and higher hydrocarbons, tend to degrade readily in oxygenated soil if they are present in low concentrations. This tendency to degrade limits the effectiveness of a soil gas investigation in geologic settings where the depth to ground water is greater than 25 feet or less than 5 feet. In the case of ground water being greater than 25 feet, the limitation is in being able to advance sampling probes to an adequate depth to detect significant amounts of hydrocarbons. As shown in Table 4.1, hydrocarbons tend to appear abruptly in the soil profile. In most geologic settings a soil gas probe must be advanced to within about 5 feet of the water table surface in order to get a reliable soil gas signal. The time required to drive soil gas sampling probes to depths greater than 20 feet tends to reduce the cost-effective nature of a soil gas investigation. At deep hydrocarbon sites, soil gas technology may only be able to delineate the distribution of soil contamination in the source area. Degradation of most volatile compounds appears to be inhibited whenever vapors are present in high concentrations. Typically high concentrations in the vicinity of leaking tanks are high enough to destroy soil bacteria and persist for long periods in shallow oxygenated soil.

Research is currently being conducted on techniques which may improve the means for remote detection of hydrocarbons in the situations described above. The occurrences of elevated levels of carbon dioxide above a dissolved plume where the primary hydrocarbons are not detectable may prove to be useful in delineating the areal extent of contamination. Preliminary work by Kerfoot et al. (1988) at the Pittman Lateral site in Nevada has indicated that this approach may be successful.

Stability of halogenated chemicals is generally related to the number and type of halogens on the molecule. Stability of the molecule increases with the number of halogens. Fluorine produces greater stability than chlorine, and chlorine produces greater stability than bromine. Fluorocarbons tend to persist even at low concentrations in the environment. As a result, they have accumulated in the atmosphere to the extent that they now pose a threat to the ozone layer (Zurer, 1987). Solvents having three or four chlorines on the molecule C_2Cl_4, C_2Cl_3H, CCl_3CH_3 and CCl_4 commonly degrade to some degree in the subsurface environment, but degradation is slow enough to have little impact on their detectability in soil gas. Dichloro- compounds [dichloroethene (DCE) and dichloroethane (DCA) isomers] are produced in the subsurface as the first degradation products of the primary chlorinated solvents. These products appear to degrade in the soil gas environment slightly faster than the primary solvents (Vogel et al., 1987). As a result, soil gas data for the dichloro- compounds is apt to be less representative of their ground-water distribution than the same data for the primary solvents. Monochlorinated vinyl chloride (C_2ClH_3), a second stage degradation product, may be the least stable chloro- compound in the soil gas environment. Vinyl chloride has been detected in soil gas associated with landfills, but seldom detected in soil gas over contaminated ground water (Table 4.2). This indicates that it is probably an unreliable indicator of ground-water contamination.

Table 4.2.   Vinyl Chloride Concentrations in Soil Gas and Ground Water (15–20 Feet to Ground Water).

Water	Soil Gas
520 μg/L	<0.005 μg/L
110 μg/L	<0.005 μg/L
510 μg/L	<0.01 μg/L
1200 μg/L	<0.005 μg/L

Interpretation

Soil gas data are normally regarded as remote or indirect indications of ground-water or soil contamination from volatile chemicals. As with other remote detection methods, the data are subject to limitations that may cause them to be misrepresentative or inaccurate at any particular location.

Most problems result from attempts to over-interpret the data. Usually this is evident when too much importance is placed on a single point or data anomaly in a very small area. Commonly investigations begin by collection of soil gas samples adjacent to wells or areas of known contamination to establish a basis for interpreting the soil gas data. The findings are commonly disappointing because the high, medium, and low concentrations in soil gas may not be measured at the same locations as high, medium, and low concentrations in ground water. Small-scale geologic and soil moisture variability typically accounts for these problems and may make the data at any given point or in a small area highly misrepresentative of all subsurface conditions. In spite of an initially poor correlation, the investigation is probably worth continuing if the contamination was detectable in at least 50% of the locations where it was known to exist. Soil gas detection of contamination is generally more successful when evaluated over a broad area and used to determine only the presence or absence of contamination in that area.

A second problem relates to the tendencies of some users to include the possible effects of short-term climate changes on the soil gas data. Typically, barometric pressure changes, recent rainfall events, and air temperature are parameters that are brought in unnecessarily. Barometric pressure changes have long been known to be responsible for only a small amount of air transport into and out of the soil. Air exchange due to barometric fluctuations is believed to be limited to the upper 1% of the thickness of the unsaturated zone (Buckingham, 1904). However, soil ventilation due to barometric changes may be important in the immediate vicinity of a borehole, where an air conduit exists into the soil.

A single rainfall event rarely has any appreciable effect on soil gas measurements. If the soils are normally unsaturated, even a heavy rain will not produce saturated conditions, except for a brief period of time (probably less than an hour) at the ground surface. However, soils consisting of fine marine sediments where the depth to water is six feet or less are typically problematical. These soils remain nearly saturated due to capillary forces drawing water upward from the water table. As a result, soil gas investigations are often not useful in these bay mud type environments.

Summary

In summary, soil gas technology is an effective tool for the delineation of subsurface volatile organic compound contamination. A well planned investigation which takes into account the effects of geologic barriers, the suitability of the compounds to the application of the soil gas technology, and reasonable interpretation of the data, will yield results that can be used to more efficiently direct a conventional soil boring and ground-water monitoring well installation program.

REFERENCES

Apgar, M. A. and Langmuir, D., 1971, Ground-Water Pollution Potential of a Landfill Above the Water Table, *Ground Water*, V. 9, no. 6, pp. 76–96.

Albrecht, K. A., Herzog, B. L., Follmer, L. R., Krapac, I. G., Griffin, R. A. and Cartwright, K., 1989, Excavation of an Instrumented Earthen Liner: Inspection of Dyed Flow Paths and Soil Morphology, *Hazardous Waste and Hazardous Materials*, V. 6, no. 3, pp. 269–279.

Ball, J. and D. M. Coley, 1986, A Comparison of Vadose Monitoring Procedures, *Proceeding of the Sixth National Symposium and Exposition on Aquifer Restoration and Ground Water Monitoring,* National Water Well Association, Dublin, OH, pp. 52–61.

Bear, Jacob, 1979, *Hydraulics of Ground Water,* McGraw-Hill, Inc., New York, NY.

Bisque, R. E., 1984, Migration Rates of Volatiles from Buried Hydrocarbon Sources through Soil Media, *Proceedings of the NWWA/API Petroleum Hydrocarbons and Organic Chemicals in Ground Water: Prevention, Detection and Restoration Conference.* National Water Well Association, Dublin, OH.

Bruce, R. R. and R. J. Luxmore, 1988, Water Retention: Field Methods, *Methods of Soil Analysis, Part 1. Physical and Mineralogical Methods,* Agronomy Monograph no. 9, American Society of Agronomy-Soil Science Society of America, Madison, WI, pp. 663–686.

Buckingham, E., 1904, *Contributions to Our Knowledge of the Aeration of Soils,* U.S. Department of Agriculture Soils Bur. Bull. 25. P. 52.

Corey, A. T., 1977, *Mechanics of Heterogeneous Fluids in Porous Media,* Water Resources Publications, Fort Collins, CO.

Crease, C. L. and Dreiss, S. J., 1985, Soil Water Samplers: Do they Significantly Bias Concentrations in Water Samples?, *Proceedings of the NWWA Conference on Characterization and Monitoring of the Vadose (Unsaturated) Zone,* Water Well Publishing Company, Dublin, OH, pp. 173–181.

Daniel, D. E. and S. J. Trautwein, 1986, Field Permeability Test for Earthen Liners, in S. P. Clemence (ed.), *Use of In-Situ Tests in Geotechnical Engineering,* American Society of Civil Engineers, New York, NY.

Davis, S. N., D. J. Campbell, H. W. Bentley and T. J. Flynn, 1985, *Ground Water Tracers.* National Water Well Association.

Dazzo, F. B. and D. F. Rothwell, 1974, Evaluation of Porcelain Samplers for Bacteriological Sampling. *Applied Microbiology,* V. 27, no. 6, pp. 1172–1174.

EPRI, 1985, Section 3: Unsaturated Zone Measurement Methods. Field Measurement Methods for Hydrogeological Investigations: A Critical Review of the Literature. EPRI EA-4301.

Evans, O. D. and G. M. Thompson, 1986, Field and Interpretation Techniques for Delineating Subsurface Petroleum Hydrocarbons Spills Using Soil Gas Analysis, *Proceedings of the NWWA/API Petroleum Hydrocarbons and Organic Chemicals in Ground Water: Prevention, Detection and Restoration Conference.* National Water Well Association, Dublin, OH.

Everett, L. G., 1980, *Groundwater Monitoring,* General Electric Company, Technology Marketing Operation, Schenectady, NY.

Everett, L. G., Schmidt, K. D., Tinlin, R. M., and Todd, D. K., 1976, *Monitoring Ground-Water Quality: Methods and Costs,* U.S. Environmental Protection Agency, Solid Waste Management Series SW-616.

Everett, L. G., McMillion, and L. A. Eccles, 1988, Suction Lysimeter Operation at Hazardous Waste Sites, *Ground-Water Contamination: Field Methods*, ASTM STP 963, A. G. Collins and A. I. Johnson, Eds., American Society for Testing and Materials, Philadelphia, pp. 304–327.

Fenn, D., E. Cocozza, J. Isbiter, O. Braids, B. Yare and P. Roux. 1977. *Procedures Manual for Ground Water Monitoring at Solid Waste Disposal Facilities*. U.S. Environmental Protection Agency. EPA/530/SW-611. 260 pp.

Green, W. H. and G. A. Ampt, 1911, Studies on Soil Physics: 1. Flow of Air and Water Through Soils, *Journal of Agriculatural Science*, V. 4, part 1, pp. 1–24.

Grover, B. L. and Lamborn, R. E., 1970, Preparation of Porous Ceramic Cups to be Used for Extraction of Soil Water Having Low Solute Concentrations, *Soil Science Society of America Proceedings*, V. 34, no. 4, pp. 706–708.

Hansen, E. A., and Harris, A. R., 1975, Validity of Soil-Water Samples Collected with Porous Ceramic Cups, *Soil Science Society of America Proceedings*, V. 39, no. 3, pp. 528–536.

Healy, R. W., M. P. deVries, and R. G. Striegl, 1986, *Concepts and Data-Collection Techniques used in a Study of the Unsaturated Zone at a Low-Level Radioactive-Waste Disposal Site near Sheffield, Illinois*, U.S. Geological Survey Water-Resources Investigations Report 85-4228, Urbana, Illinois, 35.

Hillel, D., 1980a, *Applications of Soil Physics*, Academic Press, New York, NY.

Hillel, D., 1980b, *Fundamentals of Soil Physics*, Academic Press, New York, NY.

Hillel, D., Krentos, V. D., and Strylianou, Y., 1972, Procedure and Test of an Internal Drainage Method for Measuring Soil Hydraulic Characteristics *in Situ*, *Soil Science*, V. 114, pp. 395–400.

Iwata, S., and Tabuchi, T., with Warkentin, B., 1988, *Soil-Water Interactions: Mechanisms and Applications*, Marcel Dekker, Inc., New York, NY.

Kerfoot, H. B., and C. L. Mayer, 1986, The Use of Industrial Hygiene Samplers for Soil Gas Surveying, *Ground Water Monitoring Review*, V. 6, no. 4, pp. 74–78.

Kerfoot, H. B., C. L. Mayer, P. B. Durgin and J. J. D'Lugosz, 1988, Measurement of Carbon Dioxide in Soil Gasses for indication of Subsurface Hydrocarbon Contamination, *Ground Water Monitoring Review*, V. 8, no. 2, pp. 67–71.

Levin, M. J., and Jackson, D. R., 1977, A Comparison of *in Situ* Extractors of Sampling Soil Water, *Soil Science Society of America Proceedings*, V. 41, no. 3, pp. 535–536.

Litaor, M. I., 1988, Review of Soil Solution Samplers, *Water Resources Research*, V. 24, no. 5, pp. 727–733.

Marrin, D. L. and G. M. Thompson, 1987, Gaseous Behavior of TCE lying a Contamination Aquifer, *Ground Water*, V. 25, no. 1, pp. 21–27.

McWhorter, D. and Sunada, D., 1981, *Ground water Hydrology and Hydraulics*, Water Resources Publications, Fort Collins, CO.

Morrison, R. D. and B. Lowery, 1990, Sampling Radius of a Porous Cup Sampler: Experimental Results, *Ground Water*, V. 28, no. 2, pp. 262–267.

Parizek, R. R. and Lane, B. E., 1970, Soil-Water Sampling Using Pan and Deep Pressure-Vacuum Lysimeters, *Journal of Hydrology*, V. 11, no. 1, pp. 1–21.

Peters, C. A. and R. W. Healy, 1988, The Representativeness of Pore Water Samples Collected from the Unsaturated Zone Using Pressure-Vacuum Lysimeters, *Ground Water Monitoring Review*, V. 8, no. 2, pp. 96–101.

Schmugge, T. J., T. J. Jackson, and H. L. McKim, 1980, Survey of Methods for Soil Moisture Determination, *Water Resources Research*, V. 16, no. 6, pp. 961–979.

Silkworth, D. R. and D. R. Grigal, 1981, Field Comparison of Soil Solution Samplers, *Soil Science Society of America Journal*, V. 45, no. 2, pp. 440–442.

Soilmoisture Equipment Corporation, *Operating Instructions for the Model* 1920 *Pressure Vacuum Soil Water Sampler*, Santa Barbara, Cal., 6 pp.

Stannard, D. I., 1990, Tensiometers—Theory, Construction and Use, *Ground Water and Vadose Zone Monitoring*, ASTM STP 1053, D. M. Nielsen and A. I. Johnson, Eds., American Society for Testing and Materials, Philadelphia, pp. 34–51.

Thompson, G. M. and D. L. Marrin, 1987, Soil Gas Contaminant Investigations: A Dynamic Approach, *Ground Water Monitoring Review*, V. 7, no. 3, pp. 88–93.

U.S. Environmental Protection Agency, 1988, *Design, Construction and Evaluation of Clay Liners for Waste Management Facilities*. U.S. EPA Risk Reduction Engineering Laboratory, Cincinnati, OH, EPA/530-SW-86-007F.

Vogel, T. M., C. S. Criddle and P. L. McCarty, 1987, Transformations of Halogenated Aliphatic Compounds, *Environmental Science and Technology*, V. 21, no. 8, pp. 722–734.

Watson, K. K., 1974, Some Applications of Unsaturated Flow Theory, in "Drainage for Agriculture," J. van Schilfgaard (ed.), *Agronomy*, No. 17, American Society of Agronomy, Madison, Wisconsin.

Wilson, L. G., 1980, *Monitoring in the Vadose Zone: A Review of Technical Elements and Methods*, U.S. Environmental Protection Agency, Environmental Monitoring Systems Laboratory, Office of Research and Development, Las Vegas, NV, report EPA-600/7-80-134.

Wilson, L. G., 1981, Monitoring in the Vadose Zone, Part 1: Storage Changes, *Ground Water Monitoring Review*, V. 1, no. 3, pp. 32–41.

Wilson, L. G., 1990, Methods for Sampling Fluids in the Vadose Zone, *Ground Water and Vadose Zone Monitoring*, ASTM STP 1053, D. M. Nielsen and A. I. Johnson, Eds., American Society for Testing and Materials, Philadelphia, pp. 7–24.

Wolff, R. G., 1967, Weathering of Woodstock Granite Near Baltimore, Maryland, *American Journal of Science*, V. 265, no. 2, pp. 106–117.

Wood, W. W., 1973, A Technique Using Porous Cups for Water Sampling at Any Depth in the Unsaturated Zone, *Water Resources Research*, V. 9, no. 2, pp. 486–488.

Wood, W. W., 1978, Use of Laboratory Data to Predict Sulfate Sorption During Artificial Ground-Water Recharge, *Ground Water*, V. 16, no. 1, pp. 22–31.

Zurer, P. S., 1987, Antarctic Ozone Hole: Complex Picture Emerges, *Chemical and Engineering News*, V. 65, no. 44, pp. 22–26.

5

Remote Sensing and Geophysical Methods for Evaluation of Subsurface Conditions

Richard C. Benson

INTRODUCTION

Remote sensing and geophysical methods encompass a wide range of airborne, surface, and downhole tools which provide a means of investigating hydrogeologic conditions and locating buried waste materials. Under certain conditions, some of the geophysical methods provide a means of detecting contaminant plumes.

Geophysical measurements can be made relatively quickly, thereby increasing sample density. Continuous data acquisition along a traverse line can be employed with certain techniques at speeds up to several miles per hour. Because of the greater sample density, anomalous conditions are more likely to be detected, resulting in a more accurate characterization of subsurface conditions.

Geophysical methods, like any other means of measurement, have advantages and limitations. There is no single, universally applicable geophysical method, and some methods are quite site-specific in their performance. Thus, the user must carefully select the method or methods and how they are applied to specific site conditions and project requirements.

Unlike direct sampling and analysis, such as obtaining a soil or water sample and sending it to a laboratory, the geophysical methods provide nondestructive, in situ measurements of physical, electrical, or geochemical properties of the natural or contaminated soil and rock. The success of a geophysical method depends on the existence of a sufficient contrast between the measured properties of the target and background conditions. If there is no measurable contrast, the target will not be recognized. Similarly, if a layer is sufficiently thin or if the size of the target is sufficiently small, it will not be detected.

Geophysical techniques are not new; they have been used for decades in oil and gas exploration, mineral exploration, geotechnical applications, and regional water resources development (Griffith and King, 1969; Telford et al., 1982; and Zohdy et al., 1974). Geophysical methods, as applied to hazardous waste site investigations, are somewhat different in their application because they are usually required to produce higher resolution shallow data (typically less than 100 feet or so in depth). In less than one decade (1975–1985) extensive development in geophysical field instrumentation, field methods, analytical techniques, and related computer processing has resulted in a striking improvement in our capability to provide a high resolution assessment of shallow subsurface conditions.

However, many professionals still view geophysics as a "black box" technology. This is unfortunate because the methods are based upon sound principles of physics, geochemistry, and electronics. The "black box" image simply reflects a lack of understanding the science behind the technology.

This chapter provides an overview of the various geophysical methods. The first section provides background material and identifies the three basic areas in which geophysical methods should be applied. The second section deals with airborne, surface, and downhole geophysical methods, discussing specific techniques for each. Major emphasis is placed upon the surface and downhole geophysical methods which are most commonly used for the type of applications discussed in this chapter. The chapter ends with application tables and a discussion to aid in selecting methodologies for specific field problems.

The examples of data shown within this chapter are considered to be of excellent quality. These high quality data are presented to aid the reader in understanding the geophysical methods discussed. In practice, data will often be less clear than these examples, which requires the skill of an experienced interpreter.

BACKGROUND

Traditional approaches to subsurface field investigations at hazardous waste disposal sites have often been inadequate. Site investigations have relied upon direct sampling methods such as:

- soil borings and monitoring wells for gathering hydrogeologic data and soil and water samples;
- laboratory analysis of soil and water samples to provide a quantitative assessment of site conditions; and
- extensive interpolation and extrapolation from these points of data.

This approach has evolved over many years and is commonly considered a standard one to deal with field investigations. However, there are numerous pitfalls associated with this approach, which can result in an incomplete or even erroneous understanding of site conditions. These pitfalls have been the subject of numerous papers and conferences over the past five years (Dunbar et al., 1985; Hileman, 1984; Lysyj, 1983; Perazzo et al., 1984; and Walker and Allen, 1984). They have also precipitated the tightening of ground-water monitoring regulations (EPA, 1985).

The single most critical factor we face in site evaluation work is accurately characterizing the site's hydrogeology (Benson and Pasley, 1984). If we have an accurate understanding of site hydrogeology, predicting the movements of contaminants, or designing a cleanup operation would be reasonably straightforward. If all sites had simple, horizontally stratified geology with uniform properties, site characterization would be easy. Data from just one boring would be sufficient to characterize the site. However, in most geologic settings, this will not be the case, and one must be alert to variations which can cause significant errors in site characterization.

In the design of many soil and rock sampling programs and monitoring well networks, the placement of borings and wells has been done mainly by educated guesswork. The accuracy and effectiveness of such an approach is heavily dependent upon the assumption that subsurface conditions are uniform. This approach usually assumes that regional hydrogeology and ground-water flow (as obtained from literature) is valid for the site-specific setting, and that data from a few specific borings or monitoring wells can be used to characterize the site. These assumptions are frequently invalid, resulting in nonrepresentative locations for borings and monitoring wells and erroneous generalizations from this limited information. To improve the accuracy of the site investigation, a large number of borings may be required.

Sample Density

A soil or core sample obtained from drilling may be representative of only a limited area surrounding the hole. Therefore, fractures, cavities, bedrock channels, sand lenses, and local permeable zones can easily be missed by borehole programs.

An insight to the number of discrete samples or borings that are required for accurate sampling can be obtained by considering detection probability (Benson et al., 1982). Figure 5.1a shows a target area which is 1/10 of the total site area. This target area (whose size and location are usually unknown) could be a waste burial site, a plume from a chemical spill, an old sinkhole, or a buried channel. Based upon probability calculations, the number of samples or borings required to achieve various detection probabilities are shown in Figure 5.1b (Benson and LaFountain, 1984). For example, a site to target ratio of 1/10 and a detection probability of 90% would require 16 borings spaced over a regular grid. For a smaller target, such as a narrow sand lens or fractures, the site to target area ratio will increase significantly, thus 100 to 1000 borings may be required to give a 90% confidence level in characterizing many sites, making the subsurface investigation like "looking for a needle in a haystack."

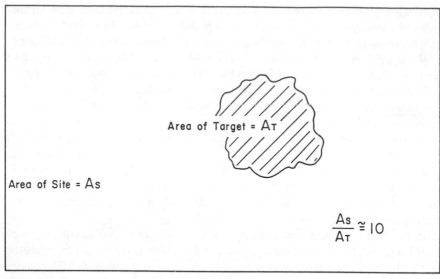

5.1a: Illustrates Site to Target Area Ratio
      As/At = 10 in this case.

Probability of Detection	As/At = 10	As/At = 100	As/At = 1000
100	16	160	1600
98	13	130	1300
90	10	100	1000
75	8	80	800
50	5	50	500
40	4	40	400
30	3	30	300

5.1b: Number of Sample points required for various As/At ratios and Probability of Detection. This table assumes uniform sample grid if a random placement is used, the number of samples must be increased by a factor of 1.6.

Figure 5.1a and 5.1b: Spatial sampling requirements.
Source: Benson and LaFountain, 1984.

It becomes obvious that achieving a good statistical evaluation of complex site conditions requires samples or borings to be placed in a close-order grid, which would reduce the site to "swiss cheese." In many cases, direct sampling alone is not sufficient to accurately characterize site conditions from a technical or cost point of view. This is the primary reason for the application of geophysical methods.

How Geophysical Methods Are Used

Data obtained from borings or monitoring wells come from a very localized area. Geophysical methods, on the other hand, usually measure a much larger volume of the subsurface (Figure 5.2). These measurements provide an average response over a large volume of subsurface conditions; providing a means of detecting subsurface conditions, such as a buried channel that a limited number of borings may miss.

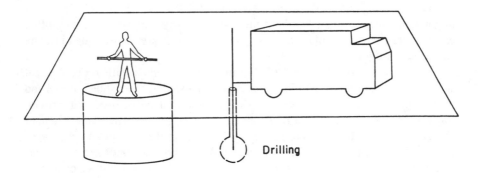

Drilling

A typical geophysical measurement integrates a larger volume of soil and rock.

Volume of soil and rock sampled by drilling is relatively small

Figure 5.2.   Comparison of volumes sampled by geophysical method and a borehole.
Source: Benson et al., 1982.

The geophysical methods to detect anomalous areas may pose potential problems. When geophysical methods are used in this manner, they are essentially anomaly detectors.

Once an overall characterization of a site has been made using geophysical methods and anomalous zones identified, a better drilling and sampling plan may be designed by:

- locating soil borings and monitoring wells to provide samples that are representative of site conditions;
- minimizing the number of samples, borings, and/or monitoring wells required to accurately characterize a site;
- reducing field investigation time and cost; and
- significantly improving the accuracy of the overall investigation.

This approach yields a much greater confidence in the final results, with fewer borings or wells, and an overall cost savings. The estimated cost of long-term sampling from monitoring wells can range from $75,000 upward over a 30 year period (1987 dollars). This is nearly two orders of magnitude greater than the cost of the original boring or monitoring well. With this in mind, it makes good sense to minimize the number of monitoring wells at a site and optimize the location of those installed. Using this approach, drilling is no longer being used for hit-or-miss reconnaissance, but is being dedicated to the specific quantitative assessment of subsurface conditions. Boreholes or wells located with this approach are called SMART HOLES because they are scientifically placed, for a specific purpose, in a specific location, based on knowledge of site conditions; eliminating much of the guess work (Benson and Pasley, 1984). While SMART HOLES might sometimes be placed without the use of geophysical methods, they often can be placed more reliably if the geophysical methods are incorporated into the subsurface program.

If borings have already been drilled or monitoring wells installed, geophysical surveys can still provide significant benefits. The location of existing borings and monitoring wells relative to anomalous site conditions can be assessed, thus providing a means of evaluating the representativeness of any existing data. Then, if additional borings or wells are needed to fill gaps in the overall site coverage, they can be accurately placed as SMART HOLES. Assessment of site conditions will often require that an area larger than the site itself be considered. Contaminant transport by ground water and the geohydrologic factors which control flow do not stop at arbitrary site boundaries or property lines. Insight into the character of the local setting is often derived from the knowledge of the broader picture. An analogy can be drawn to the use of a camera's telescopic zoom from an overall wide angle view to a close-up telescopic view of the finer details. Omitting the broad overview will often result in a number of critical gaps in information about the setting. The geophysical methods provide a means of rapid reconnaissance over larger areas, and can often be employed to obtain the big picture.

Continuous and Station Measurements

Geophysical surveys often involve making measurements of subsurface properties at discrete points over a site. That is, the instrumentation is located at a station along a survey line or a grid, and measurements are made at one point at a time. However, some techniques can provide measurement of subsurface parameters continuously as the instrumentation is moved along the survey line.

By estimating the size of geologic features or anomalies before the survey is carried out, a suitable station spacing can be selected. However, if the size estimate of the geologic feature is too large, the data will not be representative and can lead to errors in the assessment of site conditions. Continuous methods should be employed whenever possible to minimize the possibility of making such errors, to achieve maximum resolution, and to minimize project costs. This is particularly true when site conditions are suspected of being highly variable, and a small sample interval is required.

Although the continuous surface geophysical methods referred to in this document are typically limited to a depth of 50 feet or less, they are applicable to many site investigations. They can provide continuity of subsurface information which is not practically obtainable from station measurements. Continuous surface geophysical methods can be applied at speeds of 1 to 5 mph or more, resulting in a cost-effective approach for relatively shallow survey work. To illustrate the benefits of continuous measurements, a comparison between station measurements and continuous measurements is discussed below.

The lower set of data in Figure 5.3 reveals the highly variable nature of a site indicated by a continuous measurement technique. The upper set of data in Figure 5.3 shows the loss of information that can result from a limited number of station measurements and interpolating between sample points. This limited number of measurements results in a distorted set of data and leads to errors in interpretation of site conditions when target size is significantly smaller than station spacing. By running closely spaced parallel survey lines with continuous methods, subtle changes in subsurface parameters can often be mapped. Total site coverage can even be achieved if necessary.

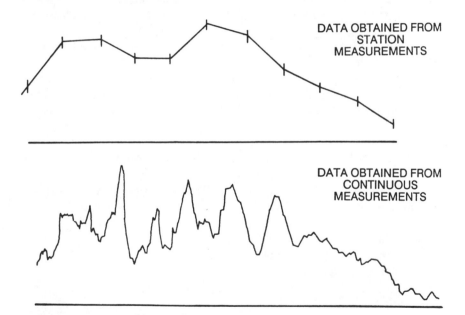

DATA OBTAINED FROM
STATION
MEASUREMENTS

DATA OBTAINED FROM
CONTINUOUS
MEASUREMENTS

**The data was obtained with an EM-34 with
a 10 meter coil spacing. The higher values
of electrical conductivity are caused by
fractures in the underlaying gypsum rock.**

Figure 5.3.   Comparison of station and continuous measurements from the same site.

The data in Figure 5.3 were obtained by surface electromagnetic measurements of subsurface electrical conductivity. The higher conductivity values indicate fractures within underlying gypsum rock. These fractures show up because they are more electrically conductive due to water content and weathering of the gypsum rock.

Site Investigation Methods Are Scale Dependent

All site investigation methods, including geophysics, are scale dependent. For example, aerial photography is an effective tool to be used in regional studies and for obtaining the big picture at a local site investigation. However, it will not provide any information about site-specific soil conditions at a depth of 10 feet. Conversely, a boring will provide information on soil conditions versus depth, but information from the boring is only valid for a very limited extent immediately around the borehole. Geophysical measurements made at the subsurface can be used to determine detailed soil conditions over a few hundred square feet or over many square miles. On the other hand, geophysical logging measurements made down a borehole will extend the measurements from the hole itself radially to a distance of 6 inches to a few feet, depending upon the method used.

Therefore, the site investigation method must be selected to suit data and project requirements (Benson and Scaife, 1987). Typically, a subsurface investigation will include measurements of the big picture (aerial photography), intermediate picture (surface geophysical measurements), and the very local details (boring and sampling data).

APPLICATIONS FOR GEOPHYSICAL MEASUREMENTS

There are three major areas for the application of geophysical methods at hazardous waste sites. They are:

- assessing hydrogeologic conditions;
- detecting and mapping contaminant plumes; and
- locating and mapping buried wastes and utilities.

Assessing Hydrogeologic Conditions

Probably the most important task of a site investigation will be characterizing hydrogeologic conditions. A variety of geophysical methods can be used to assess natural hydrogeologic conditions, such as depth to bedrock, degree of weathering, and presence of sand and clay lenses, fracture zones, and buried relic stream channels (Benson et al., 1979; 1982; 1987; Keys and MacCary, 1976). Accurately understanding the hydrogeologic conditions and anomalies can make the difference between success and failure in site characterization, because these features will often control ground water flow and contaminant transport.

Detection and Mapping Contaminant Plumes

A major objective of many site investigations is the detection and mapping of contaminant plumes. Geophysical methods can be employed in two ways to solve this problem. Some methods can be used for the direct detection of contaminants. In cases in which the contaminant cannot be detected directly, geophysical methods can be used to assess the detailed hydrogeologic conditions that control ground-water flow. Then, the location of the contaminants can be estimated and the ground-water and contaminant flow pathways can be identified (Benson et al., 1982, 1985; Cartwright and McComas, 1968; Greenhouse and Monier-Williams, 1985; McNeill, 1980).

Locating and Mapping Buried Wastes and Utilities

Geophysical methods can also be used to locate and map the areal extent, and sometimes the depth, of buried wastes in trenches and landfills (Benson et al., 1982). There are methods that can also be employed to detect buried drums, tanks, and utility lines. In many cases, the trenches associated with buried pipes and utilities will be of interest because they form a permeable pathway for contaminant migration.

Airborne, Surface, and Downhole Geophysics

There are three different modes in which geophysics can be applied: airborne, surface, and downhole. Airborne methods are usually employed to obtain a regional overview, or the big picture, of site conditions. Surface methods will provide a means of rapid reconnaissance over a larger area or a means of obtaining site-specific details. Downhole methods can be used to provide very localized details down a borehole or well (Figure 5.4).

Airborne and satellite remote sensing clearly have merits in terms of spatial coverage per unit time and cost. Imaging methods (photographic, infrared, and others) provide a "picture" of the site and the surrounding area. They give us an excellent overview of regional conditions that let us see the pieces of the puzzle totally assembled and in perspective. However, they provide little, if any, subsurface data other than those data derived by skilled interpretation. Other airborne (nonimaging) methods such as magnetics and radiometrics can provide a measure of subsurface conditions.

While surface geophysical methods yield much less spatial coverage per unit time than airborne methods, they significantly improve resolution (the ability to detect a small feature) while providing subsurface information. Sometimes, continuous data acquisition can be obtained at speeds up to several miles per hour.

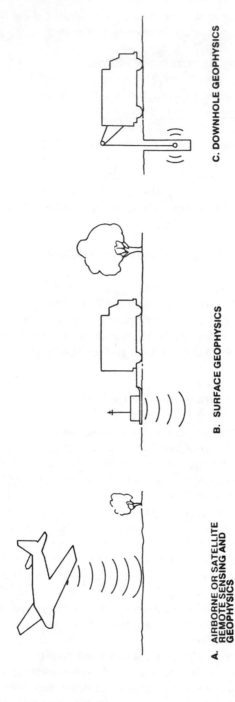

Figure 5.4. Three modes of geophysical measurements.

In certain situations total site coverage is technically and economically feasible. However, an inherent limitation of all surface geophysical methods is that their resolution decreases with depth.

The major benefit of downhole geophysical methods is that they provide detailed high-resolution data at depth around the borehole or well. Unlike surface geophysical methods, whose resolution decreases with depth, the resolution of downhole logging methods is independent of depth. In addition, most downhole methods provide continuous data along the depth of the hole. The volume sampled by downhole methods is usually limited to the area immediately around the boring (a cubic foot to a cubic yard). The cost per unit area of coverage for the downhole methods is obviously much higher than for surface methods, because all downhole techniques require a borehole or monitoring well. However, if holes are already in place or if they are to be drilled for other purposes, the overall cost of downhole logging is relatively low.

All of these approaches—airborne remote sensing and surface and downhole geophysical survey methods, have a place in subsurface investigations. Through appropriate combinations of geophysical measurements and borehole data, an accurate three-dimensional picture of subsurface conditions can be generated. The resulting understanding of subsurface conditions can then be used to develop an accurate conceptual model, which incorporates the big picture through the local details.

REMOTE SENSING AND "AIRBORNE" GEOPHYSICAL METHODS

Airborne remote sensing and geophysical methods cover a wide range of the electromagnetic spectrum, from the lower frequency airborne EM method to the very short wavelength gamma rays measured by the radiometric method. Figure 5.5 shows the range of wavelengths employed for specific measurements. The terms airborne and remote sensing, as used in this section, include measurements made from aircraft as well as from satellites.

Imaging Methods

Imaging methods are those that result in a "picture" of the surface. A wide range of aerial photos can be obtained, from those taken by hand-held 35 mm cameras, to those obtained by complex satellite sensors. Aerial photographs are a source of geological information and provide the big picture, an overview of site conditions. Aerial photos, USGS topographic maps, and USDA soil survey maps will often be the first data reviewed for a project because they provide a rapid, low-cost means of obtaining the necessary overview of site conditions.

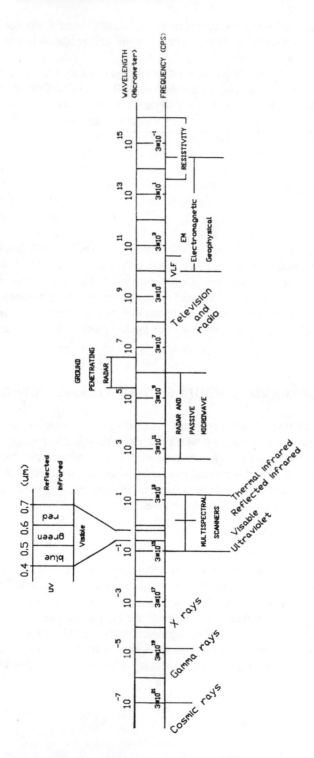

Figure 5.5.   Portions of the electromagnetic spectrum used for geophysical measurements.
Source: Technos Inc.

Large-scale aerial photos with a 19×19-inch format and a scale of 1:3,600 are commonly available from a county surveyor's or tax assessor's office, or through commercial firms. These types of photos provide a local overview of the site for project planning, and a means of accurately locating survey grids, buildings, and roads.

Standard small scale (9×9 inch; 1:24,000) black and white photos are available in stereo pairs from a number of sources, including the Soil Conservation Service (U.S. Department of Agriculture), the U.S. Geological Survey, and state agencies. These aerial photos are often available for the past years, thereby providing a historic record of site conditions. This type of photography is used to provide a very broad overview of the site and to allow photogeologic interpretation. Aerial photo interpretation can provide information on bedrock type, landforms, lineament or fracture analysis, soil texture, site drainage conditions, susceptibility to flooding, and slope of land surface.

Aside from these two relatively standard photographic formats, there are a number of other options available including high-altitude photography, color photography, false-color infrared, and a wide range of satellite imagery. Each of these can be very useful in specific applications. However, the black and white formats are the first and often the primary types used in most site characterization. They are also low in cost and readily available.

There are two other forms of imagery that are occasionally very useful for specific problems: thermal infrared and side-looking airborne radar (SLAR). Thermal infrared is different from false-color infrared in that it is a measure of the thermal response of an area measured in the infrared spectrum. The earth's surface emits radiation in the thermal infrared wavelengths. These emissions are recorded by electronic detectors compared to the reflected energy recorded by infrared film. The image (or thermogram) is presented on video or on film, or it can be stored on magnetic tape. This method can provide a means of locating springs, identifying seeps from a landfill, locating moist and dry areas, characterizing surface soil and rock, and identifying vegetation stress. It is also useful in a number of other conditions in which a difference in temperature is a characteristic feature.

Side-looking airborne radar (SLAR) is an electronic image-producing system. A radar beam is transmitted off to the side of the aircraft. The result is an obliquely illuminated view of the terrain. This oblique view enhances subtle surface features and facilitates geologic interpretation. Another important property of SLAR is that it is an active system that provides its own source of illumination in the form of microwave energy, thus imagery can be obtained either day or night, independent of cloud cover. The SLAR products commonly used for analysis are image strips and mosaics. SLAR imagery is available from the U.S. Geological Survey for selected areas in the United States.

Nonimaging Methods

Nonimaging methods do not result in picture, but provide a measurement of some parameter along the flight line of the aircraft. These methods include electromagnetic measurement [using frequencies up to a few kilohertz (KHz)], magnetic measurements, radiometric measurements of gamma radiation, and ground-pene-trating radar using frequencies of around 100 megahertz (MHz). These are referred to as nonimagery methods because as the aircraft moves along a survey line, a series of measurements are obtained rather than an image. However, by running parallel survey lines, a contour map of the measured parameters can be developed for the site. While the imaging methods provide only a measure of surface conditions, the nonimaging methods measure subsurface conditions. These methods are nor-mally applied to large areas or areas that are not easily accessible by land.

The electromagnetic method (which is described further in the section on sur-face geophysical methods) measures electrical conductivity of subsurface materials. It provides a measure of gross changes in geologic, hydrologic, and environmental conditions based upon electrical conductivity. In certain conditions this method could be used to map soil cover, locate coastal saltwater intrusion, or even map a large leachate plume.

Magnetic measurements provide a means of determining the magnetic suscep-tibility of soil and rock and therefore can provide a geologic map. The resulting maps provide an overview of the gross geologic conditions (based upon mag-netic properties) and can identify larger anomalous conditions.

Radiometric measurements provide a means of measuring the natural radio-active radiation (potassium-40 and daughter products of the uranium and thorium decay series) that is emitted from all rocks. Total measurements or spectral meas-urements can be obtained to characterize the count from specific elements. While this method has been applied to mineral exploration, a radiometric map, like a magnetic map, provides insight into the overall geologic structure and can iden-tify larger anomalous conditions.

Ground-penetrating radar, which is commonly used on the surface, has been used for limited airborne applications (this method is described further in the section on surface geophysical methods). Helicopter surveys have been successfully applied to obtain some soil, ice, snow, and permafrost thickness measurements. In general, where radar penetration is good and the site is clear of vegetation and cultural fea-tures, the method may be usable to obtain shallow profiles in these materials.

SURFACE GEOPHYSICAL METHODS

The surface methods discussed in this section are:

- Ground-Penetrating Radar
- Electromagnetics
- Resistivity
- Seismic Refraction
- Seismic Reflection
- Micro Gravity
- Metal Detection
- Magnetics

These techniques are included because they are used regularly and have proved effective for hazardous waste site assessments. A brief description of each of these surface geophysical techniques is presented in the following section.

Ground-Penetrating Radar

Ground-penetrating radar uses high-frequency electromagnetic waves from less than 100 MHz to 1000 MHz to acquire subsurface information. Energy is radiated downward into the ground from a transmitter and is reflected back to a receiving antenna. The reflected signals are recorded and produce a continuous cross-sectional picture or profile of shallow subsurface conditions.

Reflections of the radar wave occur whenever there is a change in dielectric constant or electrical conductivity between two materials. Changes in conductivity and in dielectric properties are associated with natural hydrogeologic conditions such as bedding, cementation, moisture, clay content, voids, and fractures. Therefore, an interface between two soil or rock layers that have a sufficient contrast in electric properties will show up in the radar profile (Benson et al., 1979; 1982; 1987). Figure 5.6 shows a radar record of a sand-clay interface. The water table can be detected in coarser-grained materials but not in finer-grained sediments with a large capillary boundary. Both metallic and nonmetallic buried pipes and drums can also be detected.

The vertical scale of the radar profile is in units of time (nanoseconds 10^{-9} sec). The time it takes for an electromagnetic wave to move down to a reflector

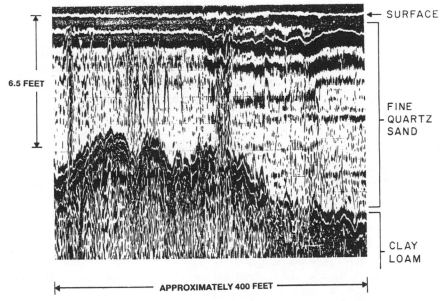

Figure 5.6.   Radar profile of quartz sand over clay. (Note the level of detail that can be obtained.)

and back to the surface is relatively short because the waves are travelling at almost the speed of light. The time scale then is converted to depth by making some assumptions about the velocity of the waves in the subsurface materials.

Depth of penetration of the radar wave is highly site-specific. The method is limited in depth by attenuation due to the higher electrical conductivity of subsurface materials or scattering. Generally, radar penetration is better in coarse, dry, sandy, or massive rock; poorer results are obtained in wet, fine-grained, clayey (conductive) soils. Data can be obtained in saturated materials if the specific conductance of the pore fluid is sufficiently low. Radar has been applied to map the sediments in fresh-water lakes and rivers. While radar penetration in soil and rock to more than 100 feet has been reported, penetration of 15 to 30 feet is more typical. In silts and clays, penetration may be limited to a few feet or less.

The continuous data produced by the radar method offers a number of advantages over some of the other geophysical methods. Continuous profiling permits data to be gathered much more rapidly, thereby providing a large amount of data. In some cases, total site coverage of an area can be obtained. Continuous radar data may be obtained at speeds of 5 to 10 mph or more. Very high lateral resolution data can be obtained by towing the antenna by hand at much slower speeds (less than 1 mph).

Radar has the highest resolution of all of the surface geophysical methods. Vertical resolution of radar data can range from less than an inch to several feet, depending upon the depth and the frequency used. A variety of antennas can be selected to cover frequencies from less than 100 MHz to 1000 MHz. Lower frequencies provide greater depths of penetration with lower resolution and the higher frequencies provide less penetration with higher resolution.

Preliminary field analysis of radar data is possible using the picture-like record. However, despite its simple graphic format, there are many pitfalls in the interpretation of radar data. Often, there are multiple bands within the data due to ringing—these may obscure layers and cause confusion in interpretation. Overhead reflections may appear on the record when not using shielded antennas (generally a problem with lower frequency unshielded antennas), and system noise can sometimes clutter up the record.

The Electromagnetic and Resistivity Methods

The electromagnetic (EM) and resistivity methods are similar in the sense that they both measure the same parameter, but in different ways. Electrical conductivity values (mhos/meter) are the reciprocal of resistivity values (ohm/meter). Electrical conductivity (or resistivity) is a function of the type of soil and rock, its porosity, and the conductivity of the fluids that fill the pore spaces. The specific conductance of the pore fluids often dominates the measurement. Both methods are applicable to the assessment of natural hydrogeologic conditions and to mapping of contaminant plumes (Griffith and King, 1969; McNeill, 1980; Telford et al., 1982; Benson, et al., 1982).

Natural variations in subsurface conductivity (or resistivity) may be caused by changes in basic soil or rock types, thickness of soil and rock layers, moisture content, and depth to water table. Localized deposits of natural organics, clay, sand, gravel, or salt-rich zones will also affect subsurface conductivity or (resistivity) values. Structural features such as fractures or voids can also produce changes in conductivity (or resistivity).

The absolute values of conductivity (or resistivity) for geologic materials are not necessarily diagnostic in themselves, but their spatial variations, both laterally and with depth, can be significant. It is the identification of these spatial variations or anomalies which enable the electrical methods to rapidly find potential problem areas (Figure 5.7).

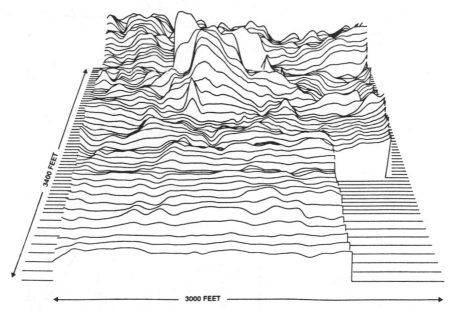

Figure 5.7. Continuous EM profile measurements show a large inorganic plume (center rear) and considerable natural geologic variation.

Because the specific conductance of the fluids in the pore spaces can dominate the measurements, detection and mapping of contaminant plumes can often be accomplished using electrical methods. Because inorganics in sufficient concentrations are often more electrically conductive than ground water, both the lateral and vertical extent of an inorganic plume can often be mapped using electrical methods. Correlation between ground-water chemistry data and results using electrical methods to map inorganics from landfills has been as good as 0.96 at the 95% confidence level (Benson et al., 1985). Electrical methods provide a means of directly mapping the extent of the inorganic contaminants in situ, obtaining

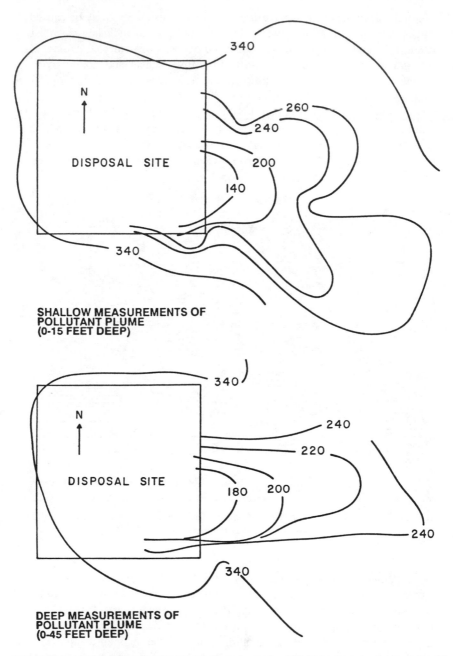

**SHALLOW MEASUREMENTS OF
POLLUTANT PLUME
(0-15 FEET DEEP)**

**DEEP MEASUREMENTS OF
POLLUTANT PLUME
(0-45 FEET DEEP)**

Figure 5.8.   Resistivity map of leachate plume from a landfill. (Values are in ohm-feet; landfill
is approximately 1 square mile.)

direction of flow, and estimating concentration gradients (Figure 5.8). These measurements can also be used for time-series measurements to obtain plume dynamics, and thus provide vital information for modeling ground-water flow (Benson et al., 1988).

If the contaminant plume consists of a mix of organics and inorganics, such as leachate from a landfill, a first approximation to the distribution of the organics can often be made by using electrical methods to map the more electrically conductive inorganics (Figure 5.8). Correlation between ground-water chemistry data for total organic carbon in a landfill leachate and results using electrical methods has been as good as 0.85 at the 95% confidence level (Benson et al., 1985).

In cases in which pure organics such as trichloroethylene (TCE) exist, electrical as well as other geophysical methods can often be used to define permeable pathways or buried channels through which the contaminants may migrate. Direct detection of hydrocarbons can sometimes be accomplished by looking for a conductivity low (resistivity high) associated with the organics. The possibility for such an anomaly exists where large amounts of hydrocarbons have been in place for a long period of time and there is a sufficient contrast in electrical values between the natural background values and the hydrocarbons. To date, this approach has had limited success.

Both EM and resistivity methods may be used to obtain data by "profiling" or "sounding." Profiling provides a means of mapping lateral changes in subsurface electrical conductivity (or resistivity) to a given depth. Profiling measurements are made by obtaining data at a number of stations along a survey line. The spacings between the measurements will depend upon the variability of the setting and upon the lateral resolution desired. At each station along the profile line, data may be obtained for one depth or a number of depths depending upon project requirements. It is useful to take at least two measurements, a shallow one and a deeper one, so that the influence of highly variable shallow soils and cultural influences can be assessed. Profiling is well suited to delineation of hydrogeologic anomalies, mapping of contaminant plumes, and location of buried material.

The sounding method provides a means of determining the vertical changes in electrical conductivity (or resistivity) correlating with soil and rock layers. In this case, the instrument is located at one location and measurements are made at increasing depths. Interpretation of sounding data provides the depth, thickness, and conductivity (or resistivity) of subsurface layers with different electrical conductivities (or resistivities) (Figure 5.9).

Electromagnetics

Two types of electromagnetic instrumentation are in use. The most common is the frequency-domain system in which the transmitter is radiating energy at all times. This system measures changes in magnitude of the currents induced

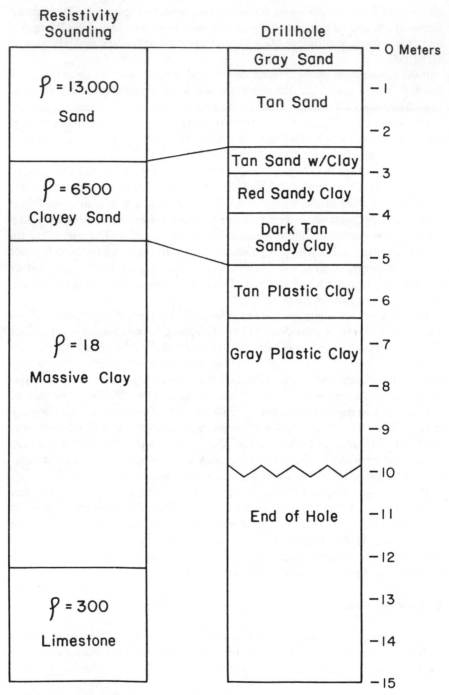

Figure 5.9.   Resistivity geoelectric section showing correlation with a driller's log. (Resistivity values are in ohm-feet.)

within the ground (McNeill, 1980). The time-domain system, in which the transmitter is cycled on and off, measures changes in the induced currents within the ground as a function of time. The frequency-domain and time-domain systems both induce currents into the ground by electromagnetic induction.

Because electromagnetic instruments do not require electrical contact with the ground, measurements may be made quite rapidly. Lateral variations in conductivity can be detected and mapped by profiling. Using commonly available frequency-domain EM instruments, profiling station measurements may be made to depths ranging from 2.5 to 200 feet.

Continuous EM profiling data can be obtained from 2.5 feet to a depth of 50 feet (Benson et al., 1982). These continuous measurements significantly improve lateral resolution for mapping small hydrogeologic features (Figure 5.3). Data can be recorded on an analog strip chart recorder or a digital data acquisition system. The excellent lateral resolution obtained from continuous EM profiling has been used to outline closely spaced burial pits, to reveal the migration of contaminants into the surrounding soils (Figure 5.7), or to delineate complex fracture patterns (Figure 5.3; Benson et al., 1982).

In addition to evaluation of natural hydrogeologic conditions and mapping of contaminant plumes, some electromagnetic instrumentation can be used to locate trench boundaries, buried wastes and drums, and metallic utility lines. Frequency-domain EM instruments provide two outputs consisting of an in-phase component and an out-of-phase component. The out-of-phase component is used to measure electrical conductivity and can also be used to locate pipes. The in-phase component is a measure of the magnetic susceptibility, and can be used to detect both ferrous and nonferrous metal. For example, using the in-phase component, a single 55 gallon steel drum can be detected at a depth of about 6 to 8 feet.

Vertical variations in conductivity can be determined by sounding. The instrumentation is placed at one location and measurements are made at increasing depths by a changing coil orientation and/or coil spacing. Data can be acquired at depths ranging from 2.5 to 200 feet by combining data from a variety of commonly available frequency-domain EM instruments. The vertical resolution of frequency-domain EM soundings is relatively poor since measurements are made at only a few depths. However, they do provide a quick means of obtaining limited vertical information. Time-domain transient EM systems, on the other hand, are capable of providing detailed sounding data to depths of 150 feet to more than 1,000 feet.

Resistivity

As with EM measurements, electrical resistivity measurements are a function of the type of soil or rock, its porosity, and the conductivity of the fluids that fill the pore spaces. The method may be used in many of the same applications as the EM method (Benson et al., 1982; Cartwright and McComas, 1968; Griffith and King, 1969; Mooney, 1980; Telford et al., 1982; Zohdy et al., 1974).

The resistivity method requires that an electrical current be passed through the ground from a pair of surface electrodes. Both direct current and switched direct current power sources are used. The resulting voltage is measured at the surface between a second pair of electrodes. This requires that metal stakes be driven into the ground or that nonpolarizing copper-copper sulfate electrodes be used. The greater the electrode spacing, the greater the depth of the measurement. Usually the depth is less than the spacing between electrodes. There are a number of electrode geometries that can be used, including the Wenner, Schlumberger, dipole, and many more. The simplest, in terms of geometry, is the Wenner Array, which consists of four electrodes, spaced equally, all in a line. The resistivity of the soil and rock is calculated based on the electrode separation, the geometry of the electrode array, the applied current, and the measured voltage.

The resistivity technique may be used for profiling or sounding, similar to EM measurements. Profiling provides a means of mapping lateral changes in subsurface electrical properties to a given depth, and is well suited to the delineation of hydrogeologic anomalies and mapping inorganic contaminant plumes (Figure 5.8).

Sounding measurements provide a means of determining the vertical changes in subsurface electrical properties. Interpretation of sounding data provides the depth, thickness, and resistivity of subsurface layers. Data can be interpreted using master curves for two to three layers (Orellana and Mooney, 1966), or computer models may be used to handle more than two to three layers (Mooney, 1980). Sounding data are used to create a geoelectric section which illustrates changes in the vertical and lateral resistivity conditions at a site. Figure 5.9 shows a geoelectric section developed from a resistivity sounding, along with a drillers log showing the correlation.

One drawback to resistivity sounding is that the array requires considerable space. For example, a Wenner array sounding (with four electrodes equally spaced) may require that the spacing between the electrodes be as much as three to four times the depth of interest. Therefore, a sounding to a depth of 100 feet could require an overall array length (from current electrode to current electrode) of 900 to 1200 feet. At many sites this space may not be available.

Comparison of Electromagnetic and Resistivity Measurements

The frequency-domain electromagnetic method is often preferred for making profiling measurements since it requires less space for a measurement to a given depth. Also, because the electromagnetic method does not require that electrodes be driven into the ground, it can be run more rapidly and is not influenced by shallow geologic noise associated with the electrodes resistivity used in measurements. On the other hand, because resistivity methods provide better vertical resolution than the frequency domain EM method, the resistivity method is commonly employed for sounding measurements. When space is limited and deep soundings are needed, there are advantages to using the time domain electromagnetics system for soundings since it requires less space than long resistivity arrays.

Electromagnetic measurements will often be affected by buried metal pipes, metal fences, nearby vehicles, buildings, power lines, etc., as are resistivity measurements. But resistivity measurements are often less sensitive to many of these problems, permitting resistivity measurements to be made near cultural metal, where electromagnetic measurements often cannot be made.

Electromagnetic and resistivity measurements from the same location will often not agree, due to the difference in the volume of material being sampled and in the differences in current distribution inherent to the two methods. Measurements will only be the same if they are made over a uniform media.

Seismic Refraction and Reflection

Seismic refraction and reflection techniques are often used; to determine the top of bedrock, determine depth of water table, assess the continuity of geologic strata, locate fractures, faults, and buried bedrock channels. The refraction method may be used to characterize the type of rock and degree of weathering based upon the seismic velocity of the rock. The seismic velocity in rock is related to its material properties such as density and hardness. Therefore, characterization of the material on the basis of seismic velocity can indicate the degree of weathering and rippability.

Seismic waves are transmitted into the subsurface by a source, which can sometimes be as simple as a sledge hammer. These waves are refracted and reflected when they pass from a soil or rock type with one seismic velocity into another with a different seismic velocity. An array of geophones placed on the surface measures the travel time of the seismic waves from the source to the geophones. The refraction and reflection techniques use the travel times of the waves and the geometry of the source-to-geophone wave paths to model subsurface conditions. The unit of time is milliseconds (10^{-3} seconds). For most refraction work, the first refracted compressional wave arrivals (P-waves) are used. For reflection work, the later arriving reflected compressional waves are used.

A seismic source, geophones, and a seismograph are required to make the measurements. The seismic source may be a simple sledge hammer or other mechanical source with which to strike the ground. Explosives may be utilized for deeper applications which require greater energy. Geophones implanted in the surface of the ground translate the ground vibrations of seismic energy into an electrical signal. The electrical signal is displayed on the seismograph, permitting measurement of the arrival time of the seismic wave and displaying the wave forms from a number of geophones. Geophone spacing can be varied from a few feet to a few hundred feet depending upon the depth of interest and the resolution needed.

Because the seismic refraction and reflection methods measure small ground vibrations, they are inherently susceptible to vibration noise from a variety of natural (e.g., wind and waves) and cultural sources (e.g., walking, vehicles, and machinery).

Seismic Refraction

The refraction method is commonly applied to shallow investigations up to a few hundred feet deep (Benson et al., 1982; Griffith and King, 1969; Haeni, 1986; Telford et al., 1982). However, with sufficient energy, surveys to a few thousand feet and more are possible. Up to three and sometimes four layers of soil and rock can normally be determined, if a sufficient velocity difference or contrast exists between adjacent layers. A typical refraction line for a shallow investigation might consist of 12 or 24 geophones set at equal spacings as close as 5 to 10 feet. Two seismic impulses at each end of the geophone array are created and their refracted waves recorded separately. The refraction survey may require a maximum source-to-geophone distance four to five times the depth of investigation.

Significantly greater source energy will be required as the depth of investigation increases. Two inherent limits to the refraction method are its inability to detect a lower velocity layer beneath a higher velocity layer and its inability to detect thin layers.

Seismic refraction work can be carried out in a number of ways. The simplest approach in terms of field and interpretation procedures can be carried out by creating two separate seismic impulses, one at each end of the geophone array. The results of this simple measurement provide two depths and thus, the dip of rock under the array of geophones. The method is described in detail by Mooney (1973).

A more detailed refraction survey can be carried out so that depths are obtained under every geophone (Figure 5.10). This survey will produce a detailed profile of the top of rock. Lateral resolution will depend upon the geophone spacing, which might range from 5 to 50 feet. This method is described in detail by Redpath (1973). The General Reciprocal Method described by Palmer (1980) will accommodate varying velocities within each layer, while calculating the depth beneath each geophone.

Seismic Reflection

By comparison, the seismic reflection survey is capable of much deeper investigations with less energy than the refraction method. While reflections have been obtained from depths as shallow as 10 feet, the shallow reflection method is more commonly applied to depths of 50 to 100 feet or more. The reflection technique can be used effectively to depths of a few thousand feet and can provide relatively detailed geologic sections (Figure 5.11). As with radar reflections, the vertical scale is measured in two-way travel time—that is, the time it takes for a wave to travel down to an interface and back up to the surface again. The time scale must then be converted to depth making some assumptions regarding seismic velocity within the strata.

There are two approaches currently used to obtain shallow seismic reflection data, both of which have evolved within the past few years. These two methods

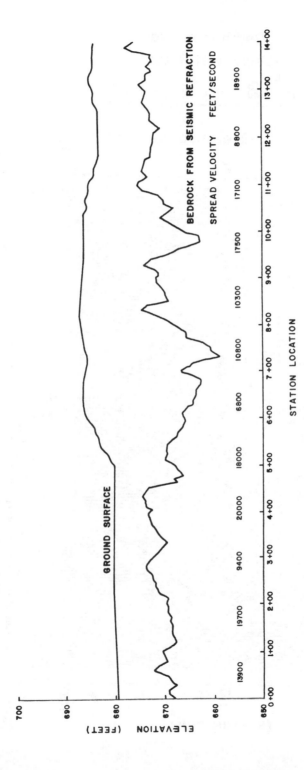

Figure 5.10. Profile of top of bedrock from seismic refraction survey. (Depth to rock determined under each geophone. Geophone spacing is 10 feet.)

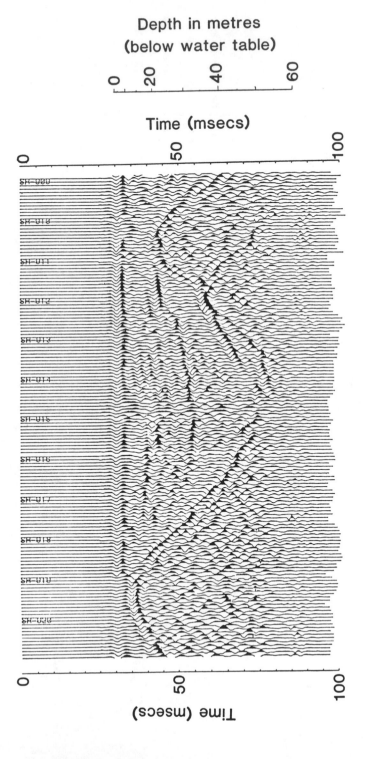

Figure 5.11.   Common offset seismic reflection data showing channel in bedrock.
Source: Dr. Jim Hunter, Geological Survey of Canada.

are the common offset method, developed by Hunter et al. (1982), and the common depth point (C.D.P.) method adapted from the oil industry by Lankston and Lankston (1983) and Steeples (1984). The common offset method uses low-cost equipment and software but has some site-specific limitations which are not inherent in the C.D.P. method. The common depth point method has fewer site-specific limitations, but is more dependent upon sophisticated hardware and software capabilities. Hardware and software for the shallow C.D.P. method have just recently become readily available (late 1988).

The shallow high resolution reflection methods discussed here attempt to utilize the highest frequencies possible (150 to 600 Hz) to improve vertical resolution, and relatively closely spaced geophones (1 to 20 feet apart) to provide good lateral resolution. Because of the need for higher frequencies, attention must be given to selection of a seismic source and its optimum coupling to soil or rock, as well as to geophone placement.

The reflection method is limited by its ability to transmit energy, particularly high frequency energy, into the soil and rock. Loose soil near the surface limits the ability of the soil system to transmit high frequency energy into and out of the rock, limiting the resolution that can be obtained. The most common limitation, however, will be that of acoustic noise caused by natural or cultural sources.

Micro Gravity

Gravity instruments respond to changes in the earth's gravitational field caused by changes in the density of the soil and rock. By measuring the spatial changes in the gravitational field, variations in subsurface geologic conditions can be determined (Griffith and King, 1969; Telford et al., 1982). There are two basic types of gravity surveys: a regional gravity survey and a local micro-gravity survey. A regional gravity survey employs widely spaced (a few thousand feet to a few miles) stations and is carried out with a standard gravity meter. These surveys are used to assess major geologic conditions over many hundreds of square miles. Micro-gravity surveys, on the other hand, have station spacings of 5 to 20 feet (typically) and are carried out with a very sensitive micro gravimeter. These surveys are used to detect and map shallow, localized geologic anomalies such as bedrock channels, fractures, and cavities.

The unit of acceleration used in gravity measurement is the gal. The earth's normal gravity is 980 gals. Micro-gravity measurements are sensitive to within a few micro gals (10^{-6} gals).

The micro-gravity survey results in a Bouguer Anomaly, which is the difference between the observed gravity values and theoretical gravity values. The Bouguer Anomaly is made up of deep-seated effects (the regional Bouguer Anomaly) and shallow effects (the local Bouguer Anomaly). It is the local Bouguer Anomaly that is of interest in micro-gravity work (Figure 5.12).

A gravimeter is designed to measure extremely small differences in the gravitational field and is a very delicate instrument. The instrument is thermostatically controlled to minimize drift caused by temperature variations. Considerable care must be taken in shipment and general field use to avoid shock to the instrument. Gravity measurements may be affected by ground noise (see Seismic Methods), winds, and temperature. To compensate for minor instrument drift throughout the day, measurements must be made at a base station every hour or so, so that drift corrections can be applied to the data. Corrections must also be made for the constantly changing earth tides, changes in elevation (to the nearest 0.01 foot), and topography. Gravity data may be presented as a profile or as a contour map, depending upon project needs.

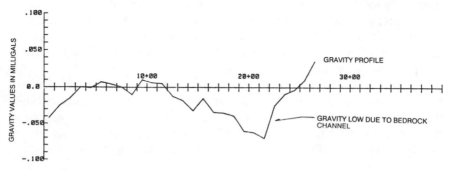

Figure 5.12.   Micro gravity profile showing bedrock channel.
Source: Technos Inc.

Metal Detection

Metal detectors are commonly used by utility and survey crews for locating buried pipes, cables, and property stakes. They can also be used for detecting buried drums and for delineating the boundaries of trenches containing metallic drums or trash (Figure 5.13; Benson et al., 1982). Metal detectors can detect both ferrous metals such as iron and steel, and nonferrous metals such as aluminum and copper.

Metal detectors have a relatively short detection range, because the detector's response is proportional to the cross-section of the target and inversely proportional to the sixth power of the distance to the target. Small metal objects, such as quart-sized containers, can be detected at a distance of approximately 2 to 3 feet. Specialized metal detectors will detect larger objects, like 55-gallon drums, at depths of 3 to 10 feet, and massive piles of 55 gallon drums may be detected at depths of up to about 15 feet. The metal detector is a continuously sensing instrument used with a sweeping motion while moving forward along a survey line. It may also be held steady while a traverse line is walked and the results are recorded on a strip-chart recorder. The area of detection of a metal detector is approximately equal to its coil size or coil spacing (typically 1 to 3 feet). Metal detectors can be affected by nearby metallic pipes, fences, cars, buildings and, in some cases, changes in soil conditions.

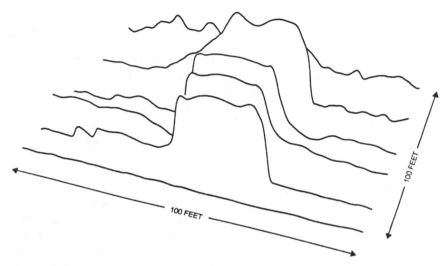

Figure 5.13.   Results of a metal detector survey to locate a burial trench.

Magnetometry

A magnetometer measures the intensity of the earth's magnetic field. As with gravity surveys, a magnetic survey can be used to map geologic conditions over large areas. This type of survey is useful for mapping regional geologic conditions. In certain geologic environments, magnetics can also be used to map depth to bedrock, channels, and fractures (Breiner, 1973; Griffith and King, 1969; Telford et al., 1982). The primary application of magnetic measurements at hazardous waste sites is in detecting buried drums, tanks, and pipes (Benson et al., 1982; Breiner, 1973). A magnetometer will only respond to ferrous metals (iron and steel) and will not detect nonferrous metals. The presence of buried ferrous metals creates a local variation in the strength of the earth's magnetic field, permitting the detection and mapping of buried ferrous metal (Figure 5.14).

Two types of magnetic measurements are commonly made: total field measurements and gradient measurements. A total field measurement responds to the total magnetic field of the earth, any changes caused by a target, natural magnetic, and cultural magnetic noise (ferrous pipe, fences, buildings, and vehicles).

The effectiveness of total field magnetometers can be reduced or totally inhibited by noise or interference caused by time-variable changes in the earth's magnetic field, or spatial variations due to magnetic minerals in the soil, steel debris, pipes, fences, buildings, and passing vehicles.

A base station magnetometer can be used to reduce the effects of natural noise by subtracting the base station values from those of the search magnetometer. This can minimize any errors due to natural long-period changes of the earth's field. Cultural noise, however, will remain a problem with total field measurements. Many of these problems can be avoided by use of gradient measurements and proper field techniques.

Gradient measurements are made by a gradiometer, which is simply two magnetic sensors separated vertically (or horizontally), usually by a few feet. Gradient measurements have some distinct advantages over total field measurements. They are insensitive to natural spatial and temporal changes in the earth's magnetic field and minimize most cultural effects; because the response of a gradiometer is the difference of two total field measurements it responds only to the local gradient. As a result, it is better able to locate a relatively small target, such as a buried drum. The disadvantage of a gradiometer is that it provides a slightly less sensitive measurement than a total field instrument.

A total field magnetometer's response is proportional to the mass of the ferrous target and inversely proportional to the cube of the distance to the target (such as a drum). A gradiometer response is inversely proportional to the fourth power of the distance to the target (such as a drum), making it less sensitive than the total field measurement. While gradiometers are inherently less sensitive than total field instruments, they are also much less sensitive to many sources of noise. Typically, a single drum can be detected at depths up to about 20 feet with a total field magnetometer, or about 10 feet with a gradient magnetometer. Massive piles of drums can be detected at depths up to 50 feet or more with a total field magnetometer, or about 25 feet with a gradient magnetometer.

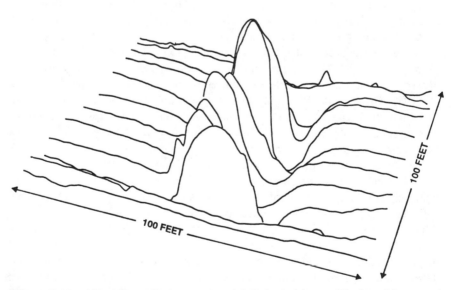

100 FEET

100 FEET

Figure 5.14. Magnetic gradient over a trench with buried drums. (The trench is approximately 20 feet by 100 feet.)

A total field or gradient proton procession magnetometer normally requires the operator to stop and take a measurement while a fluxgate gradiometer permits the continuous acquisition of data as the magnetometer is moved across the site. Continuous coverage is much more suitable for detailed (high resolution) surveys to identify local targets, such as drums, and the mapping of areas in which complex anomalies are expected.

DOWNHOLE GEOPHYSICAL MEASUREMENTS

One of the most common subsurface investigation techniques is that of sampling soil and rock at discrete intervals (typically every 5 feet) as a boring is advanced. This method provides gross information on subsurface lithology but sand lenses, fractures, or other subtle changes in geology, which can affect hydraulic conductivity, can easily go undetected. Though continuous sampling or coring can improve the description of geologic conditions, it is very costly and time consuming, and material description is somewhat subjective. Furthermore, 100% sample recovery is rarely achieved.

A number of downhole logging techniques are available for determining the characteristics of soil, rock, or fluid along the length of a borehole or a monitoring well (Keys and MacCary, 1976). These methods provide continuous, high-resolution in situ measurements that are often more representative of hydrogeologic conditions than samples obtained from borings. A number of logging techniques are available, and an adequate assessment of subsurface conditions will often require that multiple logs be used since each log responds to a different property of the soil, rock, or fluid. Some of these techniques will provide measurements from inside plastic or steel casing, and some will allow measurements to be made in the unsaturated zone as well as the saturated zone.

Downhole logging measurements can be correlated to the known geologic strata in one hole and then can be used to identify and correlate geologic strata in other holes without sampling. Thin layers and subtleties, not readily detected in soil or core samples, can often be resolved by logging. Logging can significantly improve the ability to accurately characterize and correlate strata between borings by providing high-resolution data independent of subjective interpretations of soil and rock type.

A number of soil and rock properties can be measured in situ. Values for soil and rock porosity, density, seismic velocity, and elastic moduli can be obtained to facilitate engineering design. Even more important is the ability to identify the uniformity or lack of uniformity of subsurface conditions. Downhole measurements can be used to identify permeable zones, such as sand lenses in glacial tills, weathered zones, and fractures or solution cavities in rock. The same measurements are also effective for identifying impermeable zones, such as aquitards, and assessing their continuity and integrity.

Monitoring wells that have been in place for years provide the basis for long-term chemical monitoring. For many of these wells, neither geologic logs nor installation records are available. Using downhole techniques, it is possible to obtain geologic information and well construction details. In addition, logging may be used to determine whether a problem exists with well construction and what type of remedial work, if any, is necessary to correct it.

By running nuclear logs in existing holes with steel or PVC casing, geologic strata outside the casing can be characterized. Under some conditions in an open borehole or PVC cased well, contaminants outside PVC casing can be detected by running electromagnetic induction logs. A downhole television camera can be used within cased wells to assess monitoring well conditions, or it can be used within an uncased borehole to assess the existence of fractures.

While each log is susceptible to both natural and cultural noise, borehole diameter will probably be of most concern. Most logs provide measurements within a radius of 6 to 12 inches from the hole. Therefore, as the hole diameter becomes larger, the measured results become more dominated by drilling and well construction aspects.

A description of the most commonly used logs is given below. Table 5.1 lists the conditions in which these logs can be used and some of the limitations inherent in the use of each log.

Nuclear Logs

Natural Gamma Log

A natural gamma log records the amount of natural gamma radiation that is emitted by rocks and unconsolidated materials. The chief use of natural gamma logs is the identification of lithology and stratigraphic correlation in open or cased holes above and below the water table.

The gamma-emitting radioisotopes normally found in all rocks and unconsolidated materials are potassium-40 and daughter products of the uranium and thorium decay series. Because clays and shales concentrate these heavy radioactive elements through the processes of ion exchange and adsorption, the natural gamma activity of clay and shale-bearing sediments is much higher than that of quartz sands and carbonates. Therefore, the gamma log, which indicates an increase in clay or shale content by an increase in counts per second (Figure 5.15), is useful for evaluating the presence, variability, and integrity of clays and shales. The radius of investigation for the natural gamma log is from about 6 to 12 inches (Keys and MacCary, 1976) from the borehole wall.

Gamma-Gamma (Density) Log

A gamma-gamma log is used to determine the relative bulk density of the soil or rock and to identify lithology. The log can be used in open or cased holes above and below the water table (Figure 5.16).

Table 5.1.   General Characteristics and Use of Downhole Geophysical Logs.

Downhole Log	Parameter Measured (or Calculated)	Casing Uncased/PVC/Steel			Saturated	Unsaturated	Radius of Measurement	Affect of Hole Diameter, and Mud
NATURAL GAMMA	Natural Gamma Radiation	yes	yes	yes	yes	yes	6–12 inches	moderate
GAMMA GAMMA	Density	yes	yes	yes	yes	yes	6 inches	significant
NEUTRON	Porosity Below Water Table—Moisture Content Above Water Table	yes	yes	yes	yes	yes	6–12 inches	moderate
INDUCTION	Electrical Conductivity	yes	yes	no	yes	yes	30 inches	negligible
RESISTIVITY	Electrical Resistivity	yes	no	no	yes	no	12 inches to 60 inches	significant to minimal depending upon probe used.
SINGLE POINT RESISTANCE	Electrical Resistance	yes	no	no	yes	no	near borehole surface	significant
SPONTANEOUS POTENTIAL (SP)	Voltage—Responds to Dissimilar Minerals and Flow	yes	no	no	yes	no	near borehole surface	significant
TEMPERATURE	Temperature	yes	no	no	yes	no	within borehole	—NA—
FLUID CONDUCTIVITY	Electrical Conductivity	yes	no	no	yes	no	within borehole	—NA—
FLOW	Fluid Flow	yes	no	no	yes	no	within borehole	—NA—
CALIPER	Hole Diameter	yes	yes	yes	yes	yes	to limit of senor typically 2–3 feet	—NA—

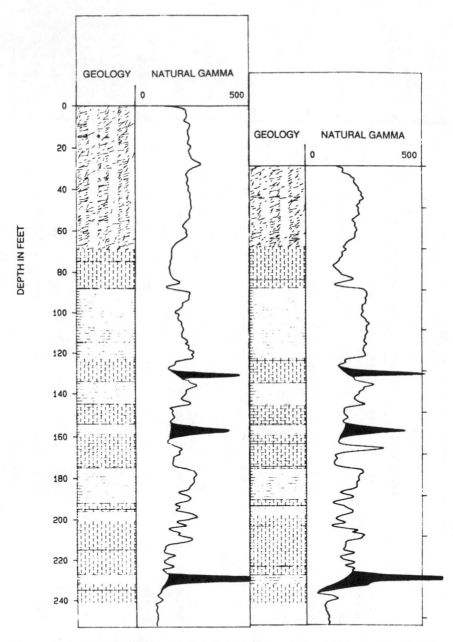

Figure 5.15.   Natural gamma logs from two nearly boreholes, 100 feet apart. (Note the characterization and correlation of the shale and limestone units.)
Source: Technos Inc.

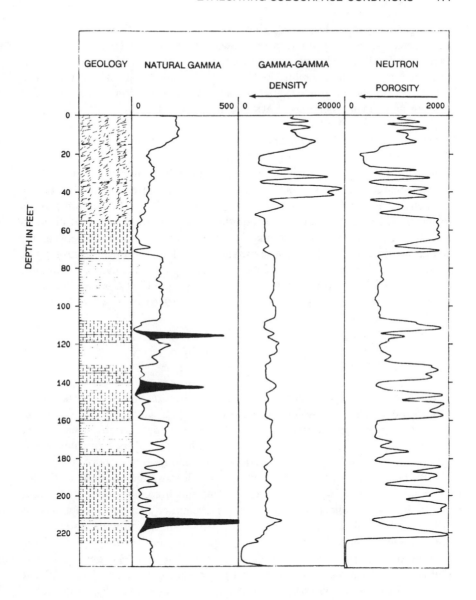

Figure 5.16.   A suite of logs from within the same borehole. (The natural gamma log provides a means of characterizing the shale. The gamma-gamma log provides a measure of density and the neutron log provides a measure of porosity within the shale/limestone units.)
Source: Technos Inc.

The gamma-gamma log is an active probe containing both a radiation source and a detector. This log provides a response, in counts per second, that is averaged over the distance between the source and the detector. The radius of investigation for the gamma-gamma log is relatively small (only about 6 inches). Therefore, borehole diameter variations and well construction factors can affect this log more than other logs (Keys and MacCary, 1976).

Neutron-Neutron (Porosity) Log

A neutron-neutron log provides a measure of the relative moisture content above the water table and porosity below the water table (Figure 5.16). It can be run in open or cased holes, above and below the water table. The neutron-neutron log is an active probe with both a radiation source and a detector. It provides a response, in counts per second, that is averaged over the distance between the source and the detector. The radius of investigation for the neutron-neutron probe is approximately 6 inches (up to 12 inches in very porous formations). Borehole diameter variations and well construction factors can affect this log, but not as severely as the density log (Keys and MacCary, 1976).

Non-Nuclear Logs

Induction Log

The induction log is an electromagnetic induction method for measuring the electrical conductivity of soil or rock in open or PVC-cased boreholes above or below the water table (similar to EM measurements made on the surface). The induction log can be used for identification of lithology and stratigraphic correlation. Electrical conductivity is a function of soil and rock type, porosity, permeability, and the fluids filling the pore spaces. Because the response of the log (in millimhos/meter) will be a function of the specific conductance of the pore fluids, it is an excellent indicator of the presence of inorganic contamination (Figure 5.17) and in some cases (when organics are mixed with inorganics or when a thick layer of hydrocarbons is present), organic contamination. Variations in conductivity with depth may also indicate changes in clay content, permeability of a formation, or fractures. An induction log provides data similar to that provided by a resistivity log (because conductivity is the reciprocal of resistivity). However, the induction log can be run without electrical contact with the formation. Therefore, the induction log can be used in both the vadose zone and the saturated zone and it can be used to log through PVC casing.

The radius of investigation for the induction log is approximately 2.5 feet from the center of the well. Because this log has a much larger radius of investigation than other logs, it is almost toally insensitive to borehole and construction effects, and as such is a good indicator of the overall soil and rock conditions surrounding the borehole.

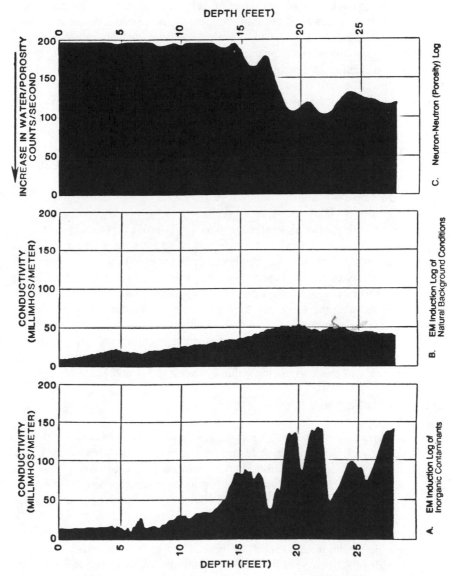

Figure 5.17. Induction and porosity logs are used to identify contaminants and permeable zones.
Source: Technos Inc.

Resistivity Log

The resistivity log measures the apparent resistivity (measured in ohm-feet or ohm-meters) of rock and soil within a borehole. Because resistivity is the reciprocal of conductivity, which is the property measured by an induction log, the resistivity log responds to and measures the same properties and features as the induction log. However, because of the need for electrical contact with the borehole wall, the resistivity log can only be run in an uncased hole filled with water or drilling fluid.

There are a number of electrode spacings or geometries that may be used for resistivity logs. The most common is the "normal" log. Short normal probes (typically an electrode spacing of 16 or 18 inches) give good vertical resolution and measure the apparent resistivity of the formation immediately around the borehole. Long normal probes (typically an electrode spacing of 64 inches) have less vertical resolution but measure the apparent resistivity of undisturbed rock within a larger radius from the borehole, similar to the induction log (Keys and MacCary, 1976).

Resistance Log

A resistance log (sometimes referred to as single-point resistance) measures the resistance (in ohms) of the earth materials lying between a downhole electrode and a surface electrode. It can only be run in uncased holes in the saturated zone. The primary uses of resistance logs are geologic correlation and the identification of fractures or washout zones in resistive rocks. The resistance log should not be confused with the resistivity log, which provides a quantitative measure of the material resistivity.

The radius of investigation of the resistance log is quite small. It is in many cases as strongly affected by conductivity of the borehole fluid as it is affected by the resistance of the surrounding volume of rock (Keys and MacCary, 1976).

Spontaneous-Potential Log

The spontaneous-potential (SP) log measures the natural potential (in millivolts) developed between the borehole fluid and the surrounding rock materials. It can only be run in uncased holes within the saturated zone. The SP voltage consists of two components. The first component results from electrochemical potential caused by dissimilar minerals. The second component is the streaming potential caused by water moving through a permeable medium.

SP measurements are subject to considerable noise from the electrodes, hydrogeologic conditions, and borehole fluids. Even though these measurements do not provide quantitative results, they have a number of applications, including:

- characterizing lithology;
- providing information on the geochemical oxidation-reduction conditions; and
- providing an indication of fluid flow.

The radius of investigation of the SP log is highly variable (Keys and MacCary, 1976).

Temperature Log

A temperature log is a continuous record of the temperature of the borehole fluid immediately surrounding the sensor as it is lowered within an open borehole. The temperature log will often indicate a zone of ground-water flow within the uncased portion of a borehole. Flow is indicated when an increase or decrease in water temperature occurs. Changes in temperature can also be used to monitor leaks in casing where damage or corrosion has occurred. A temperature log may have a sensitivity of 0.5°C or better.

Fluid Flow

There are many ways of measuring fluid flow within a borehole (Keys and MacCary, 1976). The most commonly used method is the use of an impeller-type flow meter that provides counts per second. The count rate can usually be calibrated to provide results in feet/minute or gallons/minute.

Fluid Conductivity

A fluid conductivity log provides a measurement of the specific conductance of the borehole fluids (in micromhos/centimeter). If accurate values are needed (as opposed to anomaly detection) a temperature log must also be run so that corrections can be made.

Caliper Log

A caliper log provides a record of the diameter of an open borehole or of the inside diameter of a well casing. The caliper probe consists of spring-loaded arms which extend from the logging tool so that they follow the sides of the borehole or casing.

Caliper logs are utilized to measure borehole diameter, to locate fractures and cavities in an open borehole. The caliper log can be used to determine well construction details and casing diameter. It can also be used to reveal casing deterioration due to extreme corrosion or accumulation of minerals on the interior of the well casing.

APPLICATIONS OF GEOPHYSICAL METHODS

There is no simple, exact way to select the geophysical method(s) required to solve a particular problem. Tables 5.2 through 5.4 are provided to illustrate how geophysical methods may be used to carry out assessment of hydrogeologic conditions, detecting and mapping contaminants, and locating and mapping buried wastes and utilities. However, simple tables and rules of thumb often fail when considering specific project needs and site-specific conditions and, therefore, the tables presented here should only be used as an initial guide.

Assessing Hydrogeologic Conditions

The first and often the most important task of most site investigations is the evaluation of natural hydrogeologic conditions. A description of overall hydrogeologic conditions and identification of any hydrogeologic anomalies is usually required. Knowledge of the natural anomalies, in relation to the overall setting, can ensure that drilling and sampling is done in the locations that will most likely yield information on the location or migration of a contaminant plume.

The first step in any site investigation is to obtain appropriate background literature, maps, and aerial photos so that geophysical surveys and other site work can be planned. Photo imagery is almost a necessity in any serious site investigation, to assist in planning a geophysical survey and to locate the survey grid.

Table 5.2 lists possible applications of surface geophysical methods and some of their advantages and limitations in evaluating hydrogeologic conditions. Variations in the shallow natural setting are best evaluated with ground-penetrating radar, which provides the greatest resolution. However, depth of penetration of the radar signal is highly site-specific, and is typically less than 30 feet. When silts and clays are present at the surface, penetration may be limited to only a few feet.

Even with these limitations, ground-penetrating radar can often help solve problems at a depth greater than its sensing range. For example, by looking for anomalies in shallow marker beds, or by observing shallow soil piping, shallow radar data can be used to predict the presence of cavities and fractures far beyond its range. Investigation of such Near Surface Indicators (NSI) with radar and other methods to evaluate deeper conditions is a powerful technique (Benson and Yuhr, 1987).

High-resolution seismic reflection can be used in combination with radar to provide a more complete depth profile. While this method has less resolution than radar, information can be acquired to depths of hundreds of feet or more. The reflection method is often found to be ineffective at depths shallower than 25 to 50 feet, where radar is most effective. Therefore, these two methods are quite complementary for developing detailed geologic profiles. It should be noted, however, that the cost of seismic work is considerably greater than the cost for a radar survey.

Seismic refraction and resistivity soundings provide good vertical information, though they are not capable of achieving the lateral and vertical resolution of radar, or in some cases, the vertical resolution of seismic reflection. The frequency-domain EM techniques have very good lateral resolution in the continuous mode to depths of about 50 feet, but are somewhat lacking in their capability to produce vertical detail (sounding data). Yet, the EM methods can provide some relative sounding information (i.e., thick versus thin or shallow versus deep) very quickly and more cost effectively than resistivity or seismic refraction.

Table 5.2.   Surface Geophysical Methods for Evaluation of Natural Hydrogeologic Conditions.[a]

Method	General Application	Continuous Measurements	Depth of Penetration	Major Limitations
Radar	Profiling and mapping; Highest resolution of any method	yes	to 100 feet (typically less than 30 feet)	Penetration limited by soil conditions.
EM (Frequency Domain)	Profiling and mapping; Very rapid measurements	yes (to 50 feet)	to 200 feet	Affected by cultural features (metal fences, pipes, buildings, vehicles).
EM (Time Domain)	Soundings	no	to few 1000 feet	Does not provide measurements shallower than about 150 feet.
Resistivity	Soundings or profiling and mapping	no	No limit (commonly used to a few 100 feet)	Requires good ground contact and long electrode arrays. Integrates a large volume of subsurface. Affected by cultural features (metal fences, pipes, buildings, vehicles).
Seismic Refraction	Profiling and mapping soil and rock	no	No limit (commonly used to a few 100 feet)	Requires considerable energy for deeper surveys. Sensitive to ground vibrations.
Seismic Reflection	Profiling and mapping soil and rock	no	to few 1000 feet	Shallow surveys, <100 feet are most critical. Sensitive to ground vibrations.
Micro Gravity	Profiling and mapping soil and rock	no	No limit (commonly used to a few 100 feet)	Very slow, requires extensive data reduction. Sensitive to ground vibrations.
Magnetics	Profiling and mapping soil and rock	yes	No limit (commonly used to a few 100 feet)	Only applicable in certain rock environments. Limited by cultural ferrous metal features

[a]Applications and comments should only be used as guidelines. In some applications, an alternate method may provide better results.

Probably the two best techniques to map lateral variations in soil and rock, from a speed and resolution point of view, are radar and continuous EM measurements. While radar performance is highly site-specific, the EM technique can be applied in almost any environment and can often provide deeper information, but with much less vertical resolution than radar. Continuous EM profiling measurements provide high lateral resolution and can be run at speeds from 1 to 5 mph, depending on the detail required. The rapid speed at which EM measurements can be obtained and the option of continuous profile measurements at depths of up to 50 feet makes EM the best choice for profile work under most situations.

The resistivity method can also be used for profile measurements by moving the electrode array in small increments to provide data at closely spaced intervals. This is a slow process relative to an EM survey, and resistivity data can be affected by near-surface geologic noise at the electrodes.

Sometimes, one method may work and another will fail under a given set of site conditions. For example, in many cases, resistivity measurements can be made adjacent to a chain link fence or a buried pipeline where EM measurements cannot.

In order for any geophysical method to work, there has to be a contrast in the parameter being measured. The best method is the one in which the parameters being measured have the greatest contrast and will be least influenced by site-specific conditions and noise. The final decision must be made on a site-by-site basis.

Once the surface methods have defined the 2-dimensional or 3-dimensional conditions reasonably well, boring locations can be selected. These locations should be selected to be representative of the normal background conditions at the site and to investigate any anomalies.

Generally, if there are anomalous site conditions present, including sand lenses, fractures, subtle changes in formation permeability, or geochemical anomalies, downhole logs should be run.

The drilling program should be designed to provide a means of accurately characterizing soil and rock conditions to the greatest extent possible within the budget available. If an adequate downhole logging program is used, most of the holes can be drilled without sampling. However, it is always good practice to continuously sample or core one borehole and then log it along with any other holes. This procedure provides a reference for the logging data to compare to site-specific soil samples or rock cores. The logs can then be used to extrapolate soil and rock type and other conditions to nearby boreholes.

When the appropriate logs are combined, continuous in situ logging measurements can be obtained in both the vadose zone and the saturated zone to characterize hydrogeologic conditions. Geologic formations can be identified and easily correlated from hole to hole. Relative estimates of clay content, density, and porosity can be given. Permeable sand lenses and fractures can be identified, as can impermeable clay and shale zones. In addition, the continuity of impermeable zones can be assessed. The maximum amount of data should always be obtained from each borehole because borings are often few and costly.

Natural gamma logs can be used for geologic characterization and stratigraphic correlation. For example, the natural gamma logs shown in Figure 5.15 clearly show the contrast between the limestone units (low counts) and the shale units (high counts). In this case, correlation of the stratigraphy from natural gamma logs in adjacent boreholes is easily made.

Figure 5.16 shows a suite of natural gamma, density, and porosity logs from the same borehole. The density log shows variable conditions in the overlying soil, but fairly uniform density within the shale and limestone units. In contrast, the porosity log shows considerable variation throughout both the shale and limestone units. Without calibration, these logs can be used to indicate relative changes in density and porosity. By calibrating these logs, quantitative results for density and porosity may be obtained in some situations.

Detecting and Mapping Contaminant Plumes

Table 5.3 illustrates how surface geophysical methods can be applied to mapping contaminant plumes. The fundamental approach to evaluating the direction of ground-water flow and the possible extent of a contaminant plume is by determining the hydrogeologic characteristics of the site (that is, determining the presence of pathways such as buried channels, fractures, and permeable zones).

"Direct" detection of inorganics (or organics mixed with inorganics) can be accomplished by electrical methods, including ground-penetrating radar, as shown in Table 5.3. When inorganics are present in sufficient amounts, they can be detected by electrical methods and radar. The higher specific conductance of the pore fluids acts as a tracer by which the plume can be mapped. In cases in which inorganic plumes have a very low specific conductance, or dispersed organics are encountered, they will not be detectable by electrical methods.

Where suitable penetration is possible, ground-penetrating radar can provide a means for mapping the depth to the top of and lateral extent of shallow inorganic plumes. However, because of the site-specific behavior of radar, the EM or resistivity methods are most often used. Of the two methods, EM measurements are preferred for profile work, particularly where continuous sampling can be employed. Resistivity is preferred for sounding work.

Both resistivity and EM conductivity can miss a contaminant plume if the measurements are in the wrong location. However, rapid EM profiling by either continuous or station measurements allows coverage of a site with closely spaced data. It is not unreasonable from a cost perspective to have overlapping measurements, therefore providing total site coverage using the EM profiling method.

Both resistivity and EM are capable of vertical soundings. The frequency-domain EM method provides a depth of penetration that is limited to about 200 feet and provides less resolution than the resisitivity method since measurements are made at only a few depths. The depth to which resistivity sounding data can be obtained is virtually unlimited. Depths of a few hundred feet to thousands of feet are obtainable. However, the long resistivity arrays necessary for deep measurement

Table 5.3. Surface Geophysical Methods for Mapping of Contaminant Plumes.[a]

Mapping Permeable Pathways, Bedrock Channels, etc.
The fundamental approach to evaluating the direction of groundwater flow and the possible extent of a contaminant plume is by determining the hydrogeologic characteristics of the site (see Table 5.2 for Evaluation of Natural Hydrogeologic Conditions).

Mapping of Inorganics or Mixed Inorganics and Organics
When inorganics are present in sufficient concentrations above background or organics are part of such a inorganic plume, they can be mapped by the electrical methods and sometimes radar. The higher specific conductance of the pore fluids acts as a tracer by which the plume can be mapped.

Mapping of Hydrocarbons
When sufficient hydrocarbons have been present in the soil or floating on a shallow water table, for a sufficient period of time they may sometimes be mapped by the electrical methods or by radar. Because of their low conductivities (high resistivity) they may sometimes be detected by the electrical methods. Due to changes in dielectric constant or suppression of the capillary zone they may sometimes be mapped by radar (in some situations where degradation of hydrocarbons is occurring, conductivity may increase). These applications are limited and should be treated with caution. A more reliable approach is to map natural permeable pathways (see Table 5.2 for Evaluation of Natural Hydrogeologic Conditions and Table 5.4 for Mapping of Cultural Pathways).

Radar	• Limited applications—may sometimes be used to detect shallow floaters (0 to 20 feet) to map hydrocarbons in soil. May detect thickness in some cases.
EM	• May be applicable to detect low conductivity at some sites.
Resistivity	• May be applicable to detect high resistivity at some sites.

[a]Applications and comments should only be used as guidelines. In some applications, an alternate method may provide better results.

may not be practical in some areas due to space restrictions and cultural factors. Here, the time-domain EM transient systems, which have a smaller coil size, would be the choice for measurements to depths from 150 feet to a few thousand feet or more.

In some cases, organics can be mapped because they are mixed with inorganics. Figure 5.7 shows the inorganic plume from a chemical/drum recycling center which also contains organics. Figure 5.8 shows the inorganic plume from a landfill that contains low levels of organics. The results in Figure 5.18 show an excellent comparison of EM, resistivity, and Organic Vapor Analyses (OVA) responses obtained from a mixed plume or organics and inorganics confined in a buried channel. Clearly, if inorganics are present, they should be used as an easily detectable tracer that will provide a first approximation of where the organics may be.

Some investigators have suggested that direct detection of major hydrocarbon spills can be accomplished by looking for EM conductivity lows (or resistivity highs) associated with the organics. Recent spills of petroleum products do not seem to yield a resistivity high or an EM conductivity low. However, the possibility for an anomaly exists where the product has been in the ground for some

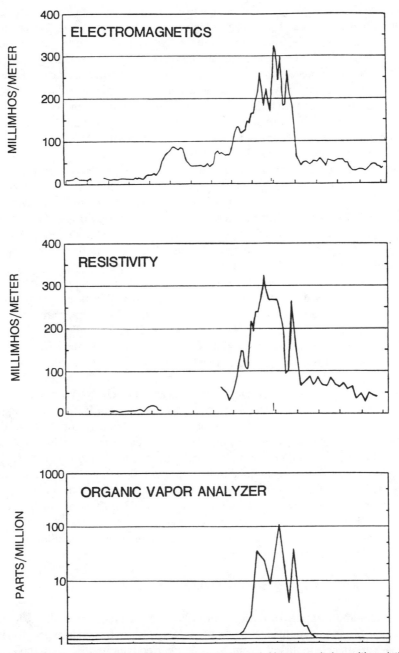

Figure 5.18.   Organic vapor profile over buried channel. Note correlation with resistivity and EM measurements. (This is an excellent example of a buried channel controlling flow and the level of correlation between organic and inorganic contaminants.)

time, where there is a substantial amount of floating product, and where the conductivity of the natural soil conditions is high enough (or the resistivity low enough) to provide a reasonable contrast between the hydrocarbons and the natural soil.

Once the spatial extent of a contaminant has been mapped by surface geophysics and after boreholes have been installed, continuous downhole logging can be used to evaluate changes in the vertical hydraulic conductivity of soil and rock, as well as the distribution of contaminants. The vertical distribution and concentration of contaminants at a site can vary significantly as a result of small local changes in hydraulic conductivity. Because hydraulic conductivity can change by more than an order of magnitude in less than a foot, it can have a significant impact on test results obtained from a monitoring well. The chemical concentration in a well may be low, average, or high, depending upon screen length and location. Two downhole logging techniques particularly well suited for hydraulic conductivity evaluations are the electromagnetic (EM) induction log (or resistivity logs—only in an open borehole) and neutron (porosity) log. Both of these logs can be run in an open borehole or within an existing PVC-cased well, either above or below the water table (Figure 5.17).

The EM induction log, shown in Figure 5.17a, indicates the presence of inorganic contaminants that have preferentially migrated within five discrete zones of increased hydraulic conductivity in the limestone. These zones are indicated by higher electrical conductivity values and range from 2 to 4 feet in thickness. The presence of high hydraulic conductivity zones detected by the EM induction method was confirmed by using a downhole television camera which visually located small cavities and fractures in each zone. Figure 5.17b shows an EM induction log of natural conditions taken in a background well. No permeable zones are indicated by the log since there are no inorganics present. Figure 5.17c shows a neutron log taken in the same background well. In this log, zones of variable porosity are revealed whether contaminants are present or not. Conditions shown by this log are also representative of conditions at the contaminated well.

An adequate assessment of conditions in a borehole often requires that more than one log be run. At this site, an EM induction log was used to identify the contaminated zones, and a neutron log was used to identify zones of increased porosity. Once conditions at a site are understood, a reliable and representative monitoring well system can be designed and data from existing monitoring wells can be more accurately evaluated.

Locating and Mapping Buried Wastes and Utilities

Location and mapping of buried wastes, utilities, drums, and tanks is a common application of geophysical methods. Table 5.4 lists the surface geophysical methods applicable to this problem. Locating buried bulk wastes where no metal is present can often be accomplished by ground-penetrating radar, if soil conditions are suitable. Often the shallow edges of trenches can be detected even in soil conditions that provide poor radar penetration. Shallow EM tools are also effective for most location problems. When metals are present, EM conductivity,

Table 5.4. Surface Geophysical Methods for Location and Mapping of Buried Wastes and Utilities.[a]

Method	Bulk Wastes w/o Metals	Bulk Wastes w/Metals	55 Gal. Drums	Pipes and Tanks
Radar	Very good if soil conditions are appropriate; Sometimes effective to obtain shallow boundaries in poor soil conditions	Very good if soil conditions are appropriate; Sometimes effective to obtain shallow boundaries in poor soil conditions	Good if soil conditions are appropriate (may provide depth)	Very good for metal and non-metal if soil conditions are appropriate (may provide depth)
EM	Excellent to depths less than 20 feet	Excellent to depths less than 20 feet	Very good (Single drum to 6–8 feet)	Very good for metal tanks
Resistivity	Good (sounding may provide depth)	Good (sounding may provide depth)	—NA—	—NA—
Seismic Refraction	Fair (may provide depth)	Fair (may provide depth)	—NA—	—NA—
Micro Gravity	Fair (may provide depth)	Fair (may provide depth)	—NA—	—NA—
Metal Detector	—NA—	Very good (shallow)	Very good (shallow)	Very good (shallow)
Magnetometer	—NA—	Very good (ferrous only; deeper than metal detector)	Very good (ferrous only; deeper than metal detector)	Very good (ferrous only; deeper than metal detector)

[a]Applications and comments should only be used as guidelines. In some applications, an alternate method may provide better results.

metal detectors, and magnetometers are the primary choices. Metal detectors and magnetometers are unaffected by most soil types or by the presence of contaminants. However, EM measurements are influenced by both variations in soil and the presence of contaminants.

To locate buried 55-gallon steel drums, the use of metal detectors, magnetometers, or the in-phase component of EM measurements are recommended. All three methods can be used to locate single 55-gallon drums as well as large piles of drums within their depth limitations.

Both the metal detector and the EM will respond to ferrous and nonferrous metals, while a magnetometer will respond only to ferrous metals. Therefore, it is necessary to assess what metals may be present in order to select the appropriate method.

While radar can be used to find drums, it will often be unable to detect a single drum if it is not oriented so that energy is reflected back to the antenna. Further more, many natural and human-made objects may have a radar response similar to that of a drum.

For small, discrete, critical targets such as a single 55 gallon drum, continuous measurements (on closely spaced lines of about 5 feet) are required to assure detection. Radar, EM equipment, metal detectors, and certain magnetometers can provide these continuous measurements. However, there may be cases in which the proximity of other metal structures may limit the use of EM to locate drums or trenches, making radar a clear choice.

Metal detectors and radar both provide reasonably good spatial resolution to pinpoint the location of a target. However, EM equipment and magnetometers do not provide the same target resolution because the shape of their response curve is broader and often more complex.

Metal detectors, EM units, and total field magnetometers are highly susceptible to interference from nearby metallic cultural features. Any of these features can produce an erroneous response that may be incorrectly interpreted as a subsurface target. Because metal detectors are relatively short-range devices, they can be operated closer to such sources of interference than can most magnetometers. Measurements made with a total field proton procession magnetometer are susceptible to interference from high magnetic gradients, natural changes in the earth's magnetic field, and nearby power lines, whereas fluxgate gradiometers do not suffer from these shortcomings.

Seismic, resistivity, magnetic, and gravity techniques may also be used to locate boundaries of larger trenches and landfills. These techniques are much slower and will provide less resolution than the previously described methods. However, they are often the only techniques that can be used to estimate the thickness of a landfill or trench. It should be noted that interpretation of such data should be done with caution by experienced personnel.

SUMMARY

All of the geophysical methods discussed are scientifically sound, and all have been proven in the field. Like other technologies, however, they may fail to provide the desired results when applied to the wrong problem or improperly used. The techniques must be matched to site-specific conditions by a person who thoroughly understands the methods and their limitations.

To improve the accuracy of site characterization, adopting a broader integrated systems approach is recommended. Geophysics is just one of many technologies that can be readily incorporated into a site investigation program. An integrated systems approach provides the benefits of both direct sampling and remote sensing techniques. Airborne and/or surface geophysical methods are generally used as initial reconnaissance tools to cover an area in a quick search for anomalous conditions. Surface geophysical methods can then be employed for a detailed assessment of site conditions. After potential problem areas have been identified, the drilling locations for borings and monitoring wells can be selected with a higher degree of confidence to provide representative samples. Analyses of soil and water samples from properly located borings or monitoring wells will then provide the necessary quantitative measurements of subsurface parameters. Downhole geophysical methods can be applied to define details of conditions with depth. This approach delivers greater confidence in the final data interpretation with fewer borings and wells, and an overall cost savings. Furthermore, the drilling operations are no longer being used for hit-or-miss reconnaissance, but rather as specific quantitative tools (SMART HOLES).

Before selecting a method or methods, the project objectives must be clear and as much as possible should be learned about site conditions. Information such as accessibility and site topography should be available. In addition, general soil and rock types and conditions, the approximate depth to water table, depth to rock, background specific conductance of ground water, should be known or estimated. If appropriate, the type of contaminant should also be defined. Finally, one should consider whether it is likely that sufficient contrast may exist in the parameters being measured. If in fact there is no contrast in the parameter being measured between one layer and another, the geophysical method will fail to provide a response. Similarly, if a layer is sufficiently thin or the size of the target is sufficiently small, the layer or target will not be detected.

The question of whether drilling or geophysics should be done first often arises. Because the results of geophysical work usually result in identification of anomalous conditions, geophysics should generally be done first so that anomalous areas can be identified for drilling and sampling.

However, if borings or monitoring wells have already been installed, geophysical surveys can still deliver increased accuracy. The location and data from existing boreholes and monitoring wells can be assessed using geophysical methods, thus providing a means of evaluating the validity of data already acquired. If additional boreholes or wells are needed to fill gaps in the data, they can be located with a high degree of confidence.

Because each geophysical technique measures a different parameter, the information from one is often complemented by that from another. The synergistic use of multiple geophysical techniques often serves to enhance site characterizations. Those familiar with traditional well logging will recognize this concept, as multiple logs are commonly obtained to aid interpretation.

It should be noted that the use of any geophysical technique depends on its specific application and on site conditions. Therefore, no single method should be expected to solve all site evaluation problems. Furthermore, geophysical technology is not in itself a panacea; its successful application is dependent upon integrating the geophysical data with other sources of information. This must be done by persons with training and experience in geophysical methodology, as well as in the broader aspects of the earth sciences. Geophysical methods do not offer a substitute for borings and wells, but provide a means to minimize the number of boreholes and wells, to assure that they are in reasonably representative locations, and to fill in the gaps between boreholes.

REFERENCES

Benson, R. C., and R. A. Glaccum, 1979. Radar Surveys for Geotechnical Site Assessment. In: Geophysical Methods in Geotechnical Engineering, Specialty Session, American Society of Civil Engineers, Atlanta, GA, pp. 161–178.

Benson, R. C., R. A. Glaccum, and M. R. Noel, 1982. Geophysical Techniques for Sensing Buried Waste and Waste Migration. Environmental Protection Agency—Environmental Monitoring Systems Laboratory, Las Vegas, NV, pp. 236.

Benson, R. C., and L. J. LaFountain, 1984. Evaluation of Subsidence or Collapse Potential Due to Subsurface Cavities. In: Sinkholes, Their Geology, Engineering and Environmental Impact, Proceedings of the First Multidisciplinary Conference on Sinkholes, Orlando, FL, pp. 201–215.

Benson, R. C., and D. C. Pasley, 1984. Ground Water Monitoring: A Practical Approach for a Major Utility Company. Fourth National Symposium and Exposition on Aquifer Restoration and Ground Water Monitoring, National Water Well Association.

Benson, R. C., M. Turner, W. Vogelson, and P. Turner, 1985. Correlation Between Field Geophysical Measurements and Laboratory Water Sample Analysis. Proceedings of the National Water Well Association/Environmental Protection Agency Conference on Surface and Borehole Geophysical Methods in Ground Water Investigations, National Water Well Association.

Benson, R. C. and J. Scaife, 1987. Assessment of Flow in Fractured Rock and Karst Environments. Proceedings of the Second Multidisciplinary Conference on Sinkholes and the Environmental Impacts of Karst, Orlando, FL.

Benson, R. C., and L. Yuhr, 1987. Assessment and Long Term Monitoring of Localized Subsidence Using Ground Penetrating Radar. Proceedings of the Second Multidisciplinary Conference on Sinkholes and the Environmental Impact of Karst, Orlando, FL.

Benson, R. C., M. Turner, P. Turner, and W. Vogelson, 1988. In Situ, Time-Series Measurements for Long-Term Ground-Water Monitoring. In: Ground Water Contamination: Field Methods, ASTM STP 963, A. G. Collins and A. I. Johnson, Eds. American Society for Testing and Materials, pp. 58–72.

Breiner, S., 1973. Applications Manual for Portable Magnetometers. Geometrics, Sunnyvale, CA, 58 pp.

Cartwright, K. and M. McComas, 1968. Geophysical Surveys in the Vicinity of Sanitary Landfills in Northeastern Illinois. Ground Water, Volume 6, pp. 23–30.

Dunbar, D., H. Tuchfeld, R. Siegel, and R. Sterbentz, 1985. Ground Water Quality Anomalies Encountered During Well Construction, Sampling and Analysis in the Environs of a Hazardous Waste Management Facility. Ground Water Monitoring Review, Vol. 5, No. 3, pp. 70–74.

Greenhouse, J. P. and M. Monier-Williams, 1985. Geophysical Monitoring of Ground Water Contamination Around Waste Disposal Sites. Ground Water Monitoring Review, Vol. 5, No. 4, pp. 63–69.

Griffith, D. H. and R. F. King, 1969. Applied Geophysics for Engineers and Geologists, Pergamon Press.

Haeni, P., 1986. Application of Seismic-Refraction Techniques to Hydrologic Studies. U.S. Geological Survey, Open File Report No. 84–746, Hartford, CT, 144 pp.

Hileman, B., 1984. Water Quality Uncertainties. Environmental Science and Technology, Vol. 18, No. 4, pp. 124–126.

Hunter, J. A., R. A. Burns, R. L. Good, H. A. MacAulay and R. M. Cagne, 1982. Optimum Field Techniques for Bedrock Reflection Mapping with the Multichannel Engineering Seismograph. In Current Research, part B, Geological Survey of Canada, Paper 82-1B, pp. 125–129.

Keys, W. S. and L. M. MacCary, 1976. Application of Borehole Geophysics to Water-Resources Investigations. Techniques of Water-Resources Investigations of the United States Geological Survey, Chapter E1.

Lankston, R. W. and M. M. Lankston, 1983. An Introduction to the Utilization of the Shallow or Engineering Seismic Reflection Method. Geo-Compu-Graph, Inc.

Lysyj, Ihor, 1983. Indicator Methods for Post-Closure Monitoring of Ground Waters. National Conference on Management of Uncontrolled Hazardous Waste Sites, Hazardous Materials Control Research Institute, pp. 446–448.

McNeill, J. D., 1980. Electromagnetic Resistivity Mapping of Contaminant Plumes, In: Proceedings of National Conference on Management of Uncontrolled Hazardous Waste Sites, Washington, DC, pp. 1–6.

Mooney, H. M., 1973. Engineering Seismology. In: Handbook of Engineering Geophysics, Vol. 1; Bison Instruments, Minneapolis, MN.

Mooney, H. M., 1980. Electrical Resistivity. In: Handbook of Engineering Geophysics, Vol. 2; Bison Instruments, Minneapolis, MN.

Orellana, E. and H. M. Mooney, 1966. Master Tables and Curves for Vertical Electrical Sounding Over Layered Structures. Interciencia, Madrid, Spain.

Palmer, D., 1980. The Generalized Reciprocal Method of Seismic Refraction Interpretation. K. B. S. Burke, Ed., Dept. of Geology, University of New Brunswick, Fredericton, N.B., Canada, 104 pp.

Perazzo, J. A., R. C. Dorrler, and J. P. Mack, 1984. Long-Term Confidence in Ground Water Monitoring Systems. Ground Water Monitoring Review, Vol. 4, No. 4, pp. 119–123.

Redpath, B. B., 1973. Seismic Refraction Exploration for Engineering Site Investigations. Technical Report E-73-4, U.S. Army Engineer Waterways Experiment Station Explosive Excavation Research Laboratory, Livermore, CA, 52 pp.

Steeples, D. W., 1984. High Resolution seismic reflections at 200 Hz. Oil and Gas Journal, pp. 86–92, Dec. 3.

Telford, W. M., L. P. Geldart, R. E. Sheriff, and D. A. Keys, 1982. Applied Geophysics, Cambridge University Press.

United States Environmental Protection Agency, 1985. RCRA Ground-Water Monitoring Technical Enforcement Guidance Document.

Walker, S. E. and D. C. Allen, 1984. Background Ground Water Quality Monitoring: Temporal Variations. Fourth National Symposium and Exposition on Aquifer Restoration and Ground Water Monitoring, National Water Well Association.

Zohdy, A. A., G. P. Eaton, and D. R. Mabey, 1974. Application of Surface Geophysics to Ground-Water Investigations. Techniques of Water-Resources Investigations of the United State Geological Survey, Chapter D1, 116 pp.

6

Monitoring Well Drilling, Soil Sampling, Rock Coring, and Borehole Logging

H. E. "Hank" Davis, James Jehn, and Stephen Smith

DRILLING METHODS

Drilling and sampling for ground-water monitoring utilizes much of the same technology used in conventional drilling and geotechnical exploration but with some very significant differences. The primary purpose of most exploration is to recover an intact physical specimen which can be tested for physical strength, assayed for mineral properties, or measured for geologic formation properties such as dip, strike, and formation continuity. For ground-water monitoring, primary consideration must be given to obtaining a sample which is representative of existing conditions and valid for chemical analyses. The sample must not be contaminated by drilling fluid or by the drilling and sampling procedures. Great care must be taken to protect the sample from contaminants at the surface and in transit to the laboratory. In cases in which the borehole itself is to be logged by geophysical or geochemical methods, it is often necessary to drill without conventional circulation fluids. The method of drilling or sampling is often site-specific, depending upon the type of logging or testing to be done.

This chapter describes most conventional methods of exploration and includes comments by the authors on the suitability of each method for ground-water monitoring investigations. Two general categories of methods are discussed: (1) methods which do not use circulation fluids, and (2) methods which require the use of circulation fluids to transport drill cuttings to the surface.

195

Drilling Without Circulation Fluids

Probing

Probing can be done with a tool as simple as a slender steel rod, 1/4- to 1/2-inch in diameter and 3 to 4 feet long, having a tee handle (see Figure 6.1). This tool is often used by backhoe operators to probe into the soil by hand in advance of any power digging to feel resistance to penetration which would indicate the presence of a subsurface obstruction. Probing can also be done—generally with a larger and longer probe—using a backhoe, drilling machine or other hydraulic powered device to push or press the probe.

Figure 6.1. Probe.

Probes can be used in some situations to locate and outline buried objects, such as underground tanks, piping, conduit, and survey markers. They can also be used to "hand probe" ahead of a borehole when resistance to advance is felt in the first 3 or 4 feet of a boring. A probe can often be used to locate and avoid accidental drilling through or digging up of a buried object.

When a probe is advanced or pushed, it forces the formation out of its path by "displacing" the soil. Thus, a probe is a simple form of displacement boring.

Displacement Boring

The word "boring" here may be a misnomer, but through common usage it has come to mean a borehole made by displacing soil material out of the path of the boring without recovery of the displaced material to the surface. The displaced material is simply forced into the sidewall of the hole.

Generally, the tool that is forced into the soil is a piston or plug type sampler (see Figure 6.2) such as the 2 inch modified California Sampler. The plug or piston, in closed (plugged) position, forces all material in its path to be displaced to the sidewall. When the tool (sampler) reaches the desired sampling depth, the plug or piston is retracted by means of a center actuating rod and the advance is resumed—this time with the sampler open and taking sample. Most material in the path of the open sampler will become part of the sample taken, but the material directly in the path of the wall of the sampler will still be displaced.

Figure 6.2. 2″ Modified California sampler.

Displacement boring is a fairly simple method of obtaining shallow samples and can be accomplished without the need for heavy equipment or circulation fluids. Depending upon the size and depth of samples needed, and the formation's resistance to penetration, it can be done entirely by hand using 35-pound hammers or with lightweight, power-operated post or picket drivers.

From a hole and sample contamination viewpoint, displacement boring is one of the cleanest methods available. The method is well suited to use for shallow borings in soft materials where the location is not accessible to heavy equipment. It is generally not as efficient as other more advanced methods, where heavy equipment is usable, or where a large number of samples must be taken.

Auger Drilling

Auger drilling utilizes a helical (spiral) tool form to convey dug material to the surface (see Figure 6.3). Mechanically, an auger consists of a long inclined plane with a fixed mechanical advantage. The digging and conveying capacity of a specific auger is directly proportional to the torque applied to rotate the auger. Auger drilling does not normally require the use of circulation fluids.

An auger is essentially a conveyer which has a drill head, or cutting bit, or combination of head and bit at its bottom end to cut the formation, which is then conveyed upward. Four basic types of auger are in use: bucket auger, digger auger, solid-stem continuous flight auger (CFA), and hollow-stem continuous flight auger. For descriptive purposes, the hollow-stem continuous flight auger is referred to simply as hollow-stem auger (HSA), despite the fact that it is a continuous flight design.

Bucket auger machines utilize an auger bucket attached to a square torque bar that passes through a ring-type drive mechanism. Generally, the auger bucket advances into the formation by a combination of dead weight and tooth cutting angle. When the bucket has advanced one or two feet, the bucket is withdrawn from the hole by means of a wire-line hoist cable attached to the top of the square torque bar. When the bucket reaches the surface it is swung to the side of the hole and the spoil is dumped out through the bottom by means of a hinge and latch device on the bucket bottom.

The square torque bar—known as a "Kelly Bar"—is often of a telescoping type which permits digging to greater depths. The solid bar is nested within one or more square tubes. Most bucket auger machines have a depth capacity of 30 to 75 feet and most are used for large-diameter holes ranging from about 16 to 48 inches. Buckets smaller than 16 inches in diameter are very rare. Most bucket auger machines are "gravity fed" and are used for vertical holes. They are not normally used to drill monitoring wells but are sometimes used to drill production wells and recovery wells. They are most commonly used to drill drain wells, caissons, and footings.

Figure 6.3.   Digger auger tool.

Digger augers, also known as "pressure diggers" are power-feed machines that can exert great downward force on the cutter head. They dig by means of a power-rotated square torque bar which is equipped with a helical cutter head on the lower end. The cutter head consists of one or more full 360-degree spirals which may have shutter plates. The shutter plates are hinged to allow entry of the material as the cutter head is rotated. They close when the loaded auger is withdrawn to the surface, thus holding the cut spoil from sliding off the auger flights into the borehole. Cutter heads without shutter plates generally consist of three or more 360-degree spirals which have a very shallow pitch so that the spoil will not slide off the cutter head when rotation is stopped.

To dig with a digger auger, the cutter head is advanced one to three feet and retracted to the surface where the spoil is "spun off." Digger machines make holes by continuing the sequence of dig—retract—spin. They can dig angle holes, and have a depth capacity of from 12 feet to about 75 feet. Boring diameter is seldom less than 12 inches and can be eight feet or more. Digger augers are sometimes used for recovery well drilling and pump-out holes, but are only rarely used for monitoring wells.

Continuous-flight augers (see Figure 6.4) consist of a plugged tubular steel center shaft or axle, around which is welded a continuous steel strip in the form of a helix. An individual auger is known as an auger or "flight," and normally

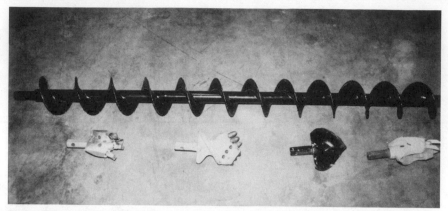

Figure 6.4.   Continuous flight auger. (Photo courtesy of Mobile Drilling Company.)

is 5 feet long, though other lengths are not uncommon. Manufacture of flights is such that, when connected to one another, the helix is continuous across the connections and throughout the depth of the borehole. Connections are normally made by means of a hex or square pin welded to the male end of a flight, which slips into a corresponding hex or square sleeve welded to the female end of a flight. Torque is transmitted from the auger drilling machine to and through the flights by the hex or square connections on each flight. Down force is transmitted by shoulder to shoulder contact of the flights. Retract force is taken by a pin or bolt inserted through a hole that has been drilled through a flat face of the hex or square connections at a 90-degree angle to the axis of the flight. Typically, this pin or bolt is known as a "U Pin" or "Drive Clip."

Auger drill heads are attached to the bottom auger flight in the same manner as flights are connected to each other. Most auger drill heads are of the field-replaceable bit type where a hardened or tungsten carbide steel inserted bit does the cutting. Other types include the "Clay Head" or "Mining Roof Bit," which consists of a one piece drill head.

Auger drill heads are generally designed to cut a hole approximately 10% greater in diameter than the actual diameter of the auger they serve. For example, the auger drill head designed to cut a 6- to 6.25-inch diameter hole is used with augers which actually measure 5.5 inches in diameter when new. The auger that actually has a 5.5-inch diameter is known as a 6-inch auger. Thus, a conventional nonhollow stem continuous flight auger is known by the nominal diameter of the drill head.

In addition to diameter, augers are specified by the pitch of the auger and the shape and dimension of the connections. The pitch is the distance along the axis of the auger which it takes for the helix to make one complete 360-degree turn. Pitch does not wear. All other exposed parts of an auger do wear and can be a source of confusion when specifying. The pitch of an auger used for vertical hole drilling will generally be 65% to 80% of the hole diameter. This "slow" pitch allows easy conveyance of spoil. An auger used exclusively for horizontal drilling will generally have a pitch equal to or greater than the hole diameter because that auger is not used to lift spoil; it only pulls spoil. Horizontal augers are known as "fast" pitch augers.

Hollow stem augers (see Figure 6.5) are a form of continuous flight auger in which the helices (also know as flights, flighting, or spirals) are wound around and welded to a tubular center stem or axle. The tubular axle is internal flush without upset, or without a thickened wall, at the ends. Connection of one section to another is by means of a thread, key/keyway, or spline-type connector. Threaded connectors transmit both torque or ram/retract force at the thread. In the key/keyway and spline systems the torque is transmitted by the spline or key/keyway; ram is transmitted by shoulder to shoulder contact and retract force is carried by the bolts. When sections are connected, the hollow-stem auger will present a smooth uniform bore throughout its length and the flighting will be continuous from top to bottom of the hole.

Hollow-stem augers are specified by the inside diameter of the hollow stem. It is *not* specified by the hole size it drills, which is the source of much confusion.

The hollow-stem auger is a conveyer, but when compared to regular continuous flight augers, the center axle (hollow stem) is much larger. This change in

Figure 6.5. Hollow stem augers. (Photo courtesy of Mobile Drilling Company.)

proportion between axle diameter and hole size makes considerable change in the conveyance characteristics of the auger, which must be accommodated by technique change on the drillers part. Essentially, this is done by more rotation, not more RPM. Some formations must be augered at very slow RPM or they do not auger at all. Heavy formations, such as adobe or "fat" clays, should be auger-drilled at 30 to 50 RPM. Good clean sand that will stand open can be successfully augered at 750 RPM.

Hollow-stem auger drill heads are generally made of two pieces: an annular outer head attached to the bottom of the lead auger and an inner pilot or center bit mounted in a plug which is removable through the center of the auger to the surface. Being able to withdraw the center bit and plug while leaving the auger in place is a principal advantage of hollow stem augers. Withdrawing the plug while leaving the auger in place provides an open, cased hole into which samplers, down-hole drive hammers, instruments, casing, anodes, wire, pipe, or numerous other items can be inserted. Replacing the center bit and plug, assuming nothing was left in the hole, allows continuation of the borehole. The center plug is normally "tripped" and held in place by drill rods, although some manufacturers offer a system of operating the center plug by wire line, thus greatly reducing the "trip time."

Hollow-stem augers are available with inside diameters, in inches of 2.5, 3.25, 3.375, 4.0, 4.25, 6.25, 6.625, 8.25, and 10.25. The most commonly used sizes are 3.25 inches and 4.25 inches for 2-inch monitoring wells and 6.625 inches for 4-inch monitoring wells. The two largest sizes, 8.25 and 10.25 inches, are finding some use in dual-purpose (monitor then pump-out) wells and larger-diameter wells in which the use of circulation fluids must be avoided. The cost of these largest auger sizes and the machines to operate them is relatively high; operating costs are also high and they are not routinely used for monitoring wells.

Drilling machines (see Figure 6.6) used to operate continuous flight augers (CFA) and hollow-stem augers (HSA) are of top-drive design where all down force is applied directly to the top of the auger. Most are designed to drill with 5-foot auger lengths, although 10-foot and 20-foot machines are becoming available. The main characteristic of these machines is relatively high torque, which is the main feature that distinguishes auger machines from rotary drills. An auger drilling machine used for simple CFA work might consist of a rotary and feed/retract system only. Some exploration consists of simply augering a hole and collecting spoil from the flights as it arrives at the surface. More often the machine also has a hoist, a driving device, and an off-hole mechanism. These features permit removal of augers and use of sampling tools without having to move the machine.

The newest design features include the ability to work through a large-bore hollow top-drive spindle directly into the HSA, thereby eliminating the need for disconnect/reconnect and on-off hole manipulation. These new features also permit continuous access to the bottom of the borehole and, in fact, will permit percussion drilling or drive sampling through the center of the HSA while the HSA is actually making hole.

Figures 6.6.   Auger drilling machine. (Photo courtesy of Mobile Drilling Company.)

Drilling with Circulation Fluids

Circulation fluids, also known as drilling fluids, are an essential element of all the methods of drilling described in the remainder of this section. A circulation fluid can be a liquid, such as water or drilling mud (water with special additives), or it can be a gas, such as air or foam (air with additives of various types). Circulation fluids are normally forced down through the drill pipe, out through the bit, and back up the annulus between the borehole wall and drill pipe. Circulation fluids function to cool and lubricate the cutting bit, stabilize the borehole, and remove cuttings.

Removal of cuttings requires a minimum up-hole velocity of the circlation fluid, which depends on the viscosity of the fluid and the size and density of the cuttings.

Minimum velocity can be estimated using a modified form of Stokes Law, or by reference to one of many available nomographs. Generally, the minimum up-hole velocity needed to transport cuttings is about 150 feet per minute for plain water with no additives and about 3,000 feet per minute for air with no additives. Additives decrease the required minimum velocity. Excessive velocities can cause hole erosion or "sandblasting," which can cause caving of the borehole.

Because air is everywhere and water has to be hauled, it is always a good idea to at least consider the possible use of air. A good rule-of-thumb says that the correct pump volume (in gpm) when multiplied by 4 will be the correct volume of air (in cfm). Primary advantage of air, for the geologist, is the quick recovery of cuttings at the surface due to the high velocity return.

The use of circulation fluids invariably involves the addition of chemicals to the borehole. Additions to circulation liquid include various types of drilling muds, most of which are a form of bentonitic clay, although some polymers are in use. Additions to circulation gas include detergents (from which foam is generated) or water (for mist). These may contain complex chemicals. Furthermore, compressed air usually contains a substantial amount of hydrocarbon lubricants, although compressed air without hydrocarbons may soon be available for drilling. Therefore, as a general rule, methods of drilling which require the use of circulation fluids in the borehole are used for contaminant exploration only when absolutely necessary—and then with great care.

Several types of specialized drilling methods that use circulation fluids are described below.

Wash Boring

A wash boring is a fairly simple, almost obsolete method of advancing a borehole, in which the formation is cut by a chopping and twisting action of a bit and washed (flushed) to the surface by circulation fluids. Equipment can be as simple as a tripod with sheave, drill pipe with bit, pump with hoses, and hoist with rope. Circulation fluid is pumped into the drill pipe and out through the bit, which is raised and dropped as it is turned back and forth.

In a wash boring, the chopping action cuts the formation and the occasional turning of the bit maintains the roundness and straightness of the hole. The cut material is washed up around the drill pipe to the surface where it is screened out, and settles out of the circulation fluid in a circulation tank or pit before the water is recirculated back through the pump into the borehole.

When wash boring, samples of cuttings are generally caught in a sieve or screen held in the return stream. Relatively undisturbed samples of unconsolidated material can be obtained by driving a sampler into the bottom of the borehole using the hoist, drill pipe, and drive hammer after the bit has been removed from the drill pipe.

Rotary Drilling

Rotary drilling is a drilling method in which drill pipe with an attached bit is continuously rotated against the face of the hole while circulation fluid is pumped through the pipe and bit to flush cuttings to the surface. It differs in principle from wash boring in that in wash boring the drill pipe is not continuously rotated, but rather is periodically turned by the driller.

Rotary drilling is usually accomplished with truck-mounted or trailer-mounted machines which carry their own pumps and operating components. A typical rotary drilling machine consists of an engine, the rotation mechanism, a feed/retract system, drum hoist(s), a cathead or driving device, an on-off hole mechanism for moving the rotation mechanism away from the drilling axis, and a pump or compressor complete with pressure hose, piping, swivel, and other equipment as necessary to circulate the drilling fluid.

Pumping and circulation with a rotary drill is no different in principle from that used in wash boring, but the rotary drill can drill holes that are larger in diameter and much deeper. The increased capacity is a result of: (1) the rotary mechanism, which causes continuous rotation of the drill pipe and the bit, and (2) the feed/retract system, which allows continuous application of down force on the bit, causing the bit to cut new formation while the circulation fluid flushes cuttings to the surface.

For effective rotary drilling, the down force on the bit should be great enough to continuously cause penetration of the formation. As a rule-of-thumb, this force should be approximately 1500 to 2500 pounds per inch of bit diameter. If the crushing strength of the formation exceeds that applicable with the drill it will normally be necessary to use a heavier, more powerful drill or change the method of drilling.

Rotating speeds of most rotary drills are in the range of approximately 30 to 250 rpm. Rotary drills are rarely used at speeds in excess of 250 rpm due to vibration, and only the newest rigs have speeds low enough (15 to 25 rpm) to permit efficient use of a "down-hole hammer," which is described later.

Specifications for drilling machines often show pull-down capabilities that are well in excess of the weight of the truck and drill combined. Although the rated pull-down may be theoretically possible, it generally can be achieved only by tying the drill down to a previously installed earth anchor. For shallow sampling or monitoring holes, the extra time and effort is usually not justified. As a general rule, and in most monitoring applications, use of pull-down in excess of that which can be applied and safely contained by the weight on the rear axle(s) of the truck is not advisable.

Rated depth capacity of a rotary drill is that length of drill pipe (stem or rod) which weighs 80% of the maximum main-hoist, single-line, bare-drum capacity. For example, a drill having a maximum main-hoist, single-line, bare-drum capacity of 10,000 pounds would have a rated depth capacity that would be equivalent to 8,000 pounds (10,000×0.80). This drill used with "NW" drill rods, which weigh 5.5 pounds per foot, would have an "NW" rated depth capacity of 1,455 feet (8,000/5.5).

Hole diameter rating of a rotary drill is based on several factors, the most significant of which are the delivered volume of the circulation pump or compressor and, in deeper holes, the delivery pressure of the pump or compressor. In rating the pump or compressor, both delivered volume and delivery pressure will be specific to the type of drilling fluid in use, and the drilling fluid to be used is often determined from advance knowledge of the diameter and depth of hole required. For example, if the driller knows that he has to place a 4.5-inch outside diameter casting with a 1-inch all round gravel pack, he also knows that he will have to drill a hole not less than 6.5 inches in diameter. He will probably choose to drill at least a 7-inch diameter hole. At some depth around 500 to 600 feet (less with water flow problems or artesian heads), the circulation fluid engineer ("mud engineer") takes on more responsibility than the driller. In the case of companies which do not have fluid engineers, the driller must have a thorough knowledge of circulation fluids.

Many holes drilled with the rotary method contain some casing which is grouted. Therefore, the hole should be drilled large enough to allow at least a 1.5-inch annulus between the casing and the wall of the hole. This space provides room for a grout pipe.

Three basic types of rotary drill are in common use: (1) stationary table, in which the rods are rotated by means of a square or splined Kelly as it passes through a fixed rotating table; (2) moving table, in which the rods are rotated by means of a square or splined Kelly as it passes through a rotating table, which is moved up or down by means of hydraulic cyclinders; and (3) top drive, in which the rotating spindle travels up or down applying feed, retract, and rotation forces directly to the top of the drill stem. Top-drive types are the most recent design, and they seem to be used more than the others in monitoring work.

Reverse Circulation

Reverse circulation is a method of rotary drilling in which the circulation fluid flows from the ground surface down the annulus around the drill pipe. The fluid carries the cuttings back to the surface inside the drill pipe. At the surface, the fluid is expelled through a swivel into the circulation pit or tank where the cuttings settle out.

Reverse circulation is especially advantageous in very large boreholes and also in those cases where the erosive velocity of conventional rotary circulation would be detrimental to the borehole wall. To increase the diameter of a hole drilled by conventional rotary methods, the capacity of the pump or compressor must be increased to maintain an adequate up-hole velocity to lift cuttings. With reverse circulation, the up-hole velocity is controlled by the size of the drill pipe, not the borehole.

Reverse circulation has few applications in monitoring work. However, one form of drilling that is often referred to as reverse circulation has significant potential for monitoring. That method is described below.

Center Stem Recovery

Center stem recovery, or CSR, is a form of rotary drilling similar to reverse circulation in which two concentric drill pipes are assembled as a unit to create a controlled annulus. For this reason, the method is often referred to as dual-tube or dual-wall reverse circulation drilling. Circulation fluid, which may consist of air, water, mist, foam, or "mud" is pumped through an outer swivel down through the annulus to the bit where it is deflected upward into the center pipe. The bit used with CSR equipment is of a design which cuts an annular ring or kerf and forces all cut material to move toward the center of the hole. Cut material is returned up through the inner pipe and swivel to the surface. The spoil may be collected as a sample or simply allowed to pile up on the ground, depending upon the purpose of the boring. CSR systems are available for both percussion and rotary type drills.

CSR drilling will find increased use in monitoring work if oil-less compressed air becomes available as a circulation fluid or if plain water, or a combination of air and water, can be tolerated.

Percussion Drilling

Percussion drilling is a form of drilling in which the basic method of advance is hammering, striking, or beating on the formation. Rotation may also be involved but, if so, it is used primarily to maintain roundness and straightness of the percussion borehole. Three basic types of percussion drilling equipment are in use: (1) cable-tool, or "churn" drilling, in which a bit, hammer, or other heavy tool is alternately raised and dropped; (2) air percussion, in which an air-actuated device with attached bit breaks the formation; and (3) air-operated casing hammer. All three methods use impact energy to break or cut the formation.

Cable-tool drilling is one of the oldest methods of drilling, and is still widely used for drilling water-supply wells. Its application in monitoring work is limited, mainly because the method is slow. Drilling rates of 10 to 20 feet per day are common. Furthermore, holes much smaller than 6 inches are impractical because of the need to use a relatively large, heavy bit. Nevertheless, the method does not use large volumes of drilling mud and allows sampling of ground water as the hole is advanced in high-yielding formations. For these reasons, the method will continue to be used in monitoring work in some circumstances.

In cable-tool drilling, the bit breaks up and pulverizes the soil or rock. Cuttings are recovered by adding water or water with additives ("mud") to the borehole in order to form a slurry which is then periodically bailed from the borehole. In unconsolidated formations, casing is advanced behind the bit. The diameter of the casing is slightly larger than the diameter of the bit, and it is equpped with a drive shoe on the lower end. Casing is driven by retrieving the bit from the hole and equipping it with drive clamps. The weight of the bit is then applied to the top of the casing to drive it into the hole.

In air percussion drilling, air is used to actuate a hammer which is operated on the end of the drill pipe. Air exhausted from the hammer is used to carry cuttings to the surface continuously as the hole is advanced. The most common form of air percussion used in exploration drilling employs a "down-hole hammer" in which the hammer and bit are operating in the bottom of the borehole. Connection to the machine is with a drill stem to the surface, providing only very slow rotation and sufficient feed or retract force for proper operation of the hammer. "Down-hole hammers" are excellent tools for making holes in hard formation where the formation will stand open without caving.

The third method of percussion drilling uses an air-operated casing hammer that is similar to a pile-driving hammer except for a hole through its axis so that drill stem can be rotated through the center of the casing hammer (Figure 6.10b). This arrangement allows drilling to proceed with the drill stem even while the casing is actually being driven. Sometimes the casing hammer drives casing while the drill stem is idle, sometimes the stem is used while the hammer is idle, and at other times both the driving and rotating systems work at the same time. The casing hammer drives (hammers) the casing only; therefore, casing must be cleaned out by the tool on the inner drill stem. Usually this tool is a rotary rock bit or a down-hole hammer.

The air-operated casing hammer and the down-hole hammer both require internal lubrication which is provided by hydrocarbon lubricants added by means of in-line oilers. This need for lubrication eliminates the combined systems from consideration for some monitoring work unless proper filtering is available and used. The casing hammer, however, does not input oil into the formation directly and therefore it can often be used if some other method of casing cleanout is used which does not add oil to the formation.

ODEX is an adaptation of the air-operated down-hole hammer. It uses a "swing-out" eccentric bit as a casing under-reamer. The percussion bit is a two-piece bit consisting of a concentric pilot bit behind which is an eccentric second bit that swings out to enlarge the hole diameter. The driller controls the swing-out by forward or reverse rotation of the drill stem. Immediately behind (above) the eccentric bit is a "drive sub" which engages a special internal shouldered drive shoe on the bottom of the ODEX casing. Thus, ODEX casing is actually "pulled" down by the drill stem as the hole is advanced. Cuttings blow up through the drive sub and stem/casing annulus to a swivel, which conducts them to a sample collector or onto the ground. ODEX is also known as TUBEX.

SIM-CAS is very similar to ODEX except that the casing is pushed down from a "pushing" head on the drill at the surface as opposed to being pulled from the bottom. The eccentric bit, drill stem sizes, hole sizes, and air requirements are essentially the same as ODEX.

Both SIM-CAS and ODEX use compressed air or foam for operation; therefore, unless proper filtration is provided, they may not be acceptable for monitoring work.

SELECTION OF DRILLING METHODS

Introduction

Selecting drilling methods for a ground-water investigation is typically a process of evaluating tradeoffs. Drilling methods that allow for quick, efficient well construction may not be well-suited for soil or rock sampling. Methods that accommodate geophysical logging are not necessarily well adapted to drilling in highly contaminated areas. Compared to production wells or mineral exploration, the geologist, engineer, or consultant usually has more options to consider when drilling monitoring wells.

A higher level of field supervision is normally required when drilling monitoring wells. Construction of monitoring wells and collection of environmental samples is a relatively new business for many drilling contractors. Old techniques and drilling short cuts may introduce contaminants into soil samples or the formation. A higher degree of accuracy is usually required in tallying lengths of casing and screen, and measuring the depth of boreholes, gravel-packed intervals, and annular seals.

In selecting a drilling method for monitoring well construction, the single most important consideration is the collection of representative ground-water, soil, or rock samples from a specified depth or interval. However, time, cost, and many other factors must be considered.

Health and Safety

At hazardous waste sites, contamination levels in the soil and subsurface may be high or may not be known. With some drilling methods or equipment, adequate worker protection is inherently more difficult than with other methods. Auger drilling or cable-tool drilling methods do not require the use of large amounts of circulation fluids to remove cuttings, and these methods generally present the lowest risk of contaminant exposure under most conditions. However, neither method is risk-free. Care must be taken to avoid contact with cuttings, dust, or any water that may have been used to assist in the removal of cuttings.

From the standpoint of health and safety, drilling methods which use air as a circulating fluid may present the greatest risk, particularly for monitoring well installation in areas where ground water is highly contaminated. In high-yield aquifers, holes drilled with air below the water table will produce water at high rates even when casing is advanced as the hole is drilled. The water can be difficult to manage, and special precautions may be necessary to minimize worker exposure.

In some situations, particularly where an underground fire is suspected or know to occur, introducing high volumes of air into a borehole should be completely avoided. Air used for drilling will accelerate the rate of burning. Augers also generate considerable heat. Therefore, where fire, heat, or explosion is a possibility, such as at a landfill or a coal mine which is on fire, drilling with a water-based drilling fluid is the preferred method.

Access and Noise

Drilling and sampling in connection with ground-water investigations is frequently carried out in urban areas where access and equipment noise are important considerations in selecting drilling methods. Most drilling equipment is mounted on trucks which have limited maneuverability, and in congested areas near factories, power plants, refineries, and manufacturing facilities, side-to-side and overhead clearances are critical to the selection of drilling methods. Auger drilling equipment is generally the smallest, lightest, and most maneuverable. Mast height is usually less than with other types of drilling rigs, and on some auger rigs the upper part of the mast can be detached. For extremely "tight" locations, skid-mounted equipment which can be winched into position is available from some drilling contractors. Electric powered "skid rigs" can be used for indoor work where adequate ventilation for engine exhaust is not available.

Rotary, cable tool, and percussion drilling equipment is usually mounted on larger trucks than auger equipment, and it requires more room to maneuver and operate. Furthermore, additional space is generally required to drill a hole once the rig is on location. Drill pipe for rotary rigs is generally at least 20 feet long, as more space is required to handle the pipe than the 5-foot augers which are customarily used with auger drilling rigs. Circulation fluids such as mud or water also require holding tanks which extend from the side or the rear of the rig.

Cable tool rigs may require guylines for lateral mast support, which limit their use in restricted locations. These guylines extend at least 15 to 20 feet to the front and rear of the mast.

Noise can be a major obstacle to monitoring well construction in populated areas. Many cities have noise ordinances which restrict the operation of heavy machinery, such as drilling equipment, to specific daylight hours. The allowable noise levels are also restricted. In municipalities which do not have noise ordinances, complaints can still be filed under provisions of nuisance ordinances.

Some drilling equipment can be modified to control noise, but for other types of equipment, noise control is impractical or impossible. Noise from impact or percussion equipment such as casing hammers and casing drivers is most difficult to control. Therefore, other drilling methods are better suited to use in residential or urban areas. Engine noise from most rigs can be muffled or shielded, although providing for noise control may increase costs. Special noise blankets are available to reduce engine or compressor noise.

Disposal of Fluids and Cuttings

Disposal of drilling fluids and cuttings is an important consideration in the selection of drilling methods. If soil and ground water are contaminated, fluids and cuttings may have to be handled as a hazardous waste. Disposal to a licensed hazardous waste landfill can add significant cost to the drilling project.

Drilling methods which use mud or water as a circulating medium produce large volumes of contaminated fluids. Using air as a circulating medium below the water table can also produce large volumes of ground water. If this water is contaminated, cost for transport and disposal may be greater than drilling costs. Where disposal of contaminated cuttings and drilling fluids is a major concern, auger or cable-tool drilling have an advantage over other drilling methods.

Lithology and Aquifer Characteristics

Lithology and aquifer characteristics are primary factors to take into consideration when selecting a drilling method. Some drilling methods are better suited to drilling certain types of soils and rocks than others, and the characteristics of the subsurface materials may preclude use of some methods altogether. Auger drilling can be used effectively only in unconsolidated and semiconsolidated materials. Augers cannot be used to efficiently drill most types of bedrock and they are also limited in their ability to penetrate boulders. However, large-diameter (24-inch and larger) bucket auger or digger auger rigs are effective in boulders.

Auger drilling equipment is also limited in its ability to drill below the water table, particularly in loose granular soils. With hollow-stem equipment, sand or gravel may flow upward inside the auger, and obtaining an adequate sample or installing monitoring well casing may be difficult. Special, retrievable, auger plugs are available for use in these materials.

Air- and mud-rotary drilling equipment can be used to advance a hole through most types of unconsolidated or consolidated materials. However, lost circulation zones are particularly troublesome unless casing is advanced as the hole is drilled. The use of dual-wall casing with center stem recovery drilling also reduces the severity of lost circulation problems.

Lost circulation zones are usually caused by fractures in rock or porous unconsolidated material. When such a zone is encountered, drilling fluids and cuttings may not return to the surface, creating several problems. When drilling with mud or water, the lost fluids and additives have to be replaced, often at considerable expense. Drilling progress can be slowed or stopped if water has to be hauled in by truck. Without circulation return, the hole is advanced "blindly," and the driller and geologist do not have cuttings to assess changes in lithology.

When a drill hole is advanced without circulation return for more than a few feet, the potential for stuck drill pipe exists. Cuttings that are not transported to the surface are carried into the lost circulation zone. When circulation is stopped to add a length of drill rod, the cuttings may fall back into the hole on top of the bit, and the entire string of drill pipe can become stuck in the hole. At depths greater than a few hundred feet, drill pipe can also get stuck by the process known as differential sticking. Drilling fluids moving through the wall of the borehole into the lost circulation zone press the drill pipe against the wall of the hole and it cannot be withdrawn.

In addition to lost circulation zones, boulders also represent a problem for drilling with conventional rotary methods. The boulders roll beneath the bit instead of being crushed. In some loose boulder zones, caving can occur to such an extent that a hole several feet in diameter can be created with an 8- or 10-inch bit.

With rotary drilling, caving and lost circulation in unconsolidated materials can be effectively overcome by advancing casing closely behind the bit. Badly caving zones or lost circulation zones can also be cemented and redrilled, and additives can be mixed with drilling fluid to attempt to bridge lost circulation zones.

When using air as a circulating fluid instead of mud or water, the anticipated aquifer yield is an important factor in selecting a drilling method. In high-yield aquifers, drilling with air rotary more than a few feet below the water table is impractical in many situations. The large volumes of water are difficult to contain and control at the surface. In unconsolidated materials, the water moving up the borehole will erode the walls unless casing is installed. However, in low-yield materials, a hole can be advanced several hundred feet below the water table without producing excessive volumes of water.

Air can be used as a circulation fluid below the water table in some high-yield aquifers with center stem recovery drilling methods. The dual-wall drill pipe that is advanced with the borehole seals off most inflow, and water can only enter the hole through the bottom. However, in extremely productive aquifers, the rate of inflow through only the end of the casing may eventually become unmanageable as the hole is advanced below the water table.

Depth of Drilling

For most monitoring well installations, depth is usually not a major consideration in selecting a drilling method. Rotary and cable tool methods have both been used to drill to depths of several thousand feet. Auger drilling is generally effective only at depths up to about 150 feet, although hollow stem augers have been used to drill to depths of more than 400 feet under favorable conditions. Bucket augers or digger augers are generally limited to depths of less than 100 feet because of the length of the kelly bar. The main limitation for drilling deep holes with continuous flight or hollow-stem auger equipment is the friction between the auger flights and the wall of the hole.

Depth limitation may be encountered with some other drilling methods under certain conditions. In dry unconsolidated materials, casing hammers are most effective at depths of less than about 200 feet. At greater depths, the penetration rate decreases because of the increased friction between the casing and the wall of the hole. Telescoping casing strings can be used to overcome this limitation to a certain extent.

Sample Type

Collection of soil or rock samples is one of the most important considerations in selecting drilling methods. For collecting relatively undisturbed samples of unconsolidated materials, auger drilling with hollow-stem augers is the method of choice. No fluids are introduced into the hole, and samples remain relatively uncontaminated by the drilling process. The center stem recovery method also allows collection of relatively undisturbed samples in unconsolidated materials, but if air is used as a circulation fluid, the chemical characteristics of samples may be altered.

When drilling with conventional rotary methods in unconsolidated materials, collecting undisturbed samples is more difficult than with an auger rig. The drill rods and bit have to be "tripped out" so that sampling equipment can be lowered into the hole, and sampled materials may be chemically altered by drilling fluids. If the hole is filled with mud or water, some invasion of the sampled interval in the bottom of the hole will probably occur.

Undisturbed and unaltered samples of unconsolidated materials are nearly impossible to obtain with a cable-tool drilling rig. Materials several feet below the bottom of the hole are affected by the impact of the bit.

In consolidated materials, coring by conventional or wireline methods is a preferred method of obtaining relatively undisturbed samples, and most conventional or wireline equipment can be used with rotary drilling equipment. Obtaining undisturbed samples of rock with most cable-tool drilling equipment is impossible, although some hybrid equipment is available which can be used for both cable-tool and rotary drilling.

Many auger drilling rigs can be used for rock coring, and some soil and rock sampling programs can be carried out using only auger drilling equipment. Hollow-stem augers are used to advance the hole to bedrock, and the augers are left in the hole to serve as temporary casing while the hole is advanced using wireline coring methods. However, this method has some serious drawbacks for monitoring well construction. The augers provide an ineffective seal, and fluids can migrate relatively freely along the outer flights while the rock is being cored. Cross-contamination by drilling fluids or contaminated ground water can take place, and initial samples from the monitoring well may be unrepresentative.

Cost

Cost should be a factor in selecting methods for carrying out a drilling program. However, in addition to the footage or hourly rates associated with drilling, other costs should be considered. For example, the cost of obtaining and analyzing water samples from monitoring wells as part of a regularly scheduled sampling program can exceed the cost of well installation after only a few rounds of samples. If excessive time is required to obtain a sample because of improper drilling or development methods, or if samples are unrepresentative because contaminants were introduced into the aquifer during the drilling process, then the use of the "cheapest" drilling method may not provide the lowest overall project cost.

The cost of drilling, soil and rock sampling, and monitoring well construction usually includes the cost of a supervisory geologist or engineer. Therefore, faster drilling methods have a cost advantage which is not included in the drilling footage rate. Cable-tool drilling, which is slower than most other methods, can result in excessive overall costs of monitoring well construction if observation is provided.

An accurate comparison of costs between different drilling methods is difficult. Under some conditions, one method may have distinct cost advantages over others, but not for other conditions. For example, under certain conditions, air rotary can be an extremely fast and inexpensive method for drilling. In dry Cretaceous-age sandstones and siltstones in the Great Plains or on the Colorado Plateau, the cost of air rotary drilling can be relatively low (several dollars per foot) for a 5-inch diameter hole, and the penetration rate can be as high as 100 feet per hour. However, in a coarse-grained alluvium in the same area, an air rotary rig would have difficulty achieving a penetration rate of 5 feet per hour.

SOIL SAMPLING

Types of Samples

Four basic sample types are involved in monitoring and exploration work.

Bulk Samples are nothing more than a shovelful or handful of material taken from the spoil (return cuttings from the borehole); hopefully, they represent the formation penetrated by drilling. When taken, these samples are usually placed in containers and transported to the laboratory. This type of sample is not widely used in ground-water monitoring and generally would be considered the least accurate of the four basic types of sample.

Representative Samples are generally taken in some form of drive or push tube, and they represent a certain depth increment. For instance, if a hollow stem auger was advanced to 10-foot depth, the center bit removed and a 2-foot long tube lowered and driven for 18 inches, the recovered sample in the tube would be "representative" of the depth 10 to 11.5 feet. A "representative" sample is also a sample in which all the constituents are present, but not necessarily in an "undisturbed" form. Thus, a representative sample for monitoring purposes is a sample taken from a specific depth increment which contains all of the constituents present in the formation at that depth interval.

"Undisturbed" Samples are very high quality samples taken under strictly controlled conditions in order to minimize structural disturbance of the sample. The goal of undisturbed sampling is to sample all constituents of the formation without altering the presampling relationships between constituents in the sample. Every effort is made to avoid disturbing the sample in the sampling process. Undisturbed samples are generally limited to geotechnical work and are rarely taken for groundwater monitoring purposes.

"In situ" is a Latin term which roughly translated means "in place." For many years the term "in situ" was used to describe the highest possible quality of sample, since it was thought impossible to take a truly undisturbed sample. Recently, and with the advent of more tests actually made in place (in situ), the term undisturbed sample is preferred and is more accurate for those samples that are removed from a borehole in a relatively undisturbed fashion. The interchangeable use of the two terms (in situ and undisturbed) should be avoided.

Composite Samples are manufactured samples. They are a blend or mix of samples or material put together by some process other than taking all of the material that they represent. They might be a mix of bulk material or a half tube from one formation mixed with a part of a tube from another formation or a small part of each sample taken from the borehole mixed in such a way as to represent the total borehole.

Types of Samplers

Solid-Barrel Samplers consist of a tubular section which screws onto a connector head and on which is screwed a drive shoe. The length is generally between 12 and 60 inches and the diameter ranges from 1 to 6 inches. The sampler is normally made out of steel or stainless steel, and it may be used with thin-walled liners which can be slid into or out of the sampler barrel. Liners may be of brass, aluminum, stainless steel, or various synthetics. Allowable liner materials are generally determined by the engineer or geologist based upon the type of material that is being investigated and the types of tests and analyses that will be conducted.

Split-Barrel Samplers (Figure 6.7) are similar to the solid-barrel samplers except that the tubular section is split longitudinally into two equal semicylindrical halves. The split-barrel sampler may be used lined or unlined. The split-barrel sampler is also referred to as a split-spoon sampler.

The split-barrel sampler is generally available in 2-, 2.5-, 3-, 3.5-, and 4-inch outside diameters and it is by far the most commonly used sampler in both groundwater monitoring and geotechnical work. For geotechnical work, the sampler is often loaded with 12 1-inch "rings" on the bottom and a 6-inch "liner" on top. For monitoring work, three 6-inch liners are generally used to collect the samples.

Thin-Wall Tube Samplers (Figure 6.8) consist of a connector head and a 30-inch or 36-inch long thin-wall steel, aluminum, brass, or stainless steel tube which

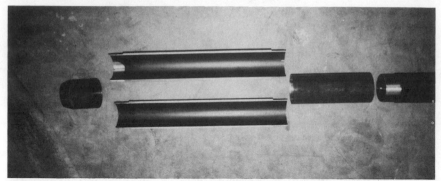

Figure 6.7. Split-barrel sampler. (Photo courtesy of Mobile Drilling Company.)

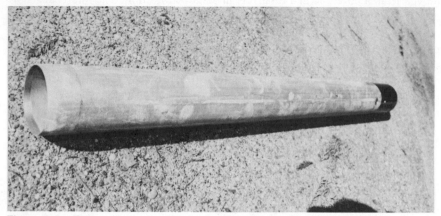

Figure 6.8. Thin-wall sampler. (Photo courtesy of Mobile Drilling Company.)

is sharpened or "rolled & reamed" at the cutting end and drilled for attachment to the connector head at the opposite end. This sampler is often referred to as a "Shelby tube" sampler. The original seamless tubing, used years ago, was made by U.S. Steel's Shelby process. Tubes today are generally made of electric welded drawn-over mandrel steel (EW-DOM) and are much stronger, straighter, and cheaper than the older type.

The thin-wall tube sampler is used primarily in soft or clayey formations where it will provide more sample recovery than the split barrel or solid barrel samplers and where relatively undisturbed samples are desired. The most commonly used thin-wall sampler tube west of the Rockies has a 3-inch outside diameter, is 30 inches long, and has a 2.81-inch cutting diameter. East of the Rockies, samplers with 2-inch outside diameters are more commonly used.

Common to all three of the samplers discussed above is the "connector head." The connector head is the connection between the sampler barrel or tube and the drill rod, hammer, or pressing sub that connects the sampler to the drill machine. For all three samplers, the connector head contains a valve (normally a ball valve) which allows air, water, or mud to escape through the connector head as the sampler is advanced to fill with soil. When sampler advance is stopped, the valve closes, and any effort of the sample to slip back upon sampler withdrawal causes a vacuum to occur in the sampler. The top of the connector head generally is machined as a box thread to accept the drill rod in use. The bottom of the connector head will be machined to allow attachment of the solid-barrel, split-barrel, or thin-wall tube.

Rotary Samplers such as the Pitcher or Denison samplers (Figure 6.9) can be very useful, exotic, and expensive when used in geotechnical work. For monitoring work, their usefulness is limited because they require the use of circulation fluids. However, there are forms of the rotary sampler in use in monitoring wherein the hollow-stem auger rotating around a nonrotating sampler does the scavenging work of circulation fluid. It is described in more detail below.

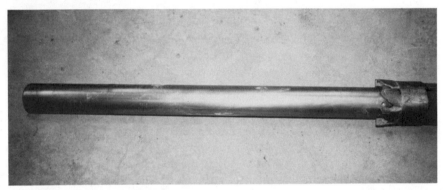

Figure 6.9.   Rotary sampler "Pitcher type." (Photo courtesy of Pitcher Drilling Company.)

Piston Samplers typically consist of a thin-wall tube, piston, and mechanisms for regulating movement between the tube and the piston. They are expensive, precisely made, and are not widely used outside of geotechnical work. A piston sampler of the fixed-piston design is shown in Figure 6.10.

To obtain a sample, the fixed piston sampler is lowered to the bottom of the hole on drill rods with a fixed piston flush with the lower end of the sampler tube. Pressurized fluid actuates the driving piston, forcing the tube into the soil while the fixed piston remains stationary.

The modified California sampler is a mechanical retractable piston sampler developed by the California Division of Highways in the 1940s. It was derived from a 2-inch California sampler that was developed in the 1920s. The design of the 2-inch sampler was in turn derived from a 1-inch plug sampler developed for agricultural use in 1915 by a student at the University of California at Davis. Split-barrel samplers are sometimes inaccurately referred to as California samplers. The confusion apparently arises from the fact that many split-barrel samplers

Figure 6.10. Piston sampler. (Photo courtesy of Techniquest Associates.)

use a liner with a 2.5-inch outside diameter, which is the same as the outside diameter of the 2-inch modified California sampler. However, the inside diameters are different.

Methods of Sampling

Samples in monitoring are generally taken from the bottom of the borehole although samples of cuttings (bulk samples) are taken from the top. For example, if the sample is to be taken from 12.5 feet to 14 feet, the hole is drilled to 12.5 feet. After that sample is taken, the borehole is advanced to the next sampling depth and the sampling process is repeated. Under some circumstances, continuous samples are obtained, wherein the sampling objective is to incrementally obtain a continuous column of the formation from top to bottom of the hole. Driving (hammering) or pressing is the preferred method for most samplers.

Driving (hammering) is the most common method of obtaining split-spoon samples. For most sampling, a 140-pound safety hammer with enclosed impact surfaces is used. The 140-pound safety hammer is connected to the sampler by drill rods and the hammer remains at the ground surface.

In-hole sampling hammers are sometimes used. They are connected directly to the sampler, lowered down the hole on wireline, and operated by a "free-fall" hoist. When the sampler has been driven to the desired depth, it is removed by hydraulic ram pullback or "bumped back" by reverse action of the hammer. When "bumping-back," losing the sample by jarring it out of the sampler is always a possibility.

With either type of hammer, a record of the number of hammer blows that are needed to advance the sampler provides useful information on the type of material that is being penetrated. Blows are usually counted for each 6-inch increment of the total drive, so for an 18-inch drive, three numbers are recorded. If the sampler cannot be advanced 6 inches with a reasonable number of blows (usually 50 or so), "sampler refusal" occurs and the sampling effort at the particular depth is terminated. If "auger refusal" has not occurred, the hole can be advanced for another attempt.

Pressing or pushing of samplers is done by using the drilling machine's hydraulic feed ram(s). For safety, a pressing sub should be used to connect the drill to the drill rods. The sub also allows the use of the machine's hydraulic retract system to withdraw the sampler from the hole. Pressing is the normal mode of advance for the thin-wall tube sampler (Shelby tube) and is sometimes used for split-barrel or solid-barrel samplers.

Rotation While Pressing is a sampling technique which uses hollow-stem augers and a sampling tube that is advanced with the augers (Figure 6.11). This method was briefly described earlier, but it is not a typical rotary method. The augers rotate, but the sampler tube is advanced without rotation. The sampler tube is lowered into the augers and retrieved from the surface, using wireline or drill rods. Typically, that sampler is lowered and locked into place so that it protrudes slightly ahead of the auger cutter head. Augering then begins in a routine manner and the sampler is pressed into the formation. After the hole has been advanced a distance equal to the length of the sampler, the augering is stopped and the full sampler is brought to the surface by wireline or drill rods, and an empty sampler is lowered into the hole. With wireline samplers, a double row ball/thrust bearing assembly in the head of the sampler prevents rotation of the sampler as the augers are rotated about the sampler. When drill rods are used to control the sampler, the rods and the sampler on the lower end of the rods are advanced without rotation at the same rate that the augers are advanced with rotation.

Figure 6.11.   Rotation while "pressing." (Photo courtesy of Mobile Drilling Company.)

ROCK CORING

Introduction

The purpose of rock coring is to obtain undisturbed (i.e., intact or unbroken) samples of consolidated, semiconsolidated, or, rarely, unconsolidated rock materials by use of traditional diamond-bit drilling methodology. It is used in subsurface exploration for dams, tunnels, and power plants, on rock foundations, mineral deposits in rock, quarry rock materials and in investigations in cement consolidated materials and other investigations where intact rock samples are desirable. Holes can be drilled in any configuration from vertical to horizontal. The diamond drilling can be used to retrieve a continuous sample, showing the characteristics of the entire interval that was drilled, or it can be used to obtain ''spot'' cores from selected intervals.

Samples obtained by diamond drilling may be tested for standard penetration permeability, porosity, and specific retention. The types of drill rigs and penetration equipment used in rock core drilling are diverse. Figure 6.12 illustrates a typical rock core sampler utilized in the rotary drilling process. Types of core tubes vary. Standard core tubes are attached to the end of drill rod and the entire string of drill rods, core tube, and bit are removed from the core hole following each core run. Wireline core tubes use an inner barrel which is extracted through the drill string with a special overshot device that is lowered into the drill string on a wireline.

Figure 6.12. Rock core sampler.

Rock core drilling with diamond bits is done with circulation fluid which is usually air or water, and loss of fluid circulation is a very common occurrence. The lack of return water may indicate that rock cuttings are accumulating either in openings in the rock mass or in the annular space between the drill rods and the inside of the core hole. If cuttings are accumulating in voids, the results of any permeability tests, geophysical logging, or other down-hole measurements may be affected. If cuttings are accumulating in the annulus, or if they cannot be retrieved, reentering the hole may be difficult.

Other coring problems commonly arise when the rock mass surrounding the drill hole is not self-supporting. Fragments of rock that protrude into the hole above the bit during drilling may retard or prevent extraction. The protruding material may also cause rapid abrasion of the sidewalls of the bit, the core barrel, or the drill rods. Sidewall caving that occurs after the rods have been withdrawn may also prevent the barrel from reaching the bottom of the hole for the start of the next run.

In general, holes that are small in diameter and are close to vertical are less susceptible to caving than large diameter holes or holes that are drilled at lower angles. The main techniques for overcoming caving are: (1) use of high density drilling mud; (2) installing casing in the unstable interval and reducing hole size below the casing; and (3) cementing the hole and redrilling after the cement has set. The choice of techniques depends on the nature of the investigation and the severity of the problem.

Another problem frequently encountered in rock coring is rapid bit wear during drilling of highly abrasive rocks. However, excessive bit wear is also common in rocks that are not particularly abrasive. Impregnated diamond bits have largely replaced surface set diamond bits. They are cheaper, far more robust, resist impact better, and handle broken formations far more effectively than surface set bits. They are made for most standard sizes and types of core barrels. They are *not* standardized between manufacturers and thus the user is advised not to attempt interchanges without consulting the maker. The individual diamond particles in an impregnated bit are actually in the matrix as opposed to setting on the matrix as is the case with a surface set bit. As an impregnated bit wears, more of the diamond is exposed to do the cutting. As a surface set bit wears, the 'hold' on the stone loosens and finally the stone drops into the formation.

Impregnated bits can be used almost anywhere that surface set bits can be used and, in fact, impregnated bits now account for about 75% of the market.

Core Losses

During diamond core drilling, core losses frequently occur. Core losses in relatively unconsolidated materials are frequent and may be caused by erosion of the core during circulation of the fluid down the core barrel. These losses can be minimized if: (1) bottom discharge bits are utilized; (2) the volume of the water is kept to the minimum that is required to remove the drill cuttings from the core hole; (3) vibration and chatter of the drill rods are minimized; (4) the speed and rotation and rate of advancement of a drill rod are controlled; and (5) appropriate core catching devices are utilized at the junction between the bit and the core barrel. Basket or spring retainers are generally used to prevent core loss.

Core losses in relatively consolidated materials can occur when the rock that is being cored is highly fractured and broken or when a fragment of the rock becomes wedged in a portion of the core bit or barrel. If a fragment becomes wedged, the only practical solution is to retrieve the core barrel and remove rock fragment at the surface. Blockages are usually apparent when the sudden increase in pressure of the drilling fluid or prevention of downward advance of the drill string is observed.

Frequently a dropped core occurs when a segment of core drags out of the barrel and back into the hole. At times this can be recovered on the next core run. If the loss is written off on an initial run and then the dropped core is picked up on a subsequent run without being noticed, an error will be introduced into the record of the depth of core. Therefore, care should be taken on every core run to determine: (1) percent recovery; (2) amount and location of core loss; and (3) actual depth of the beginning and end of the core run in each case. On runs subsequent to dropping of core, the risk of overfilling the core barrel is present.

Rock Coring Logs

The purpose of the core log is to record all relevant information obtained during the drilling and coring and to record a field description of the core. Core log forms are generally more specialized than the standard drilling log sheets and generally will contain columns for recording percent core recovery, rock quality designation (RQD), and number and orientation of fractures (Figure 6.13).

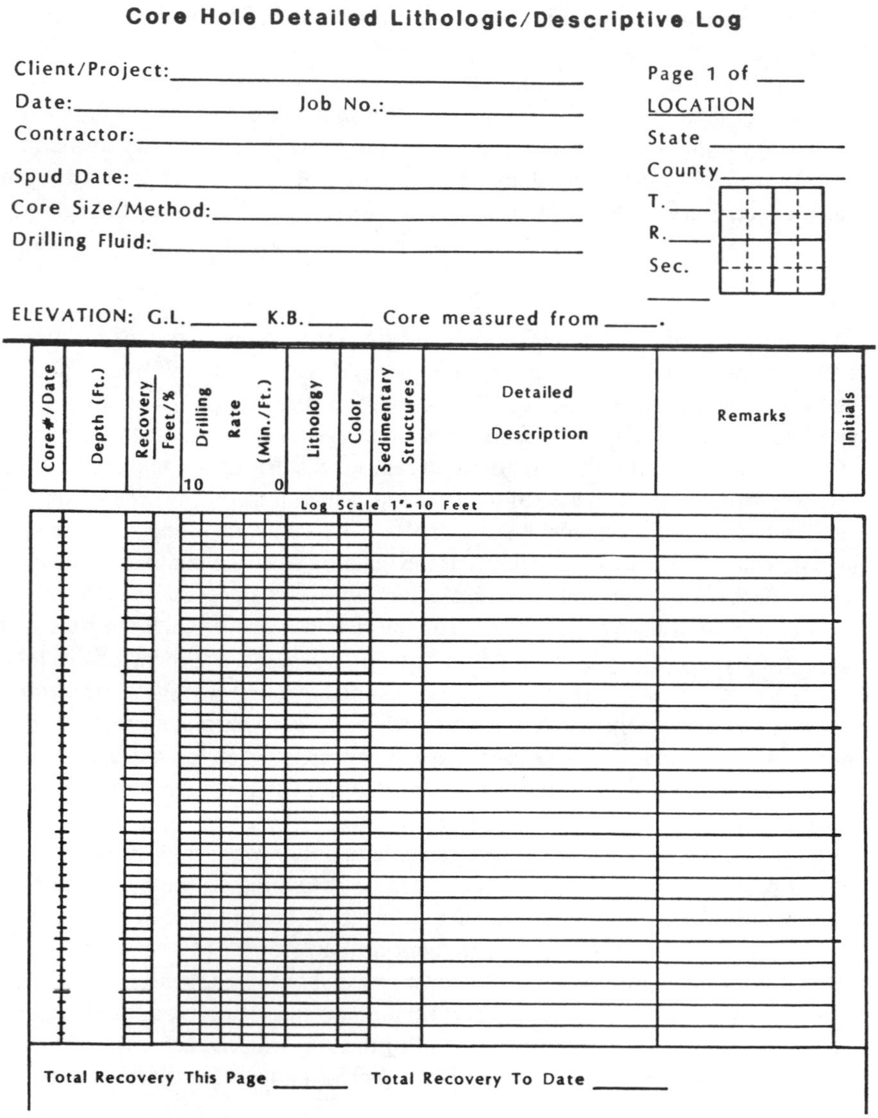

Core Hole Detailed Lithologic/Descriptive Log

Client/Project:_____   Page 1 of ____

Date:_____ Job No.:_____   LOCATION

Contractor:_____   State _____

Spud Date:_____   County_____

Core Size/Method:_____   T. ____

Drilling Fluid:_____   R. ____

Sec. ____

ELEVATION: G.L. _____ K.B. _____ Core measured from _____.

Core #/Date	Depth (Ft.)	Recovery Feet/%	Drilling Rate (Min./Ft.)	Lithology	Color	Sedimentary Structures	Detailed Description	Remarks	Initials
		10 0					Log Scale 1"=10 Feet		

Total Recovery This Page _____   Total Recovery To Date _____

Figure 6.13. Core log form.

The percent recovery should be calculated and recorded for each core run. Percent recovery is computed by dividing the length of recovered core by the length of the core run and multiplying by 100. Unrecovered core from one core run that is recovered on the subsequent run should be included in the run in which it was originally cored for the sake of recovery and other measurements. Such occurrences should be noted on the boring log.

RQD should also be calculated and recorded, for although the measurement was originally developed for geotechnical work, RQD is also a useful measurement for monitoring and sampling. RQD is calculated by measuring the total length of all sound pieces of core 4 or more inches in length, dividing this total length by the length of the core run, and multiplying the result by 100. Both the total length of all sound core pieces 4 inches or longer and the percent RQD are usually recorded on the log. Table 6.1 lists percent RQD and a qualitative description of rock. In general, the higher the RQD percent, the higher the qualitative value of the rock.

Table 6.1. Rock Quality Terms.

Percent RQD	Descriptive Rock Quality
0– 25	Very poor
25– 50	Poor
50– 75	Fair
75– 90	Good
90–100	Excellent

HANDLING PROCEDURES FOR SOIL AND ROCK SAMPLES

Introduction

The purpose of soil and rock sampling is to collect representative samples of materials in a horizon of interest. A representative sample is assured by minimizing the possibility of errors associated with each phase of the sampling process, including collecting the sample, transferring the sample from the sampler to a container, and transporting the sample to the laboratory or storage area. The need for complete, accurate, written documentation of each step of this process cannot be overstated.

Each sample should be properly labeled and identified according to a sample number, project number (if applicable), sample depth (with datum) to correspond with a sample number and depth on the borehole log. The sample should be properly sealed and packed according to the type of sample and, if practical, the sample and the container should both be labeled. This procedure will reduce the possibility of misidentification of the samples if they are removed from the container in a laboratory. Examples of soil sampling packaging are the Shelby tube, the liners of sampling devices, glass jars, conventional plastic bags, and other more elaborate containers such as core boxes, core tubes, and specialized wax containers. Some core samples may be waxed in the field for preservation of volatile organic compounds or to preserve in situ pore water volume.

The samples should be properly packaged and sealed prior to shipping and the shipping container should be appropriately labeled. Care should be taken so that the samples remain as undisturbed as possible during the shipping process. Prior to packaging and shipping, the samples frequently are photographed with information such as depth, sample number, borehole, and project number visible within the photograph. Then, if the samples get mixed up at any point during the shipping or handling process, identification may be possible through the photographs.

Chain of custody forms (Figure 6.14) are frequently utilized to provide a method for identifying every individual who has custody of samples. The sampler signs the custody of the sample to the person who is responsible for the shipping, who then signs it over to the individual responsible for storage or laboratory analysis. The chain-of-custody form indicates the samples by number, time of pickup, time of delivery, person samples were received from, person delivering samples, person receiving samples, and condition of samples at pickup and at delivery.

Rock Core

After the core has been removed from the core barrel, it should be inspected, logged, and placed into the shipping and storage containers. These may be one of several types of rigid containers with lids. Rock core is relatively fragile and represents a significant investment of time and money. It deserves special labeling, handling, and storage procedures.

Suitable containers for storing and shipping core are core boxes or rigid PVC tubes. Core boxes can be constructed from aluminum, stainless steel, wood, and reinforced plastic. Cardboard is generally not suitable except for temporary or emergency storage. Before it is placed in the core box, core may be waxed or it may be placed in a clear plastic tubing to preserve the moisture content.

As core is placed into a core box, it should be labeled by writing directly on the core. Information should include top and bottom, depth, position of core loss zones, and identification of fractures. Fractures that were made after the core was removed from the core barrel should be distinguished from fractures that are interpreted as in-place fractures.

Core boxes should also be labeled on the tops and ends. Information should include the core run number, depth of core, data, hole number, project name and number, and the total depth of core in the box. Wooden blocks are usually used as spacers inside core boxes to mark ends of core runs and the positions of core loss zones. These should also be labeled with the depth.

Core Custody Log

Person Taking Custody: _____

Date: _____

Time of Possession From: _____   To: _____

Total No. of Core Tubes Moved This Load: _____

Core Moved From: _____

                            To: _____

Person Relinquishing Custody: _____

Checked Core Being Moved (Initial): _____

Core Moved

Tube No.	Depth From:	To:
————		
————		
————		
————		
————		
————		
————		
————		
————		
————		
————		
————		
————		
————		
————		
————		
————		
————		
————		
————		
————		
————		
————		
————		
————		
————		
————		
————		
————		

Comments:

Figure 6.14. Chain of custody form.

The inside of the core box lid can also be labeled to show core orientation, depths, and other significant features. Figure 6.15 shows an example of core stored in a core box.

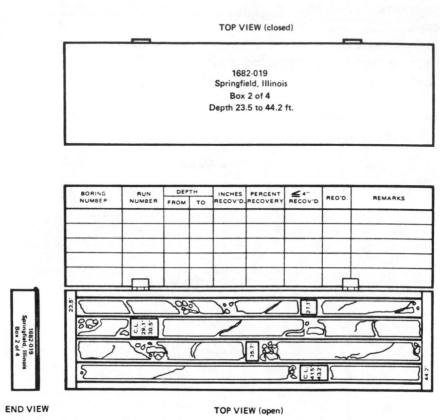

Figure 6.15.   Arrangement of rock core and labeling of rock core boxes.

Cuttings or Disturbed Samples

Procedures for handling, labeling, and packaging samples of cuttings or bulk samples are similar to those used for cores. After the lithologic characterization is completed, the samples are generally packaged and labeled according to depth, project name and number, boring or well number, location, and date. Cuttings or disturbed samples can usually be labeled on the sample container only, and the container becomes part of the label. If the sample is separated from the container, identification is usually impossible.

Cloth bags, plastic bags, and jars are the containers that are most commonly used for storing and shipping cuttings or disturbed samples labeled with the depth and the sample number. Sometimes cuttings or disturbed samples are split, the original sample is maintained in the original container, and the split sample is shipped to a laboratory for analysis. Once a sample is split, the original sample container should be identified as holding a split sample. A more detailed discussion is available as ASTM Method D-4220.

BOREHOLE LOGGING

Introduction

Borehole logs are the written reports prepared by the field geologist or engineer and the drilling contractor. These reports are prepared onsite as the boring is advanced. They provide a record of the drilling, sampling, and well construction procedures, and they are often the sole record of the significant occurrences that occurred during field work.

Every drilling, sampling, and monitoring project is unique, and numerous forms for logging are widely used. To try to describe them all is impossible. However, certain types of information are common to most projects and most boring logs. These are covered in the following sections.

Log Header Information

The log header is the top part of most borehole log forms. It usually includes, as a minimum, the project name, location, surface elevation, surface conditions, type of drill rig or pumping equipment, bit size and type, geologist's name, boring location, and project number. Figure 6.16 shows an example of a geologist's boring log. It is not necessary that the second sheet of the borehole log contain the same header information. It is imperative that the headers of each of these sheets be particular to the project. As an example, a standard drilling form will be different than a rock coring form, which will in turn be different from a pumping test or water-quality form. It may also be important to record information such as water level and datum, casing diameter, depth of hole, casing stickup, casing material, drilling methodology, and sampling methodology.

Log Completion Information

Log completion information should include the time and date that drilling started, the time and date of completion, total depth drilled, total depth cased, abandonment procedures, if any, and final water-level measurement, among other important data.

LOG OF WELL _____   SHEET _1_ OF _____

PROJECT AND LOCATION					ELEVATION AND DATUM		PROJECT NO
DRILLING AGENCY					DATE STARTED		DATE FINISHED
DRILLING EQUIPMENT				ROD LENGTH	COMPLETION DEPTH		ROCK DEPTH

		SIZE	DEPTH	SIZE	DEPTH	NO. SOIL SAMPLES	METHOD
BIT		SIZE	DEPTH	SIZE	DEPTH	NO. WATER SAMPLES	WATER LEVEL
CASING		SIZE	DEPTH	SIZE	DEPTH	FOREMAN	
		SIZE	DEPTH	SIZE	DEPTH		
SCREEN SETTING		SIZE	DEPTH	SIZE	DEPTH	INSPECTOR	
		SIZE	DEPTH	SIZE	DEPTH		

DEPTH FT.	ELEV. FT.	DESCRIPTION	CONDUC-TIVITY µ mhos/cm	DEPTH FT.	SAMPLE NO	TYPE	RECOV FT	PENETR. RESIST. BL/6 IN	CASING	REMARKS

Figure 6.16. First page of a boring log.

Sample Information

Sample information is recorded as the hole is advanced and samples are collected. The types of information that should be recorded on the boring log are:

- sampling method; for example, bulk samples of cuttings, relatively undisturbed drive samples, or core samples;

- number of blow counts required to advance a soil sampler. These are usually recorded for each 6-inch interval;
- size and type of sampler. This information may be recorded only once on the log header, but if different types of samplers are used in the same hole, the type of sampler used for each sample should be recorded;
- sample number and depth. Sample depth and all other depths recorded on the boring log should always be measured from the ground surface unless a very compelling reason to do otherwise is present. If a datum other than ground surface is used, it should be recorded;
- length of drive, push, or core run. In soils, sampling attempts are usually in the range of a few inches to about 3 feet, depending on the hardness of the soil. Core runs in rock are usually 5 to 20 feet long;
- length of the recovered sample and the calculated percent recovery. Percent recovery is an important measure of the degree to which the sample is representative of actual subsurface conditions, and most boring log forms have a separate column for recording the value;
- rock quality designation, or RQD; and
- portion of sample saved or submitted to a laboratory for analysis. All of a rock core is frequently saved because it usually stores well and is relatively expensive to obtain. Some or all of a soil sample may be discarded.

Soil and Rock Descriptions

Describing soil and rock samples in the field is an art as much as it is a science. The time available to examine each sample is usually limited, lighting conditions are variable, and it may be raining, snowing, or 100° in the shade. In contaminated areas, the field personnel may be working in protective clothing, including gloves and respirators. Nevertheless, no one is better prepared to provide sample descriptions than the field person who has witnessed the entire process of drilling and collecting the soil sample.

Several methods for classifying and describing soils or unconsolidated sediments are in relatively widespread use. The Unified Soil Classification System (USCS) may be the most popular, and the guidelines for its use are reproduced in Figure 6.17. With the USCS, a soil is first classified according to whether it is predominantly coarse-grained or fine-grained.

Coarse-grained soils are then further subdivided according to the predominance of sand and gravel. Fine-grained soils are subdivided on the basis of the liquid limit and the degree of plasticity. The accurate identification of silts and clays can be aided by the use of some simple field tests. Clay is sticky, will smear readily, and can be rolled into a thin thread even when the moisture content is low. When it is dry, clay forms hard lumps. On the other hand, silt has a low dry strength, can be rolled into threads only at a high moisture content, and a wet silt sample will puddle when it is tapped.

In some respects, rock classification can be more difficult than soil classification. A nearly infinite number of rock lithologies have been recognized and named,

GRADATION CHART

PARTICLE SIZE

MATERIAL SIZE		LOWER LIMIT		UPPER LIMIT	
		MILLIMETERS	SIEVE SIZE*	MILLIMETERS	SIEVE SIZE*
CLAY SIZE		.005		.074	#200#
SILT SIZE		.074		.42	#40#
SAND	FINE	.074		.42	#40#
	MEDIUM	.42	#40#	2.00	#10#
	COARSE	2.00	#10#	4.76	#4#
GRAVEL	FINE	4.76	#4#	19.1	3/4"
	COARSE	19.1	3/4"	76.2	3"
COBBLES		76.2	3"	304.8	12"
BOULDERS		304.8	12"	914.4	36"

* U.S. STANDARD   • CLEAR SQUARE OPENINGS

PLASTICITY CHART

FOR LABORATORY CLASSIFICATION OF FINE-GRAINED SOILS

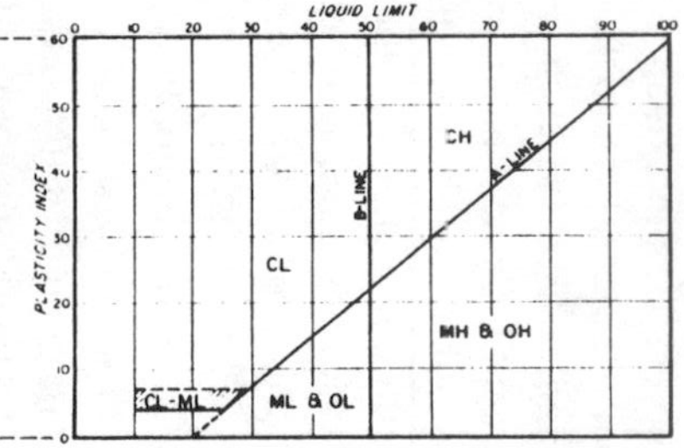

(Liquid Limit horizontal axis 0–100; Plasticity Index vertical axis 0–60. A-LINE and B-LINE shown. Regions labeled: CL-ML, ML & OL, CL, CH, MH & OH.)

SOIL CLASSIFICATION CHART

MAJOR DIVISIONS			GRAPH SYMBOL	LETTER SYMBOL	TYPICAL DESCRIPTIONS
COARSE GRAINED SOILS	GRAVEL AND GRAVELLY SOILS	CLEAN GRAVELS (LITTLE OR NO FINES)		GW	WELL-GRADED GRAVELS, GRAVEL SAND MIXTURES, LITTLE OR NO FINES
				GP	POORLY-GRADED GRAVELS, GRAVEL SAND MIXTURES, LITTLE OR NO FINES
	MORE THAN 50% OF COARSE FRACTION RETAINED ON NO. 4 SIEVE	GRAVELS WITH FINES (APPRECIABLE AMOUNT OF FINES)		GM	SILTY GRAVELS, GRAVEL-SAND-SILT MIXTURES
				GC	CLAYEY GRAVELS, GRAVEL SAND-CLAY MIXTURES
MORE THAN 50% OF MATERIAL IS LARGER THAN NO. 200 SIEVE SIZE	SAND AND SANDY SOILS	CLEAN SAND (LITTLE OR NO FINES)		SW	WELL-GRADED SANDS, GRAVELLY SANDS, LITTLE OR NO FINES
				SP	POORLY-GRADED SANDS, GRAVELLY SANDS, LITTLE OR NO FINES
	MORE THAN 50% OF COARSE FRACTION PASSING ON NO. 4 SIEVE	SANDS WITH FINES (APPRECIABLE AMOUNT OF FINES)		SM	SILTY SANDS, SAND-SILT MIXTURES
				SC	CLAYEY SANDS, SAND-CLAY MIXTURES
FINE GRAINED SOILS	SILTS AND CLAYS	LIQUID LIMIT LESS THAN 50		ML	INORGANIC SILTS AND VERY FINE SANDS, ROCK FLOUR, SILTY OR CLAYEY FINE SANDS OR CLAYEY SILTS WITH SLIGHT PLASTICITY
				CL	INORGANIC CLAYS OF LOW TO MEDIUM PLASTICITY, GRAVELLY CLAYS, SANDY CLAYS, SILTY CLAYS, LEAN CLAYS
				OL	ORGANIC SILTS AND ORGANIC SILTY CLAYS OF LOW PLASTICITY
MORE THAN 50% OF MATERIAL IS SMALLER THAN NO. 200 SIEVE SIZE	SILTS AND CLAYS	LIQUID LIMIT GREATER THAN 50		MH	INORGANIC SILTS, MICACEOUS OR DIATOMACEOUS FINE SAND OR SILTY SOILS
				CH	INORGANIC CLAYS OF HIGH PLASTICITY, FAT CLAYS
				OH	ORGANIC CLAYS OF MEDIUM TO HIGH PLASTICITY, ORGANIC SILTS
	HIGHLY ORGANIC SOILS			PT	PEAT, HUMUS, SWAMP SOILS WITH HIGH ORGANIC CONTENTS

NOTES:

1. DUAL SYMBOLS ARE USED TO INDICATE BORDERLINE CLASSIFICATIONS
2. WHEN SHOWN ON THE BORING LOGS, THE FOLLOWING TERMS ARE USED TO DESCRIBE THE CONSISTENCY OF COHESIVE SOILS AND THE RELATIVE COMPACTNESS OF COHESIONLESS SOILS

COHESIVE SOILS
(APPROXIMATE SHEARING STRENGTH IN KSF)

VERY SOFT	LESS THAN .25
SOFT	.25 TO 0.5
MEDIUM STIFF	0.5 TO 1.0
STIFF	1.0 TO 2.0
VERY STIFF	2.0 TO 4.0
HARD	GREATER THAN 4.0

COHESIONLESS SOILS

VERY LOOSE
LOOSE
MEDIUM DENSE   THESE ARE USUALLY BASED
DENSE   ON AN EXAMINATION OF SOIL
VERY DENSE   SAMPLES, PENETRATION
RESISTANCE, AND SOIL
DENSITY DATA

SAMPLES

■ INDICATES UNDISTURBED SAMPLE
⊠ INDICATES DISTURBED SAMPLE
☐ INDICATES SAMPLING ATTEMPT WITH NO RECOVERY

[INDICATES LENGTH OF CORING RUN

NOTE: DEFINITIONS OF ANY ADDITIONAL DATA REGARDING SAMPLES ARE ENTERED ON THE FIRST LOG ON WHICH THE DATA APPEAR.

Figure 6.17. Unified Soil Classification System.

and textbooks and college courses are devoted to the identification of rocks. However, for most ground-water monitoring work or contamination investigations, a graduate degree in petrography is not necessary and would probably be a hindrance. In many instances, lithology is not the most important factor in evaluating the hydrogeologic aspects of rock, athough lithology provides important clues. Features such as weathering, fractures, bedding planes, and porosity are usually more significant, and these can be easily recognized with a moderate degree of experience and training. Two excellent field guides for use in the identification and naming of rocks are the *AGI Data Sheets,* which are published by the American Geological Institute, and Compton's *Manual of Field Geology.* Both of these are standard references in most geologist's libraries. (See also Tables 6.2 and 6.3.)

Local knowledge and experience is indispensable in assisting in the accurate classification, identification, and logging of soil and rock. Whenever possible, field personnel should try to review existing logs prior to drilling in a new area. On drilling projects in areas where existing information is limited, local experts from a geological agency or a college may be willing to examine and describe samples.

Table 6.2.   Degree of Weathering.

Term	Description
Fresh	The rock shows no discoloration, loss of strength, or any other effect of weathering.
Slightly Weathered	The rock is slightly discolored, but not noticeably lower in strength than the fresh rock.
Moderately Weathered	The rock is discolored and noticeably weakened, but 2-inch diameter drill cores cannot usually be broken up by hand, across the rock fabric.
Highly Weathered	The rock is usually discolored and weakened to such an extent that 2-inch diameter cores can be broken up readily by hand, across the rock fabric. Wet strength usually much lower than dry strength.
Extremely Weathered	The rock is discolored and is entirely changed to a soil, but the original fabric of the rock is mostly preserved. The properties of the soil depend upon the composition and structure of the parent rock.

Table 6.3.   Bedding Characteristics.

Average Bed Thickness	Term
<0.001 foot	Thinly laminated
0.001 to 0.01 foot	Laminated
0.01 to 0.1 foot	Thin bedded
0.1 to 1.0 foot	Medium bedded
>1.0 foot	Thick bedded

Drilling Information

A record of drilling operations provides valuable documentation for possible contract disputes involving the driller, and a record of drilling conditions can help identify changes in subsurface conditions. A record of drilling information also helps prepare cost estimates and schedules for later drilling and sampling.

Drilling information should be recorded separately from soil or rock descriptions as the hole is advanced. A separate column should be used on the boring log form, and entries should be made at a position that indicates the depth. For emphasis, the depth for the entry may also be written with the entry. Types of information that are most useful include the following:

- changes in penetration rate, rig noise, or drilling action.
- interruptions in drilling, including breakdowns for repairs, "trips" for changing bits, interruptions to install casing, or interruptions to mix drilling fluids. Bit and casing diameter should be recorded.
- the condition of the circulation fluids and cuttings, including loss of circulation, appearance of water in cuttings from auger drilling or rotary drilling with air, and change in color and presence of odors. If water is uncounted in a hole drilled with air, the rate of water production should be estimated and recorded.
- penetration rate and pulldown. Many rotary rigs are equipped with instruments which automatically record penetration rate. If one is not available, the field person can estimate the rate in feet per minute or, in difficult drilling conditions, minutes per foot.
- changes in drilling personnel. For deep monitoring wells, drilling rigs are sometimes operated 24 hours a day, and two or three separate crews may be involved in drilling a single hole.

If a boring is converted to a monitoring well, a separate form, usually called a well completion log, is usually used to record the well construction information.

DRILLING CONTRACTS

Introduction

A contract is a binding agreement between two or more parties. A drilling contract is typically an agreement between a driller, who agrees to drill one or more wells or borings, and a customer, who agrees to pay the driller for the work. In the past, drilling customers have typically been landowners or municipalities who needed a water supply well. However, with monitoring wells, the customer may be a municipality, a consultant, the owner of the facility potentially responsible for ground-water contamination, or the state or federal government.

Thousands of wells have been drilled with nothing more than a verbal agreement and a handshake between the well driller and the customer, and many wells will continue to be drilled this way. However, because monitoring well drilling projects frequently involve the expenditure of large sums of money and require close adherence to technical specifications, some form of written agreement between the driller and the customer should be used. The written agreement may be as simple as a purchase order, or it may be a construction contract which is several hundred pages long.

Purchase orders are preprinted forms issued by a company or a government agency, and the terms of the agreement are typically printed on the back in "fine print." The purchase orders are heavily "loaded" in favor of the issuing company, and for this reason alone they should be reviewed carefully by the drilling contractor. They are not normally a completely satisfactory instrument for a drilling contract. The front of the purchase order is typically arranged in columns that are well suited to ordering specific quantities of materials or supplies, but the format is difficult to adapt to the purchase of drilling services without reference to additional technical specifications, proposals, or plans and drawings. Furthermore, the terms and conditions are usually not adequate to cover the contingencies that can arise during the construction of wells. However, in spite of their shortcomings, purchase orders are widely used because purchasing departments are comfortable with them. Preparing and executing a more formal contract can involve legal review and lawyer's fees.

The remainder of this section describes the important components of a comprehensive drilling contract such as one that might be prepared for a drilling program involving several monitoring wells. The purpose of this section is two-fold: (1) to eliminate some of the mystery involving the language in formal written contracts, and (2) to educate the driller, client, facility owner, and consultant on the important components of a standard drilling contract. The purpose of this section is not to demonstrate how to write contracts, but some of the information contained in this section should be useful to the geologist who is requested to assist in preparing a drilling contract.

The Agreement

The agreement is typically the first item which makes up the drilling contract documents. It identifies the parties, it describes the work in general terms, and it specifies the documents that are part of the contract. The first few lines of the agreement usually contain the date that the agreement is signed, and the last lines are the signature of the parties who enter into agreement.

The agreement may define the time of completion and the start date of work. It also usually states the contract sum. The agreement may include provisions for changes, termination, and method of payment. However, some of these items may be described in other documents which comprise this set of contract documents.

General Conditions

The general conditions contain the principal contract provisions that govern the parties involved. These provisions define the legal relationships between the various parties (i.e., owner, drilling contractor, subcontractors, and consultant).

The general conditions are the "fine print" of the contract. Many large companies that enter into numerous construction agreements typically have a carefully honed set of general conditions to accompany every construction contract. These conditions are usually heavily "loaded" in favor of the company which prepared them. A drilling contractor who accepts the standard general conditions issued by such a company should be prepared to consult an attorney for legal advice prior to signing the contract. Important clauses or articles in standard general conditions are described below.

- *Definitions* define the owner, the contractor, and subcontractors. These clauses describe the functions, rights, and obligations of each of these parties.
- *Indemnification* requires the contractor to indemnify and hold harmless the owner, his agents, and employees from and against claims, damages, losses, and expenses arising out of the negligence or omissions of the contractor, or the contractor's agents and employees.
- *Bonds* are required to protect the owner against the contractor's default so that in the event of the default, the wells can still be drilled.
- *Disputes* are usually required to be resolved by an arbitration process.
- *Time* for completing the project is usually stated in the formal agreement, but the method of measuring the time is usually stated in the general conditions.
- *Safety* requirements are described and the responsibility for initiating, maintaining, and supervising safety programs are usually placed on the contractor.
- *Changes* may consist of additions, deletions, or other revisions. The general conditions usually contain a clause describing the mechanism for changing the scope of work.
- *Corrections* of defective work are required under the terms of most general conditions. Defective work is corrected at the expense of the contractor.
- *Termination* provisions describe the situation under which either the owner or the contractor may terminate the contract.

Other clauses that may be included in the general conditions of the comprehensive construction contract include royalties and patents; subcontracts; permits, laws, taxes, and regulations; project representatives; warranty and guarantee; insurance; and cleanup.

Construction Drawings

Drawings are prepared in advance of bidding for those drilling projects that are competitively bid. Along with the specifications, they describe and locate the various elements of the drilling project. For most drilling projects, the drawings will consist of one or more maps showing the locations of the borings or wells, access, and utilities. In addition, construction or completion diagrams for all of the types of monitor wells to be drilled on the project should be included. These drawings will show the total depth of the well, the length of the screen or slotted section, the length of the blank casing, requirements for surface casing, and position of grouted or gravel-packed intervals.

The terms of some general conditions create parity among drawings and specifications. Other terms of general conditions establish priorities, so that one document prevails if the information between the two does not agree. For example, the drawings may show that the slot size on a well screen should be 0.020 inches, whereas the specifications state the slot size should be 0.040 inches. If the contract documents specify that the terms of the drawings prevail over the specifications, then the driller should base his bid on screen with the 0.020-inch slot size. If no priority is specifically stated, the drawings usually prevail in the event of disputes.

Specifications

Specifications provide technical information concerning materials, components, and equipment with respect to quality, performance, and results. Two types of specifications can be written. Procedure specifications specify in detail the quality, properties, and composition of materials. Procedure specifications may specifically state wall thickness for well casing, the size of gravel in the gravel pack, or the brand of well screen to be installed. On the other hand, performance specifications specify performance characteristics without stipulating the methods by which the characteristics are to be obtained. Performance specifications are not widely used for monitoring well drilling contracts.

The clause "or approved equal" is common in many procedure specifications. This clause allows the widest possible competition and permits drilling contractors to shop around for the best bargain when preparing their bid. Use of the phrase may cause controversy, however. It raises the question of how to measure and judge equality. If the clause is used, the specifications should include a statement explaining who will be responsible for approving substitute items.

Important components of the technical specifications in a drilling contract may include some or all of the following items:

- drilling methods
- driller's logs, geophysical logs, and driller's reports
- sampling methods and depth at which samples will be obtained
- casing installation
- grouting
- well screens
- gravel pack

- plumbness and alignment
- well development
- aquifer testing
- plugging requirements

Other clauses may also be included. The technical specification section is usually written by the consultant or geologist. Along with the drawings, the technical specifications play an important role in determining the final quality of the monitoring well construction. A complete set of drawings and technical specifications reduces the need for on-the-spot decisionmaking by the owner's representative. However, because of the unknown nature of the subsurface conditions on many monitoring well projects, the need for regular supervision to provide answers to construction questions cannot be completely eliminated.

Special Conditions

Some well-drilling contracts may contain special conditions that supplement and modify the general conditions. They are typically used where a set of standard general conditions for construction contracts are used for a drilling contract. Special conditions are written individually for each project and usually govern in the event of conflict with the terms in the general conditions.

The purpose of the special conditions overlaps somewhat with the purpose of the technical specifications. Important components of the special conditions document may include:

- scope and general description of work, including well size, depth, and location
- subsurface information, including lithology and depth to water table
- work schedule
- liquidated damages, or a fixed sum which is paid by the contractor for each day's delay in completion beyond an agreed-upon date
- special permits and taxes
- property boundaries
- location of existing utilities

Other Documents

Noncontractual documents such as instructions for bidders, advertisements for bids, proposals prepared by bidders, statutes, local rules and regulations, and well construction codes may or may not be specifically referenced in the contract documents. Nevertheless, they are used as guides in resolving contract disputes and in determining the legal obligations of the parties. The subsurface conditions and site inspection clauses in instructions to bidders are a frequent source of contract dispute for well-drilling contracts. In the instruction to bidders, the clauses typically require the bidder to represent that "he has visited the site and has familiarized himself with the local conditions." Disputes arise when the plans and specifications contain information concerning site conditions which differ substantially from those encountered during drilling. The information in the plans and specifications may have been obtained from earlier drilling, or it may represent a geologist's educated guess. Frequently, a clause is inserted in the general conditions which states that data containing subsurface conditions should not be relied upon for estimating drilling costs. The driller is expected to have sufficient experience in the area to judge drilling conditions and develop a realistic bid.

7

Design and Installation of Ground-Water Monitoring Wells

David M. Nielsen and Ronald Schalla

INTRODUCTION

For many waste disposal and chemical spill scenarios, installing ground-water monitoring wells to detect trace [i.e., parts per billion (ppb)] levels of both organic and inorganic contaminants in ground-water systems is a common practice. Tens or perhaps hundreds of thousands of monitoring wells have been designed and installed since the late 1970s and thousands more are installed each year, many by consultants and contractors unaware of correct monitoring well construction practices. As a result, many existing monitoring wells and some wells currently being installed have critical design flaws, or were or are being installed using methods and materials that may adversely affect subsequent ground-water sample chemical quality. The objective of most ground-water monitoring programs is to obtain "representative" ground-water samples that retain both the physical and chemical properties of the ground water. Proper ground-water monitoring well design and installation techniques are necessary to minimize potential chemical alteration of samples.

Most ground-water monitoring well design and installation problems can be traced back to a mistaken belief in a "cookbook" approach that ignores site-specific hydrogeologic, geographical, and contaminant-related conditions. The fact is that each site at which monitoring wells are installed is unique, requiring a unique design. The designer must develop well design and installation specifications that anticipate specific site conditions and accommodate changes caused by unanticipated geologic conditions.

Lack of a sufficient number of professionals adequately trained and experienced in proper monitoring well construction practices and procedures contributes to other ground-water monitoring well design and installation problems. Furthermore, modern analytical laboratory capability is now reaching the parts-per-trillion detection level, while our means of gaining subsurface access to obtain ground-water samples is comparatively crude. However, most potential sources of sample chemical alteration inherent in monitoring well design and construction can be anticipated and controlled. ·

The basic requirement for well design and installation is for workable, flexible guidelines adaptable to a wide variety of ground-water monitoring situations and usable by both consultants and contractors. One step toward developing such guidelines is identifying the most common problem areas in well design and construction. Among the most common monitoring well design flaws and installation problems are the following:

- use of well casing or well screen materials that are not compatible with the hydrogeologic environment, the anticipated contaminants, or the requirements of the ground-water sampling program, resulting in chemical alteration of samples or failure of the well;
- use of nonstandard well screen (i.e., field-slotted, drilled, or perforated casing) or incorrect screen slot-sizing practices, resulting in well sedimentation and the acquisition of turbid samples throughout the monitoring program;
- improper length and placement of the well screen so that acquisition of water-level or water-quality data from discrete zones is impossible;
- improper selection and placement of filter pack materials, resulting in well sedimentation, well screen plugging, ground-water sample chemical alteration, or potential well failure;
- improper selection and placement of annular seal materials, resulting in alteration of sample chemical quality, plugging of the filter pack and/or well screen, or cross-contamination from geologic units that have been sealed off improperly; and
- inadequate surface protection measures, resulting in surface water entering the well, alteration of sample chemical quality, or damage to/destruction of the well.

Any one or a combination of these monitoring well design and installation problems could cause a well or series of wells to be unsuitable for obtaining representative ground-water samples. Improperly designed or installed wells often must be abandoned and replaced, which is costly and time consuming. Thus, proper monitoring well design and installation practices are essential to ensure cost-effective acquisition of representative ground-water samples.

Proper design and installation of ground-water monitoring wells requires a thorough review of site-specific conditions, as well as an up-to-date knowledge of well design and installation practices and procedures. Site-specific design considerations include the following:

- purpose or objective of the ground-water monitoring program (i.e., water-quality monitoring versus water-level monitoring);
- surficial conditions, including topography, drainage, climate, seasonal variations in climate, and site access;
- known or anticipated hydrogeologic setting, including type of geology (unconsolidated/consolidated), aquifer physical characteristics (type of porosity, hydraulic conductivity), type of aquifer (confined/unconfined), recharge/discharge conditions, and ground-water/surface-water interrelationships;
- characteristics of known or anticipated contaminants, including chemistry, density, viscosity, reactivity, and concentration;
- anthropogenic influences (man-induced changes in hydraulic conditions); and
- any applicable regulatory requirements.

A unique set of site-specific design considerations exists for each site and for each individual well installation; this requires that each well be designed as a unique structure.

To develop a knowledge of proper monitoring well design practices, it is first necessary to understand the individual design components of monitoring wells and how they combine to produce the final structure—the well itself. While it is not practical to describe a "typical" monitoring well in which the design components are fixed, it is possible to describe each of the individual design components, which include the following:

- well casings
- well screens
- filter packs
- annular seals
- surface protection

These individual design components can be tailored and assembled to suit the site-specific considerations described above. Figure 7.1 illustrates the design components typical of most monitoring wells, and how they are assembled to produce a well.

Proper installation of ground-water monitoring wells requires a knowledge of state-of-the-art practices for well installation, both to avoid potential contamination of the borehole or well caused by the well construction process itself, and to permit easy access to the subsurface for ground-water sampling and water-level measurement. Proper application of joining and placement techniques for

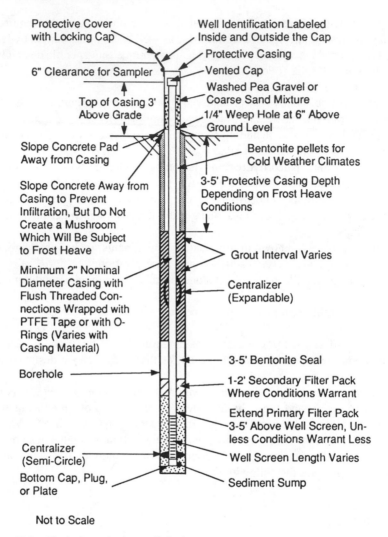

Protective Cover with Locking Cap

Well Identification Labeled Inside and Outside the Cap

Protective Casing

6" Clearance for Sampler

Vented Cap

Top of Casing 3' Above Grade

Washed Pea Gravel or Coarse Sand Mixture

1/4" Weep Hole at 6" Above Ground Level

Slope Concrete Pad Away from Casing

Bentonite pellets for Cold Weather Climates

Slope Concrete Away from Casing to Prevent Infiltration, But Do Not Create a Mushroom Which Will Be Subject to Frost Heave

3-5' Protective Casing Depth Depending on Frost Heave Conditions

Grout Interval Varies

Minimum 2" Nominal Diameter Casing with Flush Threaded Connections Wrapped with PTFE Tape or with O-Rings (Varies with Casing Material)

Centralizer (Expandable)

Borehole

3-5' Bentonite Seal

1-2' Secondary Filter Pack Where Conditions Warrant

Extend Primary Filter Pack 3-5' Above Well Screen, Unless Conditions Warrant Less

Centralizer (Semi-Circle)

Well Screen Length Varies

Bottom Cap, Plug, or Plate

Sediment Sump

Not to Scale

Figure 7.1. Typical monitoring well design components.

well casing and screen, slot-sizing procedures for screens, placement and sizing techniques for filter packs, placement procedures for annular seals, and installation of surface protective measures must all be used to ensure that a monitoring well will perform as intended.

There are some limitations to the application of the information presented in this chapter. For example, most of the monitoring well design and installation techniques described herein are derived from the experiences of the authors and others related to monitoring wells installed in unconsolidated geologic materials. While most of these techniques apply directly, or are easily adapted, to the installation of monitoring wells in bedrock, there are other techniques that may apply specifically to bedrock completions that are not described. Furthermore, nearly all of the technology developed to date pertaining to well design and installation has been intended for application to geologic materials that are considered aquifers (i.e., water-bearing geologic units that yield significant quantities of water to wells). Therefore, the techniques described herein are most effectively applied to monitoring wells constructed in these geologic materials; the authors are not aware of technology that will allow the installation of truly sediment-free monitoring wells in fine-grained (i.e., greater than 30% silt or clay by weight) geologic materials.

SELECTION OF MONITORING WELL
CASING AND SCREEN MATERIALS

The purpose of casing in a ground-water monitoring well is to provide a means of access from the surface to some point in subsurface saturated geologic materials. Casing prevents the collapse of geologic materials into the borehole and allows access to the groundwater, by means of a well screen generally affixed to the subsurface end of the casing, for determinations of ground-water quality and piezometric head.

Historically, the selection of casing material for water-supply wells and other types of wells focused on the material's structural strength, ease of handling, and durability in long-term exposures to natural subsurface conditions. When ground-water samples taken from monitoring wells began to be chemically analyzed at the parts-per-billion level, however, the focus shifted to the potential impact that casing materials may have on the chemical integrity, or "representativeness" of the ground-water samples. Additionally, durability in long-term exposures to potentially hostile man-induced chemical environments became a real concern. The selection of appropriate materials for monitoring well casing and screen must consider all of these factors.

Unique site-specific and logistical factors should be the controlling criteria in the selection of monitoring well casing and screen materials. Site-specific factors include the geologic environment, natural geochemical environment, anticipated well depths, and types of contaminants present or anticipated. Logistical factors include well drilling method, ease of handling and cleaning, and cost (for materials and shipping). Because no single casing or screen material is applicable over the wide range and variety of natural and man-induced site-specific conditions, it is critical that these conditions be evaluated thoroughly before selecting a material for monitoring well casing and screen. The selection of monitoring well casing and screen materials must be based on the ability of three primary casing characteristics—physical strength, chemical resistance, and chemical interference potential—to meet site-specific conditions.

Requirements of Casing and Screen Materials

Physical Strength of Well Casing and Screen

Monitoring well casing and screen must have the structural strength to withstand the forces exerted on them by the surrounding geologic materials and the forces imposed on them during installation (Figure 7.2). The material should be able to retain its structural integrity for the expected duration of the monitoring program, under both natural and man-induced subsurface conditions. The three components of casing and screen structural strength are tensile strength, compressive (column) strength, and collapse strength. Each property must be evaluated for a particular application. Relative strengths of stainless steel and PVC casing are presented in Tables 7.1 and 7.2, respectively. The values in these tables are for materials available from one manufacturer, though relative strengths of materials supplied by other manufacturers should be similar, except for the tensile strength of the casing joint, which varies with the coupling design. A comparison of the relative tensile and collapse strengths of small-diameter casing, for five common materials used in monitoring wells, is presented in Table 7.3. The weight per unit length, which is used along with tensile strength to calculate maximum permissable string length for a casing material, is presented in Table 7.4.

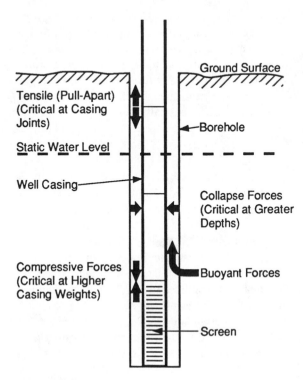

Figure 7.2. Forces exerted on a monitoring well casing and screen during installation.

Table 7.1.   Material Strength Data for Type 304 Stainless Steel.[a]

Nominal Size	O.D. (in.)	I.D. (in.)	Wt (lb/ft)	Strength			
				Collapse (psi)	Tensile (lb)	Column (lb)[b]	Joint Tensile (lb)
2-in. schedule 40 casing	2.375	2.067	3.653	3,526	85,900	6,350	15,900
2-in. schedule 5 casing	2.375	2.245	1.604	986	37,760	3,000	15,900
2-in. wire-wound screen	2.375	1.900	4.0	1,665	10,880	810	15,900
4-in. schedule 40 casing	4.500	4.026	10.790	2,672	254,400	69,000	81,750
4-in. schedule 5 casing	4.500	4.334	3.915	315	92,000	26,800	81,750
4-in. wire-wound screen	4.500	4.000	6.0	249	16,320	4,500	81,750
5-in. schedule 40 casing	5.563	5.047	14.6	2,231	343,200	145,490	91,500
5-in. schedule 5 casing	5.563	5.345	6.4	350	148,800	66,660	91,500
5-in. wire-wound screen	5.560	5.030	4.8	134	38,600	13,040	91,500
6-in. schedule 40 casing	6.625	6.065	19.0	1,942	444,800	270,000	94,500
6-in. schedule 5 casing	6.625	6.407	7.6	129	178,400	113,660	94,500
6-in. wire-wound screen	6.620	6.090	5.5	176	54,000	19,170	94,500

[a]Information provided by Johnson Filtration Systems Inc.
[b]For all column calculations, the span = 20 ft, hinged at one end, and fixed at other end.

The tensile strength of a casing or screen material, defined as the load required to pull the casing apart, is the most significant strength-related property of casing or screen materials. Tensile strength varies according to casing composition, manufacturing technique, and the type of casing joint used; it is closely related to the strength of the parent material as well as the casing dimensions (diameter and wall thickness). For a monitoring well installation, the casing material selected should have as a minimum enough tensile strength to support its own weight when suspended from the surface in an air-filled borehole. The maximum theoretical installation depth can be calculated by dividing the tensile strength for a given casing material by the linear weight of the casing. In most cases, the casing will encounter water in the borehole during installation; the buoyant force of the water increases the length of casing that can be suspended in the borehole by a factor

Table 7.2.   Material Strength Data for PVC.[a]

Nominal Size	O.D. (in.)	I.D. (in.)	Wt (lb/ft)	Collapse (psi)	Tensile (lb)	Column (lb)[b]	Joint Tensile (lb)
2-in. schedule 40 casing	2.375	2.067	0.64	307	7,500	90	1,800
2-in. schedule 80 casing	2.375	1.939	0.88	947	9,875	125	1,800
2-in. wire-wound screen	2.375	1.875	0.8	99	1,800	25	1,800
4-in. schedule 40 casing	4.500	4.026	1.9	158	22,200	1,030	6,050
4-in. schedule 80 casing	4.500	3.826	2.6	494	30,850	1,375	6,050
4-in. wire wound screen	4.620	4.000	1.7	79	2,250	150	6,050
5-in. schedule 80 casing	5.563	4.813	3.9	324	42,780	2,940	6,050
5-in. wire-wound screen	5.560	4.810	2.5	79	4,610	307	6,050
6-in. schedule 80 casing	6.625	5.761	5.4	292	58,830	5,760	4,000
6-in. wire-wound screen	6.620	5.680	3.7	87	5,770	552	4,000

[a]Information provided by Johnson Filtration Systems Inc.
[b]For all column calculations, the span = 20 ft, hinged at one end, and fixed at other end.

Table 7.3.   Comparative Strengths of Well Casing Materials.[a]

Material	Casing Tensile Strength (lb) 2-in. nominal	Casing Tensile Strength (lb) 4-in. nominal	Casing Collapse Strength (lb/in.^{2}) 2-in. nominal	Casing Collapse Strength (lb/in.^{2}) 4-in. nominal
Polyvinylchloride (PVC)	7,500	22,000	307	158
PVC casing joint[b]	2,800	6,050	300	150
Stainless steel (SS)[c]	37,760	92,000	896	315
SS casing joint[b]	15,900	81,750	No data	No data
Polytetraflouroethylene (PTFE)	3,800	No data	No data	No data
PTFE casing joints[b]	540	1,890	No data	No data
Epoxy fiberglass	22,600	56,500	330	250
Epoxy casing joints[d]	14,000	30,000	230	150
Acrylonitrile-butadiene-styrene (ABS)	8,830	22,000	No data	No data
ABS casing joints[d]	3,360	5,600	No data	No data

[a]Information provided by E. I. du Pont de Nemours & Company, Wilmington, DE.
[b]All joints are flush-threaded.
[c]Stainless steel casing materials are schedule 5 with schedule 40 joints; other casing materials (PVC, PTFE, epoxy, ABS) are schedule 40.
[d]Joints are not flush-threaded, but are a special type that is thicker than schedule 40.

Table 7.4.   Weight per Unit Length and Weight Ratios of Well Casing Materials (2-in. nominal).

| Material | Weight by Schedule Number (lb/ft) | | | | Approximate Weight Ratios[a] |
	#5	#10	#40	#80	
PVC	—	—	0.65	0.91	1.3
Stainless steel	1.62	2.06	3.65	5.07	3.4[b]
PTFE	—	—	1.50	1.90	3.0
Epoxy fiberglass	—	—	0.50	—	1.0
ABS	—	—	0.60	—	1.2

[a]Weight ratio is obtained by multiplying each schedule 40 weight by 2.
[b]Schedule 5 casing with schedule 40 couplings for a 10-ft length of pipe.

that depends primarily on the specific gravity of the casing material. The tensile strength of the casing joints is more important than the tensile strength of the casing itself, because the joints are usually the weakest points in the casing string. Therefore, joint strength is more commonly used to determine the maximum axial load that can be placed on a casing string.

The compressive or column strength of a casing or screen material is defined as the load required to deform the material by compressing it. The properties of the casing or screen material, specifically the yield strength and stiffness, are more significant in determining compressive strength than are the dimensional parameters, although casing wall thickness is also important.

Another significant strength-related property of casing and screen materials is collapse strength, or the capability of a casing to resist collapse caused by any and all external loads to which it is subjected, both during and after installation. The collapse strength of a casing material is determined principally by dimensional parameters; most notably, the collapse strength of a piece of casing is proportional to the cube of its wall thickness. Therefore, a small increase in wall thickness provides a significant increase in collapse strength. Casing and screen are most susceptible to collapse during installation, when the casing string has not yet been confined and restrained by the placement of filter pack or annular seal materials around it. Once a casing string is properly installed and confined, its resistance to collapse is enhanced so that collapse is no longer a concern (NWWA/PPI, 1980).

Among the external loadings on casing that may contribute to casing collapse are the following:

- net external hydrostatic pressure produced when the static water level outside the casing is higher than that on the inside;
- unsymmetrical loads on the casing resulting from uneven placement of backfill materials;
- uneven collapse of unstable formation materials;
- weight of grout on the outside of a partially water-filled casing; and
- forces associated with well development that produce large differential pressures on the inside and outside of the casing.

Of these, only the first, external hydrostatic pressure, can be predicted and calculated with any accuracy. To provide sufficient margin against possible collapse by all normally anticipated external loadings, a casing material is selected so that its resistance to collapse is greater than that required to resist external hydrostatic pressure alone. Generally, a safety factor of at least two is recommended (NWWA/PPI, 1980). In well installations in difficult drilling or geologic conditions, a larger safety factor should be employed.

Except for joint strength, all of the strength characteristics of a piece of casing are reduced when the casing is cut to produce slotted well screen. Continuous-slot wire-wrap well screen varies in strength depending on the configuration of the vertical columns and the wire-wrap screen, and the type of material.

Chemical Resistance of Casing and Screen Materials

Materials used for monitoring well casing and screen must be durable enough to withstand potential chemical attacks from either natural chemical constituents or contaminants in ground water; in particular, metallic casing materials should be resistant to corrosion (galvanic or electrochemical) and plastic casing materials should be resistant to chemical degradation. Because the extent to which chemical attacks occur is primarily dependent on the presence and concentration of certain chemical constituents in ground water, the casing material can be selected only with a knowledge of existing or anticipated ground-water chemistry. Not only may natural or man-induced ground water chemistry affect the structural integrity of monitoring well casing or screen, but by-products of casing deterioration also may adversely affect the chemistry of water samples taken from monitoring wells.

Chemical Interference from Casing and Screen Materials

Materials used for monitoring well casing and screen must not assimilate chemicals from ground water by adsorption on to the material surface or by absorption into the material matrix or pores. Loss of chemicals from a ground-water sample may create so-called "false negatives," which produce the false impression that chemical constituents are not present, or are present below their actual concentration in solution. Conversely, these materials must not desorb or leach chemical constituents from the well casing or screen into the ground water to be sampled. The addition of leached or desorbed chemicals to a ground-water sample may produce "false positives," which indicate possible ground-water contamination when, in fact, none is present. Therefore, in the selection of monitoring well materials, the potential interactions between casing or screen materials and the natural and man-induced geochemical environment must be carefully considered.

Types of Casing and Screen Materials

Casing used in monitoring wells could conceivably be made of almost any rigid tubular material, although experience dictates that the choices are limited to only a few materials. Casing materials typically used in ground-water monitoring wells can be categorized into four general types:

- thermoplastic materials, including polyvinyl chloride (PVC) and acrylonitrile butadiene styrene (ABS);
- fluoropolymer materials, including polytetrafluoroethylene (PTFE), tetrafluoroethylene (TFE), fluorinated ethylene propylene (FEP), perfluoroalkoxy (PFA), and polyvinylidine fluoride (PVDF);
- metallic materials, including carbon steel, low carbon steel, galvanized steel, and stainless steel (particularly types 304 and 316); and
- fiberglass-reinforced materials, including fiberglass-reinforced epoxy (FRE) and fiberglass-reinforced plastic (FRP).

Each of these materials has physical and chemical characteristics that influence its use in site-specific hydrogeologic and contaminant-related conditions. PVC, ABS, PTFE, low-carbon and galvanized steel, stainless steel and FRE casing and screen materials are discussed in greater detail in the following sections; other materials are infrequently used in monitoring wells, or too little practical application-related information is available on which to base decisions on selection of these materials for use in ground-water monitoring wells.

Thermoplastic Materials

Thermoplastics are man-made materials that consist of varying formulations of plastics that can be formed and reformed repeatedly; they are softened by heating and harden upon cooling. This characteristic allows thermoplastics to be molded or extruded into rigid well casings.

Nearly all thermoplastic well casing is one of two materials: polyvinyl chloride (PVC) or acrylonitrile butadiene styrene (ABS). The strength, rigidity, and temperature resistance characteristics of both of these materials are generally sufficient to allow well casings and screens made from them to withstand the typical stresses of handling, installation, and loading. In addition, rigid, hardened thermoplastics offer complete resistance to galvanic and electrochemical corrosion, high resistance to abrasion, high strength-to-weight ratios, light weight, durability in most natural ground-water environments, low maintenance, partial flexibility, workability, and low cost, making them ideal for many monitoring well applications.

Rigid PVC well casing is produced by combining polyvinyl chloride resin with various types of stabilizers, lubricants, pigments, fillers, and processing aids. The amounts of these additives can be varied to produce different PVC plastics with properties tailored to specific applications. PVC used for well casing is composed of a rigid hardened (unplasticized) polymer formulation (PVC Type 1) that has high tensile, compressive and collapse strength, and good chemical resistance except to low molecular weight ketones, aldehydes, and chlorinated solvents (Barcelona et al., 1983).

ABS plastics are produced from three monomers—acrylonitrile, butadiene, and styrene. The ratio of the components and the way in which they are combined can be varied to produce plastics with a wide range of properties. Acrylonitrile contributes rigidity, impact strength, hardness, chemical resistance, and heat resistance; butadiene contributes impact strength; styrene contributes rigidity, gloss, and ease of manufacturing (NWWA/PPI, 1980). ABS used for well casing is a rigid, strong unplasticized polymer formulation that has good heat resistance and impact strength.

Strength-related characteristics of thermoplastic materials. Typical physical properties of PVC and ABS thermoplastic well-casing materials are provided in Table 7.5. Included in this table are minimum tensile strength and compressive strength values for several cell classes of each material, as required under ASTM standard specification D-1784 (for PVC) and D-1788 (for ABS). Dimensions, hydraulic collapse pressure, and unit weight of PVC and ABS well casing are provided in Tables 7.6 and 7.7 respectively.

In comparison to metallic materials, the tensile, compressive, and collapse strength of thermoplastic materials is relatively low. With respect to tensile strength, the light weight of thermoplastics offsets the low strength so that for most installations of thermoplastic well casings, the axial loading is not a limiting factor. Assuming a dry borehole and a safety factor of two (equivalent to a tensile strength of half the value given in Table 7.5), the theoretical maximum permissible string length for thermoplastic materials is in excess of 4,000 feet. This maximum string length is for fully cured, solvent-cemented connections under conditions of short-term loading only. In monitoring well applications, the use of solvent-cemented joints is discouraged because of the high potential for sample chemical alteration; threaded joints are much more commonly used. For threaded connections, the maximum string length is reduced substantially and, for long-term loading, the maximum permissible string length is further reduced. Though the degree of reduction of maximum permissible string length depends on the type of joint used, it can be expected that flush-joint, threaded connections will reduce the theoretical maximum permissible string length by 30% to 70%, to about 1200–2000 feet.

Because their specific gravities are not much higher than that of water (1.4 for PVC, 1.3 for ABS), thermoplastic well casings are relatively light when immersed in water. The buoyant force of water for thermoplastics is therefore very high, increasing the maximum string length by about 40% for that portion of the casing in contact with water.

Table 7.5. Typical Physical Properties of Thermoplastic Well Casing Materials at 73.4°F.[a]

Property	ASTM Test Method	ABS Cell Class, per D-1788		PVC Cell Class, per D-1784	
		434	533	12454-B&C	14333-C&D
Specific gravity	D-792	1.05	1.04	1.40	1.35
Tensile strength, lb/in^2	D-638	6,000[b]	5,000[b]	7,000[b]	6,000[b]
Tensile modulus of elasticity, lb/in^2	D-638	350,000	250,000	400,000[a]	320,000[b]
Compressive strength, lb/in^2	D-695	7,200	4,500	9,000	8,000
Impact strength, izod, ft-lb/inch notch	D-256	4.0[b]	6.0[b]	0.65	5.0
Deflection temperature under load (264 psi), °F	D-648	190[b]	190[b]	158[b]	140[b]
Coefficient of linear expansion, in/in-°F	D-696	5.5×10^{-5}	6.0×10^{-5}	3.0×10^{-5}	5.0×10^{-5}

[a]Source: NWWA/PPI, 1980.
[b]These are minimum values set by the corresponding ASTM Cell Class designation. All others represent typical values.

Table 7.6. Dimensions, Hydraulic Collapse Pressure and Unit Weight of PVC Well Casing.

| Outside Diameter (inches) | | SCH# | Wall Thickness (in.) | SDR | Weight in Air (lbs/100 feet) | | Weight in Water (lbs/100 feet) | | Hydraulic Collapse Pressure (psi) | |
Nominal	Actual				PVC 12454	PVC 14333	PVC 12454	PVC 14333	PVC 12454	PVC 14333
2	2.375	SCH80	0.218	10.9	94	91	27	24	947	758
		SCH40	0.154	15.4	69	66	20	17	307	246
2½	2.875	SCH80	0.276	10.4	144	139	41	36	1110	885
		SCH40	0.203	14.2	109	105	31	27	400	320
3	3.500	SCH80	0.300	11.7	193	186	55	48	750	600
		SCH40	0.216	16.2	143	138	41	36	262	210
3½	4.000	SCH80	0.316	12.6	235	227	67	59	589	471
		SCH40	0.226	17.7	172	176	49	43	197	158
4	4.500	SCH80	0.337	13.3	282	272	80	70	494	395
		SCH40	0.237	19.0	203	196	58	51	158	126
4½	4.950		0.248	20.0	235	226	67	58	134	107
			0.190	26.0	182	176	52	46	59	47
5	5.563	SCH80	0.375	14.8	391	377	112	98	350	280
		SCH40	0.258	21.6	276	266	79	69	105	84
6	6.625	SCH80	0.432	15.3	538	519	154	134	314	171
		SCH40	0.280	23.7	358	345	102	89	78	62

Table 7.7. Dimensions, Hydraulic Collapse Pressure and Unit Weight of ABS Well Casing.

| Outside Diameter (inches) | | SCH# | Wall Thickness (in.) | SDR | Weight in Air (lbs/100 feet) | | Weight in Water (lbs/100 feet) | | Hydraulic Collapse Pressure (psi) | |
Nominal	Actual				ABS 434	ABS 533	ABS 434	ABS 533	ABS 434	ABS 533
2	2.375	SCH80	0.218	10.9	71	70	3.4	2.7	829	592
		SCH40	0.154	15.4	52	51	2.5	2.0	269	192
2½	2.875	SCH80	0.276	10.4	108	107	5.1	4.1	968	691
		SCH40	0.203	14.2	82	81	3.9	3.1	350	250
3	3.500	SCH80	0.300	11.7	145	144	6.9	5.5	656	468
		SCH40	0.216	16.2	107	106	5.1	4.1	229	164
3½	4.000	SCH80	0.318	12.6	176	175	8.4	6.7	515	368
		SCH40	0.226	17.7	129	128	6.1	4.9	173	124
4	4.500	SCH80	0.337	13.3	211	209	10.0	8.0	432	308
		SCH40	0.237	19.0	152	151	7.2	5.8	138	98
5	5.563	SCH80	0.375	14.8	294	291	14.0	11.2	306	218
		SCH40	0.258	21.6	207	205	9.8	7.9	92	66
6	6.250	SCH80	0.432	15.3	404	400	19.2	15.4	275	196
		SCH40	0.280	23.7	268	266	12.8	10.2	69	49

Chemical resistance and chemical interference characteristics of thermoplastic materials. With respect to chemical resistance, thermoplastic well-casing materials are superior in some respects to metallic materials because they are nonconductors and thus totally immune to electrochemical or galvanic corrosion. In addition, they are resistant to biological attack, and to chemical attack by soil, water, and other naturally occurring substances present in the subsurface. The resistance of PVC and ABS to common hazardous materials applies for most acids, oxidizing agents, salts, alkalies, oils, and fuels (NWWA/PPI, 1980). Even after long-term (6 months) immersion in common types of gasoline containing high concentrations of aromatic hydrocarbons (benzene, toluene, ethylbenzene, and xylenes), rigid PVC did not exhibit any swelling or other alteration effects (Schmidt, 1987). Thermoplastics are, however, susceptible to chemical attack by high concentrations of certain organic solvents. These solvents can produce an effect called solvation, the physical degradation of the plastic. Solvent cementing of thermoplastic well casing is based on solvation, which occurs in the presence of very high concentrations of specific organic solvents. If these solvents, which include tetrahydrofuran (THF), methyl ethyl ketone (MEK), methyl isobutyl ketone (MIBK), and cyclohexanone, are present in high enough concentrations (i.e., parts-per-thousand or percentage concentrations) they could be expected to chemically degrade thermoplastic well casing to some degree. In general, the chemical attack on the thermoplastic polymer matrix will increase as the organic content of the solution with which it is in contact increases.

Barcelona et al. (1983) list the groups of chemical compounds that may cause degradation of the thermoplastic polymer matrix and/or the release of compounding ingredients which otherwise would remain in the solid material. These chemical compounds include low molecular weight ketones, aldehydes, amines, chlorinated alkenes, and alkanes. Unfortunately, there is currently a significant lack of information regarding critical concentrations of these chemical compounds at which deterioration of the thermoplastic material is significant enough to affect either the structural integrity of the material or ground-water sample chemical quality. Little published information exists regarding the performance of these well casing materials under field conditions, though experience indicates that PVC is suitable in monitoring applications in which low (i.e., parts per billion or parts per million) levels of most organic constituents are present.

Extensive research has been conducted in the laboratory (specifically on water supply piping) to evaluate vinyl chloride monomer migration from new and old PVC pipe. The data generated all support the conclusion that, under conditions in which PVC is in contact with water, the level of trace vinyl chloride migration from PVC pipe walls is extremely low compared to residual vinyl chloride monomer (RVCM) in PVC pipe walls. Since 1976, when the National Sanitation Foundation (NSF) established an RVCM monitoring and control program for PVC pipe used in potable water supplies and well casing, process control of RVCM levels in PVC pipe has improved markedly. According to Barcelona et al. (1983), the level of RVCM allowed in NSF-certified PVC products (less than or equal to 10 ppm RVCM) limits potential leached concentrations of vinyl chloride monomer to less than 1 part per billion; leachable amounts of vinyl chloride monomer should decrease as RVCM levels in products continue to be reduced. Though a potential for chemical analytical interference exists even at the low parts per billion levels at which vinyl chloride monomer may be found in a solution in contact with PVC, the significance of this interference is not currently known. However, no documented instances of residual vinyl chloride monomer occurrence in ground-water samples are known.

With few exceptions, plasticizers are not added to PVC formulations used for well casing because the casing must be a rigid material. If plasticizers were added, their levels would not be expected to exceed 0.01% (Barcelona et al., 1983). By contrast, flexible PVC tubing may contain from 30% to 50% plasticizers by weight. The presence of these high levels of plasticizers in flexible PVC tubing has been documented by several researchers (Junk et al., 1974; Barcelona et al., 1983, 1985b) to produce significant chemical interference effects. However, because rigid, hardened PVC well casing contains no plasticizers, no plasticizer-induced chemical interference problem should exist for PVC well casing.

Rigid PVC may contain other additives, primarily stabilizers, at levels approaching 5% by weight. Some representative chemical classes of additives that have been used in the manufacture of rigid PVC well casing are listed in Table 7.8. Boettner et al. (1981) determined through a laboratory study that several of the PVC heat-stabilizing compounds, notably dimethyltin and dibutyltin species, could potentially leach out of rigid PVC at very low (sub-parts-per-billion) levels. This leaching was found to decrease dramatically over time. Factors that influenced the leaching process in this study included solution pH, temperature and ionic composition, and exposed surface area and surface porosity of the pipe material.

Table 7.8. Representative Classes of Additives in Rigid PVC Materials Used for Pipe or Well Casing.[a]

(Concentration in wt. %)	
Heat Stabilizers (0.2–1.0%)	**Fillers (1–5%)**
Dibutyltin diesters of lauric and maleic acids	$CaCO_3$
Dibutyltin bis (laurylmercaptide)	diatomaceous earth
Dibutyltin-β-mercaptopropionate	clays
di-n-octyltin maleate	**Pigments**
di-n-octyltin-S,S'-bis isoctyl mercaptoacetate	TiO_2
di-n-octyltin-β-mercaptopropionate	carbon black
Various other alkyltin compounds	iron and other metallic oxides
Various proprietary antimony compounds	**Lubricants (1–5%)**
	stearic acid
	calcium stearate
	glycerol monostearate
	montan wax
	polythylene wax

[a]*Source:* Barcelona et al., 1983.

It is currently unclear what impact, if any, the leaching of low levels of stabilizing compounds may have on analytical interference.

In addition to setting a limit on RVCM, the National Sanitation Foundation (NSF) has set specifications for certain chemical constituents in PVC formulations. The purpose of these specifications, outlined in NSF Standard 14, is to control the amount of chemical additives in both PVC well casing and pipe used for potable water supply. The maximum contaminant levels permitted in a standardized leach test on NSF-approved PVC products are given in Table 7.9. Most of these levels correspond to those set by the Safe Drinking Water Act for chemical constituents in water covered by the National Interim Primary Drinking Water Standards. Only PVC products that carry either the "NSF wc" (well casing) or "NSF pw" (potable water) designation have met the specifications set forth in Standard 14; only these products should be used for casing or screen in monitoring wells. Other non-NSF-listed products may include in their formulation chemical additives not addressed by the specifications, or may carry levels of the listed chemical parameters higher than those permitted by the specifications. As an example, even though neither lead nor cadmium have been permitted as compounding ingredients in U.S.-manufactured, NSF-listed PVC well casing since 1970, PVC manufactured in other countries may be stabilized with lead or cadmium compounds that have been demonstrated to leach from the PVC (Barcelona et al., 1983).

Table 7.9.   Chemical Parameters Covered by NSF Standard 14 for Finished Products[b] and in Standard Leach Tests.

Parameter	Maximum Contaminant Level (mgL^{-1})	
Antimony (Sb)	0.05[a]	0.05[a]
Arsenic (As)	0.05	
Barium (Ba)	1.0	
Cadmium (Cd)	0.01	
Chromium (Cr)	0.05	
Lead (Pb)	0.05	
Mercury (Hg)	0.002	
Phenolic substances	0.05[a]	
Residual vinyl chloride monomer[b] (RVCM)	10[a]	
Selenium (Se)	0.01[a]	
Tin (Sn)	0.05	

Source: Barcelona et al., 1983.
[a]Not covered under National Interim Primary Drinking Water Regulations.
[b]Total residual after complete dissolution of polymer matrix.

Tabulated values are the maximum levels permissible in NSF-listed products after standardized leach testing in weakly acidic aqueous solution. [Carbonic acid solution with 100 mgL^{-1} hardness as $CaCO_3$ with 0.5 mgL^{-1} chlorine; pH 5.0 to 0.2; and surface to solution ratio of 6.5 cm^{2}mL^{-1}]

In other laboratory studies of leaching associated with PVC well-casing material Curran and Tomson (1983) and Parker and Jenkins (1986) determined that little or no leaching occurred. In the former study, it was found that, in testing several different samples (brands) of rigid PVC well casing, trace organics either were not leached or were leached only at the sub-parts per billion level. In the latter study, which was conducted using ground water in contact with two different brands of PVC, it was concluded that no chemical constituents were leached at sufficient concentrations to interfere with reversed-phase HPLC analysis for low parts-per-billion levels of 2,4,6-trinitrotoluene (TNT), hexahydro-1,3,5-trinitro-1,3,5,7-tetrazocine (HMX), or 2,4-dinitrotoluene (DNT) in solution. In a more recent study (Parker et al., 1990) PVC casings sorbed lead and leached cadmium in small quantities over several hours. The study by Curran and Tomson (1983) confirmed previous field work at Rice University (Tomson et al., 1979) that suggested that PVC well casings did not leach significant amounts (i.e., at the sub-parts-per-billion level) of trace organics into ground-water samples.

Another potential area for concern with respect to chemical interference effects is the possibility that some chemical constituents could be sorbed by (either adsorbed onto or absorbed into) PVC well-casing materials. Miller (1982) conducted a laboratory study to determine whether several plastics, including rigid PVC well casing, exhibited any tendency to sorb potential contaminants from solution. Under the conditions of his test, Miller found that PVC moderately adsorbed tetrachloroethylene and adsorbed lead, but did not adsorb trichlorofluoromethane, trichloroethylene, bromoform, 1,1,1-trichloroethane, 1,1,2-trichloroethane, or chromium. In this experiment, sorption was measured weekly for six weeks and compared to a control; maximum sorption of tetrachloroethylene occurred at 2 weeks. Although Miller (1982) attributed the losses of tetrachloroethylene and lead strictly to adsorption, the anomalous behavior of tetrachloroethylene compared to that for other organics of similar structure (i.e., trichloroethylene) is not explained. In a follow-up study to determine whether or not the tetrachloroethylene could be desorbed and recovered, only a small fraction of the tetrachloroethylene was recovered. Thus, whether strong adsorption or some other mechanism (i.e., enhanced biodegradation in the presence of PVC) accounts for the difference is not clear. In the laboratory study of Parker and Jenkins (1986), it was found that significant losses of TNT and HMX from solution occurred in the presence of PVC well casing. A follow-up study to determine the mechanism for the losses led them to attribute the losses to increased microbial degradation rather than to adsorption. These results raise questions regarding whether losses found in other laboratory or field studies which did not consider biodegradation as a loss mechanism could, in fact, be attributed to biodegradation rather than to either adsorption or absorption. Only additional research into this question will provide a suitable answer.

In another laboratory study, Reynolds and Gillham (1985) found that adsorption of selected organics (specifically 1,1,1-trichloroethane, 1,1,2,2-tetrachloroethane, bromoform, hexachloroethane, and tetrachloroethylene) onto PVC and other polymeric casing materials could be a source of bias to ground-water samples collected from water standing in the well bore. PVC was found to adsorb four of the five compounds studied (all except 1,1,1-trichloroethane), but it was demonstrated that the *rate* of adsorption was sufficiently slow that adsorption bias would likely not be significant for the adsorbed compounds if well purging and sampling were to take place in the same day.

A field study conducted to determine the potential for sorption of low concentrations (about 100 ppb) of volatile aromatic hydrocarbons onto various casing materials demonstrated that there was no significant difference in adsorption among PVC, PTFE, and stainless steel (Sykes et al., 1986).

With few exceptions, the work that has been done to determine chemical interference effects of PVC well casing (whether by leaching of chemical constituents from or sorbing to PVC) has been conducted under laboratory conditions. Furthermore, in most of the laboratory work the PVC has been exposed to a solution (usually distilled, deionized, or "organic-free' water) over prolonged periods of time (several days to several months), thus allowing the PVC an extended period of time in which to exhibit sorption or leaching effects. While this may be comparable to a field situation in which ground water was exposed to the PVC well casing as it may be between quarterly or monthly sampling rounds, few studies consider the fact that prior to sampling, the well casing is usually evacuated (purged) of stagnant water residing in the casing between sampling rounds. Thus, the water that would have been affected by any sorption or leaching effects (if they were present at all) would ideally have been removed and replaced with aquifer-quality water that is eventually obtained as a sample considered "representative" of existing ground-water conditions. Because the sample is generally taken immediately after the purging of stagnant water in contact with the casing, it will have had a minimum of time (i.e., seconds or fractions of a second) with which to come in contact with casing or screen materials and thus be affected by sorption or leaching effects. Because of this, Barcelona, et al. (1983) and Reynolds and Gillham (1985) suggest that the potential sample bias effects due to adsorptive interactions with well casing materials may be discounted. Barcelona et al. (1983) correctly point out that these effects are far more critical in sample transfer and storage procedures employed prior to sample analysis.

Fluoropolymer Materials

Fluoropolymers are man-made materials consisting of different formulations of monomers (organic molecules) that can be either molded by powder metallurgy techniques or extruded while heated. Fluoropolymers are technically included among the plastics, but possess a unique set of properties that distinguish them from other plastics. Fluoropolymers are nearly totally resistant to chemical attack, even by extremely concentrated and aggressive acids (i.e., hydrofluoric, nitric, sulfuric) and organic solvents. They are also resistant to biological attack, oxidation, weathering and ultraviolet radiation, have a broad useful temperature range (up to 287°C), a high dielectric constant, a low coefficient of friction, antistick properties, and a greater coefficient of thermal expansion than most other plastics and metals.

There exists a variety of fluoropolymer materials that are marketed under a number of different trademarks. Descriptions of some of the more popular fluoropolymers, with the trademarks they carry, follow; basic physical properties of each are included.

Polytetrafluoroethylene (PTFE) was discovered by E. I. DuPont de Nemours in 1938 and was available only to the U.S. government until the end of World War II. Four principal physical properties of polytetrafluoroethylene are (Hamilton, 1985):

1. extreme temperature range—from $-400°F$ ($-240°C$) to $+550°F$ ($+287°C$) in constant service;
2. outstanding electrical and thermal insulation;
3. lowest coefficient of friction of any solid material;
4. almost completely chemically inert, except for some reaction with halogenated compounds at elevated temperatures and pressures.

In addition, PTFE is flexible without the addition of plasticizers, and is fairly easily machined, molded, or extruded. Because PTFE never melts, molding and extruding the material poses some difficulties not encountered with melt-processible materials. Despite this, PTFE is by far the most widely used and produced fluoropolymer. Trade names, manufacturers, and countries of origin of PTFE and other fluoropolymer materials are listed in Table 7.10. Typical physical properties of the various fluoropolymer materials are described in Table 7.11.

Fluorinated ethylene propylene (FEP) was also developed by E. I. DuPont de Nemours and is perhaps the second most widely used fluropolymer. It duplicates nearly all of the physical properties of PTFE except the upper temperature range, which is 100°F (38°C) lower. Production of FEP finished products is generally faster because FEP is melt-processible, but raw material costs are higher.

Perfluoroalkoxy (PFA) combines the best properties of PTFE and FEP, but its cost is substantially higher than either.

Table 7.10. Trade Names, Manufacturers, and Countries of Origin for Various Fluoropolymer Materials.

Fluoropolymer Product	Trade Name	Manufacturer	Country of Origin
PTFE (or TFE)—Polytetrafluoroethylene (or Tetrafluoroethylene)	Teflon	DuPont	USA, Japan, Holland
	Halon	Allied	USA
	Fluon	ICI	UK, USA
	Hostaflon	Hoechs	W. Germany
	Polyflon	Daikin	Japan
	Algoflon	Montedison	Italy
	Soriflon	Ugine Kuhlman	France
FEP—Fluorinated Ethylene Propylene	Neoflon	Daikin	Japan
	Teflon	DuPont	USA, Japan, Holland
PFA—Perfluoroalkoxy	Neoflon	Daikin	Japan
	Teflon	DuPont	USA, Japan, Holland
PVDF—Polyvinylidene Fluoride	Kynar	Penn Walt	USA
CTFE—Chlorotrifluoroethylene	Kel-F	3M	USA
	Diaflon	Daikin	Japan

Table 7.11.   Typical Physical Properties of Various Fluoropolymer Materials.

Properties	Units	Method	TFE	FEP	PFA	E-CTFE	CTFE
Tensile strength @ 73°F	psi	D638–D651	2500–6000	2700–3100	4000–4300	7000	4500–6000
Elongation @ 73°F	%	D638	150–600	250–330	300–350	200	80–250
Modulus @ 73°F							
Tensile	psi	D638	95,000–115,000	95,000	240,000	206,000	238,000
Flexural	psi	D790	70,000–110,000	250,000	95,000–100,000	240,000	$1.5\text{–}3.0 \times 10^5$
Elasticity in tension		D747	58,000				
Flexural strength @ 73°F	psi	D790	Does not break	Does not break	Does not break	7000	8500
Izod impact strength (½ × ½ in. notched bar)							
@ +75°F	ft lb/ in. of notch	D256	3.0	No break	No break	No break	5.0
@ –65°F	ft lb/ in. of notch		2.3	2.9			
Tensile impact strength							
@ +73°F	ft lb/ sq in.	D1822	320	1020			
@ –65°F	ft lb/ sq in.		105	365			

continued

Table 7.11. Continued.

Properties	Units	Method	TFE	FEP	PFA	E-CTFE	CTFE
Compressive stress @ 73°F	psi	D695	1700				4600–7400
Specific gravity		D792	2.14–2.24	2.12–2.17	2.12–2.17	1.68	2.10–2.13
Coefficient of fraction static and kinetic against polished steel			0.05–0.08	0.06–0.09	0.05–0.06	0.15–0.65	0.2–0.3
Coefficient of linear thermal-expansion	/°F	D696	5.5×10^5	5.5×10^5	6.7×10^5	14×10^5	2.64×10^5

Source: Norton Chemware Catalog.

Polyvinylidene fluoride (PVDF) is tougher and has a higher abrasion resistance than other fluropolymers and is resistant to radioactive environments. It has a lower upper temperature limit than either PTFE or PFA, but is comparable in cost to both.

Care should be exercised in the use of trade names to identify fluoropolymers. Some manufacturers use one trade name to refer to several of their own different materials. For example, DuPont refers to several of its fluoropolymer resins as Teflon®, although the products referred to have different physical properties and different fabricating techniques. Although the materials are frequently interchangeable in service under most conditions, this may not always be the case.

Strength-related characteristics of fluoropolymer materials. As Dablow et al. (1986) point out, several strength-related properties of fluoropolymers (PTFE in particular) must be taken into consideration during the monitoring well design process, including: (1) pull-out resistance (tensile strength) of flush-joint threaded couplings; (2) compressive strength of the well screen; and (3) flexibility of the casing string. As with other materials, the tensile strength of fluoropolymer casing joints is the limiting factor affecting the length of casing which can be supported safely in a dry borehole. According to Dablow et al. (1986), experimental work conducted by DuPont indicates that one type of PTFE threaded joint will resist a pull-out load of approximately 900 pounds. With a safety factor of two, two-inch schedule 40 PTFE well casing, with a weight of approximately 1.2 pounds per foot, should be able to be installed to a depth of about 375 feet. Barcelona et al. (1985a) suggest that the recommended string length not exceed 320 feet. More recent information available from DuPont (Ferguson, personal communication, 1987) suggests that the pull-out load or tensile strength of a schedule 40 PTFE flat, square-threaded joint is 540 pounds. This would indicate that the maximum string length should not exceed 225 feet, assuming a safety factor of two.

Compressive strength of fluoropolymer well casings, and particularly well screens (slotted well casing), is also a recognized problem area. A low compressive strength (compared to thermoplastics) may lead to failure of the fluoropolymer casing at the threaded joints, where the casing is weakest and the stress is greatest. Additionally, according to Dablow et al. (1986), the "ductile" behavior of PTFE (which is a cold-flowable material) under compressive stress has resulted in the partial closing of screen slots with a consequent reduction in well efficiency in some fluoropolymer wells. Dablow et al. (1986) outline several design considerations that can be employed to minimize this problem. The first would be to specify a larger screen slot size than indicated by sieve analyses. In compressive strength tests conducted by DuPont to determine the amount of deformation in PTFE well screens, it was determined that a linear relationship exists between applied stress and the amount of screen deformation. This relationship is graphically presented in Figure 7.3. Ideally, from this graph, the anticipated screen slot size deformation can be determined and included in screen design by calculating the load, and adding anticipated screen slot deformation to the screen slot size determined by sieve analysis. Unfortunately, this leaves the well designer with the problem

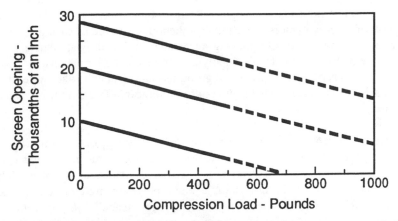

Figure 7.3. Results of short-term static compression tests on Teflon screen (from Dablon et al., 1986).

of deciding the precise time at which well development should be performed. If it is performed too soon, large quantities of sediment may enter the well, and may potentially lodge in the screen slots, thus reducing the efficiency of the well. If development is attempted too late, the slots may have closed to the point at which development becomes very difficult or is ineffective because of the lack of screen open area. Additionally, the data depicted on the graph are valid for short-term loading conditions only. It can be expected that under long-term loading, the screen slots will deform (i.e., close) further, though the degree of deformation is not known.

Another installation consideration for fluoropolymer well casing and screen to minimize compressive stress problems, as noted by Dablow et al. (1986), is to keep the casing string suspended in the borehole, so that the casing is in tension, and backfill the annulus around the casing string while it remains suspended. The intent is to reduce compressive stress by supplying support on the outer wall of the casing and screen. This could only be accomplished successfully in relatively shallow wells, in which the long-term tensile strength of the fluoropolymer casing was sufficient to withstand tensile stresses imposed on the casing by long-term suspension in the borehole. Additionally, continuous suspension of casing in the borehole would be impossible with hollow-stem auger well installations.

The third strength-related area of concern with respect to installation of fluoropolymer well casing is the extreme flexibility of the casing string, which causes the casing to become bowed and nonplumb when a load is placed on it. This may result in difficulties, following well installation, in installing pumps or other sampling devices, or in obtaining accurate water levels from these wells. Dablow et al. (1986) suggest three means of resolving this problem, including: (1) suspending the casing string in the borehole during backfilling (as discussed above); (2) using casing centralizers, or; (3) inserting a rigid PVC or steel pipe temporarily inside the fluoropolymer casing during backfilling.

Finally, while not truly a strength-related problem, there exists the antistick property of fluoropolymers, which makes them undesirable as casing materials because of the difficulty in achieving an annular seal with neat cement grout. The failure of grout to bond to fluoropolymers results in the formation of a "micro annulus," or a gap between casing and grout, that can channel surface water down the outside of the casing. This may result in alteration of ground-water sample chemistry.

Chemical resistance and chemical interference characteristics of fluoropolymer materials. For construction of ground-water monitoring wells, fluoropolymers possess several commonly perceived advantages over thermoplastics and metallic materials. For example, fluoropolymers are almost completely inert to chemical attack, even by extremely aggressive acids (i.e., hydrofluoric, nitric, sulfuric, and hydrochloric) and organic solvents. In addition, fluoropolymers have been thought to neither adsorb nor absorb chemical constituents from solution. Recent studies by Reynolds and Gillham (1985) and Gillham and O'Hannesin (1989a) indicate, however, that PTFE is prone to sorption of selected organic compounds, specifically 1,1,1-trichloroethane, 1,1,2,2-tetrachloroethane, hexachloroethane, and tetrachloroethene. A fifth organic compound studied by Gillham and O'Hannesin, bromoform, was not sorbed by PTFE. An observation of particular note made by Reynolds and Gillham (1985) was that tetrachloroethene was strongly and rapidly absorbed by PTFE so that significant reductions in concentration occurred within minutes of exposure to a solution containing the aforementioned organic compounds. These results indicate that PTFE may not be as chemically inert as previously thought. Perhaps the best application of PTFE casing is in situations in which concentrations of organic solvents at high (i.e., parts-per-thousand) levels *and* highly corrosive conditions would preclude the use of either thermoplastic or metallic materials. Recent findings of a study using mixed metals and artificial ground water showed no significant alteration of metal concentrations (excluding mercury) for periods as long as 72 hours.

Metallic Materials

Metallic well casing and screen materials available for use in monitoring wells include carbon steel, low carbon steel, galvanized steel, and stainless steel. Well casings made of any of these metallic materials are stronger, more rigid, and less temperature sensitive than thermoplastic, fluoropolymer, or fiberglass-reinforced casing materials. The strength and rigidity characteristics of metallic casing materials are sufficient to meet virtually any subsurface condition encountered in a ground-water monitoring situation. However, all metallic materials are subject to corrosion, a chemical resistance and chemical interference problem that may also affect casing strength in long-term exposures to certain subsurface geochemical environments.

Corrosion of metallic well casings and screens can both limit the useful life of the monitoring well installation and result in ground-water sample analytical bias. It is important, therefore, to select both casing and screen that are fabricated of corrosion-resistant materials.

Corrosion is defined as the weakening or destruction of a metallic material by chemical action. Several well-defined forms of corrosive attack on metallic materials have been observed and defined. In all forms, corrosion proceeds by electrochemical action, and water in contact with the metal is an essential factor. The forms of corrosion typical in environments in which well casing and screen materials are installed include (Johnson Division, 1966):

1. general oxidation or "rusting" of the metallic surface, resulting in uniform destruction of the surface with occasional perforation in some areas;
2. selective corrosion or loss of one element of an alloy (i.e., dezincification), leaving a structurally weakened material;
3. bi-metallic corrosion, caused by the creation of a galvanic cell at or near the juncture of two different metals;
4. pitting corrosion, or highly localized corrosion by pitting or perforation, with little loss of metal outside of these areas; and
5. stress corrosion, or corrosion induced in areas where the metal is highly stressed.

To determine the potential for corrosion of metallic materials, it is first necessary to determine natural geochemical conditions. The following list of indicators of corrosive conditions can help recognize potentially corrosive conditions (modified from Johnson Division, 1966):

1. low pH—if ground water pH is less than 7.0, water is acidic and corrosive conditions exist;
2. high dissolved oxygen content—if dissolved oxygen content exceeds 2 parts per million, corrosive water is indicated;
3. presence of hydrogen sulfide (H_2S)—presence of H_2S in quantities as low as 1 part per million can cause severe corrosion;
4. total dissolved solids (TDS)—if TDS is greater than 1000 parts per million, the electrical conductivity of the water is great enough to cause serious electrolytic corrosion;
5. carbon dioxide (CO_2)—corrosion is likely if the CO_2 content of the water exceeds 50 parts per million; and
6. chloride ion (Cl-) content—if Cl- content exceeds 500 parts per million, corrosion can be expected.

Combination of these corrosive agents generally increases the corrosive effect. To date, however, no data exist on the expected life of steel well-casing materials exposed to natural subsurface geochemical conditions, perhaps because the range of subsurface conditions is so wide and unpredictable.

Carbon steels were produced primarily to provide increased resistance (compared to iron) to atmospheric corrosion; achieving this increased resistance requires that the material be subjected to alternately wet (saturated) and dry (open to the atmosphere) conditions. In most monitoring wells, water fluctuations, which may be extreme during sampling, are usually not sufficient in either duration or occurrence to provide these conditions, so corrosion is a frequent problem. The difference between the corrosion resistance of carbon and low-carbon steels is negligible under conditions in which the materials are buried in soils or in the saturated zone, so both materials may be expected to corrode approximately equally. Corrosion products include iron and manganese (and trace metal) oxides as well as various metal sulfides (Barcelona et al., 1983). Under oxidizing conditions, the princpal corrosion products are solid hydrous metal oxides, while under reducing conditions, high levels of dissolved metals (principally iron and manganese) can be expected (Barcelona et al., 1983). While the electroplating process of galvanizing (application of a zinc coating) somewhat improves the corrosion resistance of either carbon or low-carbon steel, in many environments the improvement is only slight and short-term. The products of corrosion of galvanized steel include iron, manganese, zinc, and cadmium (Barcelona et al., 1983). These constituents can be contributed to ground-water samples in wells in which galvanized steel casing or screen are used.

Clearly the presence of corrosion products represents a high potential for alteration of ground-water sample chemical quality. The surfaces on which corrosion occurs also present potential sites for a variety of chemical reactions (i.e., formation of organometallic complexes) and adsorption to occur. These surface interactions can cause significant changes in dissolved metal or organic compound concentrations in ground-water samples. According to Barcelona et al. (1983), even flushing the stored water from the well casing prior to sampling may not be sufficient to minimize this source of sample bias because the effects of the disturbance of surface coatings or accumulated corrosion products in the bottom of the well would be difficult, if not impossible, to predict. On the basis of these observations, the use of carbon steel, low-carbon steel, and galvanized steel in wells used for ground-water quality monitoring should be discouraged in most natural geochemical environments. These materials may, however, be well suited to use in piezometers or monitoring wells that are used strictly for water-level monitoring or for sampling ground water for constituents other than metals or organic compounds.

On the other hand, stainless steel performs very well in most corrosive environments, particularly under oxidizing conditions. In fact, stainless steel requires exposure to oxygen in order to attain its highest corrosion resistance. Oxygen combines with part of the stainless steel alloy to form a thin, invisible protective film on the surface of the metal (Johnson Division, 1966). As long as the film remains intact, the corrosion resistance of stainless steel is very high.

Several different types of stainless steel alloys are available for use in monitoring wells; the most common are Type 304 and Type 316, which are part of the 18-8 or 300 series of stainless steels. Both types of stainless steel are available in low-carbon forms, designated by an "L" after the number (i.e., 304L), which are more easily welded than the normal carbon types. Table 7.12 describes

Table 7.12. Dimensions, Hydraulic Collapse and Burst Pressure and Unit Weight of Stainless Steel Well Casing.

Nom. Size (in.)	Schedule No.	Outside Diameter (in.)	Wall Thickness (in.)	Inside Diameter (in.)	Internal Cross-Sectional Area (sq. in.)	Internal Pressure (psi)		External Pressure (psi)	Weight (lb/ft)
						Test	Bursting	Collapsing	
2	5	2.375	0.065	2.245	3.958	820	4.105	.896	1.619
	10	2.375	0.109	2.157	3.654	1.375	6.884	2.196	2.063
	40	2.375	0.154	2.067	3.356	1.945	9.726	3.526	3.087
	80	2.375	0.218	1.939	2.953	2.500	13.766	5.418	5.069
2½	5	2.875	0.083	2.709	5.761	865	4.330	1.001	2.498
	10	2.875	0.120	2.635	5.450	1.250	6.260	1.905	3.564
	40	2.875	0.203	2.469	4.785	2.118	10.591	3.931	5.347
3	5	3.500	0.083	3.334	8.726	.710	3.557	.639	3.057
	10	3.500	0.120	3.260	8.343	1.030	5.142	1.375	4.372
	40	3.500	0.216	3.680	7.389	1.851	9.257	3.307	7.647
3½	5	4.000	0.083	3.834	11.540	.620	3.112	.431	3.505
	10	4.000	0.120	3.760	11.100	.900	4.500	1.081	5.019
	40	4.000	0.226	3.548	9.887	1.695	8.475	2.941	9.194
4	5	4.500	0.083	4.334	14.750	.555	2.766	.315	3.952
	10	4.500	0.120	4.260	14.250	.800	4.000	.845	5.666
	40	4.500	0.237	4.026	12.720	1.580	7.900	2.672	10.891
5	5	5.563	0.109	5.345	22.430	.587	2.949	.350	6.409
	10	5.563	0.134	5.295	22.010	.722	3.613	.665	7.842
	40	5.563	0.258	5.047	20.000	1.391	6.957	2.231	14.754
6	5	6.625	0.109	6.407	32.220	.494	2.467	.129	7.656
	10	6.625	0.134	6.357	31.720	.606	3.033	.394	9.376
	40	6.625	0.280	6.065	28.890	1.288	6.340	1.942	19.152

Source: ARMCO, Inc. Stainless Steel Pipe Specifications.

dimensions, hydraulic collapse pressure, burst pressure, and unit weight of stainless steel casing.

Type 304 stainless steel is perhaps the most practical from a corrosion resistance and cost standpoint. As indicated in Table 7.13 (modified from Johnson Division, 1966), Type 304 is composed of slightly more than 18% chromium and more than 8% nickel, with about 72% iron and not more than 0.08% carbon. The chromium and nickel content give the Type 304 alloy excellent resistance to corrosion; its low carbon content improves its weldability. Table 7.13 demonstrates that Type 316 stainless steel is compositionally similar to Type 304 with two exceptions—a 2–3% molybdenum content and a higher nickel content (replacing the equivalent percentage of iron). This compositional difference gives Type 316 stainless steel an improved resistance to sulfur-containing species as well as sulfuric acid solutions (Barcelona et al., 1983) so it performs better under reducing conditions than Type 304. According to Barcelona et al. (1983), Type 316 stainless steel is less susceptible to pitting or pinhole corrosion caused by organic acids or halide solutions. However, they also point out that for either formulation of stainless steel, long-term exposure to corrosive conditions may result in corrosion and the subsequent chromium or nickel contamination of ground-water samples. Insoluble halogen and sulfur compounds may also form as a result of corrosion of stainless steel.

Table 7.13.   Composition of Stainless Steel Well Casing/Screen Materials.

Chemical Component[a]	SS 304	SS 316
Carbon	0.08	0.08
Manganese	2.00	2.00
Phosphorous	0.04	0.045
Sulfur	0.03	0.03
Silicon	0.75	1.00
Cromium	18.0–20.0	16.0–18.0
Nickel	8.0–11.0	10.0–14.0
Molybdenum	–	2.0–3.0
Iron	remainder	remainder

[a]All chemical components measured in percentage.

The resistance to corrosion of both types of stainless steel can be improved by treatment with nitric acid and potassium dichromate solutions. These treatment processes, usually done at steel mills or factories, are referred to as Mil-Spec or QQ-P-35C. The passivation of stainless steel substantially increases corrosion resistance in ground water with high concentrations of halides in solution. Passivation may also be important for reducing adsorption of certain radionuclides if a site being monitored contains mixed waste (Raber et al., 1983). Another study showed Type 304 stainless steel casings sorbed arsenic and lead while leaching chromium and cadmium (Parker et al., 1990). This study also showed that Type 316 casings also sorbed arsenic, lead, and chromium. At actual field sites Type 304 and 304L stainless steel wells may leach chromium in quantities of 10 to

30 parts per billion for more than a year (Smith, 1988; Smith et al., 1989; Schalla et al., 1988).

Fiberglass-Reinforced Epoxy Materials

Fiberglass-reinforced epoxy casing combines the best characteristics of metallic and thermoplastic casing materials. It is nearly as strong as stainless steel casing (see Table 7.3) but it weighs about the same as PVC. Although fiberglass-reinforced epoxy casing is not in widespread use in monitoring wells, its light weight, high strength, and durability in hostile chemical environments make it desirable.

Published data are limited regarding adsorption of chemicals onto fiberglass-reinforced epoxy casing and leaching of constituents from the casing, but those data that are available indicate that this material is relatively inert in most monitoring environments (Hunkin et al., 1984). Some adsorption of volatile organic compounds occurs with casing and screen materials composed of commercially available epoxy fiberglass, but this adsorption does not have a significant effect on water samples if proper purging and sampling are performed (Gillham and O'Hannesin, 1990).

Coupling Procedures for Joining Casing

Rigid monitoring well casing and screen is produced in various lengths (usually 5, 10, or 20 feet) that are joined by various coupling methods during installation. Only a limited number of methods are available for joining lengths of casing or casing and screen together. The joining method used is dependent on the type of casing and casing joint type used. Irrespective of which joining technique is used, a uniform inner casing diameter should be maintained in monitoring well installations; a uniform outer casing diameter is also commonly desirable. Inconsistent inner diameters result in problems in using tight-fitting downhole equipment (i.e., development tools, borehole geophysical tools, sampling or purging devices), while an uneven outer diameter creates potential problems (i.e., bridging) with filter pack and annular seal placement. The latter problems tend to promote vertical water migration in the annular space between the borehole and the casing at the casing/annular seal interface to a greater degree than is experienced with uniform outer diameter casing (Morrison, 1984). Figure 7.4 illustrates some common types of joints used for assembling lengths of casing.

Joining Thermoplastic Casing

There are two basic methods of joining sections of thermoplastic well casing: solvent cementing (using slip joints) and mechancial joining (using threaded joints). In solvent cementing, a solvent primer is generally used to clean the two pieces of casing to be joined and a solvent cement is then spread over the cleaned surface areas. The two sections are assembled while the cement is wet, allowing

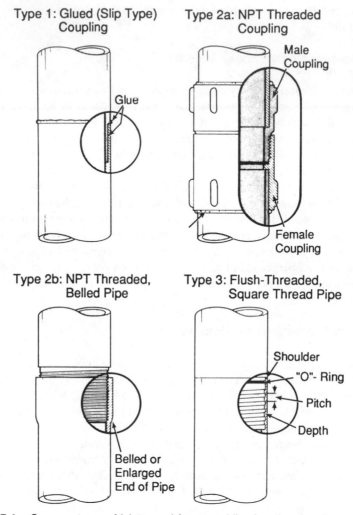

Figure 7.4. Common types of joints used for assembling lengths of casing.

the active solvent agent(s) to penetrate and soften the two casing surfaces that are joined. As the cement cures, the two pieces of casing are fused together; a residue of chemicals from the solvent cement remains at the joint. There are many different formulations of solvent cement for thermoplastics, but most cements consist of two or more of the following organic chemical constituents: tetrahydrofuran (THF), methyl ethyl ketone (MEK), methyl isobutyl ketone (MIBK), cyclohexanone, and dimethylformamide (Sosebee et al., 1983).

Clearly, the cements used in solvent welding, which are themselves organic chemicals, could produce some impact on the integrity of ground-water samples. Sosebee et al. (1983) demonstrated that the aforementioned volatile organic solvents do, in fact, contaminate ground-water samples collected from monitoring wells in which PVC adhesives are used to join well casing sections, masking the presence of these and other compounds in ground water. Barcelona et al. (1983) note that even minimal solvent cement application is sufficient to result in consistent levels of primer/cement components above 100 parts per billion in groundwater samples despite proper well development and flushing prior to sampling, and that these effects may persist for months after well construction, even after repeated attempts to develop the wells. Dunbar et al. (1985) cite a case in which THF was found at low levels (10 to 200 ppb) in samples taken from several PVC monitoring wells in which PVC solvent cement was used, more than two years after the wells were installed. In samples from adjacent monitoring wells in which threaded PVC casing was used, no THF was found, prompting the conclusion that the THF concentrations were a relict of solvent cement use in well construction. All of these results point to the fact that solvent cementing is not appropriate for use in joining sections of thermoplastic casing used in ground-water monitoring wells used for determinations of ground-water quality.

The most common method of mechanical joining of thermoplastic materials is by threaded connections. Molded and machined threads are available in a variety of thread configurations, including acme, buttress, standard national pipe thread (NPT), and flat square threads. Casing buyers should be aware of the fact that most manufacturers have their own thread type and that threaded casing from one manufacturer will probably be incompatible with casing from another manufacturer. If threads do not match and a joint is forced, it is likely that the joint will fail or leak during or after casing installation.

Flush-joint threaded casing is by far the most popular type of casing. Flush-joint threaded casing has male and female threaded ends that, when threaded together, produce a union internally and externally equal in diameter to the casing sections that are joined. This internal and external flush joint is designed to be free of gaps or irregularities at the connection when hand tightened. The smooth interior surface permits unobstructed insertion and removal of development and sampling equipment. This smooth surface is significant because the clearance provided for most equipment is usually less than 0.3 in., and sharp edges could cause plastic-coated cables, chambers on pumps, and surge blocks to snag and become temporarily or permanently lodged in the casing. The flush exterior surface permits easy installation and completion using steel casing or hollow-stem augers.

Effective seals are needed at casing joints to prevent neat cement or bentonite grout or contaminated water (i.e., cross-contamination in multiple completion wells) from entering the well. Some flush-threaded coupling designs include O-rings to ensure an effective coupling. For casing that does not come with O-rings, O-rings can be ordered directly from O-ring manufacturers in a greater variety of materials and sizes than are available from casing suppliers. The most common elastomer bases are nitrile (Buna-N), neoprene, ethylene propylene, butyl, Viton, polyacrylate, polysulfide, silicone, and fluorosilicone. Of these, the best choices for monitoring wells usually would be either nitrile, ethylene propylene, or Viton; however, materials should be selected on the basis of chemical resistance, temperature stability, strength, and other properties important to stand up to site conditions. Low tensile strength and the lack of tear and abrasion resistance make silicone and fluorosilicone poor choices for most applications. Some common ground-water contaminants and the relative suitability of elastomer materials are presented in Table 7.14. The O-ring elastomers are listed in decreasing order of preference for each chemical medium. Additional information on compatibility of elastomers for many more chemical media is available from O-ring manufacturers.

Table 7.14.   O-Ring Elastomer Materials Suitable for Common Chemical Media.

Chemical Media	Elastomers[a]	Chemical Media	Elastomers[a]
Arsenic (acidic)	N, E, V, C	Potassium dichromate	N, E, V, C
Benzene	V	Salt water	N, E
Carbon tetrachloride	V	Stoddard solvent	N, V, P
Chrome plating solutions	E, V, B	Tetrahydrofuran	E
Creosote	N, V, P	Toluene	V
Dry-cleaning fluids	V	Trichloroethane	V
Gasoline	N, V	Trichloroethylene	V
Hydrazine	E, B	Trinitrotoluene (TNT)	V
JP4 (MIL-J5624)	N, V	Type I fuel (MIL-S-3136)	N, V
Lacquer solvents	S	Type II fuel (MIL-S-3136)	V, N
Methylene chloride	E, B	Vinyl chloride	Metal
Methyl ethyl ketone	E, B	Xylene	V
Phenol	V	Zinc salt solutions	N, E, V, C

[a]Elastomer codes: B, butyl; N, nitrile (Buna-N); C, neoprene (chloroprene); E, ethylene propylene; V, fluorocarbon ("Viton"); P, polyacrylate; S, polysulfide.

In lieu of using O-rings, couplings can be sealed using Teflon tape; however, Teflon tape is difficult to put on threaded couplings in cold wet weather. Gaskets or O-rings, particularly those installed at the factory, can save time during installation and ensure a good seal, which is not always true when Teflon tape is wrapped on threads.

Most current designs of the flush-joint threaded coupling have square threads, which are easier to screw together, less likely to become cross threaded, and easier to unscrew than the NPT V-shaped threads.

Revised ASTM Standard F-480 (1988) specifies a particular type of joint for thermoplastic casing materials (Figure 7.5). The ASTM F-480 joint features interlocking angled faces, an O-ring seal, a two-threads-per-inch square-form thread and an easy-starting lead (Foster, 1989). The thread is quick and easy to field assemble, is reusable, and the O-rings are readily available in inert materials. The joint, after it is made up, will resist external forces of at least 25 p.s.i. Most manufacturers now produce the F-480 joint, and it is available in PVC, stainless steel and FRE.

Because all joints in a monitoring well casing should be watertight, the extent of tightening of joints should comply with manufacturers' recommendations. Caution should be exercised in tightening joints in thermoplastic casing, as overtightening could lead to structural failure of the joint (NWWA/PPI, 1980).

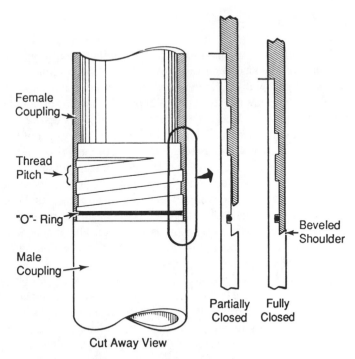

Female
Coupling

Thread
Pitch →{

"O"- Ring

Male
Coupling

Beveled
Shoulder

Partially   Fully
Closed   Closed

Cut Away View

Figure 7.5. ASTM standard flush-threaded coupling.

When using thermoplastic well casing, the ASTM F-480 threaded joint is preferred; the problems associated with the use of solvent primers and cements are thus avoided. Casing with threads machined or molded directly onto the pipe (without use of larger diameter couplings) provides a flush joint between both inner and outer diameters and is thus best suited to use in monitoring well construction. Though this type of joint reduces the tensile strength of the casing string compared to a solvent-cemented joint, in shallow (i.e., less than 200 feet) monitoring well installations this is not a critical concern.

Some manufacturers of flush-joint threaded casing provide sections of casing in exact lengths (i.e., compensated lengths), so that each length when threaded to the next is exactly 5.0 or 10.0 feet in length. This eliminates the potential for screening the wrong interval during placement of the well screen and casing. Less exact casing lengths can vary as much as 0.5 feet, and typically are from 0.05 to 0.2 feet short of the specified dimension.

Joining Fluoropolymer Casings

Because fluoropolymers are inert to chemical attack or solvation even by pure solvents, solvent welding cannot be used with fluoropolymers. As with thermoplastics, threaded joints are preferred.

Joining Metallic Casing

There are generally two options available for joining metallic well casings: resistance welding or threaded joints (either with or without couplings). Welding produces a casing string with a relatively smooth inner and outer diameter while threaded joints may or may not, depending upon whether or not couplings are used. With welding, it is generally possible to produce joints that are as strong as, or stronger than, the casing, thus enhancing the tensile strength of the casing string. The disadvantages of welding include greater assembly time, the difficulty of properly welding casing in the vertical position, enhancement of corrosion potential in the vicinity of the weld, and the danger of ignition of potentially explosive gases which may be encountered in some monitoring situations.

Because of these disadvantages, threaded joints are much more commonly employed in monitoring well installations completed with metallic casing and screen. Threaded joints provide inexpensive, fast, and convenient connections, and greatly reduce potential problems with chemical resistance or interference (due to corrosion) and ignition of explosive gases. One disadvantage to using threaded joints is that the tensile strength of the casing string is reduced. Another is that threaded, coupled joints increase the probability of bridging of materials installed in the annulus. Because strength requirements for small-diameter wells typical of monitoring installations are not as critical, and because of the high initial tensile strength of metallic casings, the reduced tensile strength at the threaded joint usually does not pose a significant problem.

Most of the stainless steel casing produced for use in monitoring wells is thin-walled casing in which threads cannot be machined. The joints in thin-walled (i.e., schedule 5 or 10) stainless steel casing are usually made in a short (two- to three-inch) length of schedule 40 pipe that is welded onto the end of the thin-walled casing at the factory. While the outer diameter of this type of threaded joint is flush, the inner diameter is not. The inside diameter of the joint in this type of casing is thus the effective inside diameter of the casing, which may be critical in the use of some downhole equipment. Additionally, because the inner joint between the thin-walled casing and the schedule 40 joint is not smooth, the use of some development techniques (i.e., use of a surge block) may be restricted, and extra care should be exercised so sampling devices are not damaged during installation or removal from the well.

Joining Fiberglass-Reinforced Epoxy Casing

Several types of joints that do not require the use of cements or other chemicals are available for use with fiberglass-reinforced epoxy casing. The most common joint for fiberglass casing used in monitoring wells is the flat square thread joint that has a uniform inside diameter. As with stainless steel, the thin-walled nature of most fiberglass casing does not allow joints to be machined directly onto the casing. With fiberglass, however, the end of the casing (i.e., the portion at the joint) is belled outward to make it thicker, and the threads are machined into the thicker, belled portion of the casing. This slightly increases the outside diameter of the casing at the joint, but because the transition from thin-walled casing to thicker-walled joint is smooth, the likelihood of bridging of materials installed in the annulus is reduced substantially compared to joints made with couplings.

Factors Affecting Selection of Well Casing Diameter

The diameter of most ground-water quality monitoring wells is smaller than that of water wells or wells used in water-resource studies. The advantages and disadvantages of small-diameter versus large-diameter wells have been debated (Rinaldo-Lee, 1983; Schalla and Oberlander, 1983; Schmidt, 1982, 1983; Voytek, 1983). Important reasons for using large-diameter wells include determining large-scale aquifer characteristics (i.e., transmissivity, storativity) and boundary conditions of high-yield aquifers via pumping tests. However, where such high-yield conditions do not exist, the reasoning has been that small-diameter wells are better (Voytek, 1983). Small-diameter wells are less expensive because: (1) smaller quantities of materials are installed; (2) drilling costs per foot are lower because borehole diameters are smaller and less-costly drilling methods can be used; and (3) the quantities of potentially contaminated drill cuttings, well-development-related water and sediment, and purge water that may require disposal at an approved hazardous waste disposal site are much lower.

Care should be exercised in the selection of casing diameter. While casing outside diameters are standardized, variations in wall thickness (i.e., casing schedule) result in variations in casing inside diameters. As illustrated by Table 7.15, two-inch nominal casing is a standard 2.375 inches outside diameter; wall thickness varies from 0.065 inches for schedule 5 to 0.218 inches for schedule 80. This means that inside diameters for two-inch nominal casing vary from 2.245 inches for schedule 5 thin-walled casing (typical of stainless steel, for example) to only 1.939 inches for schedule 80 thick-walled casing (typical of PVC). This factor must be taken into consideration when determining the proper casing size for a particular monitoring program.

The diameter of the casing for a monitoring well is dependent on the purpose of the well, and on the need for the well to accommodate downhole equipment. Additional casing diameter selection criteria include: drilling or well installation method used, anticipated depth of the well and associated strength requirements, ease of well development, volume of water required to be purged prior to sampling, rate of recovery of the well after purging, and cost.

Downhole Equipment

A variety of downhole equipment may be utilized in a monitoring well, including well development tools, borehole geophysical tools, pumps for conducting pumping tests, water-level measuring devices, and purging and sampling devices. In general, large-diameter wells (i.e., those greater than 4-inch inside diameter)

Table 7.15. Outside Diameter, Wall Thickness, and Inside Diameter of Well Casing.

Casing Size (Nominal)	Outside Diameter (Standard)	Wall Thickness				Inside Diameter			
		Sch 5	Sch 10	Sch 40	Sch 80	Sch 5	Sch 10	Sch 40	Sch 80
2 in	2.375	0.065	0.109	0.154	0.218	2.245	2.157	2.067	1.939
3 in	3.500	0.083	0.120	0.216	0.300	3.334	3.260	3.068	2.900
4 in	4.500	0.083	0.120	0.237	0.337	4.334	4.260	4.026	3.826
5 in	5.563	0.109	0.134	0.258	0.375	5.345	5.295	5.047	4.813
6 in	6.625	0.109	0.134	0.280	0.432	6.407	6.357	6.065	5.761

can be developed by a wider variety of methods with commercially available development tools than can small-diameter wells; most well development tools in current use have been designed for larger diameter wells than are used for most ground-water monitoring programs. The same applies to borehole geophysical equipment—only wells 4-inch or greater in inside diameter will accommodate most borehole geophysical tools. Some slim-hole well logging tools are available, but the parameters they measure are generally not applied in monitoring studies.

Only wells 4-inch inside diameter or larger can accommodate pumps of sufficient capacity to conduct a pumping test in most formations. For wells exceeding several hundred feet in depth, 8-inch inside diameter wells may be advisable (Schmidt, 1982). While wells less than 4-inch inside diameter are not themselves suitable for conducting pumping tests, they may be used as observation wells during a pumping test. In materials of low hydraulic conductivity in which pumping tests would not be successfully applied, small-diameter wells will meet the requirements of other hydraulic testing methods such as slug tests or bail tests.

For most water-level measuring methods (chalked tapes, electric probes, etc.) small-diameter wells are adequate; some piezometers used strictly for measuring water levels may, in fact, be smaller than 1-inch inside diameter. However, only large-diameter wells (4-inch inside diameter or greater) will accommodate the installation of most float-type continuous water-level recorders.

With respect to purging and sampling devices, a greater variety of high-capacity pumping devices is available for large-diameter wells than for small-diameter wells. Recently, devices for both purging and sampling from small-diameter wells (i.e., wells of 2-inch nominal diameter) have become widely available, but most of these are limited to relatively low pumping rates (i.e., less than 1.5 gallons per minute at 50 feet of pumping lift). While low pumping rates are usually required for sampling, higher rates (e.g., 3 to 5 gallons per minute) are commonly desired for purging.

Well-Drilling Method

While almost any diameter well can be installed with drilling methods such as air or mud rotary or cable tool, a wider variety of well-installation methods is available for small-diameter wells. For example, generally only wells of 4-inch nominal diameter or less can be successfully installed through a hollow-stem auger. Additionally, small-diameter wells can be more easily driven or jetted into place.

Anticipated Well Depth and Casing Strength

For shallow monitoring wells, the strength characteristics of nearly all diameters of all casing materials are adequate. The greater strength requirements for deeper well installations—to prevent well casing failure, severe bends in the casing, and difficulties in well construction procedures—favor larger diameter, thicker-walled casing.

Ease of Well Development

With regard to ease of well development, smaller diameter wells usually take less time to develop because smaller volumes of water and sediment are involved. However, because the number of development methods that can be used in smaller diameter wells is limited, and because the development effects are more subdued in smaller diameter wells, the well development process may not be as effective in these wells as it is in larger diameter wells.

Purge Volume

Because it is usually necessary to remove water standing in a well prior to taking a ground-water sample, the volume of water required to be purged from the well must be considered. Because volume increases significantly with well diameter, excessive storage volume in large-diameter wells can increase the time required for sampling by increasing the volume of water that must be evacuated (purged) from the well prior to sampling. For example, the purge volume of a 4-inch nominal diameter well is four times the purge volume for a 2-inch nominal diameter well. Additionally, with increased volume, purging costs also increase, especially if the purged water is contaminated and thus must be treated or contained, and later disposed as a hazardous material. However, in situations in which the saturated thickness of the monitored zone (and thus the column of water in the well) is large, where well depths are great, and where formation hydraulic conductivities are high, a pump capable of removing large volumes of water rapidly may be desirable to purge the well. Typically, a casing of at least 4-inch inside diameter is required to accommodate a high-capacity submersible pump.

Rate of Recovery

Rate of recovery of the well after purging is also an important consideration. In formations with very low hydraulic conductivities, the greater storage volume of a large-diameter well can result in a much slower recovery of water levels after purging. Small-diameter wells are usually preferred in low hydraulic conductivity formations because the well volume is small and because the time of recovery is directly proportional to well volume, which increases with well diameter. In a formation with a given hydraulic conductivity, it takes less time for a small-diameter well to recover when a slug of water is removed than for a large-diameter well (Rinaldo-Lee, 1983).

Unit Cost of Casing Materials and Drilling

Unit (per foot) costs of both casing materials and drilling increase as well diameter increases. Depending upon the material selected, unit costs for a 4-inch or 6-inch nominal diameter casing can be two to ten times the unit cost for a

2-inch inside diameter casing (Richter and Collentine, 1983). Several small-diameter wells can commonly be constructed for the same price as a single large-diameter well of the same depth, though this is not always the case, particularly with deep wells. The collection of a greater amount of data over a large area can sometimes be accomplished with small-diameter wells for the same price.

The highly variable geologic and contaminant conditions found at sites at which monitoring wells are constructed, and the varied purposes for which monitoring wells are installed, dictate that a flexible approach to selection of well diameter is required, and that no one well diameter is optimum.

Keeping Casing Centered in the Borehole

It is important to install the casing string (casing and screen) so that it is centered in the borehole. The primary reason for this is to ensure that, if the well is to be artificially filter-packed, the filter-pack material fills the annular space evenly around the screen. Perhaps equally important is ensuring that annular seal materials fill the annular space evenly around the casing.

Centering can be accomplished effectively in one of several ways. For wells that are installed in cohesive materials in which the borehole stands open during well construction, centering guides (also called centralizers) can be affixed to the casing string before or as the string is installed. Centralizers, which are typically expandable metallic (i.e., stainless steel) or plastic (i.e., PVC) devices that attach to the outside of the casing and are adjustable along the length of the casing, are generally attached immediately above the well screen (and, in some cases, at the bottom of the well screen) and at 10- or 20-foot intervals along the casing to the surface. It is easier to measure the annular depth during placement of the filter pack and grout without entangling the measuring tapes if factor attached, semicircular shaped centralizers are used rather than expandable centralizers. Centralizers may also be used in this manner in boreholes that are temporarily cased (i.e., wells drilled by cable-tool or air rotary with casing hammer). The centralizers should be adjusted to a diameter just less than that of the borehole or temporary casing, so that the casing is kept an equal distance from all sides of the borehole. Installed in this manner, the casing string will remain centered in the borehole.

A casing string can also be installed centered in the borehole (or nearly so) without centralizers if it is installed through the center of a hollow-stem auger. The auger acts as a centering device, maintaining a relatively consistent annular space on all sides of the casing. Because the inside of the hollow-stem auger is always somewhat larger than the casing string installed within it, it is possible for the casing to be installed slightly off-center, with the casing resting against one side of the auger. However, this should not materially affect the well installation.

Sediment Sumps

It is a common practice in water-supply wells to install a sediment sump or trap (also called a "tailpipe"), or a piece of blank casing installed below the well screen, to collect sediment either brought into the well during development or carried into the well by continued pumping over time. Some contractors have carried this practice over to the installation of monitoring wells. Other monitoring well designers have suggested that a sump installed below the screen would allow for the collection of samples of dense, nonaqueous phase liquids (DNAPLS).

In a properly installed filter-packed monitoring well, very little sediment should be developed into the well, whereas in a naturally developed well, a great deal of sediment may be brought into the well during development. Though this may argue for the use of a sediment sump, Yu (1989) points out that sediment brought into the well during development or well purging should be removed prior to ground-water sampling to avoid the phenomenon of chemicals sorbing onto and then desorbing from the sediments that may have collected in the sump. Furthermore, Yu (1989) contends that the two suggested uses of the sump are mutually exclusive (i.e., if the sump traps sediment, it cannot trap DNAPLS), and that if it does trap DNAPLS, it would be difficult to purge without leaving residue at the bottom of the well that would contaminate future samples taken from the well. With these apparent problems in mind, the use of a sediment sump is probably not appropriate for most monitoring well applications.

Cleaning Requirements for Casing and Screen Materials

During the production of any casing or screen material, certain chemical substances may be used to assist in the extrusion, molding, machining, or stabilization of the casing material. For example, during the production of steel casing, considerable quantities of oils and solvents are used in various manufacturing stages, as during the machining of threads. In the manufacturing of PVC well casing, a wax layer can develop on the inner wall of the casing; additionally, protective coatings of natural or synthetic waxes, fatty acids, or fatty acid esters may be added to enhance the durability of the casing (Barcelona et al., 1983). All of these represent potential sources of chemical interference, and must be removed prior to installation of the casing in a borehole. If trace amounts of these materials remain adhered to the casing after installation, they may affect the chemical integrity of samples taken from those wells.

Careful preinstallation cleaning of casing materials is essential to avoid potential chemical interference problems from the presence of such substances as cutting oils, cleaning solvents, lubricants, threading compounds, waxes, or other chemical residues. For PVC, Curran and Tomson (1983) suggest washing the casing with a strong laboratory-grade nonphosphate detergent solution (Liquinox) and then rinsing with water before installation. Barcelona et al. (1983) suggest the same procedure for all casing materials. To accomplish the removal of some cutting oils, lubricants, or solvents, it may be necessary to steam-clean casing materials or employ a high-pressure hot water wash. Care should also be taken to ensure that casing materials are protected from contamination while they are onsite awaiting installation in the borehole. This is usually done by providing a clean storage area away from any potential contaminant sources (air, water, or soil) or using plastic sheeting spread on the ground for temporary storage adjacent to the work area.

Factory cleaning of casing or screen in a controlled environment by standard detergent washing, rinsing, and air-drying procedures is superior to any cleaning efforts attempted in the field. Factory cleaned and sealed casing and screen can be certified by the supplier. Individually wrapped sections in a common shipping crate are easiest to keep clean and to install.

Costs of Casing Materials

As Scalf et al. (1981) point out, the dilemma for the field investigator is the relation between cost and accuracy. PVC is approximately 1/10 the cost and Type 304 stainless steel is approximately 1/3 the cost of equivalent diameter fluoropolymer materials, which may be a major consideration when a ground-water monitoring project may entail the installation of a large number of wells, or deep wells. On the other hand, if the particular compounds of interest in a monitoring program are also components of the lower-cost casing, and analytical sensitivity is in the parts-per-billion range, the data generated on contaminants detected which are potentially attributable to the casing may be suspect.

In many situations, it may be possible to use the lowest-cost casing material without compromising accuracy. For example, if the contaminants to be monitored are already defined and they do not include chemical constituents that could potentially leach from or sorb onto the lower-cost well casing (as defined by laboratory or field studies), it should be possible to use the less expensive casing without the worry of compromising analytical accuracy. Or, as Scalf et al. (1981) suggest, wells constructed of "less than optimum" materials might be used with a reasonable level of confidence for sampling if at least one identically constructed well was available in an uncontaminated part of the monitored formation to provide ground-water samples as "blanks" for comparison.

Hybrid Wells

The preceding discussions on chemical interference are all related to situations in which the casing or screen material is in contact with the saturated portion of the subsurface. For materials that are not in contact with the saturated zone, the arguments regarding chemical attack and sorption/leaching phenomena are generally not valid. Thus, it may be possible to utilize less chemically resistant or less chemically inert casing materials in the unsaturated zone, coupled with a more chemically resistant/inert material in the saturated zone. It should be noted, however, that seasonal or more frequent (i.e., tidal or pumping-influenced) variations in ground-water levels must be taken into account when determining the depth at which the more resistant or inert material should be used. In addition, it should be noted that some types of casing material may not be compatible. For example, the joining of two different types of steel casing (i.e., galvanized and stainless steel) may result in the creation of a galvanic cell and subsequent corrosion at the joint because the two metals have different electromotive potentials. Thus, materials should be considered for compatibility prior to installation.

Because of the high cost of fluoropolymer well casing and its inherent strength limitations and installation/handling difficulties, several authors have proposed the use of "hybrid" monitoring wells, or wells constructed of fluoropolymer casing or screen in the saturated zone, with a less-expensive, stronger, more easily handled material in the unsaturated zone. This type of arrangement would take advantage of the best properties of both materials; the material perceived as more chemically inert would be in contact with ground water to be sampled, and the other material would supply the necessary rigidity, strength, and ease of handling necessary for a deep, straight well installation. Pettyjohn et al. (1981), for example, suggest the use of fluoropolymer casing and screen in the saturated zone with galvanized steel casing in the unsaturated zone; Dablow et al. (1986) suggest the use of stainless steel casing in the unsaturated zone and fluoropolymer casing and screen in the saturated zone. A more practical combination of materials would probably be PVC for use in the unsaturated zone and fluoropolymer in the saturated zone. However, difficulties in finding compatible threaded joints and the strength problems with fluoropolymers noted earlier would seem to make the installation of so-called hybrid wells impractical or at least difficult.

DESIGN OF MONITORING WELL INTAKES— WELL SCREEN AND FILTER PACK

Proper design of a hydraulically efficient monitoring well in unconsolidated geologic materials and in certain types of poorly consolidated geologic materials requires that a well screen be placed opposite the zone to be monitored and surrounded by materials that are coarser and of higher hydraulic conductivity than natural formation material. This allows ground water to flow freely into the well from the adjacent formation while minimizing or eliminating the entrance of fine-grained formation materials (clay, silt, fine sand) into the well. The well can thus provide adequate quantities of ground-water samples that are free of suspended solids. Sediment-free water samples allow for dramatically shortened filtering times or the elimination of sample filtering, and thus greatly reduce the potential for sample bias or interference.

Much of the technology applied to the design and installation of monitoring wells has been derived from the water well industry. It should be noted that while production or water-supply wells and monitoring wells are similar in many ways, there are some distinct differences between the two types of wells. For example, one significant difference between monitoring wells and production or "water" wells is that the intake section, or screen, of monitoring wells is often purposely situated in a zone of poor quality water and/or poor yield. The quality of water entering a monitoring well can range from drinking water to a hazardous waste or leachate. In contrast, production wells are normally designed to obtain water from highly productive zones containing good quality water. The screen of a monitoring well often extends only a short length to obtain water from, or to monitor conditions within, an individual water-bearing unit. Water wells are often designed to obtain water from multiple water-bearing strata. Although there are usually differences between the design and function of monitoring wells and water wells, water wells are sometimes used as monitoring wells and vice versa.

There are two basic types of wells and well intake designs for wells installed in unconsolidated or poorly consolidated geologic materials—naturally developed wells, and wells with an artificially introduced filter pack. In a naturally developed well, the well screen is in direct contact with formation materials, and the filter pack is produced through proper well development. In an artificially filter-packed well, coarse-grained filter-pack material is installed in the annular space between the well screen and the formation to prevent formation material from entering the well. In both types of wells, the effective diameter of the well is increased by surrounding the well intake with an envelope of coarse material with higher hydraulic conductivity than the natural formation material.

Naturally Developed Wells

In a naturally developed well, formation materials are allowed to collapse around the well screen after it has been installed in the borehole. A high hydraulic conductivity envelope of coarse materials (filter pack) is developed adjacent to the well screen in situ by removing the fine-grained materials from the natural formation materials during the well development process.

As described by Johnson Division (1966), the envelope of coarse, graded material created around a well screen during the development process can be visualized as a series of cylindrical zones. In the zone immediately adjacent to the well screen, development removes all particles smaller than the screen openings, leaving only the coarsest material in place. Slightly farther away, some medium-sized grains remain mixed with coarse materials; beyond that zone, the material grades back to the original character of the water-bearing formation. By creating this succession of coarse, graded zones around the screen (Figure 7.6), development stabilizes the formation so that no further movement of fine-grained materials should take place, and the well should yield sediment-free water.

The decision on whether or not a filter pack can be naturally developed is generally based on geologic conditions, specifically the grain-size distribution of natural formation materials in the zone to be monitored. Naturally developed wells are generally recommended in situations in which natural formation materials are relatively coarse-grained, permeable, and uniform in grain size. Grain-size distribution must be determined by conducting a mechanical (sieve) analysis of a sample or samples taken of the formation materials from the intended screened interval. For this reason, the importance of obtaining accurate formation samples cannot be overemphasized.

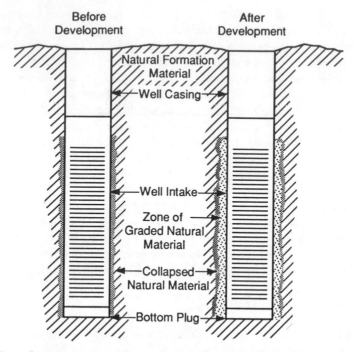

Figure 7.6. Development of filter pack around a naturally developed well.

After a sample(s) of formation material is(are) sieved, a plot of grain size vs cumulative percentage of sample retained on each sieve is made (Figure 7.7). Based on this grain-size distribution, and specifically upon the effective size and uniformity coefficient of the formation materials, well screen slot sizes are selected. The effective size is equivalent to the sieve size that retains 90% (or passes 10%) of the formation material (Figure 7.8); this is termed the D10 size. The uniformity coefficient is the ratio of the sieve size that retains 40% (or passes 60%, the D60 size) of the formation material to the effective size (Figure 7.9). A naturally developed well can normally be justified if the effective grain size of the formation material is greater than 0.01 inch and the uniformity coefficient is greater than 3.0 (Johnson Division, 1966). While these criteria are usually applied to production water wells, they are also applicable to monitoring wells.

Well Screen Design in a Naturally Developed Well

Proper sizing of monitoring well screen slot size is perhaps the most important aspect of monitoring well design. There has been an unfortunate tendency among some monitoring well designers to install a "standard" or common screen slot size (i.e., 0.010-inch slots) in every well, regardless of formation characteristics. As Williams (1981) points out, this can lead to difficulties with well development and poor well performance.

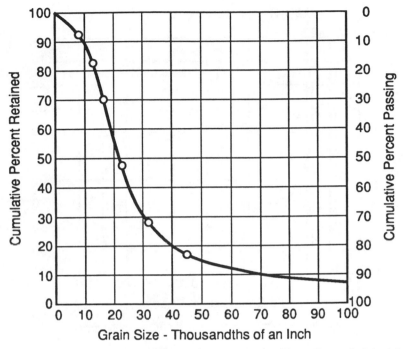

Figure 7.7.   Plot of grain size vs. cumulative percentage of formation material retained on sieves.

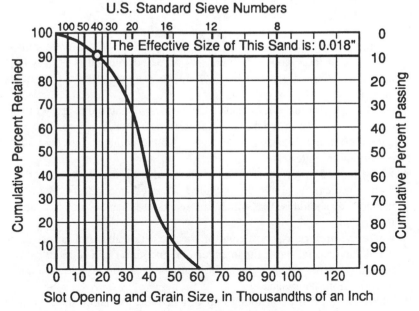

Figure 7.8.   Determining effective size of formation materials.

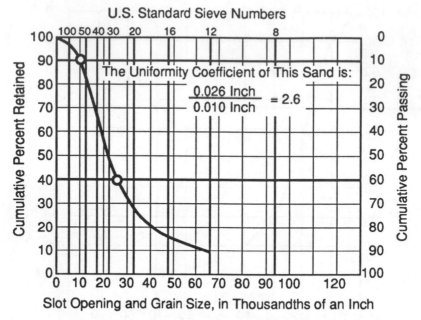

Figure 7.9. Determining uniformity coefficient of formation materials.

Well screen slot sizes are generally selected based upon the following criteria, which were developed primarily for production wells but, with some modifications, also apply to monitoring wells:

- where the uniformity coefficient of the formation material is greater than 6 and the material above the intended screened interval is noncaving, the slot size should be that which retains no less than 40% of formation materials.
- where the uniformity coefficient of the formation material is greater than 6 and the material above the intended screened interval is a readily caving material, the slot size should be that which retains no less than 50% of formation materials.
- where the uniformity coefficient of the formation material is less than 3 and the material above the intended screened interval is noncaving, the slot size should be that which retains no less than 50% of formation materials.
- where the uniformity coefficient of the formation material is less than 3 and the material above the intended screened interval is a readily caving material, the slot size should be that which retains no less than 60% of formation materials.
- where an interval to be monitored has layered formation materials of differing sizes and gradations, and where the 50% (or average) grain

size of the coarsest layer is less than 4 times the 50% size of the finest layer, the slot size should be selected on the basis of the finest layer. Otherwise, separate screened sections should be sized for each zone.

The slot size determined from a sieve analysis is seldom that of commercially available screen slot sizes (Table 7.16), so the nearest smaller standard slot size is generally used. In most monitoring wells, because optimum yield from the well is not as critical to achieve as it is in production wells, screens are usually designed to have smaller slot sizes than indicated by the above design criteria so that much less than 60% of the formation material adjacent to the well will be pulled into the well during development.

Table 7.16. Typical Commercially Available Slotted Casing Slot Widths.

0.006	0.016	0.040
0.007	0.018	0.050
0.008	0.020	0.060
0.010	0.025	0.070
0.012	0.030	0.080
0.014	0.035	0.100

Source: NWWA PPI, 1980.

Installing a naturally filter-packed well is advantageous in formations comprised of coarser materials, particularly if mud rotary drilling is used. The absence of an artificial filter pack allows for maximum effectiveness for developing the formation and for removing the finer drill cuttings and drilling fluids from the borehole in the screened interval. Perhaps the biggest drawback for naturally developed wells may be the time required for well development to remove fine-grained formation material. Because the design of the well screen may allow up to 60% of the formation materials near the well screen to enter the well, development can often be a long, drawn-out process. Increased development has other disadvantages: (1) it delays installation of the overlying well seal; (2) it may require the handling of large quantities of contaminated sediment and water; and (3) it allows settlement adjacent to the screen, which may result in the invasion of overlying sediments. Also, unless a formation is fairly coarse grained, developing a natural filter pack for a monitoring well is difficult, because the small diameter of most monitoring wells limits the size and capacity of development equipment, the short screen length limits the rate of water withdrawal, and the removal of formation fines from the well is sometimes difficult.

Artificially Filter-Packed Wells

Artificially filter-packed wells are produced when the natural formation materials surrounding the well intake are deliberately replaced by coarser material introduced from the surface. In much of the literature describing production well design (i.e., Johnson Division, 1966), the term "gravel pack" is used to describe the artificial material added to the borehole to act as a filter. Because the term "gravel" is classically used to describe large-diameter granular material and because nearly all coarse material emplaced artificially in monitoring wells is coarse to medium sand-sized material and not gravel, the use of the term "sand-pack" or "filter-pack" is preferred. True gravel-sized material is rarely used as filter pack in a monitoring well.

The artificial introduction of coarse material into the annular space between a centrally positioned well screen and the borehole wall serves a variety of purposes. As with a naturally developed filter pack, the primary purpose is to work in conjunction with the well screen to filter out fine materials from the formation adjacent to the well. In addition, the artificial filter pack stabilizes aquifer materials in order to avoid excessive settlement of materials above the well intake during well development. The introduction of material coarser than natural formation materials also results in an increase in the effective diameter of the well and in an accompanying increase in the amount of water that flows toward and into the well.

Artificial filter packs have been used extensively by water well contractors to construct efficient, large-diameter wells to provide water for irrigation and for municipal and industrial uses. Monitoring wells serve a different purpose and thus have different filter pack design requirements. Because monitoring wells are designed to serve as sampling ports in an aquifer, they are typically smaller in diameter and screened in only a portion of the aquifer. Furthermore, the design properties for monitoring well filter packs are more important than those for water well filter packs, because the disturbance of water chemistry and hydrology must be minimized.

Water well design practices have historically been used in the design of well screens and filter packs for monitoring wells because available design texts, which pertain almost exclusively to water well technology (Driscoll, 1986; Johnson Division, 1966; Campbell and Lehr, 1973; Anderson, 1971), do not fully or adequately address the subject of well screen and filter pack requirements for monitoring wells. Information on monitoring well design appears in only a limited number of recent research papers. Monitoring wells and water wells do have the following filter pack design principles in common:

- filter packs are installed to create a permeable envelope around the well-screen; and
- the selection of the filter pack grain size should be based on the grain size of the finest layer to be screened.

The decision regarding whether an artificial filter pack should be utilized in the construction of a monitoring well is based primarily on geologic considerations. There are several geologic situations in which the use of an artificial filter pack material may be required for a monitoring well:

1. where the natural formation material is unconsolidated and consists primarily of uniformly fine-grained sand, silt, or clay-sized particles;
2. where a long screened interval is required, and/or the well screen spans highly stratified geologic materials of widely varying grain sizes;
3. where the formation in which the intake is to be placed is poorly cemented, such as a friable sandstone; and
4. where the formation is a fractured or solution-channeled rock in which particulate matter is carried through large fractures or solution openings.

The use of an artificial filter pack in a fine-grained formation material allows the screen slot size to be considerably larger than if the screen were placed in the formation material without the filter pack. This is particularly true where fine-grained sands, silts and clays predominate in the zone of interest, and small enough slot sizes in well screens to hold out formation materials are impractical. The larger screen slot size afforded by an artificial filter pack thus allows for the collection of adequate volumes of sediment-free samples, and results in both decreased head loss and increased well efficiency.

Filter packs are especially well suited to use in extensively stratified formations, in which thin layers of fine-grained materials alternate with layers of coarser materials. In such a geologic environment, it is often difficult to precisely determine the position and thickness of each individual stratum and to choose the correct position and slot size for a well screen. Completing the well with an artificial filter pack, sized to suit the finest layer of a stratified sequence, resolves the latter problem and helps to ensure that the well will produce water free of suspended sediment.

Quantitative criteria exist with which decisions can be made concerning whether a natural or an artificial filter pack should be used in a production well (Campbell and Lehr, 1973; Johnson Division, 1966; Williams, 1981; U.S. EPA, 1975). In production wells, the use of an artificial filter pack is recommended in situations where the effective grain size of the natural formation materials is smaller than 0.010 inches and the uniformity coefficient is less than 3.0. For monitoring wells, California Department of Health Services (1985) recommends a different approach and suggests that an artifiical filter pack be employed if a sieve analysis of formation materials indicates that a screen slot size of 0.020 inches or less is required to retain 50% of the natural material.

Economic considerations may also affect decisions concerning the appropriateness of an artificial filter pack. Costs associated with filter packed wells are generally higher than those associated with naturally developed wells, primarily because specially sized and washed sand must be purchased and transported to the site. Additionally, larger boreholes are required for artificially filter packed wells (minimum 6-inch-diameter borehole for a 2-inch-inside diameter well; minimum 8-inch borehole for a 4-inch well). The increased costs for larger diameter wells are particularly significant if drilling is done by hollow stem augers or cable tool rigs.

Filter Pack and Well Screen Design for Artificially Filter-Packed Wells

To achieve the purposes of creating a high hydraulic conductivity envelope around the well screen, holding back formation materials and not interfering with ground-water chemistry, the filter pack must have certain characteristics. While some of these characteristics are the same as those desirable for production wells, monitoring wells have unique requirements. Artificial filter pack design factors for monitoring wells include: (1) filter pack grain size and well screen slot size; (2) filter pack grain size distribution properties (i.e., uniformity coefficient, effective size, kurtosis, and skewness); (3) filter pack grain shape properties (i.e., roundness, sphericity); (4) filter pack dimensions (thickness and length); and (5) filter pack material type. When an artificial filter pack is dictated by sieve analysis or by geologic conditions, the filter pack grain size and well screen slot size are generally designed at the same time.

Filter pack grain size and well screen slot size. Filter pack grain size and well screen slot size are determined by the grain-size distribution of the formation material. The filter pack, which is the interface with the formation, is the principal hydraulic structure of the well and is designed first. The first step in designing the filter pack is to obtain samples of the formation intended to be monitored and perform sieve analyses on the samples, the same procedure followed for sizing well screen slot sizes in a naturally developed well. The filter pack grain size is then selected on the basis of the finest formation materials present.

Although design techniques vary, all have in common a ratio to establish a grain size differential between the formation materials and filter pack materials. Generally this ratio refers to either the average or D50 (50% passing) grain size of the formation material or the 30% passing size (D30) of the formation material. For example, Walker (1974) and Barcelona et al. (1985a) recommend the use of a filter pack grain size that is three to five times the average (50% passing; D50) size of the formation materials. On the other hand, U.S. EPA (1975) and Johnson Division (1966) recommend that filter pack grain size be selected by taking the D30 grain size of the formation materials and multiplying it by a factor of between 3 and 6 (Figure 7.10), with 3 used if the formation is fine and uniform, and 6 used if the formation is coarser and nonuniform. In both cases, the uniformity coefficient of the filter pack materials should be between 1.0 and 2.5, and the gradation of the filter material should form a smooth and gradual size distribution when plotted. In uniform formation materials, both of these approaches to filter pack material sizing will give more or less similar results; however, in coarse, poorly sorted formation materials, the average grain size method may be misleading and should be used with discretion (Williams, 1981).

Because many formations have uniformity coefficients from 3 to 10 or higher, the formation's coarsest particles will probably be more coarse than the coarsest particle in the filter pack if these design criteria are used. The formation's finest

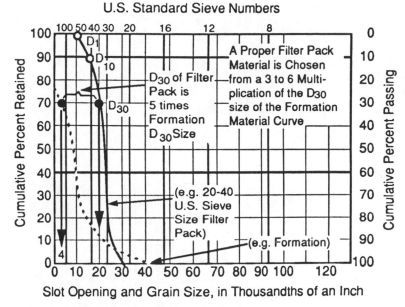

Figure 7.10. Artificial filter pack design criteria.

particles may be three to six times finer than the finest particles in the filter pack. The multiplier of three to six mentioned above is based on the assumption that, for uniform filter packs, the largest pore space will be one-third to one-sixth the average filter-pack particle size. Because retaining the bulk of formation particles is very important for monitoring wells, the same multipliers can be applied more conservatively for less uniform formations (i.e., 6 to 10) using the D10 (10% passing) size for selecting filter packs (Table 7.17).

It is possible to use a filter pack grain size that is too fine. For example, if a 0.020-inch slotted well screen is used with its appropriate (based on the D_1 size in Table 7.17) 10- to 20-mesh filter pack in a very coarse formation (i.e., coarse gravel with cobbles or boulders), the filter pack will be lost to the formation during development. The void left by the lost sand could result in the settlement of finer overlying formations, or the plastic flow of bentonite grout annular seal into the screened interval.

The size of well screen slots can only be selected after the filter pack grain size is specified. In production wells, slot size is generally chosen on the basis of its ability to hold between 85% and 100% of the filter pack materials (U.S. EPA, 1975). In monitoring wells it is desirable for the well screen to retain 90% to 100% of filter pack materials (Figure 7.11), because development is generally done after the well has been completed, and it is important to avoid excessive settling of the materials surrounding the well. To retain 99 to 100% of the

Table 7.17.   Recommended (Achievable) Filter Pack Characteristics for Common Screen Slot Sizes.

Size of Screen Opening [mm (in.)]	Slot No.	Sand Pack Mesh Size	1% Passing Size (D_1) (mm)	Effective Size (D_{10}) (mm)	30% Passing Size (D_{30}) (mm)	Range of Uniformity Coefficient	Roundness (Powers Scale)	Fall Velocities[a] (cm/s)
0.125 (0.005)	5	40–140	0.09–0.12	0.14–0.17	0.17–0.21	1.3–2.0	2–5	6–3
0.25 (0.010)	10	20–40	0.25–0.35	0.4–0.5	0.5–0.6	1.1–1.6	3–5	9–6
0.50 (0.020)	20	10–20	0.7–0.9	1.0–1.2	1.2–1.5	1.1–1.6	3–6	14–9
0.75 (0.030)	30	10–20	0.7–0.9	1.0–1.2	1.2–1.5	1.1–1.6	3–6	14–9
1.0 (0.040)	40	8–12	1.2–1.4	1.6–1.8	1.7–2.0	1.1–1.6	4–6	16–13
1.5 (0.060)	60	6–9	1.5–1.8	2.3–2.8	2.5–3.0	1.1–1.7	4–6	18–15
2.0 (0.080)	80	4–8	2.0–2.4	2.4–3.0	2.6–3.1	1.1–1.7	4–6	22–16

[a]Fall velocities in centimeters per second are approximate for the range of sand pack mesh sizes named in this table. If water in annular space is very turbid, fall velocities may be less than half the values shown here. If a viscous drilling mud is still in the annulus, fine sand particles may require hours to settle.

filter pack, the well screen slot size should be approximately equal to the D_1 size of the filter pack (Table 7.17, Figure 7.11). To retain 90% of the filter pack use the D_{10} sizes shown in Table 7.17.

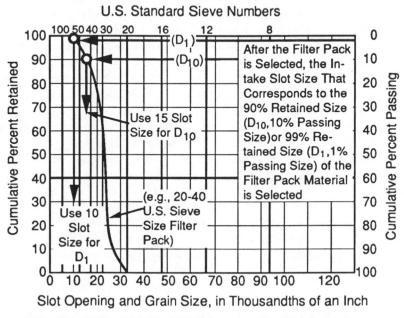

Figure 7.11. Selecting well intake slot size based on filter pack grain size.

Filter pack grain size distribution properties: uniformity coefficient, kurtosis, and skewness. Two types of artificial filter packs are in common use in production wells and in monitoring wells—the uniform filter pack and the graded filter pack. Uniform filter packs are generally preferred to graded packs for monitoring wells. Graded packs are more susceptible to the invasion of formation materials at the formation/filter pack interface, resulting in a partial filling of voids between grains and reduced hydraulic conductivity. With a uniform filter pack, the fine-grained formation materials can travel between the grains of the filter pack and be pulled into the well during development, thereby increasing formation permeability while retaining the highly permeable nature of the filter-pack material.

Futhermore, the filter pack should be composed of uniformly sized particles because the slots in a given well screen have uniform openings. Ideally, the uniformity coefficient [the 60% passing (D60) size divided by the 10% passing (D10) size (effective size)] of the filter pack should be as close to 1.0 as possible (i.e., the D60 size and the D10 size should be nearly identical), though this is rarely achieved. A low uniformity coefficient is very desirable, particularly if the tails of the particle-size distribution curve are also uniform (i.e., mesokurtic or platykurtic). The importance of kurtosis and skewness for filter packs is discussed by Schalla and Walters (1990).

Commercial filter pack materials that meet uniformity coefficient requirements typically have suitable kurtosis and skewness values. These characteristics, which describe the distribution of particle sizes, are important because during installation in the borehole, particles falling at terminal velocity in the quiescent fluid surrounding the well screen will be influenced by a number of variables, including fluid and particle density, fluid viscosity, and particle diameter, shape, and surface roughness. Assuming that all variables except particle size are constant, the fall velocity for sand-sized particles composed of quartz will be approximately proportional to the square root of the particle diameter (Simons and Senturk, 1977). For example, a coarse sand grain 4 mm (0.156 inches) in diameter falls twice as fast as a medium sand grain 1 mm (0.039 inches) in diameter. Approximate fall velocities for sand-sized particles are shown in Table 7.17.

If a coarse/medium sand filter pack material were placed in a well annulus through a few feet of water, the particles would segregate according to size, with the finest particles at the top of the screen and the coarsest particles at the bottom (Figure 7.12). During well development, fine formation particles would pass through the filter pack and well screen at the bottom and become lodged in the slots, as detailed in the bottom window in Figure 7.12. At the top of the screen, most, or possibly all, of the filter pack material would be lost during development if the screen were designed to retain 85% of the filter pack, because the finest 15% of the filter pack material would be concentrated at the top of the screened interval (see upper window in Figure 7.12). If sand were added in increments, a series of fine layers alternating with coarse layers would form at intervals along the length of the screen.

The problem with particle segregation would be eliminated through the use of a uniform filter pack, if the design of the filter pack and well screen allowed less than 5% of the filter-pack material to pass through the screen slots, rather than the 85% retained (i.e., 15% passing) criterion for production water wells (Johnson Division, 1966; Dablow et al., 1986). With a more uniform filter pack, the segregation problems with the coarse fraction would be diminished, because the tail of the coarse portion of the distribution curve would consist of a higher percentage of smaller, coarse-grained particles.

Uniformity of the filter-pack material is important because the screen slot size is uniform and the particles in a filter pack that are not uniform will segregate during placement through water, and will not be distributed evenly. Through the use of elaborate and time-consuming circulation processes, the particles in a nonuniform filter pack can be distributed somewhat more evenly (Campbell and Lehr, 1973). However, using uniform filter pack materials is more cost effective and does not require the introduction of fluids into the wellbore. Grain-size distribution characteristics, their ideal values, and desirable ranges for filter pack materials are shown in Table 7.17.

Filter pack grain shape properties: roundness and sphericity. Roundness and sphericity are important parameters for filter pack design because particles that

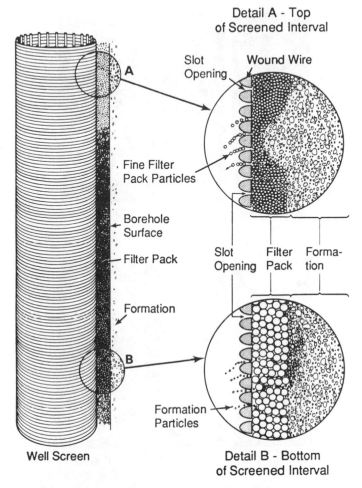

Figure 7.12. Particle segregation in a nonuniform filter pack.

are less round and less spherical tumble and oscillate as they fall through water. Tumbling and oscillating slow the rate at which the particles fall. Sand grains generally become less rounded and spherical as the particle size decreases. The greater angularity of the smaller grains tends to slow them, thus increasing the difference between the velocities of the large and small particles and increasing the likelihood of segregation of the particles.

Particle roundness (i.e., the curvature of the edges of a particle) is also important because minor changes in roundness can increase the potential for sand bridging during well development and well screen clogging. A common method used to define particle roundness is the Powers scale (based on the visual comparison of particles to photographic charts), which ranges from 1 (very angular) to 6 (well rounded) (Figure 7.13) (Powers, 1953). The potential for bridging and well screen

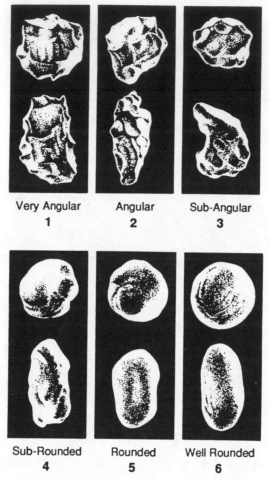

Figure 7.13. Terms for degree of rounding of grains as seen with a hand lens. After Powers, M.C., 1953, *Journal of Sedimentary Petrology,* v. 23, p. 118. Courtesy of the Society of Economic Paleontologists and Mineralogists.

clogging is particularly high between very angular (Powers scale 1) and suban-gular (Powers scale 2 to 3). Roundness values of between 4 and 6 on the Powers scale are preferred for monitoring well filter pack sand material.

Sphericity can define quantitatively how nearly equal are the three dimensions of a particle (Schalla and Walters, 1990). Sphericity, like roundness, can also reduce the potential for sand bridging in monitoring wells. However, sphericity is rarely included in well specifications that define the shape of sand particles, and numeri-cal values are infrequently mentioned by producers of filter pack materials.

Filter pack dimensions. With respect to filter pack length, the pack should gener-ally extend from the bottom of the well screen to at least 3 to 5 feet above the top of the well screen. This serves to account for any settlement of filter pack material that may occur during well development and allows a sufficient "buffer" between the well screen and the annular seal above.

The filter pack must be at least thick enough to surround the well screen completely, but thin enough to minimize resistance caused by the filter pack to the flow of fine-grained formation material and water into the well during development. To accommodate the filter pack, the well screen should be centered in the borehole and the annulus should be large enough and approximately symmetrical to preclude bridging and uneven placement of filter pack material around the screen. A thicker filter pack does not materially increase the yield of the well, nor does it reduce the amount of fine material in the water flowing to the well (Ahrens, 1957). In fact, thicker filter packs prevent the effective development of the formation and the removal of residual materials (i.e., drill cuttings, drilling fluid) produced by the installation of the borehole. Most references in the literature (U.S. EPA, 1975; Johnson Division, 1966; Williams, 1981; Walker, 1974 and others) suggest that a filter pack thickness of around 3 inches is suitable, though thinner (i.e., 2-inch) filter packs have also proved successful in monitoring wells.

Filter pack materials. In order to provide for the minimum potential for alteration of ground-water sample chemistry, the materials comprising the filter pack in a monitoring well should be as chemically inert as possible. For example, Barcelona et al. (1985a) suggest that filter pack materials should be composed primarily of clean quartz sand or glass beads. The individual grains of the filter pack materials should consist of less than 5% nonsiliceous material (Paul et al., 1988). Siliceous (quartz) material is preferred because it is nonreactive under nearly all subsurface conditions and it is generally readily available. In no case should filter pack materials comprised of crushed stone be utilized because of the irregular nature of the particles and the potential for chemical alteration of ground water that would come in contact with the filter pack material. This can occur as a result of the exposure of fresh surfaces of reactive minerals in the rock, on which chemical reactions can occur; this is particularly true regarding the use of crushed limestone. Limestone ($CaCO_3$) may significantly raise the pH of water with which it comes in contact; this, in turn, can affect the presence of other chemical constituents. Thus, such material is not appropriate for use as a filter pack in a monitoring well.

Filter pack materials should be washed, dried, and packaged at the factory in 100-lb. bags, which equal approximately one cubic foot. It is recommended that bags with plastic (i.e., polyethylene) liners sandwiched between paper be requested to minimize bag breakage and sand loss and contamination during shipping or during storage in wet weather.

Methods of Filter Pack Installation

Several methods of emplacing artificial filter packs in the annular space of a monitoring well are available, including gravity (free-fall) emplacement, emplacement by tremie pipe, reverse circulation emplacement, and backwashing.

Placement of filter packs by gravity or free-fall (i.e., dumping sand down the annulus) can be accomplished only in relatively shallow wells, with an annular space greater than two inches, where the potential occurrence of bridging or segregation of the filter pack material are minimized. Bridging can result in the occurrence of large unfilled voids in the filter pack, or in the failure of filter pack materials to reach their intended depth. Segregation of filter pack materials can result in a well that consistently produces sediment-laden samples; this problem is particularly likely to occur in deep wells in which a shallow static water level is present in the borehole. In this situation, as the sand falls through the column of water, the greater drag exerted on smaller particles due to their greater surface-area-to-weight ratio causes finer grains to fall at a slower rate than coarser grains. Thus, coarser materials end up comprising the lower portion of the filter pack, and finer materials make up the upper part. This is usually not a problem in the emplacement of truly uniform filter packs, where the uniformity coefficient is between 1.0 and 2.5 (Johnson Division, 1966), but in most cases, placement by gravity or free-fall is not recommended. Another drawback to gravity emplacement is that formation materials along the borehole wall can become incorporated into the filter pack, thus contaminating the filter pack and reducing its effectiveness.

With the tremie pipe emplacement method, the filter pack material is introduced through a rigid or partially flexible tube or pipe via gravity directly to the interval adjacent to the well screen (Figure 7.14), thus eliminating the potential for bridging. Initially, the pipe is positioned so that its end is at the bottom of the casing-borehole annulus. The filter pack material is then poured down the tremie or slurried into the tremie with water and the tremie is raised periodically so that the filter pack material can fill the annular space around the intake. The preferred minimum diameter of a tube used for a tremie pipe is 1½ inches; larger diameter pipes are advisable for filter pack materials that are coarse-grained or characterized by a uniformity coefficient exceeding 2.5 (California DOHS, 1985).

If the filter pack is being installed in a temporarily cased borehole (i.e., via cable tool), the temporary casing is pulled back progressively to expose the screen as the filter pack material builds up around the well screen. The same approach is recommended for hollow-stem auger installations in noncohesive unconsolidated formations. Raising the temporary casing (or the auger) before beginning filter-pack placement is undesirable because formation materials may enter the borehole and cave against the well screen. This may result in a well that produces sediment-laden samples. Pulling the casing back one or two feet at a time, while adding filter-pack material, is a safer, more conservative approach in noncohesive unconsolidated formations that are prone to caving than is the technique of filling the space between the temporary casing and the well screen and then pulling back the temporary casing to expose the entire screen length at one time. In consolidated or well-cemented formations, or in cohesive unconsolidated formations (i.e., predominantly clay or silt), the temporary casing or hollow-stem auger can usually be raised well above the filter pack prior to filter pack emplacement. The progress of the work should be continually checked with a

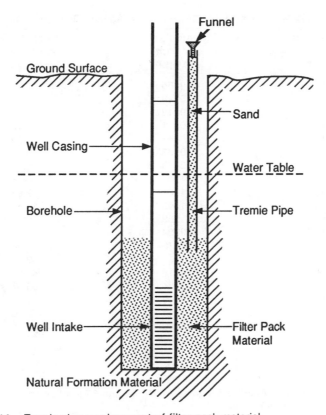

Figure 7.14. Tremie-pipe emplacement of filter pack material.

weighted measuring tape, accurate to the nearest 0.1 foot, to determine when the filter pack has reached the desired height in the borehole. The minimum volumes of sand that are needed can be determined using Tables 7.18 and 7.19. Table 7.20 provides a specific set of values (e.g., density, packaging, and volume yield) for commonly used filter pack and other annular fill materials.

For deep wells (i.e., greater than 250 feet) and nonuniform filter-pack materials (uniformity coefficient greater than 2.5), a variation of the standard tremie method, employing a pump to pressure-feed the materials into the annulus, is suggested by California DOHS (1985).

In the reverse circulation method, a sand and water mixture is fed into the annulus around the well screen, and a return flow of water passes into the well screen. The water is then pumped up through the casing to the surface (Figure 7.15). The filter pack material is generally introduced into the annulus at a maximum rate of 1.5 cubic feet per minute to allow for an even distribution of material around the screen (Johnson Division, 1966). Because of the potential for alteration of sample chemical quality posed by the introduction into the borehole of water from a surface source, this method is infrequently used for water-quality monitoring wells.

Table 7.18. Calculating Minimum Filter Pack and Annular Seal Volumes.

Diameter of Casing or Hole (in.)	Cubic Feet per Foot of Depth	Cubic Meters per Meter of Depth ($\times 10^{-3}$)
1	0.0055	0.509
1.5	0.0123	1.142
2	0.0218	2.024
2.5	0.0341	3.167
3	0.0491	4.558
3.5	0.0668	6.209
4	0.0873	8.110
4.5	0.1104	10.26
5	0.1364	12.67
5.5	0.1650	15.33
6	0.1963	18.24
6.5	0.2304	21.42
7	0.2673	24.84
8	0.3491	32.43
9	0.4418	41.04
10	0.5454	50.67
11	0.6600	61.31
12	0.7854	72.96
14	1.0690	99.35
16	1.389	129.09

Table 7.19. Minimum Annulus Volumes for 2-, 4-, and 6-Inch Casings in Boreholes of 4 to 16 Inches.

O. D. of Casing Inside I. D. of Borehole (in.)	Cubic Feet per Foot of Depth	U.S. Gallons per Foot of Depth	Liters per Meter of Depth	Cubic Meters per Meter of Depth
2.5 in 4	0.053	0.398	4.59	0.005
2.5 in 6	0.162	1.213	14.00	0.014
2.5 in 8	0.315	2.356	27.19	0.027
2.5 in 10	0.511	3.825	44.13	0.044
2.5 in 12	0.751	5.620	64.84	0.065
4.5 in 6	0.086	0.642	7.41	0.007
4.5 in 8	0.239	1.785	20.60	0.021
4.5 in 10	0.435	3.254	37.55	0.038
4.5 in 12	0.675	5.049	58.26	0.058
4.5 in 14	0.959	7.170	82.74	0.083
4.5 in 16	1.279	9.564	110.36	0.110
6.5 in 8	0.119	0.888	10.25	0.010
6.5 in 10	0.315	2.356	27.19	0.027
6.5 in 12	0.555	4.151	47.90	0.048
6.5 in 14	0.839	6.273	72.38	0.072
6.5 in 16	1.159	8.666	100.00	0.100

Table 7.20.   Volume Calculation Information for Materials Commonly Used in Monitoring Well Construction.

Material Description	Density	Packaging Unit	Yield per Packaging Unit
Colorado silica sand[a] 8–12 mesh	98.5 lb/ft^3	100-lb bag	1.01 ft^3
Colorado silica sand 10–20 mesh	92.9 lb/ft^3	100-lb bag	1.07 ft^3
Colorado silica sand 20–40 mesh	89.0 lb/ft^3	100-lb bag	1.12 ft^3
Colorado silica sand 40–100 mesh	86.6 lb/ft^3	100-lb bag	1.15 ft^3
Sodium bentonite[b] 8 mesh	70 lb/ft^3	100-lb bag	1.43 ft^3
Sodium bentonite 8 mesh	70 lb/ft^3	50-lb bag	0.71 ft^3
Sodium bentonite − 10 + 100 mesh	69 lb/ft^3	100-lb bag	1.45 ft^3
Sodium bentonite 200 mesh	58–59 lb/ft^3	100-lb bag	171 ft^3
Volclay[c] grout	9.4 lb/gal	52-lb bag	3.50 ft^3
Volclay ¼-in. pellets	80 lb/ft^3	50-lb bucket	0.62 ft^3
Pre-mix concrete	136 lb/ft^3	90-lb bag	0.66 ft^3

[a]Colorado Silica Sand, Inc., 3250 Drennan Industrial Loop, P.O. Box 15615, Colorado Springs, CO 80935, (303) 390-7969.
[b]Wyo-Ben, Inc., 3044 Hesper Road, P.O. Box 1979, Billings, MT 59103, (406) 652-6351.
[c]American Colloid Co., Water/Mineral Division, 5100 Suffield Ct., Skokie, IL 60077, (312) 966-5720.

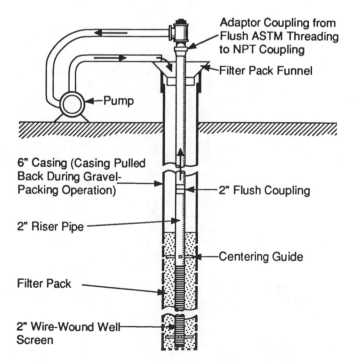

Figure 7.15.   Reverse-circulation emplacement of artificial filter pack materials.

Backwashing filter pack material into place is done by allowing sand to free-fall down the annulus while concurrently pumping clean fresh water down the inside of the casing, through the well screen, and back up the annulus (Figure 7.16). This allows for the placement of a more uniform filter pack, because the coarser materials settle out and remain in place while the finer materials will be washed back up the annulus. Backwashing is a particularly effective method of filter pack emplacement in noncohesive heaving sands and silts, but it is also effective in cohesive, noncaving geologic materials. It is not commonly used in water-quality monitoring wells because of its potential for alteration of ground-water sample chemical quality.

Secondary Filter Packs

The main purpose of the secondary filter pack is to prevent neat cement or bentonite grout or other annular seal materials from migrating into the primary

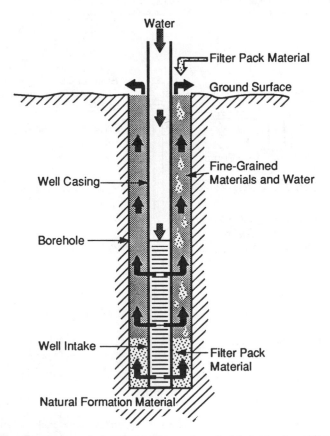

Figure 7.16. Emplacement of artificial filter pack materials by backwashing.

filter pack or from infiltrating down to the screened interval of the well, which would partially or totally seal the well from the formation to be monitored, and alter the quality of ground-water samples taken from the well. The necessity for a secondary filter depends on the particle-size distribution of the formation. It must be composed of well-graded (not uniform), preferably positively skewed sand, with the coarsest fraction equal to the 90% retained size of the filter pack, and with less than 2% by weight passing the No. 200 mesh sieve. Although the particles in the secondary filter need not be as uniform and round as those in the primary filter pack, they should have mineralogical characteristics similar to, or at least compatible with, those of the primary filter pack.

Three important surfaces must be considered in designing the secondary filter pack: the bottom surface (or the interface with the primary filter pack), the top surface (or the interface with the grout or other annular seal materials), and the outer surface (or the interface with the formation).

For a properly designed filter pack/secondary filter interface, the D10 (10% passing) size of the secondary filter must be larger than the voids (interstices) in the primary filter pack. This prevents the fine materials of the secondary filter pack from invading the primary filter pack. Therefore, the D10 size of the secondary filter should be between one-third and one-fifth the D10 size of the primary filter pack. Referring to Table 7.17, a primary filter pack of 8- to 12-mesh sand would require a secondary filter pack consisting of a layer of 20- to 40-mesh sand with 40- to 140-mesh sand on top.

At the secondary filter pack/grout interface, the filter pack material should be as fine as possible, so that neat cement or bentonite grout will not invade the secondary filter pack significantly, and will not invade the primary filter pack at all. Although the particles need to be fine grained, they should not be so fine grained that the time required for them to settle is significantly influenced by fluid viscosity or minor turbulence caused by the placement of the secondary filter pack. The smallest particle size of the secondary filter should be no larger than U.S. Standard Mesh No. 140 and no smaller than No. 200 to reduce invasion of grout and minimize settling time of the finest fraction of the secondary filter pack. This recommendation is based not on the size of the neat cement or bentonite particles in the grout slurry but on the viscosity of the grout (which should have a Marsh funnel viscosity of at least 80 seconds) and on the height of the grout column above the sand pack, which can exert a significant hydraulic head on the top of the secondary filter pack material.

The secondary filter should extend at least 1 foot above the top of the filter pack. The upper half of the secondary filter should be in contact with a lithologic layer of equal or lower permeability and thickness to prevent the slurry or sealant from migrating around the secondary filter and into the coarser filter pack.

Types of Well Screens Suitable for Monitoring Wells

The hydraulic efficiency of a well screen depends primarily upon the amount of open area available per unit length of screen. While hydraulic efficiency is of secondary concern in monitoring wells, increased open area in monitoring well screens also permits easy flow of water from the formation into the well and allows for effective development of the well. The amount of open area in a well screen is controlled by the type of well screen.

A number of different types of screens are available for use in production wells; several of these are also suitable for use in monitoring wells. Commercially manufactured well screens similar to those used for production wells are recommended for use in monitoring wells even though hydraulic efficiency is not a primary concern because of the stricter quality control followed by most commercial screen manufacturers. Hand-slotted or drilled casings should never be used as monitoring well screens because of the poor control over screen slot size, the lack of open area, and the fact that hand sawn or drilled openings provide fresh surfaces for sorption, leaching, or other chemical problems to occur. Likewise, perforated casings, produced by either the application of a casing knife or a perforating gun to installed casing down-hole, are not recommended because screen openings cannot be closely spaced, the percentage of open area is low, the opening sizes are highly variable, and opening sizes small enough to control fine materials are impossible to produce. Additionally, perforation tends to hasten corrosion attack on metal casing, because the jagged edges and rough surfaces of the perforations are susceptible to selective corrosion.

Commercially manufactured well screens, including louvered screen, bridge-slot screen, machine-slotted well casing, and continuous-slot wire-wound screen are available for use in production wells, but the latter two types of screens (Figure 7.17) predominate by far in monitoring wells, probably because they are the only types available in small (i.e., 2-inch and 4-inch nominal) diameters, which are more commonly used for monitoring wells.

Slotted well casing is often used as a well screen in monitoring wells. Such a well screen is fabricated from standard well casing, into which horizontal (i.e., circumferential) slots of predetermined widths are cut at a regular vertical spacing (typically 0.25 or 0.125 inches) by machining tools. Slotted well casing can be manufactured from thermoplastic (PVC), fluoropolymer, stainless steel, and fiberglass-reinforced epoxy materials and is available in diameters ranging from ¾-inch to 16-inch. Table 7.16 lists the most common slot widths for slotted casing. The slot openings are designated by numbers that correspond to the width of the opening in thousandths of an inch. A number 10 slot, for example, refers to an opening of 0.010 inches, while a number 40 slot refers to an opening of 0.040 inches.

The slots in slotted well casings are a consistent width for the entire wall thickness of the casing, which can result in significant clogging of the screen when irregularly shaped formation particles are brought through the screen during well development and sampling. This, in turn, can result in declining performance (with respect to well efficiency) of the well over time. Furthermore, in most slotted casing, the length of the slot on the inside of the casing is much less than the apparent length of the slot as viewed from the outside. The effect of this is

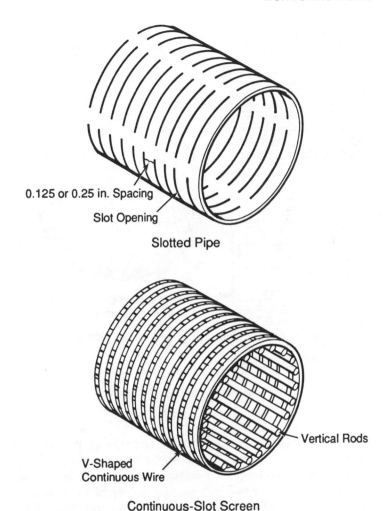

0.125 or 0.25 in. Spacing

Slot Opening

Slotted Pipe

Vertical Rods

V-Shaped
Continuous Wire

Continuous-Slot Screen

Figure 7.17. The two primary types of screen used in monitoring wells.

that the effective open area of the screen is significantly reduced. These disadvantages are overcome through the use of continuous slot, wire-wound screen.

The continuous-slot screen is manufactured by winding cold-drawn wire, approximately triangular or V-shaped in cross section, spirally around a circular array of longitudinally arranged rods (Figure 7.18). At each point where the wire crosses a rod, the two pieces are securely joined by welding, creating a one-piece rigid unit (Johnson Division, 1966). Continuous-slot screens can be fabricated of any metal that can be resistance-welded, including bronze, silicon red brass, stainless steel (304 and 316), galvanized and low-carbon steel, and any thermoplastic that can be sonic-welded, including PVC and ABS.

VERTICAL
CROSS-SECTION

HORIZONTAL
CROSS-SECTION

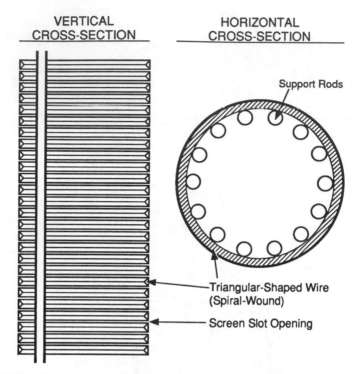

Figure 7.18.   Close-up cross-section of continuous-slot wire-wound screen.

The slot openings of continuous-slot screens are produced by spacing the successive turns of the wire as desired. This configuration provides significantly greater open area per given length and diameter than is available with any other screen type. For example, for two-inch nominal diameter PVC well screen, the open area ranges from about 4 square inches per foot of screen for the 0.006-inch slot size to more than 26 square inches per foot of screen for the 0.050-inch slot size (Table 7.21). The percentage of open area in a continuous-slot screen is often more than twice that provided by standard slotted well casing in the smaller slot sizes (0.010 and 0.020 inches), even if the slots in the slotted casing are placed 0.125 inches apart, rather than the standard 0.25 inches (Table 7.22). Continuous-slot screens also provide a wider range of available slot sizes than any other type of screen and have slot sizes that are accurate to within ± 0.002 inches. The continuous-slot screen is also more effective in preventing formation materials from becoming clogged in the openings during well development. The triangular-shaped wire is wound so that the slot openings between adjacent wraps of the wire are V-shaped, with sharp outer edges; the slots are narrowest at the outer face of the screen and widen inwardly. This allows particles slightly smaller than the openings to pass freely into the well without wedging in the opening, making these screens nonclogging.

Table 7.21. Intake Area (Square Inches Per Lineal Foot of Intake) for Continuous-Slot Wire-Wound Screen.

Screen Size (in.)	6-Slot (0.006)	8-Slot (0.008)	10-Slot (0.010)	12-Slot (0.012)	15-Slot (0.015)	20-Slot (0.020)	25-Slot (0.025)	30-Slot (0.030)	35-Slot (0.035)	40-Slot (0.040)	50-Slot (0.050)
1¼ PS	3.0	3.4	4.8	6.0	7.0	8.9	10.8	12.5	14.1	15.6	18.4
1½ PS	3.4	4.5	5.5	6.5	8.1	10.2	12.3	14.2	16.2	17.9	20.1
2 PS	4.3	5.5	6.8	8.1	10.0	12.8	15.4	17.9	20.3	22.4	26.3
3 PS	5.4	7.1	8.8	10.4	12.8	16.5	20.0	23.2	26.5	29.3	34.7
4 PS	7.0	9.0	11.3	13.5	16.5	21.2	25.8	30.0	33.9	37.7	44.5
4 Spec	7.4	9.7	11.9	14.2	17.2	22.2	27.1	31.3	35.5	39.7	46.8
4½ PS	7.1	9.4	11.7	13.8	17.0	21.9	26.8	31.0	35.2	39.4	46.5
5 PS	8.1	10.6	13.1	15.5	19.1	24.7	30.0	34.9	39.7	44.2	52.4
6 PS	8.1	10.6	13.2	15.6	19.2	25.0	30.5	35.8	40.7	45.4	54.3
8 PS	13.4	17.6	21.7	25.7	31.5	40.6	49.3	57.4	65.0	72.3	85.6

Source: Johnson Filtration Systems Inc. Environmental Equipment Catalog.

The maximum transmitting capacity of screens can be derived from these figures. To determine GPM per ft of screen, multiply the intake area in square inches by 0.31. This is the maximum capacity of the screen under ideal conditions with an entrance velocity of 0.1 feet per second.

Table 7.22. Comparison of Screen Open Area (%) for Continuous-Slot Screen and Slotted Pipe.

Screen Diameter (in.)	Continuous Slot[a]		Slotted Pipe[b]	
	10 Slot	20 Slot	10 Slot	20 Slot
2	7.6	14.4	2.9–5.1	5.5–9.4
4	6.8	12.7	2.4–4.3	4.6–7.8
6	5.3	10.0	2.0–3.6	3.9–6.6

Source: Information provided by Johnson Filtration Systems Inc.
[a]Data are for PVC; in stainless steel, the open area will be twice as great.
[b]Because pipe slotting is performed in many different ways, a range from low to high is given.

The louvered (shutter-type) screen has openings in the form of louvers that are manufactured in solid wall metal tubing by stamping with a punch outward against dies that limit the size of the openings (Helwig et al., 1984). The number and sizes of openings that can be made depends on the series of die sets used by individual manufacturers. Because a complete range of die sets is impractical, the opening sizes of commercial available screens is somewhat limited. Additionally, because of the large blank spaces that must be left between adjacent openings, the percentage of open area on louvered intakes is limited. Also the shape of the louvered openings is such that the shutter-type intakes cannot be used successfully for naturally developed wells, so their use is confined almost exclusively to artificially filter-packed wells (Johnson Division, 1966). This type of screen is available only in large-diameter metallic well screens and is rarely used in monitoring applications, though it is useful for recovery wells.

Bridge-slot screen is manufactured on a press from flat sheets or plates of metallic material that are rolled into cylinders and seam-welded after being perforated. The slot is usually vertical, and produces two parallel openings longitudinally aligned to the well axis. Normally, 5-foot sections of bridge-slot screen are available, and these can be welded into larger screen sections if desired. The chief advantages of this type of screen are reasonably high open area, minimal frictional head losses, and low cost. One important disadvantage is its low collapse strength caused by the presence of a large number of vertically oriented slots. Bridge-slot screen is usually installed in filter-packed production wells; its use in monitoring wells is limited because it is only produced in diameters 6 inches and larger and because it is available only in metallic materials.

The screen is the part of the monitoring well that is most susceptible to corrosion and/or chemical degradation, and provides the highest potential for sorption or leaching phenomena to occur. Screens have a larger surface area of exposed material than casing, are placed in a position designed to be in contact with potential contaminants (the saturated zone), and are placed in an environment where reactive materials are constantly being renewed by flowing water. To avoid corrosion, chemical degradation, sorption, and leaching problems, the materials from which screens are made are selected using the same guidelines as for casing materials.

Monitoring Well Screen Open Area

The hydraulic performance of well screens has been stated as being independent of screen design (Clark and Turner, 1983), provided that the open area of the screen exceeds a threshold of about 10%. This effective limit of about 10% is probably related to the minimum open area of rhomboidal packing of spherical particles, which is 9.2%. However, other studies indicate that screen design is important and that the thresholds are lower (Jackson, 1983; Bikis, 1979). Others (Ahmad et al., 1983) suggest that screens with open areas of 8% to 38% do not differ significantly in regard to total drawdown, even in fluvial deposits rich in silt and clay. The importance of screen entrance velocity as a design criterion for reducing head loss and entrance velocities has been overemphasized in production wells and is not applicable in monitoring wells. However, a recent standard reference book (Driscoll, 1986) on water wells indicates that the open area in a well screen should be at least equal to the effective porosity of the formation material and the filter pack. This important guideline, which is also valid for monitoring wells, would mean that screens with open areas of 10% to 30% would be needed in many monitoring situations and would require the exclusive use of continuous-slot wire-wound well screens. It is important that the well screen open area equal or exceed the formation's effective porosity, so that the screen is not the limiting factor in formation hydraulic testing.

In choosing between types of well screen, another factor to consider is the speed and effectiveness of well development. Field experience with both large- and small-diameter wells indicates that screens with a high percentage of open area greatly reduce the time and effort required for well development (Schalla, 1986). Similar findings on the importance of the percentage of open area have been reported by others (Clark and Turner, 1983; Ericson et al., 1985). A high percentage of open area is particularly important where smaller slot sizes and fine-grained filter packs must be used to retain the bulk of the formation sediments.

Well Screen Length

The process of selecting monitoring well screen lengths is largely subjective, the primary variable being the purpose of the well. In most monitoring situations, wells are installed to double as ground-water sampling points to monitor water quality, and piezometers to monitor water levels or hydraulic head in a short, discrete interval. To accomplish this objective, well screens generally must be short (i.e., between 2 and 10 feet), rarely equaling or exceeding 20 feet in length (Figure 7.19). Shorter screens provide more specific information about vertically distributed water quality, hydraulic head, and flow in the monitored formation. On the other hand, if the objective of the well is to monitor for the gross presence of contaminants in an aquifer, a longer screen, perhaps designed to monitor the entire thickness of the aquifer, might be selected. This type of well would provide both an integrated water sample and an integrated hydraulic head measurement, and would thus serve only as a screening tool.

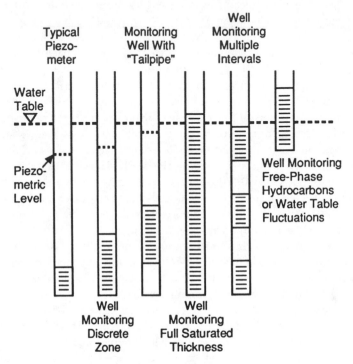

Figure 7.19. Screen length variability in monitoring wells.

Dual-Screen Well Intakes

For formations requiring less than a 0.010-inch slot size, a dual well screen or Channel Pack system (Figure 7.20) should be considered to minimize the migration of fines into the well by reducing flow velocity and gradient at the formation/filter pack interface. Basically, a dual well screen is fabricated by welding a small-diameter 0.008-inch slot, continuous-slot screen inside a slightly larger diameter, 0.008-inch slot screen and packing the space between the screens with a very uniform filter pack of 40 to 60 mesh-size sand. This dual screen/filter pack system is installed in the well like an ordinary well screen, which is then packed with filter pack, or it can be installed inside well casing as shown in Figure 7.20 without a filter pack. This system minimizes the velocity of materials moving into the well during sampling, but also inhibits effective well development.

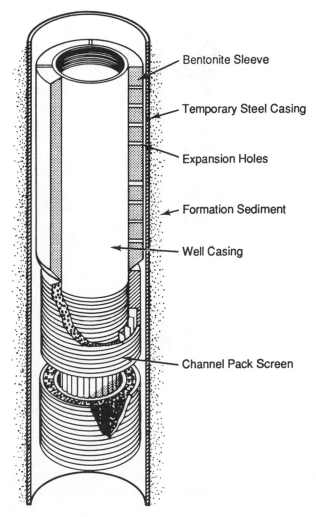

Figure 7.20.   Dual well screen/filter pack (channel-pack) arrangement.

MONITORING WELL ANNULAR SEAL DESIGN AND INSTALLATION

Any annular space that is produced as the result of the installation of well casing in a borehole provides a potential channel for downward movement of water and/or contaminants, unless the annulus is sealed. In any casing/borehole system, there are several potential pathways for water and contaminants to follow (Figure 7.21). One pathway is through the sealing material; if the material is not properly formulated and installed, or if it cracks, shrinks, or deteriorates after emplacement, vertical high-permeability channels could cause significant migration

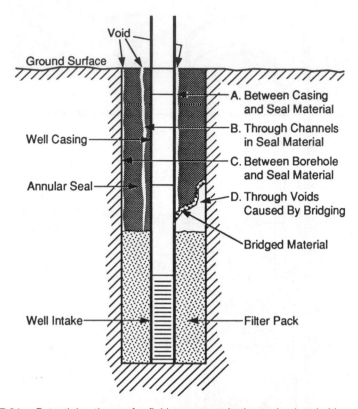

Figure 7.21. Potential pathways for fluid movement in the casing-borehold annulus.

of formation water or contaminants from zones of high hydraulic head to zones of lower hydraulic head. Because casings are relatively smooth, another potential pathway exists between the casing and sealing material. This pathway could occur because of several reasons, including temperature changes of the casing and sealing materials (principally neat cement) during the curing or setting of the sealing material, swelling and shrinkage of the sealing material while curing or setting, or poor bonding between the sealing material and the casing (Kurt and Johnson, 1982). A third pathway, resulting from improper emplacement of seal materials (i.e., from bridged annular seal materials) may also exist. All of these pathways can be anticipated and avoided with proper annular seal formulation and placement methods. Because monitoring wells are often located close to, or within areas affected by contaminants, an adequate annular seal is especially critical to both the protection of ground-water quality and to the integrity of the well.

The annular seal in a monitoring well (i.e., the sealing material placed above the filter pack in the annulus between the borehole and the well casing) serves several purposes: (1) to provide protection against surface water and potential contaminants infiltrating from the ground surface down the casing/borehole annulus; (2) to hydraulically and chemically seal off discrete sampling zones; and (3) to prohibit vertical migration of water, which may be of a different quality, in the casing/borehole annulus from one aquifer to another or from zones of high hydraulic head to zones of lower hydraulic head. Such vertical movement can cause cross contamination, which can greatly influence the representativeness of ground-water samples or cause an anomalous hydraulic response of the monitored zone, resulting in incorrect conclusions regarding formation water quality and distorted maps of potentiometric surfaces. The annular seal around the casing also increases the life of the casing by protecting it against exterior corrosion or chemical degradation, and may also provide an element of structural integrity. A satisfactory annular seal should completely fill the annular space and envelop the entire length of the well casing (exclusive of the well screen) to ensure that no vertical migration occurs within the borehole.

Annular Seal Materials

The annular seal may be composed of several different types of permanent, stable, low hydraulic conductivity materials, including bentonite (pelletized, granular, powdered, or sleeves) neat cement grout, and variations of both. The most effective seals are obtained by using expanding or nonshrinking materials that will not pull away from either the casing or the borehole after curing or setting. Bentonite, nonshrinking (also termed "expanding") neat cement, or neat cement with shrinkage-compensating additives are among the most effective materials for this purpose (Barcelona et al., 1983, 1985a; Calhoun, 1988; Johnson et al., 1980). If the casing/borehole annulus is backfilled with other material (i.e., recompacted drill cuttings, sand, or borrow material), a low hydraulic conductivity annular seal cannot be ensured, and the borehole may then act as a conduit for vertical migration of potentially contaminated ground water. This is especially true regarding the use of drill cuttings, because recompacted drill cuttings may interfere with the water chemistry of the water samples collected from the well.

Bentonite

Bentonite, a hydrous aluminum silicate composed principally of the clay mineral montmorillonite, expands significantly when hydrated; the expansion is caused by the incorporation of water molecules into the clay lattice. Expansion of bentonite in water can be on the order of 8 to 10 times the volume of dry bentonite. On hydrating, the bentonite forms an extremely dense clay mass, which sets up as a thixotropic gel with an in-place hydraulic conductivity typically in the range of 1×10^{-6} cm/s. Under favorable water-quality conditions, the bentonite expands sufficiently to provide a tight seal between the casing and the adjacent formation material.

Bentonite used for the purpose of sealing the annulus of monitoring wells is generally sodium bentonite (montmorillonite), which is more widely used because of its availability; calcium bentonite is also available, though less common. Bentonite is available as pellets, granules, chips or chunks, powder, or sleeves. Bentonite pellets are uniformly shaped and sized, and simply consist of compressed, moistened sodium montmorillonite powder; they expand at a relatively slow rate (compared to other forms of bentonite) when exposed to fresh water. Granules, chips, and chunks are irregularly shaped (angular to subangular) and sized (¼ to ¾ inches in size) particles of montmorillonite that expand at a much faster rate than pellets when exposed to fresh water. Powdered bentonite is the pulverized material produced by the processing plant after mining. While both pelletized and granular bentonite may be emplaced in dry form, powdered bentonite is generally made into a grout slurry to allow its emplacement as an annular seal. The bentonite sleeve, which is moistened, compressed bentonite made into a form that will fit around a well casing, can be placed when the casing and screen are set in the borehole by sliding it down the casing after placing the filter pack. The outer diameter of the casing must be consistent (i.e., no couplings), and casings centralizers cannot have been used in wells in which bentonite sleeves are used.

A bentonite grout is generally prepared by mixing dry bentonite powder into fresh water in a ratio of approximately 15 pounds of bentonite to 7 gallons of water, to yield one cubic foot of grout. Some bentonite powder will require the addition of either a premix additive to condition the makeup water, or a polymer additive (organic or inorganic) to delay wetting of the bentonite and prevent premature set-up, allowing adequate working time. Once the grout is mixed, it should remain workable for 15 to 30 minutes (longer if extenders are used). During this time, the grout is pumped through a tremie pipe, using a positive displacement mud or grout pump (generally either a centrifugal, diaphragm, piston, or moyno-type pump), to its intended position in the annulus. Some thick bentonite grouts without polymer additives will swell very quickly into nonpumpable gel masses, making it impossible to emplace the slurry as a seal.

The key to the success of bentonite grouts lies in the percentage of solid material in the grout. While bentonite used for drilling fluids is generally a high-viscosity mixture with a low solids content, just the opposite is desirable for bentonite used for annular seal material—high solids, low viscosity. While a solids content of greater than 30% is optimum, most mixtures are in the range of 20–25%. A high volume of solids can be created through the use of inhibitors used to retard the swelling of the bentonite while suspending it as a colloid in solution. The inhibitor is a viscosifying agent that coats the bentonite, allows it to partially hydrate and, after placement, disperses to allow the bentonite to absorb additional water and swell to create a positive seal. Another means of creating a high-solids grout is to blend various grades of sodium and calcium montmorillonite. Still another means of creating a high-solids grout is to premix a very high quality bentonite drilling fluid from a pure powdered bentonite and then add a granular bentonite, just prior to placing the mixture via a tremie pipe. A 9.4 pounds-per-gallon mixture, which is desirable for grouting, can be achieved in this manner. Blended bentonites have been engineered to build less viscosity and produce a grout with a higher weight. In some bentonite mixtures, an activator, magnesium oxide, intensifies the gelling of a mixture of sodium and calcium bentonite, helping the grout build a matrix with extracted solids to create better gel strength (Calhoun, 1988). This type of mixture has about a 60-minute working time, but it does not develop as strong a gel as pure sodium bentonite grouts.

Because it is a clay mineral, bentonite possesses a high cation exchange capacity (CEC). This allows the bentonite to "trade off" cations that make up the chemical structure of the bentonite (principally Na, Al, Fe, and Mn) with cations that exist in the aqueous solution (i.e., ground water) that is in contact with (and, in some cases, hydrates) the bentonite. The bentonite may take up cations from or release cations to the aqueous solution, depending on the chemistries of both the bentonite and the solution and on the pH and redox potential of the aqueous solution. In addition to having a high CEC, bentonite will generally set up with a moderately high pH (i.e., from 8.5 to 10.5). Thus, bentonite may have an impact on the quality of the ground water with which it comes in contact, particularly with regard to pH and metallic ion content. If a bentonite seal is placed too close to the top of the well intake, the potential exists for obtaining samples of water that are not truly representative of aquifer quality water. Because of this, the recommended practice is to place the bentonite sealing material no closer than 3 to 5 feet above the top of the well screen, or to use a secondary filter pack above the primary filter pack in an artificially filter packed well.

Other chemical considerations include the potential presence of additives (i.e., organic or inorganic polymers) in the bentonite product. The chemistry of the specific bentonite product that is used must be scrutinized closely to ensure that the bentonite does not contain an additive that could affect the representiveness of ground-water samples.

The use of bentonite as a sealing material depends on its efficient hydration following emplacement; hydration requires the presence of water (of sufficient quantity and quality) within the geologic materials penetrated by the borehole. Generally, efficient hydration will occur only in the saturated zone. Bentonite may therefore not be appropriate for use in the vadose zone, where sufficient moisture generally is not available during all times of the year to effect proper (complete) hydration of the bentonite. This is especially true in areas with arid climates and thick vadose zones. Under certain water quality conditions, notably in water with a high total dissolved solids (TDS) content (i.e., greater than 5000 ppm) or a high chloride content, the swelling of bentonite is inhibited. The degree of inhibition is dependent upon the type of bentonite used and the level of TDS or chloride in the water, but the end result is that the bentonite may not swell to its anticipated volume, and an effective annular seal may not be provided.

Recent studies were conducted to determine the effects of some organic solvents and other chemicals (i.e., xylene, acetone, acetic acid, aniline, ethylene glycol, methanol, and heptane) on hydrated clays, including bentonite (Anderson et al., 1982; Brown et al., 1983). This research has demonstrated that bentonite and other clays may lose their effectiveness as low-permeability barrier materials in the presence of concentrated solutions of selected chemical substances. These studies have demonstrated that the hydraulic conductivity of bentonite and other clays subjected to high concentrations of organic acids, basic and neutral polar organic compounds, and neutral nonpolar organic compounds may increase by several orders of magnitude due to desiccation and dehydration of the clay material, thus potentially manufacturing a conduit for vertical migration within boreholes in which bentonite is used as a sealing material. Villaume (1985) points to possible attack on, and loss of, integrity of bentonite seals due to dehydration and shrinkage of the clay by petroleum hydrocarbons in the free-product phase. Thus, where these chemical conditions exist in the subsurface, bentonite may not perform as an effective seal material, and another material should be chosen.

In summary, factors that should be considered in evaluating the use of bentonite as a sealant include:

1. position of the static water level in a given borehole, taking into account seasonal and other natural and man-induced fluctuations.
2. ambient water quality (especially with respect to total dissolved solids content and chloride content); and
3. types and potential concentrations of contaminants expected to be encountered in the subsurface.

Because the usefulness of bentonite is, for all practical purposes, limited to the saturated zone (to avoid problems if desiccation and creation of high-permeability conduits), bentonite seals are generally employed most effectively as thin (3–5 foot) annular seals just above the filter pack, or as seals only in the saturated portion of the borehole. In most cases in which static water levels are more than a few feet beneath the ground surface, another annular seal material, usually neat cement, is more commonly used to seal the remainder of the annulus.

Neat Cement

Neat cement is a mixture of portland cement (ASTM C-150) and water (without the addition of aggregate or sand), in the proportion of 5 to 6 gallons of clean water per bag (94 lb or 1 cubic foot) of cement. Cement mixtures with more than about 6 or 7 gallons of water per bag of cement may develop voids which contain only water and may generate "free water," which contains very high concentrations of soluble mineral matter from the cement, and which can adversely affect water quality in the well for prolonged periods of time. Five types of portland cement are produced: Type I, for general use; Type II, for moderate sulfate resistance or moderate heat of hydration; Type III, for high early strength; Type IV, for low heat of hydration; and Type V, for high sulfate resistance (Moehrl, 1964). ASTM Types I, II, and III correspond to API (American Petroleum Institute) Classes A, B, and C, respectively. Type I portland cement is by far the most widely used cement in ground-water related work. In cold-weather climates, cements with air-entraining agents are generally preferred due to their water tightness and freeze-thaw resistance. Air-entrained cements are designated with an "A" after the ASTM cement Type (i.e., Type IA). ASTM Standard C-845 describes three types of expanding cement, Types K, M, and S, which have characteristics similar to Type I or Type II portland cement, but which contain shrinkage-compensating additives. These have also proven useful in ground-water monitoring well construction.

Portland cement mixed with water in the proportions above yields a slurry that weighs from 14 to 15 lb/gal. A typical 14-lb/gal Type I neat cement slurry would have a mixed volume of about 1.5 cubic foot/sack and a set volume of about 1.2 cubic foot/sack. The volumetric shrinkage would be about 17%, and the porosity of the set cement about 54% (Moehrl, 1964). The setting time for such a cement mixture would range from 48 to 72 hours (depending primarily on water content).

When cement shrinkage occurs, the cement may pull away from either the casing or the borehole wall, thus destroying the integrity of the seal and opening channels or pathways for contaminant migration. Most of the problems associated with shrinkage (and other problems, including long set times) can be corrected through the use of additives or the use of shrinkage-compensated cements. While the latter option is preferred, using additives is a viable and more commonly employed approach.

A variety of additives may be mixed with the cement slurry to change the properties of the cement. The more common additives, their ranges of proportions (measured as percent by volume), and their effects on the cement mixture include the following:

1. bentonite (3%–8%), which improves the workability of the cement slurry, reduces the slurry weight and density, and produces a lower unit cost sealing material. Normally, bentonite is prehydrated before it is mixed with neat cement, because if it is added dry, it tends to lump and it will not hydrate properly. Bentonite also reduces the set strength of a seal, lengthens the set time considerably, and has been shown to be chemically incompatible with many cements (Calhoun, 1988). During setting, cement releases Ca^{+2} and OH ions, which cause flocculation of the bentonite, reducing its ability to swell. Cement-bentonite mixtures have thus fallen out of favor in monitoring well installations;

2. calcium chloride (1%–3%), which provides for an accelerated setting time and a higher early strength, particularly useful features in cold climates. This additive also aids in reducing the amount of slurry that might enter into zones of coarse material, which in turn avoids bridging of the seal;

3. gypsum (3%–6%), which produces a quick-setting, very hard cement that expands upon setting. The high cost of gypsum as an additive, however, limits its use to special operations;

4. aluminum powder (1%), which also produces a stronger, quick-setting cement that expands upon setting, thus providing a tighter seal (Ahrens, 1970);

5. fly ash (10%–20%), which increases sulfate resistance and early compressive strength;
6. hydroxylated carboxylic acid, which retards setting time and improves workability without compromising set strength; and
7. diatomaceous earth, which reduces slurry density, increases water demand and thickening time, and reduces set strength.

Water used to mix neat cement should be clean fresh water free of oil, soluble chemicals, silt, organic material, alkalies and other contaminants, and the total dissolved mineral content should be less than 500 ppm; a high sulfate content is particularly undesirable (Campbell and Lehr, 1973). Water with a high chloride content can either accelerate or retard the setting time of cement slurries, depending upon the type of cement used. Inorganic materials (sulfates, hydroxides, carbonates and bicarbonates) can also accelerate setting time. If too much water is used (i.e., more than 6 or 7 gallons), excessive shrinkage will occur upon setting, which means that the annulus will not be completely filled after the grouting operation. The voids in the annulus may not be seen from the surface, but they will be present along the length of the casing (Kurt, 1983).

When a slurry is mixed with excessive water, the water that does not adhere to or react with the cement particles tends to move upward and settle in the cement column, leading to channels and free water pockets behind the casing. This, in turn, creates zones of reduced permeability. The less dense or the greater the permeability of the cement, the less resistant is the cement to attack. Proper water-to-cement ratios are far more important than strength to the success of a cement seal. Documentable quality control should be implemented for water-to-cement ratios. Control of these ratios can be achieved by gaging the water tanks and by continuously weighing the slurry, which will help to assure the desired properties of the set cement. Although slurry volumes are increased by increasing the water-to-cement ratios, so is the permeability of the cement. Because protection of ground-water quality is one of the most important reasons for cementing wells, increasing the water-to-cement ratio is not a recommended design alternative (Kurt and Johnson, 1982). Proper mix water ratios should be adhered to as part of a documentable quality control program. Preferably a mud balance, Marsh Funnel, or some type of viscosity meter is used to determine if proper ratios have been achieved. A slurry with too much water may indeed create a permanent water-quality problem which may lead to the abandonment of the monitoring well (Williams and Evans, 1987).

The mixing of neat cement grout may be accomplished manually or with a mechanical mixer. Mixing must be continuous so that the slurry can be emplaced without interruption. Prolonged mixing should be avoided because it disrupts the hydration process and reduces the ultimate strength and quality of the cement. The grout should be mixed to a fairly stiff consistency and immediately pumped via a tremie pipe to its intended position in the annulus. The types of pumps recommended for use with neat cement grout include reciprocating (piston) pumps, diaphragm pumps, centrifugal pumps, or moyno pumps, all commonly used by well drilling contractors.

Neat cement, because of its chemical nature [lime (calcium carbonate), alumina, silica, magnesia, ferric oxide, and sulfur trioxide], is a highly alkaline substance (pH from 10 to 12), and thus introduces the potential for altering the pH of water which it contacts. The alteration of pH, in turn, can affect other chemical constituents in the water. In addition, because the emplaced mixture is a slurry and because (generally) it is emplaced in a column which imparts a high hydraulic pressure, it may tend to infiltrate into the coarse materials that comprise the primary filter pack around the monitoring well screen if it is placed directly on top of the primary filter pack. This is particularly true of thinner slurries (i.e., those mixed with more than 6 gallons of water per sack of cement). The cement infiltration problem can be aggravated in this situation if well development is attempted before the cement has completely set.

All of these problems can have a severe and persistent effect on both the performance of the monitoring well (in terms of yield) and the integrity of samples taken from the monitoring well. Direct placement of a thin grout on top of the primary filter pack, with subsequent infiltration, results in the plugging of the filter pack (and potentially the well screen) with cementitious material upon setting. Additionally, the presence of high pH cement within or adjacent to the filter pack could cause anomalous pH readings in subsequent water samples taken from the well. Dunbar et al. (1985) reported an incident attributed to this phenomenon, in which several wells completed in this manner in low-permeability geologic materials consistently produced samples with a high pH (greater than pH 9) for 2.5 years, despite repeated attempts at well development. Neat cement should thus not be emplaced directly on top of the filter pack in a monitoring well. It has been suggested in this chapter and by others that a very fine-grained secondary filter pack, from 1 to 2 feet thick, be placed atop the filter pack material before emplacement of the neat cement grout to eliminate the grout infiltration potential (Ramsey and Maddox, 1982; Barcelona et al., 1985a). A bentonite seal 2 to 5 feet thick would accomplish the same purpose but would require additional time to allow the bentonite to hydrate before cement placement. Either of these procedures should minimize the impairment of well performance and the potential effects of chemical interference caused by the proximity of neat cement to the well screen.

Another potential problem related to the use of neat cement as an annular sealing material concerns the heat generated by the cement as it sets. When water is mixed with any type of portland cement, a series of spontaneous chemical reactions called hydration reactions occur. If allowed to continue to completion, these reactions transform the cement slurry into a rigid, solid material. As the hydration reactions progress and the cement cures, heat is given off as a by-product. This heat is known as the heat of hydration (Troxell et al., 1968). The rate of generation of heat of hydration is a function of curing temperature, time, cement chemical composition, and the presence of chemical additives (Lerch and Ford,

1948). Generally, the heat of hydration is of little concern; however, if large volumes of cement are used or if the heat is not rapidly dissipated (as it would not be in a borehole because of the insulating properties of geologic materials), relatively large temperature increases may occur (Verbeck and Foster, 1950; Molz and Kurt, 1979; Jackson, 1983) and may compromise the structural integrity of some types of well casing, notably PVC casing. PVC characteristically loses strength and stiffness as the temperature of the casing increases. Because the collapse pressure resistance of a casing is proportional to material stiffness, a sufficient rise in casing temperature, coupled with the increased hydrostatic pressure due to the placement of the cement slurry, could cause unanticipated casing failure (Johnson et al., 1980). However, because the boreholes for most monitoring wells are not much larger than the well casing (compared to production wells), a heat of hydration high enough to damage PVC casing is generally not created. Moreover, because many monitoring wells are shallow, excessive hydrostatic pressures caused by the presence of grout outside the casing, are generally not created.

The use of setting time accelerators such as calcium chloride, gypsum, or aluminum powder can increase the heat of hydration, causing casings to become very hot while the grout is curing, and resulting in increased potential for casing failure.

Several methods can minimize heat of hydration. Adding agents such as bentonite or diatomaceous earth to the grout mix to retard the setting time will reduce peak temperatures. Other methods for retarding the setting time include adding inert materials such as silica sand to the grout, circulating cool water inside the casing during grout curing, and increasing the water/cement ratio of the grout mix (Kurt, 1983). The latter option, however, results in increased shrinkage and decreased strength upon setting.

Methods of Installation of Annular Seal Materials

Bentonite

Bentonite may be emplaced as a dry solid or as a grout. Generally, only pelletized, granular, chip, or chunk bentonite may be emplaced dry; powdered bentonite is usually mixed with water at the surface to form a grout and is then emplaced in the casing/borehole annulus.

In relatively shallow (i.e., 50 foot) monitoring wells with sufficient annular space (i.e., more than 3 inches) on all sides of the casing, dry bentonite may be emplaced by the gravity (free-fall) method in which the bentonite is poured from ground surface. If the gravity method is used, the bentonite should be tamped with a tamping rod after emplacement to prevent bridging of pellets or granules. In deeper wells, especially where static water levels are shallow, emplacing dry bentonite via the gravity method introduces both a very high potential for bridging and the likelihood that sloughing material from the borehole wall will be included in the seal. If bridging occurs, the bentonite may never reach its intended destination in the annulus. If sloughing material is included in the seal, "windows" of high-permeability material may develop. Either situation results in an ineffective annular seal.

In wells more than about 50 feet deep, granular bentonite may be mixed with water and conveyed through a tremie pipe from the surface directly to its intended depth in the annulus; pelletized bentonite is not effectively installed through tremie pipes. A pipe with an inside diameter of 1.5 to 2 inches should be used with granular bentonite to avoid bridging of the bentonite and subsequent clogging of the tremie. Because granular bentonite hydrates very quickly, it is usually difficult to work with unless its hydration is slowed by an organic polymer wetting agent. However, the use of organic polymers may impact the chemical integrity of ground-water samples later taken from the well, and is not recommended.

Successful use of bentonite grout as an effective well sealant can be achieved only by the proper mixing, pumping, and emplacement methods. As discussed previously, bentonite powder is generally mixed with water in a batch mixer, and the resulting grout is pumped using a positive displacement pump. The slurry should never be emplaced down the annulus using the gravity or free-fall method because the grout would either be diluted by water in the borehole, or the bentonite would segregate out of the grout. In either case, the grout would not form an effective annular seal.

Bentonite grout should be pumped under positive pressure through a side-discharge tremie pipe down the annular space (Figure 7.22). All hoses, tubes, pipes, water swivels, and other passageways through which the grout must pass should have a minimum inside diameter of 1 inch; a 1.5 inch tremie pipe is preferred. At the bottom of the tremie pipe, the discharge should be angled to the side, so the grout is not directed under pressure directly onto the top of the filter pack. If a bottom discharge tremie pipe is used, it may cause severe erosion of the filter pack or result in grout being injected directly into the filter pack. The side-discharge tremie pipe should be run to the bottom of the annular space (i.e., just above the filter pack or the level to which noncohesive material has collapsed in the borehole) and should be left there during emplacement, so that the grout fills the annulus from the bottom up. This will allow the grout to displace ground water and any loose formation materials in the annular space. The tremie can be removed after the grout has been emplaced to its intended level in the annulus.

Bentonite emplaced as a grout is already hydrated to some degree before emplacement, but its ability to form a tight seal depends upon addition and constant hydration after emplacement. Unless the geologic materials in which the grout is emplaced are saturated, thus supplying sufficient moisture to maintain the hydrated state of the bentonite, the seal may desiccate and crack, affecting its integrity. The setting of the bentonite sealing material generally requires 24 to 72 hours, during which time the grout mass builds gel strength. Bentonite grouts do not become rigid and develop structural strength as does neat cement. Well development should not be attempted until the bentonite has fully set.

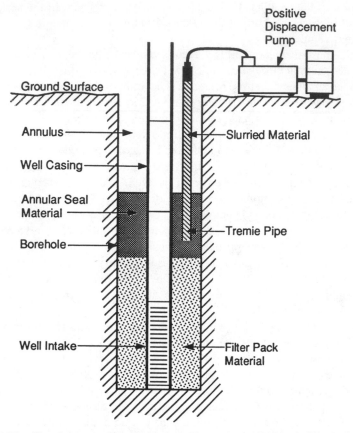

Figure 7.22. Tremie pipe emplacement of annular seal material (slurried bentonite or neat cement).

Because of the potential for chemical interference with ground-water samples posed by the moderately high pH and high cation exchange capacity of bentonite, it is recommended that a bentonite seal be placed well above the top of the well intake (i.e., at least 3 to 5 feet), and that a secondary filter pack be used on top of the primary filter pack. This should result in a minimal impact on ground-water sample integrity.

Neat Cement

As with a bentonite grout, a neat cement grout must be properly mixed, pumped, and emplaced to ensure an effective annular seal. Neat cement should be emplaced in the annulus only by free-fall (gravity) when there is adequate clearance (i.e., at least 3 inches) between the casing and the borehole, when the annulus is dry, and when the bottom of the annular space to be filled is clearly visible from the surface and not more than 30 feet deep (U.S. EPA, 1975). Allowing a neat cement grout to free-fall through standing water in the annulus introduces a high potential for the mixture to be diluted, to segregate or to bridge, after it reaches the level of standing water and before it reaches its intended depth of emplacement. In addition, in its free fall down the annulus, the grout may pick up sloughing material of high permeability from the walls of the borehole, causing a breach in the seal.

For neat cement grout, the emplacement method of choice is the tremie pipe (see Figure 7.22). Assuming the annular space is large enough, a tremie pipe with a minimum inside diameter of 1.5 inches should be inserted in the annulus to within a few inches of the bottom of the space to be sealed; the tremie pipe discharge port should be located on the side of the bottom end of the pipe. Grout is then pumped through the tremie pipe, discharging at the bottom of the annular space and flowing upward around the casing until the annular space is completely filled. This procedure allows the grout to displace ground water and loose formation materials ahead of the grout, thus minimizing both contamination and dilution of the slurry, which can reduce its bonding strength. This procedure also minimizes potential bridging of the grout with formation material. The tremie pipe may be moved upward as the slurry is emplaced or be left in place at the bottom of the annulus until grouting is completed. However, the end of the tremie pipe should always remain in the emplaced grout without allowing air spaces. After emplacement, the tremie pipe should be removed immediately to avoid the possibility of the grout setting around the pipe, causing difficulty in removing the pipe or creating a channel in the grout as the pipe is removed. To avoid the formation of cold joints, the grout must be emplaced in one continuous mass before initial setting of the cement or before the mixture loses its fluidity. The curing time required for Type I portland cement to reach its maximum strength is a minimum of 48 to 72 hours, though setting accelerators (additives) can reduce this time significantly.

SURFACE COMPLETION FOR MONITORING WELLS: PROTECTIVE MEASURES

Two types of surface completions are common for ground-water monitoring wells: the above-ground completion, which is preferred wherever practical, and the flush-to-ground-surface completion, which may be required under some site conditions. The primary purposes of either type of completion are to prevent surface runoff from entering and infiltrating down the annulus of the well, and to protect the well from accidental damage or vandalism.

Surface Seals

Whichever type of completion is selected for any given well, a surface seal of neat cement or concrete should surround the well casing and fill the annular space between the casing and borehole at the surface. Cement used for surface seals in cold-weather climates should be resistant to freeze-thaw induced crack-ing. The use of air-entrained cements is generally preferred, and can substan-tially reduce damage to or desturction of surface seals. The surface seal may be an extension of the annular seal installed above the filter pack, or a separate seal emplaced atop the annular seal. Because the annular space is generally larger and the surface material adjacent to the borehole is more highly disturbed from drilling at the surface than at depth, the surface seal will generally extend from 1 to 2 feet away from the well casing at the surface; the seal will usually taper down to the size of the borehole within a few feet of the surface. Some well in-stallers prefer to mound the cement surface seal around the well casing or pro-tective casing to allow for shrinkage of the cement, and provide a gentle slope away from the well that discourages surface runoff from entering the wellbore. The mound, however, should be limited in size and slope so that access to the well is not impaired. For climates in which alternating freezing and thawing are expected, the cement surface seal should be extended below frost depth to pre-vent potential well damage caused by frost-heaving, and the cement pad at the surface should not extend beyond the diameter of the annular seal. In some states, well installation regulations developed for water supply wells, but also applied to monitoring wells, require that the cement surface seal extend to greater depths (i.e., 10 feet or more) to ensure sanitary protection of the well.

Above-Ground Completions

In an above-ground completion (Figure 7.23), a protective casing is set into the cement surface seal while it is still wet and uncured, and is installed around the well casing. The protective casing discourages unauthorized entry into the well, prevents damage from contact with vehicles, and in the case of wells in which PVC well casing is installed, protects the casing from degradation caused by direct exposure to sunlight (i.e., from photodegradation in the presence of ultraviolet rays). The protective casing should be made of corrosion-resistant materials (i.e., stainless steel or aluminum), although it is more common to use carbon steel which may or may not be painted to inhibit corrosion. As with the well casing, the protective casing should be thoroughly cleaned before installa-tion to ensure removal of any chemicals or coatings. The inside diameter should be large enough to allow easy access to the well casing, including allowances for the size of the inner casing cap. The protective casing may be fitted with a locking cap, in which case it should be installed to provide adequate clearance (2 to 4 inches) between the top of the in-place inner well casing cap and the bot-tom of the protective casing's locking cap, when the cap is in the locked posi-tion. During installation, the protective casing should be positioned and maintained in a plumb position. It is usually installed so that approximately half of the pro-tective casing is anchored into the cement surface seal, and half extends above the seal to protect the well casing.

Like the inner well casing, the outer protective casing should be vented to pre-vent the accumulation and entrapment of potentially explosive gases, and to allow

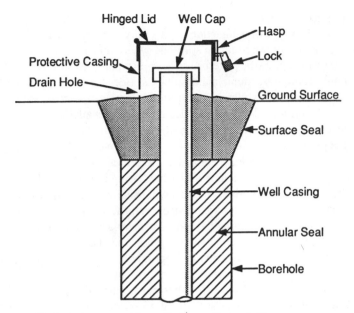

Figure 7.23. Typical aboveground monitoring well completion.

water levels in the well to respond to barometric and hydraulic pressure changes. Additionally, the outer protective casing should have a drainhole installed just above the top of the cement level in the space between the protective casing and the well casing to allow the drainage of any water accumulating in this space. This is particularly critical in cold climates, where the freezing of water trapped between the inner well casing and the outer protective casing can cause the inner casing to buckle or fail.

A case-hardened steel lock is generally installed on the casing cap to provide well security, but weather-caused corrosion is likely to cause the locking mechanism to jam in time. Because lubricants (i.e., graphite, petroleum-based sprays, and silicon) provide a potential source of ground-water sample chemical alteration, their use in lubricating locks or for freeing corroded locking mechanisms is not recommended. Rather, the use of some type of protective measure to shield the lock from the elements (i.e., a plastic covering) should be considered.

In high-traffic areas such as parking lots or in areas where heavy equipment may be working, additional protection may be afforded an aboveground completion by the installation of "bumper guards" (brightly painted posts of wood, steel, or some other durable material set in cement) within 3 or 4 feet of the well.

Flush-to-Ground-Surface Completions

In a flush-to-ground-surface completion, a protective structure such as a utility vault or meter box is generally set into the cement surface seal before it has cured, and is installed around a well casing that has been cut off below grade (Figure 7.24). This type of completion is generally used in high-traffic areas such as streets, parking lots, and service stations, where an aboveground completion would severely disrupt traffic patterns, or in areas where a flush-to-ground-surface completion is required by municipal easements. Because of the potential for surface runoff to enter the wellbore or the below-grade protective structure, this type of completion must be carefully designed and installed. For example, the bond between the cement surface seal and the protective structure, and the seal between the protective structure and its removable cover must be watertight. Use of an expanding cement should ensure that the cement surface seal bonds tightly to the protective structure; installing a flexible O-ring or gasket at the point where the cover fits over the protective structure will usually suffice for sealing the protective structure. In areas of significant street runoff, additional safeguards, such as building a low, gently sloping mound of cement around the protective structure and placing the structure slightly above grade, may be necessary to discourage entry of surface runoff. It is often very important to provide a means to lock the inner well casing to restrict access to the well, because a lock may be impractical on the outer protective enclosure. Furthermore, particularly in areas in which other below-grade enclosures exist (i.e., at a service station), it is important to identify the monitoring well to set it apart from other subsurface structures (i.e., underground storage tank fill pipes), so that the well is not mistaken for something else.

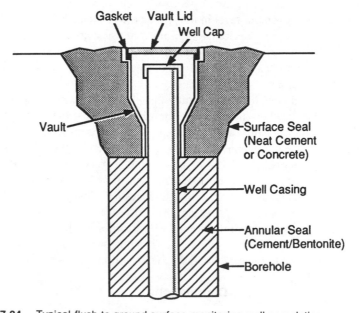

Figure 7.24. Typical flush-to-ground-surface monitoring well completion.

REFERENCES

Ahmad, M. U., E. B. Williams and L. Hamdan, 1983, Commentaries on Experiments to Assess the Hydraulic Efficiency of Well Screens, *Ground Water,* Vol. 21, No. 3, pp. 282–286.

Ahrens, T. P., 1957, Well Design Criteria: Part One; *Water Well Journal,* Vol. 11, No. 4, pp. 13–30.

Ahrens, T. P., 1970, Basic Considerations of Well Design: Part III; *Water Well Journal,* Vol. 24, No. 3, pp. 47–51.

Anderson, K. E., 1971, *Water Well Handbook,* Missouri Water Well and Pump Contractors Association, Inc., Rolla, MO.

Anderson, D. C., K. W. Brown and J. W. Green, 1982, Effect of Organic Fluids on Permeability of Clay Soil Liners; Land Disposal of Hazardous Waste: Proceedings, U.S. Environmental Protection Agency Report #EPA-600/9-82-002.

Barcelona, M. J., J. P. Gibb and R. A. Miller, 1983, A Guide to the Selection of Materials for Monitoring Well Construction and Ground Water Sampling; Illinois Department of Energy and Natural Resources, Water Survey Division, Champaign, Illinois, SWS Contract Report #327, 78 pp.

Barcelona, M. J., J. P. Gibb, J. A. Helfrich and E. E. Garske, 1985a, A Practical Guide for Ground Water Sampling; Illinois Department of Energy and Natural Resources, Water Survey Division, Champaign, Illinois, SWS Contract Report #374, 95 pp.

Barcelona, M. J., J. A. Helfrich and E. E. Garske, 1985b, Sample Tubing Effects on Ground Water Samples; *Analytical Chemistry,* Vol. 5, pp. 460–464.

Bikis, E. A., 1979, A Laboratory and Field Study of Fiberglass and Continuous-Slot Screens, *Ground Water,* Vol. 17, No. 1, p. 111 (Abstr.)

Boettner, E. A., G. L. Ball, Z. Hollingsworth and R. Aquino, 1981, Organic and Organotin Compounds Leached from PVC and CPVC Pipes; U.S. Environmental Protection Agency Report #EPA-600/1-81-062, 102 pp.

Brown, K. W., J. W. Green and J. C. Thomas, 1983, The Influence of Selected Organic Liquids on the Permeability of Clay Liners: Land Disposal of Hazardous Waste: Proceedings, U.S. Environmental Protection Agency Report #EPA-600/9-83-018, pp. 114–125.

California Department of Health Services (DOHS), 1985, The California Site Mitigation Decision Tree; California Department of Health Services, Toxic Substances Control Division Draft Working Document.

Calhoun, D. E., 1988, Sealing Well Casings: An Idea Whose Time Has Come; *Water Well Journal,* Vol. 42, No. 2, pp. 25–29.

Campbell, M. D. and J. H. Lehr, 1973, *Water Well Technology;* McGraw-Hill Book Company, New York, NY, 681 p.

Campbell, M. D. and J. H. Lehr, 1975, Well Cementing, *Water Well Journal,* Vol. 29, No. 7, pp. 39–42.

Clark, L. and P. A. Turner, 1983, Experiments to Assess the Hydraulic Efficiency of Wells, *Ground Water,* Vol. 21, No. 3.

Curran, C. M. and M. B. Tomson, 1983, Leaching of Trace Organics Into Water From Five Common Plastics; *Ground Water Monitoring Review,* Vol. 3, No. 3, pp. 68–71.

Dablow, J. F., D. Persico and G. R. Walker, 1986, Design Considerations and Installation Techniques for Monitoring Wells Cased with Teflon PTFE; Proceedings: ASTM Symposium on Field Methods for Ground Water Contamination Studies, ASTM Special Technical Publication #963, pp. 199–205.

Driscoll, F. G., 1986, *Ground Water and Wells* (2nd edition), Johnson Division, UOP, Inc., St. Paul, MN.

Dunbar, D., H. Tuchfeld, R. Siegel and R. Sterbentz, 1985, Ground Water Quality Anomalies Encountered During Well Construction, Sampling and Analysis in the Environs of a Hazardous Waste Management Facility; *Ground Water Monitoring Review,* Vol. 5, No. 3, pp. 70–74.

Ericson, W. A., J. E. Brinkman and P. S. Dar, 1985, Types and Usages of Drilling Fluids Utilized to Install Monitoring Wells Associated with Metals and Radionuclide Ground Water Studies, *Ground Water Monitoring Review,* Vol. 5, No. 1, pp. 30–33.

Ferguson, R., 1987, E. I. DuPont de Nemours, Inc., personal communication with D. M. Nielsen.

Foster, S., 1989, Flush-Joint Threads Find a Home, *Ground Water Monitoring Review,* Vol. 9, No. 2, pp. 55–58.

Gillham, R. W. and S. F. O'Hannesin, 1989, Sorption of Aromatic Hydrocarbons by Materials used in Construction of Ground Water Sampling Wells; Proceedings: ASTM Symposium on Standards Development for Ground Water and Vadose Zone Monitoring Investigations, ASTM Special Technical Publication# 1053, pp. 108–124.

Hamilton, Hugh, 1985, Selection of Materials in Testing and Purifying Waters; *Ultra Pure Water,* Jan./Feb. 1985, 3 pp.

Helwig, O. J., V. H. Scott and J. C. Scalmanini, 1984, *Improving Well and Pump Efficiency;* American Water Works Association, Denver, CO, 46 pp.

Hunkin, G. G., T. A. Reed and G. N. Brand, 1984, Some Observations on Field Experiences with Monitoring Wells, *Ground Water Monitoring Review,* Vol. 4, No. 1, pp. 43–45.

Jackson, P. A., 1983, A Laboratory and Field Study of Well Screen Performance and Design, *Ground Water,* Vol. 12, No. 6, pp. 771–772 (Abstract).

Johnson Division, U.O.P., 1966, *Ground Water and Wells;* Edward E. Johnson, Inc., St. Paul, MN, 440 pp.

Johnson, R. C., Jr., C. E. Kurt and G. F. Dunham, Jr., 1980, Well Grouting and Casing Temperature Increases; *Ground Water,* Vol. 18, No. 1, pp. 7–13.

Junk, G. A., H. J. Svec, R. D. Vick and M. J. Avery, 1974, Contamination of Water by Synthetic Polymer Tubes; *Environmental Science and Technology,* Vol. 8, No. 13, pp. 1100–1106.

Kurt, C. E., 1983, Cement-Based Seals for Thermoplastic Water Well Casings; *Water Well Journal,* Vol. 37, No. 1, pp. 38–40.

Kurt, C. E. and R. C. Johnson, Jr., 1982, Permeability of Grout Seals Surrounding Thermoplastic Well Casing; *Ground Water,* Vol. 20, No. 4, pp. 415–419.

Lerch, W. and C. L. Ford, 1948, Long-Time Study of Cement Performance in Concrete, Chapter 3-Chemical and Physical Tests of the Cements; *Journal of the American Concrete Institute,* Vol. 19, No. 8.

Miller, G. D., 1982, Uptake and Release of Lead, Chromium and Trace Level Volatile Organics Exposed to Synthetic Well Casings; Proceedings; Second National Symposium on Aquifer Restoration and Ground Water Monitoring, National Water Well Association, Worthington, Ohio, pp. 236-245.

Moehrl, K. E., 1964, Well Grouting and Well Protection; *Journal of the American Water Works Association,* Vol. 56, No. 4, pp. 423–431.

Molz, F. J. and C. E. Kurt, 1979, Grout-Induced Temperature Rises Surrounding Wells; *Ground Water,* Vol. 17, No. 3, pp. 264–269.

Morrison, R. D., 1984, *Ground Water Monitoring Technology:* Procedures, Equipment and Applications; Timco Mfg., Inc., Prairie Du Sac, Wisconsin, 111 pp.

National Water Well Association and Plastic Pipe Institute, 1980, *Manual on the Selection and Installation of Thermoplastic Water Well Casing;* National Water Well Association, Worthington, Ohio, pp. 64.

Parker, L. V. and T. F. Jenkins, 1986, Suitability of Polyvinyl Chloride Well Casings for Monitoring Munitions in Ground Water; *Ground Water Monitoring Review,* Vol. 6, No. 3, pp. 92–98.

Parker, L. V., A. D. Hewitt, and T. F. Jenkins, 1990, Influence of Casing Materials on Trace-Level Chemicals in Well Water. *Ground Water Monitoring Review,* Vol. 10, No. 2, pp. 146–156.

Paul, D. G., C. D. Palmer and D. S. Cherkauer, 1988, The Effect of Construction, Installation, and Development on the Turbidity of Water in Monitoring Wells in Fine-Grained Glacial Till, *Ground Water Monitoring Review,* Vol. 7, No. 1, pp. 73–82.

Pettyjohn, W. A., W. J. Dunlap, R. Cosby, and J. W. Keeley, 1981, Sampling Ground Water for Organic Contamination, *Ground Water,* Vol. 19, No. 2, pp. 180–189.

Powers, M. C., 1953, A New Roundness Scale for Sedimentary Particles, *Journal of Sedimentary Petrology,* Vol. 23, pp. 117–119.

Raber, E., J. Garrison and V. Oversby, 1983, The Sorption of Selected Radionuclides on Various Metal and Polymeric Materials, *Radioactive Waste Management and the Nuclear Fuel Cycle,* Vol. 4, No. 1, pp. 41–52.

Ramsey, R. H. and G. E. Maddox, 1982, Monitoring Ground Water Contamination in Spokane County, Washington; Proceedings: Second National Symposium on Aquifer Restoration and Ground Water Monitoring, National Water Well Association, Worthington, Ohio, pp. 198–204.

Reynolds, G. W. and R. W. Gillham, 1985, Adsorption of Halogenated Organic Compounds by Polymer Materials Commonly Used in Ground Water Monitoring, in Proceedings, Second Canadian/American Conference on Hydrogeology, National Water Well Association, Dublin, OH, pp. 125–132.

Richter, H. R. and M. G. Collentine, 1983, Will My Monitoring Wells Survive Down There? Installation Techniques for Hazardous Waste Studies; Proceedings: Third National Symposium on Aquifer Restoration and Ground Water Monitoring National Water Well Association, Worthington, Ohio, pp. 223–229.

Rinaldo-Lee, M. B., 1983, Small vs. Large Diameter Monitoring Wells: *Ground Water Monitoring Review,* Vol. 3, No. 1, pp. 72–75.

Scalf, M. A., J. F. McNabb, W. J. Dunlap, R. L. Crosby and J. Fryberger, 1981, *Manual of Ground Water Quality Sampling Procedures:* National Water Well Association, Dublin, Ohio.

Schalla, R., 1986, A Comparison of the Effects of Rotary Wash and Air Rotary Drilling Techniques on Pumping Test Results, in Proceedings of the Sixth National Symposium and Exposition on Aquifer Restoration and Ground Water Monitoring, National Water Well Association, Dublin, OH, pp. 7–26.

Schalla, R. and P. L. Oberlander, 1983, Variation in the Diameter of Monitoring Wells, *Water Well Journal,* Vol. 37, No. 5, pp. 56–57.

Schalla, R., R. W. Wallace, R. L. Aaberg, S. P. Airhart, D. J. Bates, J. V. M. Carlile, C. S. Cline, D. I. Dennison, M. D. Freshley, P. R. Heller, E. R. Jensen, K. B. Olsen, R. G. Parkhurst, J. T. Reiger, and E. J. Westergard. 1988, Interim Characterization Report for the 300 Area Process Trenches, PNL-6716, Pacific Northwest Laboratory, Richland, Washington, p. 677.

Schalla, R., W. H. Walters, 1989, Rationale for the Design of Monitoring Well Screens and Filter Packs, Proceedings: ASTM Symposium on Standards Development for Ground Water and Vadose Zone Monitoring Investigations, ASTM Special Technical Publication #1053, pp. 64–75.

Schmidt, G. W., 1987, The Use of PVC Casing and Screen in the Presence of Gasolines on the Ground Water Table, *Ground Water Monitoring Review,* Vol. 7, No. 2, p. 94.

Schmidt, K. D., 1982, The Case for Large-Diameter Monitor Wells; *Water Well Journal,* Vol. 36, No. 12, pp. 28–29.

Schmidt, K. D., 1983, How Representative Are Water Samples Collected From Wells? Proceedings of the Second National Symposium on Aquifer Restoration and Ground Water Monitoring, National Water Well Association, Dublin, OH, pp. 117–128.

Simons, D. B. and F. Senturk, 1977, Sediment Transport Technology, *Water Resources Publications,* U. S. Geological Survey, Fort Collins, CO, p. 807.

Smith, R. M., 1988, Resource Conservation and Recovery Act Ground-Water Monitoring Projects for Hanford Facilities: Progress Report for the Period April 1 to June 31, 1988, PNL-6675/UC-11, 41, Pacific Northwest Laboratory, Richland, WA.

Smith, R. M., D. J. Bates, and R. E. Lundgren, 1989, Resource Conservation and Recovery Act Ground-Water Monitoring Projects for Hanford Facilities: Progress Report for the Period January 1 to March 31, 1989, PNL-6957/UC-11, 41, Pacific Northwest Laboratory, Richland, WA.

Sykes, A. L., R. A. McAllister and J. B. Homolya, 1986, Sorption of Organics by Monitoring Well Construction Materials, *Ground Water Monitoring Review,* Vol. 6, No. 4, pp. 44-47.

Sosebee, J. B., P. C. Geiszler, D. L. Winegardner and C. Fisher, 1983, Contamination of Ground Water Samples With Poly (Vinyl Chloride) Adhesives and Poly (Vinyl Chloride) Primer From Monitor Wells; Proceedings: ASTM Second Symposium on Hazardous and Industrial Solid Waste Testing, ASTM Special Technical Publication #805, pp. 38-49.

Tomson, M. B., S. R. Hutchins, J. M. King and C. H. Ward, 1979, Trace Organic Contamination of Ground Water: Methods for Study and Preliminary Results; Third World Congress on Water Resources, Mexico City, Mexico, Vol. 8, pp. 3701-3709.

Troxell, G. E., H. E. Davis and J. W. Kelly, 1968, *Composition and Properties of Concrete;* McGraw-Hill Book Co., New York, NY.

U. S. EPA, 1975, Manual of Water Well Construction Practices; Report No. EPA-570/9-75-001, 156 p.

Verbeck, G. J. and C. W. Foster, 1950, Long-Time Study of Cement Performance in Concrete with Special Reference to Heats of Hydration; Proceedings, American Society for Testing and Materials, Vol. 50.

Villaume, James F., 1985, Investigations at Sites Contaminated with Dense Non-Aqueous Phase Liquids (DNAPLS); *Ground Water Monitoring Review,* Vol. 5, No. 2, pp. 60-74.

Voytek, J. E., Jr., 1983, Considerations in the Design and Installation of Monitoring Wells, *Ground Water Monitoring Review,* Vol. 3, No. 1, pp. 70-71.

Walker, William H., 1974, Tube Wells, Open Wells and Optimum Ground Water Resource Development; *Ground Water,* Vol. 12, No. 1, pp. 10-15.

Williams, C. and L. G. Evans, 1987, Guide to the Selection of Cement, Bentonite and Other Additives for Use in Monitor Well Construction; Proceedings, First National Outdoor Action Conference, National Water Well Association, Dublin, OH, pp. 325-343.

Williams, Ernie B., 1981, Fundamental Concepts of Well Design; *Ground Water,* Vol. 19, No. 5, pp. 527-542.

Yu, J. K., 1989, Should We Use A Well Foot (Sediment Trap) in Monitoring Wells? *Ground Water Monitoring Review,* Vol. 9, No. 2, pp. 59-60.

8

Monitoring Well Post-Installation Considerations

Curtis A. Kraemer, James A. Shultz, and James W. Ashley

Following the installation of a monitoring well or monitoring well system, there are several issues that should be considered to secure the well system's integrity and identity. This chapter addresses several important monitoring well post-installation considerations including: monitoring well development, surveying, well identification, report of construction details, maintenance, rehabilitation, and abandonment. In general, monitoring well development is important because it is the activity performed in the well to remove fines (silt, clay, fine sand) and drilling fluids from the filter pack and the natural formation in the vicinity of the monitoring well. This is done to provide maximum efficiency and hydraulic communication between the well and the adjacent natural formation, to assure that future aquifer test results are of maximum value, and to ensure that that representative ground-water samples may be collected in the future. Surveying of monitoring wells to a common datum is necessary to obtain accurate water-level data and to allow the correlation of stratigraphic horizons from well to well and the subsequent development of hydrogeologic cross-sections. Proper well identification and complete reporting of monitoring well construction details are necessary. Monitoring well maintenance and rehabilitation are necessary to maximize the life of a well system. Monitoring well abandonment is required to mitigate the potential for an unused, unnecessary, or malfunctioning well to become a vertical conduit for contaminant migration.

MONITORING WELL DEVELOPMENT

In general, all drilling methods create at least some amount of clogging, coating, and/or compaction of the borehole wall and the natural formation materials (unconsolidated sediment or rock) immediately adjacent to the borehole wall, resulting in localized reduction of formation hydraulic conductivity. Where highly permeable sediment is encountered, there can be a significant loss of drilling fluid to the adjacent natural formation. Additionally, during the installation of filter-packed monitoring wells, fines from the adjacent natural formation can mix with filter pack material as it is placed around the well screen.

The goal of monitoring well development is to remove fines (silt, clay, fine sand) and drilling fluid residue from the gravel pack and the natural formation in the vicinity of the screened interval in the monitoring well (for wells completed in unconsolidated sediment or incompetent rock) or open borehole (for wells completed in competent rock). Additionally, well development results in the settlement and stabilization of the material adjacent to the well screen. The well development process is composed of: (1) the application of sufficient energy in a monitoring well to create ground-water flow reversals (surging) in and out of the well and the gravel pack or natural formation to release and draw fines into the well; and (2) pumping to draw drilling fluids out of the borehole and adjacent natural formation, along with the fines that have been surged into the well. The development of monitoring wells should be performed as soon as practical after the well has been installed and the annular seal materials have cured. Ideally, this activity results in creating maximum well efficiency and hydraulic communication between the well and the adjacent natural formation, further maximizing the value of the data from future aquifer tests and the representativeness of ground-water samples to be collected from the well. Additionally, the removal of fines during development minimizes the potential for clogging and damaging of pumping equipment during future pumping tests and/or purging prior to sampling.

In general, much of the published literature regarding well development is confined to the development of water-supply wells. Water-supply well screens are generally 6 inches in diameter or larger and have a large percentage of open area to extract a large volume of water from an aquifer. In contrast, monitoring wells are not installed to provide a maximum yield, but rather for the collection of water level data and representative samples of ground water from specific depth intervals. Furthermore, monitoring wells may be installed in poorly yielding geologic materials that may not be considered aquifers. Many monitoring wells are constructed with 2- to 4-inch diameter slotted PVC screens which have a small percentage of open area. Standard water-supply well development methods are not feasible in these wells, at least not without some modification. The purpose of this chapter is to provide specific well development guidelines for monitoring wells. In addition, water-quality problems associated with monitoring well development may have to be considered. Contamination present in the discharged development water can create a disposal problem. Discharge of such water to the ground surface vs containment (e.g., in drums or tanks for appropriate future disposal) must be addressed on a site-by-site basis, along with the related issue of health and safety considerations for field personnel.

Another factor which can impact the proper completion of monitoring well development is time/cost. Monitoring wells are generally installed because of a regulatory requirement to investigate potential or known ground-water contamination; they do not generate revenue like online water supply wells (e.g., $/1,000 gallon of water use). Monitoring well development is usually charged as an hourly fee and wells installed in low-yielding formations (very fine sand, silt, or clay) can require considerable time for recovery once water is evacuated from the well casing. Well development activity can be quite lengthy and is often left until all of the monitoring wells at a site have been installed. The well development process is usually performed with portable equipment that can be handled by one or two people.

Ideally, a monitoring well should be developed to allow for the collection of turbidity-free, representative samples of ground water for chemical analysis. However, there is often a cost-benefit tradeoff during well development after much of the visible turbidity appears to have been removed. If a monitoring well is not completely developed so that representative samples of the ground water can be collected for analysis (assuming there has been no cross-contamination during installation of the wells), then all the money spent installing the monitoring well system and analyzing ground-water samples may result in analytical data that are not truly representative of actual subsurface conditions at a particular site. These analytical data can form the basis for selection of site remediation methods, and must be representative of site subsurface conditions. Therefore, the cost/benefit ratio appears to strongly favor proper development of the wells. Proper installation of a monitoring well, including pumping/bailing to remove accumulated drilling fluids and fines from the well casing during installation or as soon as it has been installed, will minimize the amount of development required. Monitoring well development should ideally continue until: (1) visibly clear water is discharged during the active portion of development; and (2) the total volume of water discharged from the well is at least equal to the estimated volume of fluid lost to the natural formation during drilling and well installation.

Monitoring well development is in part accomplished by actively agitating (surging) the water column in a well, forcing water back and forth through the well screen and the gravel pack or natural formation to release fines (silt, clay, fine sand) from the formation and bring them into the well. This material is then removed, along with drilling and well installation fluids, by pumping. Passive well development by only pumping (especially at low rates) will not effectively remove fines or drilling fluids that have moved into the adjacent natural formation. Although visibly clear water may eventually be discharged as a result of such pumping, the next activity that creates a surge in the well (a pumping test, purging prior to sampling, or sampling with a bailer) can release considerable turbidity.

Because of their intended use, monitoring wells (unlike water supply wells) spend most of their time in a dormant (unpumped) condition. Therefore, there is no activity in a monitoring well to continue removal of small amounts of fines over an extended period of time. No matter how complete the development of a monitoring well appears to be at the time of development, there is a high probability (especially for wells completed in fine-grained formations) that future introduction of pumps or bailers for testing or sampling will create a surge rendering the water produced from the well somewhat turbid. It is, therefore, imperative to adequately develop monitoring wells to minimize future turbidity problems and the possible need for redevelopment.

It is important to note that a recent field investigation by Paul et al. (1988) indicates that surging during well development increased the turbidity of monitoring wells installed in fine-grained glacial till at two sites located in Wisconsin. The well development study consisted of surging and bailing some wells, and only bailing other wells. Although the method of well installation was found to have an impact on turbidity increase, wells installed by the same method, but whose development included surging, yielded water samples with between 3 and 100 times greater turbidity than water samples from wells that were simply bailed. The conclusion of this study was that well surging should be avoided for monitoring wells completed in fine-grained till because: ''(1) it substantially increases the turbidity of water samples; (2) it does not significantly improve hydraulic well response; and (3) it adds unnecessary cost to the overall sampling program.''

During well installation, extraneous materials (e.g., grout, bentonite, sand) may inadvertently be dropped into the well, or during air-rotary drilling of bedrock, cuttings may remain on the borehole wall above the water table. Therefore, monitoring well development should also include the rinsing of the entire well cap and the interior of the well casing or open bedrock borehole above the water table using only water from the well. The purpose of this operation is to provide a well casing free of extraneous materials and potential contaminants. This washing should be conducted before or during development—not after development.

Because monitoring wells are installed at least in part to obtain representative samples for water-quality analysis, no foreign material or fluids should be introduced during development which could alter the existing subsurface chemical conditions including, for example, dispersing agents, acids, and disinfectants. Further, the addition of water to aid in development is not recommended. The introduction of foreign water to a monitoring well(s) has the potential to add contaminants not previously present at a site, or to potentially dilute contamination present in the vicinity of a monitoring well. However, if it is necessary to add water, then the source of the development water must be of good quality and documented by laboratory analysis for at least the analytical parameters of the site-specific monitoring program. Under such conditions, the amount of water discharged from the well during development should at least equal the amount of water added.

The selection of an appropriate method or combination of methods for development of a monitoring well should include consideration of the following factors:

- drilling and well installation method employed—hollow-stem auger, rotary wash, cable tool, etc.
- condition at the bottom of the well casing—capped or plugged? How?
- well casing and screen diameter
- screen length and percent of open area
- depth to water
- height of the water column within the well
- character of the natural formation (e.g., silt, gravel, bedrock)
- site accessibility
- type of personnel to perform the development
- type of equipment available
- type of suspected or known contaminants present
- need for appropriate disposal of the discharged water
- health and safety requirements
- appropriate regulatory agency approval of selected method(s)

The following sections provide guidance for six general methods applicable to the development of 2- to 4-inch diameter monitoring wells. Development methods for larger diameter monitoring wells could include combining or modifying the methods presented herein and/or the use of standard methods for water-supply wells, including: (1) single- and double-pipe air-lift, including equipment decontamination between wells; (2) mechanical surging and pumping using dedicated or decontaminated surge blocks or swabs; or (3) overpumping and rawhiding, including pump and discharge pipe decontamination between wells. These three general methods can generate significant volumes of discharged contaminated water, which could create containment and logistical disposal problems. Additionally, the performance of these three general methods are well documented in the literature, for example USEPA (1975) and Driscoll (1986).

Six monitoring well development methods and guidelines for their performance are presented in the following sections. The six methods include:

- Development by surging/pumping with compressed air
- Development by centrifugal pump
- Development by submersible pump
- Development by valved and air-vented surge plunger
- Development by bailer
- Manual development

These methods can be used individually or in any combination appropriate to complete development of a monitoring well depending upon site conditions. Except as noted, the descriptions of these methods are based upon experience of the author and have been expanded from EA Engineering, Science, and Technology, Inc. (1985).

Development by Surging/Pumping with Compressed Air

Generally, development by surging and pumping with compressed air will work for most monitoring wells no matter what the depth to water, assuming sufficient air pressure and volume. A 100–150 cubic feet per minute (CFM) air compressor is generally sufficient for development of from 2- through 4-inch diameter monitoring wells. To develop a well by this method, there must be at least 20% submergence of the air discharge line. For example, if a monitoring well has been installed to a depth of 100 feet below grade, there must be at least a 20-foot column of water in the well and preferably more, assuming that the water level will be drawn down during development. Drawdown of the water level will create a lower percent of submergence than would be calculated using the static water level. Care must be taken during the development of deep monitoring wells and wells completed in stratified gravel deposits so that air is not forced into the formation adjacent to the screened interval or open bedrock borehole. Air forced into the saturated zone of the formation reduces permeability, and can affect the quality of the ground water to be sampled in the future. Therefore, consideration should be given to other methods of development for deep monitoring wells and wells completed in stratified gravel deposits.

It is imperative that the compressed air discharge line of the air compressor include a functioning oil/air separator filter. The effectiveness of such a filter should be checked before and after each well is developed. Such a check can be performed by placing a clean white cloth over the air discharge, opening the discharge valve fully, and then checking the cloth for oil staining. If staining is observed, the problem must be corrected before well development is attempted.

Under no circumstances should the high pressure hose supplied with a rented air compressor be placed within a well. Such hoses are often sheathed with synthetic rubber and have probably laid on the ground at many previous job sites, and thus absorbed a variety of contaminants. New, fresh lengths of flexible polyethylene pipe provide a reasonable alternative for use as the air discharge line for each well developed. This pipe is relatively inexpensive and can be attached to the air compressor using a Chicago fitting, an insert fitting, and a threaded hose clamp. All connections must be securely attached. The flexible polyethylene pipe should be stored in large plastic bags until used, to mitigate contamination. Additionally, the pipe must be handled with new, clean gloves for each well, and must not be allowed to touch the ground.

Compressed air discharging at about 100 psi can be dangerous and must be handled carefully. All connections must be tightly secured. The pressure should be increased slowly so as not to blow the polyethylene pipe out of the well. To aid in keeping the pipe in the well, a length of decontaminated 3/4-inch diameter steel pipe can be attached as weight to the end of the polyethylene pipe. Alternatively, if the well is of large enough diameter, the end of the polyethylene pipe can be fitted with a metal deflector or a short length of steel pipe bent into a J-shape. Both of these attachments will discharge the compressed air directly upward through the water column, and thus keep the polyethylene pipe from being blown out of the well. It is important that these attachments be decontaminated before use in each well. The method and extent of decontamination can vary depending upon the contaminant present at a site. The top of the monitoring well should be fitted with an appropriate combination of decontaminated couplings, a T- or Y-shaped section and elbows so that the discharged (air-lifted) water does not geyser into the air, but is directed to the ground surface or an appropriate containment vessel (drums or tanks). The air line is placed in a well through a standard sanitary well seal that has been set in one branch of the T- or Y-shaped extension to the well. The air-lifted water is discharged through the other branch and elbow fittings to the directed discharge point.

Development should begin at the bottom of a well to remove potentially accumulated fines (silt, clay, fine sand), working up to the top of the screen (or bedrock borehole) and then back down to the bottom in increments of 5 feet or less and as many times as necessary. Development should consist of alternate surging and continuous air-lift pumping at each interval until the discharged water appears to be clear during surging.

Development by Centrifugal Pump

Monitoring well development using a centrifugal pump can be performed only if the depth to water is less than about 20 feet below ground surface, and can be effective for wells of very low yield. Because operation of a centrifugal pump depends upon suction, all fittings on the intake side of the pump must be air tight.

A good, inexpensive choice for the suction line into the well is 1/2- to 3/4-inch diameter flexible, polyethylene pipe. A new, unused length should be used for each well. The pipe should be stored in large plastic bags until ready for use, and should be handled with new clean gloves for each well. Additionally, the pipe must not be allowed to become contaminated by touching the ground. The end of the pipe that will be placed into the well should be cut off at an angle to minimize the potential of becoming quickly plugged in potentially accumulated silt at the bottom of the well. Additionally, this end of the pipe should be fitted with one or more large steel washers which are large enough to fit over the polyethylene pipe, but small enough to fit into the well. If the washers are re-used, they must be decontaminated before use in each well (see Manual Development section). The washers should be held in place a few inches from the end of the polyethylene pipe by two decontaminated hose clamps tightened by a screwdriver. The washer(s) will act as a plunger (surge block) when repeatedly raised and lowered 1–2 feet within various portions of the screened interval or open bedrock borehole. Such surging will force ground water back and forth through the well screen or bedrock fractures/joints. Simultaneously, pumping of the centrifugal pump will remove the turbid water and drilling fluids drawn into the well. Well development should begin at the bottom of the well to remove potentially accumulated fines (e.g., silt, clay, fine sand), working up to the top of the saturated portion of the screen (or bedrock open borehole) and back down, repeatedly surging and pumping at intervals of 5 feet or less, as many times as necessary, until the discharged water appears to be clear during surging. For monitoring wells installed in very low-yielding formations (silt, clay), the well may quickly pump dry. Therefore, there can be considerable time between each surging and pumping cycle of development. Refer also to the earlier portion of this chapter regarding the appropriateness of surging wells completed in fine-grained till deposits as investigated by Paul et al. (1988).

Development by Submersible Pump

Monitoring well development using a submersible pump can be performed in a wide variety of depth-to-water conditions. The major limiting factors include: (1) well diameter, (2) impeller construction material, and (3) type and concentration of contaminants present in the ground water. Small (less than 4-inch) diameter wells can be developed using specialty pumps. However, the presence of silt, clay, and fine sand, and/or some organic compounds in the water to be pumped can quickly clog or damage pumps with plastic impellers, and damage the bladder of squeeze-type pumps. Small diameter moyno-type, screw (progressing cavity) pumps appear to be more durable under these harsh conditions. Submersible pumps that are a nominal 4-inch diameter or larger are available with more durable, stainless steel impellers. Schalla (1986) reports that the general method of well development by submersible pump, including the use of the pump as a surge block, "proved to be the most successful technique for developing 4-inch diameter wells" for an investigation in northeastern Alabama.

The method of development by submersible pump is similar to the method using a centrifugal pump. The submersible pump must be decontaminated prior to use in each monitoring well. A good, inexpensive choice for the discharge line from the pump is 1/2- to 3/4-inch diameter flexible, polyethylene pipe. A new, unused length should be used for each well. The flexible pipe should be stored in large plastic bags or other protective means until ready for use, and should be handled with new, clean gloves for each well. Additionally, the pipe must not be allowed to become contaminated by touching the ground. The pump itself can act like a surge block when raised and lowered within the well interval to be developed. However, one or more decontaminated steel washers can be attached to the flexible discharge pipe just above the pump to increase effectiveness of the surging action. Once the pump has been placed in the monitoring well, it is repeatedly raised and lowered 1–2 feet within various portions of the screened interval (or saturated bedrock open borehole). Such surging will force ground water back and forth through the well screen (or bedrock fractures). Simultaneously, pumping of the submersible pump will remove the turbid water, fines, and drilling/well installation fluids drawn into the well. Well development should begin at the top of the saturated portion of the screened interval (or bedrock open borehole) to prevent sand locking of the pump within the well casing or borehole. Repeated surging and pumping at intervals of 5 feet or less should be performed to the bottom of the well and back up to the top of the saturated portion of the well screen or bedrock open borehole zone, as many times as necessary, until the discharged water appears to be clear during surging.

Development by Valved and Air-Vented Surge Plunger

Schalla and Landick (1986) report the development and evaluation of valved and air-vented surge plungers for the development of 2-inch diameter monitoring wells. Their study indicated the following important factors in development of the small-diameter surge plunger: length of the cylinder, sufficient weight of the plunger to overcome buoyancy and resistance, the number of water ports, and the number and size of air-vent ports. Schalla and Landick report the advantages of this device (particularly when compared to air-lifting) to include: (1) auxillary equipment (e.g., an air compressor) and tools are not required; (2) air is not introduced into the formation as long as surging is performed above the screened interval; (3) setup, shutdown, and decontamination can be performed in minutes by one person; (4) discharge of hazardous fluids can be piped directly and safely into drums using a "T" bypass; and (5) large volumes of water can be removed in a short period of time. For details of this method, the reader is referred to Schalla and Landick (1986).

Development by Bailer

Monitoring well development by bailer can be particularly useful for wells completed in very low yield formations. Bailers used for this method should be dedicated, decontaminated, and bottom-fill type with dedicated, decontaminated line with which to lower the bailer into a well. This method is very labor-intensive and, depending upon the volume of the bailer used, it may be appropriate to rig a tripod and pulley to aid in lifting the full bailer from a well. Each time the bailer is introduced and removed from a well, it will impart some surging action. Bailing should be performed throughout the screened interval or saturated portion of a bedrock open borehole. The method should be continued until the discharged water is visibly clear. Paul et al. (1988) report that for wells completed in fine-grained glacial till, development by bailing only resulted in wells with relatively less turbidity than development by surging and bailing.

Manual Development

The manual development method for a monitoring well can be used when the water level is too deep for a centrifugal (suction) pump, or where there is less than 20% submergence so air-lift with compressed air cannot be used, or a submersible

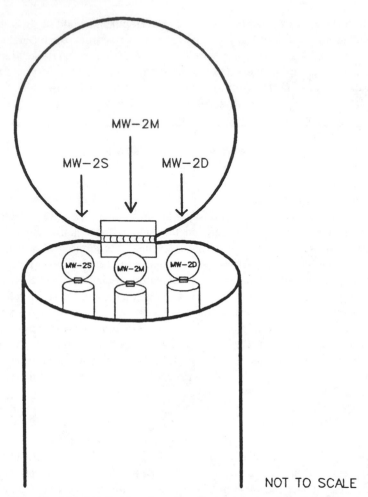

NOT TO SCALE

Figure 8.1.   Flexible polyethylene pipe with a foot-valve placed on the end and a washer attached above the foot-valve for surging. Used in manual development method for monitoring well.

pump of appropriate diameter and construction (metal impellers) is not available. This method can be effective in low-yielding wells, but is slow and requires considerable physical effort. This method consists of flexible polyethylene pipe with a foot-valve placed on the end of the pipe which will be placed in the well and a washer attached just above the foot-valve for surging (attached as noted in earlier section) (Figure 8.1). The foot-valve keeps water from flowing back out of the pipe once it has entered. By quickly and repeatedly raising and lowering the pipe about 1–2 feet, the water column is surged and water is forced into and up the pipe until it finally discharges from the other end of the pipe at ground surface. The inclusion of a downhole pitcher-type pump can reduce the physical effort required to only the surging operation. As with the previously described well development methods, new flexible pipe must be used in each well. The pipe must be carefully handled so it is not contaminated. Any re-used attachments (washers, foot valves, etc.) must be decontaminated before use in each well. Finally, development should begin from the bottom of the well to remove potentially accumulated fines (silt, clay, fine sand), working up to the top of the saturated portion of the screen (or bedrock open borehole) and back down as often as necessary to obtain discharged water that is visibly clear (if possible).

Decontamination

It is essential that every effort be made to avoid outside contamination and the cross-contamination of monitoring wells. This can best be done by ensuring that all equipment to be introduced into a well is clean. The level of effort for decontamination is a site- and project-specific issue to be resolved individually for each project. The resolution of the decontamination issue and the rationale for selection of site-specific development protocols must be made prior to installation of the monitoring well(s). At a minimum, it is recommended that reusable downhole equipment for developing monitoring wells should be steam-cleaned or washed with methanol and rinsed with clean water prior to use at each well. The reader is referred to Chapter 15 for a detailed presentation of equipment decontamination.

SURVEYING

Surveying is a necessary part of a ground-water monitoring program. The locations of monitoring wells must be surveyed so they may be accurately plotted onto maps that will be used to develop and interpret hydrogeologic data. Similarly, the well elevations must be surveyed to help assure accurate water level measurements. In many cases surveying will be required by the local, state, or federal agency directing the monitoring program or by the party for whom the program has been implemented. Usually the degree of accuracy and the reference datum, as well as whether the surveying must be performed under the supervision of a licensed surveyor, are set forth in the regulatory requirements.

In ground-water monitoring programs without specific survey requirements, there is still a need for a minimum amount of surveying. Well locations must be plotted onto a site plan to graphically represent their relative location on the site. For a small program on a small site, a sketch of the site showing the locations relative to site landmarks, as determined with a tape measure, may be adequate. Likewise, for a small program on a very large site, the approximate locations could be plotted onto a United States Geologic Survey (U.S.G.S.) quadrangle topographic map (usually 7½-minute quadrangles). As the requirement for accuracy in the ground-water monitoring program increases, the survey accuracy should also increase. Because small programs often mushroom into large programs, it may be useful to accurately survey well locations as part of a small program and minimize the problems that could occur as the project grows. Classifications and standards for vertical and horizontal control have been developed by the Federal Geodetic Control Committee and consist of First, Second, and Third Order Surveys with further divisions into classes (National Oceanic and Atmospheric Administration, 1974); First Order, Class I, is the most accurate and Third Order, Class II, is the least accurate. The details of survey accuracy are presented in Table 8.1; methods for surveying with varying levels of accuracy are described by Moffitt and Bouchard (1975).

Monitoring well locations are normally surveyed as accurately as possible. On small to moderately sized sites (up to 100 acres) the ideal accuracy would be plus or minus one linear foot. On large sites (greater than 100 acres) the ideal accuracy would be plus or minus two linear feet. These numbers are based on the author's experience.

Monitoring well elevations must be surveyed to a greater degree of accuracy. Without accurate elevations for monitoring wells, the water-level data and interpretation (discussed in Chapter 9) are subject to error. The accuracy of the elevation survey is usually to the nearest 0.01 foot. The elevations should be surveyed using a common datum. The most commonly used datum is the National Geodetic Vertical Datum (N.G.V.D.), established by the National Geodetic Survey (N.G.S.), which is part of the National Oceanic and Atmospheric Administration (NOAA). The N.G.S. has benchmarks (permanent landmarks of known position and elevation) throughout the United States from which the elevations of monitoring wells can be surveyed. The locations of some of the benchmarks are shown on 7½-minute quadrangle U.S.G.S. topographic maps. The location of the nearest benchmark can be obtained by contacting either the local or national U.S.G.S. office. It is not critical that the N.G.V.D. datum be used. However, it is useful if the water-level data obtained from one program are to be compared with data from other programs. Many industrial plants use their own datum and have established their own benchmarks; this is satisfactory and very often the relative difference between the plant datum and N.G.V.D. is known, allowing the plant elevations to be converted to N.G.V.D. elevations. If it is not practical to reference the elevations to a known datum (due to budget constraints, time, distance, etc.), an assumed datum can be used. For an assumed datum, an arbitrary point is selected (usually one of the monitoring wells) and assigned an elevation. The elevations of the other monitoring wells are then surveyed with reference to this one point. This allows the water levels of the individual monitoring wells to be compared to each other (they are surveyed to a common datum) but they cannot be compared with data from other programs (due to the use of a different datum).

Elevations of the protective casing (with the cap off or hinged back), the well casing, and the ground surface should be surveyed for each monitoring well. Water-level data may be measured from either the top of the protective casing or the top of the well casing, depending on the well construction details or the training of the field personnel. The possibility of measuring from a point without

Table 8.1. Standards for the Classification of Geodetic Control and Principal Recommended Uses.[a]

Horizontal Control

Classification	First-Order	Second-Order Class I	Second-Order Class II	Third Order Class I	Third Order Class II
Relative accuracy between directly connected adjacent points (at least)	1 part in 100,000	1 part in 50,000	1 part in 20,000	1 part in 10,000	1 part in 5,000
Recommended uses	Primary National Network. Metropolitan Area Surveys. Scientific Studies	Area control which strengthens the National Network. Subsidiary metropolitan control.	Area control which contributes to, but is supplemental to, the National Network.	General control surveys references to the National Network. Local control surveys.	

Vertical Control

Classification	First-Order Class I	First-Order Class II	Second-Order Class I	Second-Order Class II	Third Order
Relative accuracy between directly connected points or benchmarks (standard error)	0.5 mm $\sqrt{K}$	0.7 mm $\sqrt{K}$	1.0 mm $\sqrt{K}$ (K is the distance in kilometers between points.)	1.3 mm $\sqrt{K}$	2.0 mm $\sqrt{K}$
Recommended uses	Basic framework of the National network and metropolitan area control. Regional crustal movement studies. Extensive Engineering projects. Support for subsidiary surveys.		Secondary framework of the National Network and metropolitan area control. Local crustal movement studies. Large engineering projects. Tidal boundary reference. Support for lower order surveys.		Small-scale topographic mapping. Establishing gradients in mountainous areas. Small engineering projects. May or may not be adjusted to the National Network.

[a]Source: Classification, Standards of Accuracy, and General Specifications of Geodetic Control Surveys, U.S. Department of Commerce, National Oceanic and Atmospheric Administration, National Ocean Survey, Rockville, Maryland, February 1974.

a known elevation is eliminated if the elevations are measured for both reference points. The ground surface elevation is usually surveyed to the nearest 0.1 foot. If the top of either the protective casing or well casing is not level (not the same elevation all around the top), a clearly visible mark should be made (generally with an indelible marker) to indicate at what point on the casing the elevation was measured, so that water levels will be measured consistently from the same point. This will help eliminate the possibility of water-level measurement errors.

WELL IDENTIFICATION

Identifying a well by placing its number on the protective casing will assure that the location of water-level measurements and ground-water samples will be recorded correctly. The easiest way to permanently identify each well is to put the well number on the protective well casing in a visible location. Identification can be placed either on the inside or the outside of the protective casing; both have advantages and disadvantages. Placing the well identification on the outside provides the following advantage:

- Identification can be noted while approaching the well;
- So long as the identification is not on a removable cap, the identification will not be mixed with others, particularly at nested or clustered wells.

Placing the well identification on the outside has the following disadvantages:

- Identification may be removed or altered by vandalism;
- Identification may become illegible due to fading (if painted) or weathering (heavy oxidation if punched).
- If the identification is on the cap itself (not permanently attached), there could be a mix-up with other caps, particularly at nested or clustered wells.

The advantages of placing the identification on the inside of the protective casing is as follows:

- Identification cannot be vandalized;
- Identification should remain legible;
- If the identification is not on a removable cap, the identification will not be mixed with others, particularly at nested or clustered wells.

The disadvantages of placing the identification on the inside of the protective casing are as follows:

- Identification cannot be noted while approaching the well;
- If the indentification is placed on a removable cap it could get mixed with others, particularly at nested or clustered wells.

It is advisable, therefore, that the well identification be placed on both the inside and outside of the protective casing.

Well identifications are usually marked with either paint or a metal punching tool. Paint can be easily seen, but can also fade easily or be chipped off. A common technique that helps to keep the painted identification visible is to paint black numbers or letters on a white background. Punched identification numbers are sometimes difficult to read, and may become illegible due to oxidation. A combination of both is recommended, particularly if the monitoring program is to last more than one year.

Confusion can be easily avoided when the identification is marked on a removable cap by simply placing the identification number on the protective casing as well. A common identification problem arises when there are multiple wells within one protective casing. If the protective cap is hinged and therefore secure along one point, the inside of the cap can be used as a guide, as shown in Figure 8.1. This will only work where the cap is secured at one point. If there is a removable cap, some other system must be used. One solution is for each well casing to have a cap that is removable but still secured to the casing by string or wire. This way the caps can be removed, but cannot get mixed up with each other.

REPORTING WELL CONSTRUCTION DETAILS

Because the results of a ground-water monitoring program can be affected by the details of the monitoring well construction, the construction information should be reported in detail. Many states have regulations requiring the submittal of boring logs and monitoring well construction details. A telephone call to a state environmental agency should be able to clear up any reporting requirements that might be necessary.

Boring logs and monitoring well construction details are usually presented in full detail as an appendix to a report. There is usually a significant amount of information to be reported, and it would distract from the report if placed within the text. The well construction details that should be reported include:

- approximate diameter of borehole
- protective casing elevation
- well casing elevation
- ground surface elevation
- diameter, schedule, and material type of casing
- diameter, schedule, and material type of screen
- length of screen
- screen slot size
- length of casing (to allow determination of top and bottom screen elevations)
- bottom of borehole
- length of filter pack

- description of filter pack (generally referring to mineral content and grain size)
- length of bentonite seal(s)
- description of bentonite (pellets vs slurry, manufacturer, any special additives)
- length of grout seal
- description of grout seal (cement/water ratio, cement/bentonite/water ratio)
- location and description of casing and screen stabilizers, if used
- length and description of concrete surface seal
- total length and buried length of protective casing
- diameter and material type of protective casing
- any construction difficulties

The most common way of presenting the details is schematically; a figure depicting the monitoring well placed into a borehole with the appropriate backfill materials at the correct depths and the specifics presented on the figure. Two very common ways of presenting these details are shown in Figures 8.2 and 8.3. The graphic in Figure 8.2 is not to scale (vertically) and therefore the reader must take some time to determine the exact length of the screen, sand backfill, bentonite pellet seal, and so forth. Additionally, the reader must go to a separate figure that presents the boring log information, and compare it to the well construction details. This is a time-consuming process and very often confusing the the reader. However, it does present the necessary information to the reader and is a "form" graphic that can be completed very quickly and inexpensively. The graphic in Figure 8.3 is to scale (vertically), therefore the reader can very quickly see the relative lengths of screen, casing, filter pack, bentonite seal, and other well construction components. The specifics on the lengths and diameters of the well construction are also presented on the graphic. In addition, the reader is simultaneously presented with the boring log so the borehole lithology may be visually compared with the monitoring well construction. This allows the reader to note if the well is screened above the water table, throughout an entire aquifer, a portion of an aquifer, or through several aquifers, and whether the bentonite seals have been properly placed.

If the number of monitoring wells is relatively small, the well construction details can also be presented in tabular form. Table 8.2 is typical and includes columns for many of the details presented on the previous page. This form of presentation is usually not as complete as the graphic presentation, but can be useful in the text of the report, while referring the reader to the appendix for more details. This tabular presentation can be used to focus on specific well construction details that an author may want to highlight.

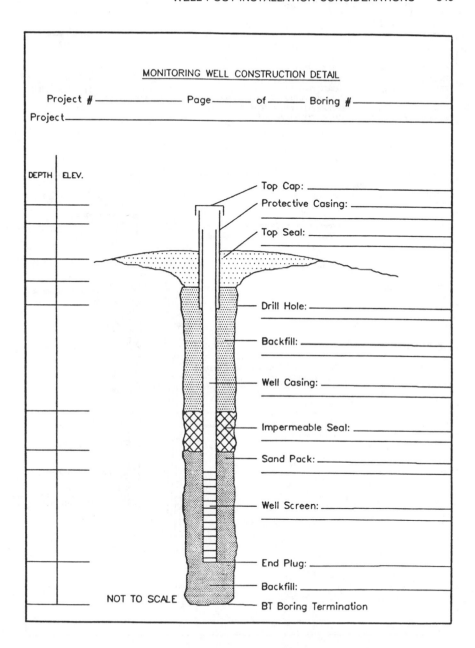

Figure 8.2.   Monitoring well construction detail (not to scale).

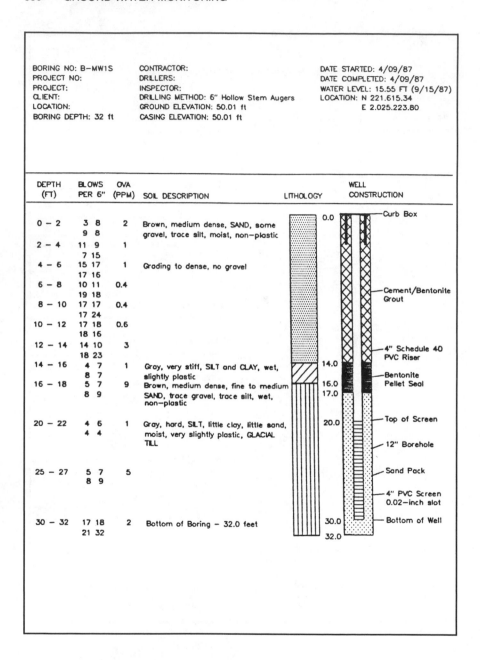

BORING NO: B—MW1S
PROJECT NO:
PROJECT:
CLIENT:
LOCATION:
BORING DEPTH: 32 ft

CONTRACTOR:
DRILLERS:
INSPECTOR:
DRILLING METHOD: 6" Hollow Stem Augers
GROUND ELEVATION: 50.01 ft
CASING ELEVATION: 50.01 ft

DATE STARTED: 4/09/87
DATE COMPLETED: 4/09/87
WATER LEVEL: 15.55 FT (9/15/87)
LOCATION: N 221.615.34
            E 2.025.223.80

DEPTH (FT)	BLOWS PER 6"	OVA (PPM)	SOIL DESCRIPTION	LITHOLOGY	WELL CONSTRUCTION
0 – 2	3 8 / 9 8	2	Brown, medium dense, SAND, some gravel, trace silt, moist, non—plastic	0.0	Curb Box
2 – 4	11 9 / 7 15	1			
4 – 6	15 17 / 17 16	1	Grading to dense, no gravel		
6 – 8	10 11 / 19 18	0.4			Cement/Bentonite Grout
8 – 10	17 17 / 17 24	0.4			
10 – 12	17 18 / 18 16	0.6			
12 – 14	14 10 / 18 23	3			4" Schedule 40 PVC Riser
14 – 16	4 7 / 8 7	1	Gray, very stiff, SILT and CLAY, wet, slightly plastic	14.0	Bentonite Pellet Seal
16 – 18	5 7 / 8 9	9	Brown, medium dense, fine to medium SAND, trace gravel, trace silt, wet, non—plastic	16.0 / 17.0	
20 – 22	4 6 / 4 4	1	Gray, hard, SILT, little clay, little sand, moist, very slightly plastic, GLACIAL TILL	20.0	Top of Screen / 12" Borehole
25 – 27	5 7 / 8 9	5			Sand Pack
					4" PVC Screen 0.02—inch slot
30 – 32	17 18 / 21 32	2	Bottom of Boring – 32.0 feet	30.0 / 32.0	Bottom of Well

Figure 8.3.   Monitoring well construction detail (to scale).

Table 8.2. Typical Table of Well Construction Details.

Well No.	Top of Casing Elevation (N.G.V.D.)	Length of Casing (ft)	Top of Screen Elevation (N.G.V.D.)	Length of Screen (ft)	Bottom of Screen Elevation (N.G.V.D.)	Water Level Elevation 9/11/87 (N.G.V.D.)	Water Level Elevation 10/15/87 (N.G.V.D.)
MW-1	344.03	14.5	329.53	10.0	319.53	327.08	326.83
MW-2	341.72	15.0	326.72	10.0	316.72	322.64	322.97
MW-3	341.17	15.0	326.17	10.0	316.17	323.19	323.42

MONITORING WELL MAINTENANCE AND REHABILITATION

Many ground-water monitoring programs are designed and implemented to collect only a few rounds of samples or water level data. After the program has served its purpose, the monitoring wells are no longer used and usually forgotten. For short-duration monitoring programs, there is little need for either well maintenance or rehabilitation, although some wells require rehabilitation shortly after their installation because of improper construction. Ground-water monitoring programs that are implemented for more than one year should include provisions for basic well maintenance and rehabilitation. An ongoing well maintenance and rehabilitation program is usually less costly than replacing the monitoring wells periodically.

Well maintenance consists of both periodically checking well performance and conducting routine procedures to preserve the well. Well maintenance needs will vary from site to site, and in some cases, from well to well. The checks that would be necessary for the maintenance of a well include the following:

- visibility (keeping the well easy to find)
- well location access (maintaining a path to the well location)
- lock removal (keeping it from rusting or other corrosion)
- protective cap removal (maintaining the hinge or well cap threads with antioxidizing agents)
- upkeep of the well identification
- concrete surface seal (checking for loss of integrity of the seal, such as cracks in concrete)
- subsurface bentonite seal (checking for any unusual ground-water quality problems)
- well screen and filter pack (confirming that the well efficiency or performance has not changed significantly)

Almost all well maintenance checks can be made during a ground-water sampling event. The checks that are based on water-quality data cannot be made until the laboratory results are completed, but the others can be made at the well location either visually or by comparing new data with previous data. If a well maintenance check indicates a problem, rehabilitation to correct the problem should be initiated.

Well rehabilitation should be initiated if well performance has been reduced significantly or if one of the well maintenance checks indicates that rehabilitation will alleviate the problem. The rehabilitation method should return well performance to its original level. The most common problem for monitoring wells is silting of the filter pack and/or well screen. This occurs when fine-grained materials (clays and silts) within the screened formation migrate into the filter pack and/or well screen and significantly reduce the well efficiency. Rehabilitation for this type of problem most often consists of redevelopment of the well. Several methods of well development are presented earlier in this chapter; these methods would be used to redevelop the monitoring well. Some other problems that might possibly affect the monitoring well performance are:

- the loss of a subsurface seal
- the breaking of the well screen
- the incrustation of the well screen
- the corrosion of the well screen

If, after reviewing the well sampling data, a change in the well performance is noted, an effort should be made to determine the problem. Silting of the well is easily diagnosed by a significant drop in the well yield, accompanied by cloudy water and/or a measurable amount of silt accumulating within the well screen. In these cases, redevelopment of the well is in order. The breaking of the screen would be diagnosed by similar conditions except that coarser material from the sand backfill and/or aquifer (providing that the screened formation consists of coarser materials) would accumulate at the bottom of the well. Incrustation of the well screen can be diagnosed by a significant drop in the well yield, with the well maintaining a sediment-free yield. The loss of a subsurface seal can be diagnosed by changes in water quality. However, wells that have lost their subsurface seals cannot be rehabilitated; they must be replaced.

The rehabilitation of a broken well screen usually consists of placing an intact, but smaller diameter screen inside the damaged screen. If the monitoring well is shallow, the smaller diameter screen may be threaded to the same diameter casing that then comprises a new monitoring well placed inside the old monitoring well. If the well is deep or of a small diameter, a smaller diameter screen may be placed inside the damaged screen and sealed to the existing casing. A rubber or lead packer may be placed just above the smaller diameter screen and wedged into the existing casing to form a seal between the screen and the casing.

Well incrustation can significantly reduce the yield of the well, and might also impact the chemical quality of samples collected from the well. There are three general types of incrustation:

1. precipitation of carbonates (or sulfates) of calcium and magnesium
2. precipitation of iron and manganese compounds
3. build-up of slime produce by iron bacteria or other slime-forming organisms

Well incrustation can be easily remedied by first determining the type of incrustation and then removing it with a combination of chemical treatment and redevelopment. The effect of chemical treatment on water-quality results from a monitoring well may be a problem and should be considered before this type of rehabilitation is attempted. Detailed discussions of the rehabilitation of incrusted wells can be found in Gass et al. (1982), Helweg et al. (1982), Campbell and Lehr (1973), and Driscoll (1986).

In some cases in which rehabilitation is required, particularly where a monitoring well has a small diameter and is relatively shallow, it may be more cost efficient to simply replace the monitoring well. However, when wells are relatively deep and/or constructed of more expensive materials, even expensive rehabilitation may be warranted. A down-the-hole television camera can be used to provide a videotape in wells with inside diameters as small as two inches. The pictures can reveal the nature and significance of a problem, the specific location of the problem, and whether or not the problem can be rehabilitated. However, the down-the-hole television cannot show if, or where, there is a problem outside of the well, such as a loss of a subsurface seal.

MONITORING WELL AND BORING ABANDONMENT

Proper monitoring well (and/or boring) abandonment is probably one of the most important post-construction elements of a ground-water monitoring well program. Proper well and boring abandonment can help reduce the threat of cross-contamination, alteration of sampling results, and potential liabilities. Some monitoring wells may simply become outdated (Bergren et al., 1988) and need to be abandoned. Frequently, little thought is given to the subject of abandonment when a well is first constructed. Just as frequently, no one follows up with an abandonment plan when the use of the well has been concluded. Perhaps most critical of all may be the immediate and proper abandonment of soil borings, test holes, and improperly constructed or located monitoring wells. It may even be justified to temporarily abandon a monitoring well which is no longer in an active program or which may be subject to site damage.

Objectives

The two principal objectives of an abandonment program should be to restore the original borehole and to prevent movement of fluids (cross-contamination) between formations or sampling zones. A monitoring well or boring should never become a conduit for contaminating previously uncontaminated aquifers. The well or boring abandonment program should be as well planned and well documented as the original well or boring installation program.

Planning for Abandonment

Before abandoning a monitoring well or boring, it is important to address six key elements. First, any state, federal, or local regulations that may control abandonment of the monitoring well or boring must be taken into account. Table 8.3 lists the agency in each state that oversees well or boring abandonment, and notes whether or not the state has abandonment regulations or guidelines.

Table 8.3. Status of State Well Abandonment Requirements.

State	Statewide Well Abandonment Laws[a]	State Agency with Program Responsibility
Alabama	Yes	Environmental Mgmt. Department 205-271-7790
Alaska	No	Department of Environmental Conservation 907-465-2600[b]
Arizona	Yes	Water Resources Department 602-255-1581
Arkansas	Yes	Water Well Construction Com. 501-666-8379
California	No	Water Resources Department 916-445-7314[b]
Colorado	Yes	Water Resources Division 303-866-3581
Connecticut	Yes	Consumer Protection Department 203-566-3275
Delaware	Yes	Natural Resources Department 302-736-4793
Florida	Yes	Environmental Regulation Department 904-488-3601[c]
Georgia	Yes	Water Well Stds Council 404-656-3214
Hawaii	Yes	Land and Natural Resources Department 808-548-7539
Idaho	Yes	Water Resources Department 208-334-4291
Illinois	Yes	Department of Mines and Minerals 217-782-7756
Indiana	No	Department of Natural Resources, Div. of Water 317-232-4160[d]
Iowa	Yes	Natural Resources Department 515-281-8690[e]
Kansas	Yes	Health and Environment Department 913-862-9360
Kentucky	Yes	Environmental Protection Department 502-564-3410
Louisiana	Yes	DOTD, Water Resources Section 504-379-1434
Maine	No	Drinking Water Office 207-289-5678[b,f]

continued

Table 8.3. Continued.

State	Statewide Well Abandonment Laws[a]	State Agency with Program Responsibility
Maryland	Yes	Department of the Environment 301-225-5850[c]
Massachusetts	No	Environmental Quality Eng. Department 617-292-5770[b]
Michigan	Yes	Public Health Department 517-335-8300
Minnesota	Yes	Health Department 612-623-5339
Mississippi	Yes	Natural Resources Department 601-961-5200
Missouri	Yes	Natural Resources Department 314-364-1752
Montana	Yes	Board of Water Well Contractors, Nat. Res. and Conservation Department 406-444-6643
Nebraska	Yes	Department of Health 402-471-2541
Nevada	Yes	Water Resources Div. 702-885-4380
New Hampshire	Yes	Water Well Board 603-271-3406
New Jersey	Yes	Bureau of Water Allocation, Environmental Protection Department 609-984-6831
New Mexico	Yes	Water Rights Div. 505-827-6120[g]
New York	Yes	Health Department 518-474-2011[h]
North Carolina	Yes	Natural Resources and Community Dev. Department 919-733-3221
North Dakota	Yes	Health Department 701-224-2354
Ohio	Yes	Health Department 614-466-1390
Oklahoma	Yes	Water Resources Board 405-271-2555
Oregon	Yes	Water Resources Department 503-378-8455
Pennsylvania	Yes	Environmental Resources Department 717-787-5828
Rhode Island	No	Water Resources Board 401-277-2217[b]
South Carolina	Yes	Health and Env. Control Department 803-734-5331
South Dakota	Yes	Water and Natural Res. Department 605-773-3352
Tennessee	Yes	Ground Water Protection Div. 615-741-0690
Texas	Yes	Texas Water Commission 512-463-8273
Utah	Yes	State Engineers Office, Water Rights Div. 801-533-6071 or 801-533-7217
Vermont	Yes	Environmental Conservation Department 802-244-5638
Virginia	Yes	State Water Control Board 804-257-0056
Washington	Yes	Department of Ecology 206-459-6045
West Virginia	Yes	Health Department 304-348-2981
Wisconsin	Yes	Natural Resources Department 608-267-7649
Wyoming	Yes	State Engineer's Office 307-777-7354

Source: National Water Well Association, Water Well Journal, September 1987.
[a]There may be requirements at other levels of government regarding well abandonment. A check should be made with all appropriate agencies.
[b]For information purposes only. No regulatory authority.
[c]Central contact. Individual water management districts have primary responsibility.
[d]Well abandonment procedures will become effective January 1, 1988.
[e]Central contact. Local boards of health have primary responsibility regarding abandonment of non-public and non-regulated water wells.
[f]Public safety laws are in effect and may be applicable.
[g]In certain declared basins.
[h]In certain areas of the state.

A second important consideration is the type of well or borehole to be abandoned and its design and construction. If it is a monitoring well, was it properly constructed, is it a monitoring well in which the construction is unknown or inadequately documented, is it a monitoring well in which the construction is known to be inadequate, or is it a well that has failed or been damaged? Improperly constructed or damaged monitoring wells will need more extensive abandonment procedures than monitoring wells that have been properly constructed with suitable materials and have an adequately grouted annular space. All test borings and exploration holes should be abandoned immediately after installation. Even some properly constructed monitoring wells may have to be abandoned by destructive methods because of their design. If they have dedicated permanent sampling equipment or are multiprobe or multilevel, they may not be amenable to normal open borehole abandonment methods. Some of the advantages of some of these special monitoring well designs may be detriments at the time of abandonment.

A third consideration in abandonment is the hydrogeologic environment. Were any low hydraulic conductivity formations penetrated by the well or borehole? What is the minimum hydraulic conductivity of the materials penetrated? What is the depth to the zone of saturation? Are the overburden materials cohesive, granular, or interbedded layers of material? If the well or boring reached or penetrated bedrock, what was the nature of the bedrock? Was it porous, fractured, cavernous, or massive?

A fourth consideration in planning for abandonment is the chemical environment around the borehole or well. Most abandonment procedures, including those recommended in this handbook, recommend or require the use of grouting materials, including sodium bentonite and portland cement. These materials react unfavorably in highly acidic or alkaline environments (Smith, 1976; Williams and Evans, 1987). Unlike a monitoring well which may have a life of 30 years, a properly abandoned well or boring must be designed to remain sealed forever. If the chemical environment is going to adversely affect the abandonment material, a different material or procedure may have to be utilized. Not only is information such as a good boring log or well log important, but sample analytical results from the well or boring are important because they offer clues to possible interference with the setting up or curing of abandonment materials and to the potential future breakdown of these materials.

The fifth planning consideration is disposal of potentially contaminated materials removed from the borehole or monitoring well. These materials may include soil or rock material, pumps or samplers, casing, screens, or the entire monitoring well if the well has to be destructively removed and the resulting borehole backfilled with grout. The handling of contaminated material should receive the same care as when the monitoring well or boring was first constructed. This subject is dealt with more fully elsewhere in this book.

The final planning consideration for abandonment of monitoring wells or borings is determining the type of equipment and the quantity of grouting or sealing materials that will be needed. In some types of abandonment, the equipment required may be very simple and the amount of grouting material required may be small. Conversely, wells constructed in cavernous limestone formations or other lost circulation zones, or which require full destruction and removal of all well construction materials, may require significant amounts of grouting material. Where full destruction of the well is needed, drilling equipment capable of drilling a borehole up to 1½ times the size of the original borehole will be needed. Monitoring projects involving only the installation of probes or borings should also have adequate equipment and materials to properly close and seal these holes promptly upon completion of their use.

Location and Inspection

An important step before proceeding with abandonment is to confirm the location of the monitoring well scheduled for abandonment. Was sufficient information generated at the time of construction of the well to reidentify the site even after abandonment? If not, sufficient locational information should be generated. If the site does not have significant magnetic interference, it may be desirable to place a permanent magnetic underground marker in the well after it is abandoned so that the site can be relocated in the future.

Another important pre-abandonment step that should be considered is an inspection of the well itself. A TV camera survey can help to confirm the condition, depth, and the construction details of the well. Even a visual inspection using a flashlight or sunlight reflected from a mirror may be helpful. If the monitoring well must be removed by a destructive method and the well is greater than 100 feet in depth, a hole deviation survey should be conducted (Bergren et al., 1988).

Abandonment Materials

To be effective, abandonment materials must fill the space provided and have an effective hydraulic conductivity less than the original materials through which the well or boring was constructed. Two principal grouting materials, neat cement and high solids bentonite, best meet the needs for abandonment.

Type I portland cement (ASTM C-150) is the material most frequently used to make a neat cement abandonment material. When mixed with no more than 5 or 6 gallons of water per 94-pound bag of cement and placed in an environment with at least 80% relative humidity, Type I cement will continue to hydrate, and may not shrink. The same mix placed under water will actually expand (Kosmatka and Panarese, 1988). However, in most other environments, portland cement will shrink upon setting.

To control shrinkage and/or improve pumpability, it is necessary to use additives. The most common additive is bentonite. Commonly, between 2% and 6% bentonite, by weight, is prehydrated with the mix water. Table 8.4 lists mixing ratios per 94-pound bag or portland cement with 0% to 6% bentonite, and the resulting slurry properties. One benefit of the addition of bentonite is improvement of pumpability of the neat cement (Smith, 1976). Pumpability can also be enhanced and the amount of mix water reduced by 12% to 30% by use of superplasticizers (Kosmatka and Panarese, 1988). A compressive strength of 500 psi for the set neat cement appears to be adequate when used as a grout or abandonment material (Smith, 1976). Neat cement can also be modified by adding clean sand and, for larger holes, by adding gravel not larger than 0.5 inch in diameter.

In addition to shrinkage problems that may occur if too much water is added to the cement mix, the cement may also segregate with development of channels and pockets of free water (Coleman and Corrigen, 1941). This problem occurs particularly in deep holes with limited diameter. The permeabilities of cements are also significantly increased with increased water-to-cement ratios (Williams and Evans, 1987).

Table 8.4. Recommended Slurry Properties of Portland Cement with Bentonite.

Percent Bentonite	Type of Cement[a]	Maximum Water Requirements		Slurry Weight		Slurry Volume
		gal/sk	ft³/sk	lb/gal	lb/ft³	ft³/sk
0	I or II	5.2	0.70	15.6	117.0	1.18
0	III	6.3	0.84	14.8	110.7	1.32
2	I or II	6.5	0.87	14.7	110.0	1.36
2	III	7.6	1.02	14.1	105.5	1.51
4	I or II	7.8	1.04	14.1	105.0	1.55
4	III	8.9	1.19	13.5	101.0	1.69
6	I or II	9.1	1.22	13.5	101.0	1.73
6	III	10.2	1.36	13.1	98.0	1.88
8	I or II	10.4	1.39	13.1	98.0	1.92
8	III	NA	NA	NA	NA	NA

Source: After Halliburton Co., Cementing Tables, 1975.
[a]A.S.T.M. Type I, II, and III used instead of API A, B, and C.

The second common abandonment material is sodium bentonite (API-13A). The principal asset of this material is its ability to rapidly hydrate to 10 or more times its original volume. Not only may it be used as a high solids pumpable grout, it may also be used in dry form. By adding a quart of polymer to 100 gallons of water it is possible to then add 1.5 to 2 pounds of granular bentonite per gallon and gently pump the resulting mixture. A mixture weighing 9.25 to 9.4 pounds per gallon can be achieved by this method. The polymer delays the hydration of the bentonite, permitting approximately a 20-minute working time (Gaber and Fisher, 1988).

Where the use of a polymer may be of concern, a similar mixture can be obtained by mixing 40 pounds of high-quality nonpolymerized powdered or granular bentonite per 100 gallons of water to form a drilling mud. The bentonite is stirred in just prior to placing the mixture (Gaber and Fisher, 1988). Again, a 9.25 to 9.4 pound per gallon abandonment material with a solids content of 20% to 22% can be obtained using this mixture.

A recent modification of grouting and abandonment materials is the development of a proprietary mixture of sodium and calcium bentonite in which 2.1 pounds of bentonite is mixed per gallon of water, followed by the addition of a magnesium oxide compound (Bertane, 1986). This mixture has about a 2-hour working time, but does not develop as strong a gel as the pure sodium bentonite mixes (Gaber and Fisher, 1988). The long-term performance of this material is also not yet known. A second proprietary mixture has also been developed by another bentonite manufacturer. This mixture also uses a retarder and is NSF (National Sanitation Foundation) approved.

Some attempts have also been made to use thickened or "heavy" drilling mud. Unfortunately, this mixture does not contain enough solids to form an effective abandonment material. Heavy drilling mud may also be the most prone of the fluid-based abandonment materials to the problems of consolidation noted earlier.

Another alternative to pumped abandonment material is the use of granular, chunk, or tablet forms of bentonite. Where there is little water in the borehole or well, dry bentonite may be an excellent choice. It may also be mixed with sand or gravel in some applications, which helps to ensure full and rapid emplacement. The dry bentonites are probably the most practical choice for abandoning borings and test holes. With careful placement, even the potential problem of bridging can be overcome.

The use of native materials for abandonment, such as puddled clay or fine sand, should only be considered in situations where the risk of contamination is effectively nonexistent or can be controlled by careful placement of regular abandonment materials.

In many cases the materials allowed for use in abandonment are controlled by state well or boring abandonment rules or regulations.

Placing Abandonment Materials

The most effective way to place the abandonment material is by tremie pipe. Therefore, the basic equipment for most abandonment procedures should be the same as for tremie grouting the original well. A grout mixer or other method of mixing the grout is needed, along with a positive-displacement pump to deliver the mixture with positive pressure to the bottom of the hole. The mixing of a granular bentonite slurry requires special precautions, however. Mixing of this slurry should be done with a blade- or paddle-type mixer with recirculation (Gaber and Fisher, 1988). A moyno pump with its screw-type pumping action is an ideal grouting pump. Other positive-displacement pumps such as the air diaphram pump and the typical piston-type mud pump may also be used.

The tremie pipe should be at least 1 inch to 1.5 inches in diameter to allow for adequate flow of bentonite or neat cement grout. A combination of rubber hose and short lengths of threaded rigid plastic or metal pipe can work well. Enough hose is needed to run from the grout pump to the well or boring with the pipe used down the well or borehole. If the end of the pipe is to be used to check placement of material, it may be best to install a "T" at the end of the pipe to allow grouting material to discharge from the side of the pipe. During placement of the grout, the tremie pipe should always remain submerged in the grouting material.

When abandoning with a mixture of neat cement and sand, the previously described tremie setup will work effectively. Placement of cement without sand, however, will require increasing the size of the tremie pipe and special selection of the pump. When the hole is large enough to permit use of concrete for abandonment, it may be feasible to use a conductor pipe. The concrete can then be fed by gravity. However, the end of the conductor pipe, like the tremie pipe, should remain submerged during placement of the concrete.

Another recently reported use of the conductor pipe (Mason, 1988) is for the placement of bentonite pellets for abandonment of water-filled, uncased holes. The conductor pipe is suspended a short distance off the bottom and water is circulated down the pipe and back up the hole. The pellets can then be added gradually to the circulating water. The pellets drop out below the pipe and fill the hole. The conductor pipe must be raised at regular intervals to prevent plugging by the pellets. It is also important to use only clean pellets so that a drilling mud is not formed by the recirculating water. Some replacement of the water may be necessary.

Direct gravity placement of abandonment material can also be used satisfactorily. All of the cement-based materials may be placed directly in shallow holes which are free of water and in some similar deeper holes which have sufficient open diameter to prevent bridging. Granular, chunk, and pellet bentonite may be used in a similar manner.

With special precautions, chunk and pellet bentonite may even be used to abandon deep water-filled boreholes or wells which have sufficient open diameter to prevent bridging. All the fines must be removed to prevent the development of a drilling mud which would prevent proper settlement of the bentonite chunks. It is also important to limit the rate of filling to prevent bridging. An effective setup (Figure 8.4) utilizes a chute formed of 1/4-inch mesh screen. The bentonite is poured down the chute at a rate of not more than 20 pounds per minute. Equal amounts of coarse sand or fine rounded stone can also be added at the bottom of the chute if a 100% bentonite fill is not required.

Procedures of Abandonment

After review of the type of monitoring well or borehole to be abandoned, a decision can be made on the type of abandonment to be performed. Boreholes and properly constructed monitoring wells are relatively simple to abandon. This is particularly true when removal of the well casing or screen is not necessary. However, it is important to recheck all steps in the abandonment planning process before proceeding. One must be sure of what may be required by the state, and be sure that the procedure and abandonment materials will do the required job.

Figure 8.4. Filling a large diameter well with chunk bentonite. Bentonite is being poured over ¼ " mesh to separate the fines.

The following checklist outlines standard abandonment procedures:

1. Check to be sure that the well or borehole is free of debris;
2. Remove any dedicated equipment that will interfere with abandonment;
3. Determine the depth of the well or borehole with a sounding weight and line;
4. If there is a screen, it may be desirable to fill the length of the screen with fine sand;
5. Proceed to place abandonment material (Figure 8.5); and
6. In any large-diameter or deep hole, sound the hole at pre-set intervals to ensure a complete fill.

The surface finish is frequently determined by the site or by state rules or regulations. A preferred final treatment is to cut all casing below land surface, cap and fill the excavation with abandonment material, and then surface spread native material. Many state regulations require the final seal to be done with cement or concrete. Location measurements and survey markers placed just below the ground surface should be made or emplaced if appropriate.

Figure 8.5. Filling a 2″ monitoring well with granular bentonite.

The other principal type of abandonment is destructive abandonment. In addition to all the requirements of simple abandonment is the need to destroy the casing, grout, screen, and any in-place sampling devices. How this abandonment is carried out will vary greatly depending on construction of the well, purpose or use of the well, site conditions, and drilling equipment available.

Where PVC or similar plastic casing has been used and the well is free of pumps and internal obstructions, a pilot bit inserted in the casing may be used as a guide in reaming out the casing and grout (Bergren et al., 1988). Steel-cased monitoring wells will generally have to be over-reamed and removed. A large-diameter, heavy-duty auger may be the most satisfactory drilling equipment. An air- or mud-rotary drill with a roller bit may also be satisfactory. If all grout has not been removed with the first pass, a larger reaming bit will need to be used with all the remaining material circulated out of the hole.

It is important to remember that if contamination was detected in the original well or boring, appropriate health and safety requirements should be maintained. Well construction and surrounding soil materials may also need to be handled and disposed of specially.

In some cases it may be desirable to fill the well with neat cement grout if that will assist in its destructive removal. Wells with stainless steel casing present particularly difficult abandonment problems, as the casing cannot be easily drilled out. The hollow-stem auger is probably the only practical solution in this case. Another important consideration is keeping the hole open after drilling until the hole is filled with abandonment material. Once the well has been destructively removed, abandonment can proceed as previously outlined.

A final type of abandonment which may be appropriate in some cases is temporary abandonment. In this case the screen or zone of production should be filled with medium to fine sand. The rest of the well should then be filled with a bentonite material. If the well needs to be put back into service, the bentonite and sand can be flushed out of the well with water and disposed of. If the well is accidentally destroyed, the potential for ground-water contamination will have been forestalled.

Records and Reports

The last step in a proper abandonment program is the development of appropriate records and reports. The records should identify the method of abandonment, the materials used, and the quantity and mixed weight of all materials. The records should clearly establish the location of the abandoned well or boring. It should be noted if a magnetic survey marker was placed at the well site. Finally, if any materials were disposed of offsite, the location and disposition should be noted. It may also be desirable to save and store samples of mixed abandonment material.

Finally, it is important to file required reports with the appropriate state agency. A similar or more detailed report and maps should be filed with the client.

REFERENCES

American Colloid Company, Technical Data Sheets on VOLCLAY Grout. Skokie, IL.

American Petroleum Institute, 1983. API Specifications for Oil-Well Drilling-Fluid Materials: API Spec. 13A.

American Society of Testing and Materials, Standard Specifications for Portland Cement, Designation: C150-81.

Bergren, C. L., Janssen, J. L. and Heffner, J. D., 1988. Abandonment of Groundwater Monitoring Wells at the Savannah River Plant, Aiken, South Carolina. Proceeding of the Focus Conference on Eastern Regional Ground Water Well Association. Dublin, OH. pp. 215–220.

Bertane, M. (1986). The Use of Grout for the Sealing of Monitoring Wells. American Colloid Company, Skokie, IL.

Campbell, M. D., and J. H. Lehr, Water Well Technology, McGraw-Hill Book Company, New York, 1973.

Coleman, R. J. and G. L. Corrigan, 1941. Fineness and Water-to-Cement Ratio in Relation to Volume and Permeability of Cement. Petroleum Technology. Tech. Pub. No. 1266, pp. 1–11.

Driscoll, F. G., Ed., Groundwater and Wells, Johnson Division, St. Paul, MN, 2nd Ed., 1986.

EA Engineering, Science, and Technology, Inc., 1985. Manual of Standard Operating Procedures for Geotechnical Services, GTS-201.

Gaber, M. S. and Fisher, B. O., 1988. Michigan Water Well Grouting Manual. Ground Water Control Section, Michigan Department of Public Health. Lansing, MI. 83 pp.

Gass, T. E., T. W. Bennet, J. Miller, and R. Miller, Manual of Water Well Maintenance and Rehabilitation Technology, the National Water Well Association. Ada, OK, Robert S. Kerr Environmental Research Laboratory, National Environmental Research Center, U.S. Environmental Protection Agency, 1982.

Halliburton Company, 1975. Halliburton Cementing Tables. Little's. Duncan, OK. 501 pp.

Helweg, O. J., V. H. Scott, and J. C. Scalmanini, Improving Well and Pump Efficiency, American Water Works Association, 1982.

Kosmatka, S. H. and W. C. Panarese, 1988. Design and Control of Concrete Mixtures. Portland Cement Association. Skokie, IL, 205 pp.

Leonard, R., 1985. Dry Grout Driven Method: New Approach to Grouting. Ground Water Age, Vol. 19, No. 7, pp. 19–23.

Mason, Carl, Personal communication, 1988.

Moffit, H., and H. Bouchard, Surveying, Harper and Row, New York, 1975.

N. L. Baroid/NL Industries, Benseal—Product Information. Houston, TX.

National Oceanic and Atmospheric Administration, Classification, Standards of Accuracy, and General Specifications of Geodetic Control Surveys, U.S. Department of Commerce, 1974.

Paul, D. G., C. D. Palmer, and D. S. Cherkauer. 1988. The Effect of Construction, Installation, and Development on the Turbidity of Water in Monitoring Wells in Fine-Grained Glacial Till. Ground-Water Monitoring Review, Vol. 8, No. 1. National Water Well Association.

Riewe, T., 1987. Memorandum: Results from the 1st Phase of UW Research Project on Bentonite Seals. Wisconsin Deparment of Natural Resources.

Schalla, R. 1986. A Comparison of the Effects of Rotary Wash and Air Rotary Drilling Techniques on Pumping Test Results. Pacific Northwest Laboratory. Presented at the Sixth National Symposium and Exposition on Aquifer Restoration and Ground Water Monitoring, 19–22 May 1986. National Water Well Association.

Schalla, R. and R. W. Landick. 1986. A New Valved and Air-Vented Surge Plunger for Developing Small Diameter Monitor Wells. Ground Water Monitoring Review, Vol. 6, No. 2. National Water Well Association.

Smith, D. K., 1976. Cementing. Society of Petroleum Engineers Monograph Series, Volume 4. Society of Petroleum Engineers of AIME. Dallas, TX.

U.S. Environmental Protection Agency. 1975. Manual of Water Well Construction Practices. Office of Water Supply. EPA-570/9-75-001.

Williams, C. and L. G. Evans, 1987. Guide to the Selection of Cement, Bentonite and Other Additives for Use in Monitoring Well Construction. Proceedings: First National Outdoor Action Conference on Aquifer Restoration, National Water Well Association. Ground Water Monitoring and Geophysical Methods. Dublin, OH, pp. 325–343.

9

Acquisition and Interpretation of Water-Level Data

Matthew G. Dalton, Brent E. Huntsman, and Ken Bradbury

INTRODUCTION

The Importance of Water-Level Data

The acquisition and interpretation of water-level data are essential parts of any ground-water monitoring program. When translated into values of hydraulic head, water-level measurements are used to determine hydraulic head distributions, which in turn are used to assess ground-water flow velocities and directions within a three-dimensional framework. When referenced to changes in time, water-level measurements can reveal changes in ground-water flow regimes brought about by natural or human influences. When measured as part of an in situ well or aquifer pumping test, water levels provide information needed to evaluate the hydraulic properties of ground-water systems.

Water Level and Hydraulic Head Relationships

Hydraulic head is the driving force for ground-water movement, and varies both spatially and temporally in ground-water systems. A piezometer is a monitoring device designed to measure hydraulic head at a discrete point in a ground-water system. Figure 9.1 shows water-level and hydraulic head relationships at a simple vertical standpipe piezometer(A). The piezometer consists of a hollow vertical casing with a short screen open at point P. The piezometer measures

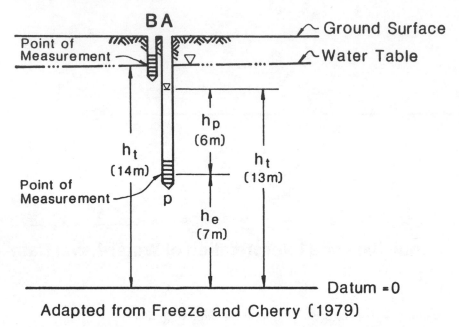

Figure 9.1.   Hydraulic head relationship at a field piezometer.

total head at point P. *Total head* (h_t) has two components, *elevation head* (h_e) and *pressure head* (h_p).

$$h_t = h_e + h_p$$

Elevation head (h_e) refers to the potential energy that ground water possesses by virtue of its elevation above a reference datum. Elevation head is caused by the gravitational attraction between water and earth. On Figure 9.1, the elevation head (h_e) at point P is 7 meters.

Pressure head (h_p) refers to the force exerted on water at the measuring point by the height of the static fluid column above it (in this discussion we neglect atmospheric pressure). On Figure 9.1, the pressure head (h_p) at point P is 6 meters. Note that h_p is measured *inside* the piezometer, and corresponds to the distance between point P and the water level in the piezometer.

Total head (h_t) is the sum of elevation head (h_e) and pressure head (h_p). The total head at point P on Figure 9.1 is 7m + 6m = 13m relative to the datum.

The water level in piezometer A is lower than the water level (water table) measured in piezometer B. The difference in elevation between the water table piezometer (B) and the water level in the deeper piezometer (A) corresponds to the hydraulic gradient between the two. In this case there is a downward vertical gradient because total hydraulic head decreases from top to bottom.

Hydraulic Media and Aquifer Systems

The "classic" definition of an *aquifer* as "a water-bearing layer of rock that will yield water in a usable quantity to a well or spring" (Heath, 1983) was developed to address water supply issues, but is less useful for describing materials in terms of modern ground-water monitoring. Today, ground-water monitoring (including well installation, water-level measurement, and water-quality assessment) occurs in hydrogeologic media ranging from low hydraulic conductivity shales, clays, and granites to high hydraulic conductivity sands and gravels. The term "aquifer" (in ground-water monitoring) is being used as a relative term to describe any and all of these materials in various settings.

Aquifers are also generally classified based on where a water level lies with respect to the top of the geologic unit. Figure 9.2 shows an example of layered hydrogeologic media forming both *confined* and *unconfined* aquifers. The confined aquifer is a relatively permeable unit, bounded on its upper surface by a relatively lower permeability layer. Hydraulic head in the confined aquifer is described by a *potentiometric surface,* which is an imaginary surface representing the distribution of total hydraulic head (h_t) in the aquifer, which is higher in elevation than the physical top of the aquifer.

The sand layer in the upper part of Figure 9.2 contains an *unconfined aquifer* which has the *water table* as its upper boundary. The water table is a surface corresponding to the top of the unconfined aquifer where total hydraulic head

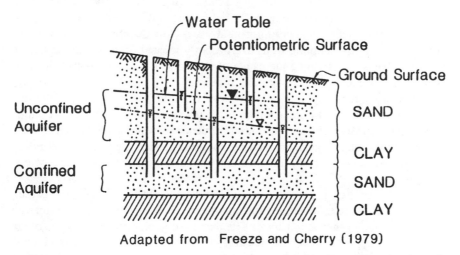

Adapted from Freeze and Cherry (1979)

Figure 9.2.   Unconfined aquifer and its water table confined aquifer and its potentiometric surface.

is zero relative to atmospheric pressure or the hydrostatic pressure is equal to the atmospheric pressure.

Notice that water levels in the piezometers on Figure 9.2 vary with the depth and position of the well. This variation corresponds to the variation of total hydraulic head throughout the saturated system. Hydraulic head often varies greatly in three dimensions over small areas, and the design and placement of water-level monitoring equipment is critical for a proper understanding of the ground-water system.

DESIGN FEATURES FOR WATER LEVEL MONITORING SYSTEMS

The primary use of ground-water level (hydraulic head) data from wells or piezometers is assessment of ground-water flow directions and hydraulic gradients. The design of ground-water monitoring systems usually considers requirements for both water-level monitoring and ground-water sampling. In many cases, both needs can be accommodated without installing separate systems. However, to collect acceptable water-level data, certain requirements need to be met which may not always be consistent with the requirements for collecting ground-water samples. For example, additional wells may be required to fully assess the con-figuration of a water table or potentiometric surface over and above what might be required to collect ground-water samples. Conversely, the design of wells to collect ground-water samples may differ from wells which are used solely to col-lect ground-water level data.

Water-level monitoring data are generally collected during two phases of a monitoring program. The initial phase is when the site to be monitored is being characterized to provide data to design a monitoring system. The second phase is when water-level data are being collected as part of the actual monitoring pro-gram to assess whether changes in ground-water flow directions are occurring, and to confirm that sampling wells are properly located. These latter data also provide a basis to determine the cause of such changes and to assess whether changes to the monitoring system are required.

To design a water-level monitoring system, a detailed understanding of the site geology is necessary. The site geology is the physical structure in which ground water flows, and as such has a profound influence on water level data. It is very important that reliable geologic data be collected so that the water-level moni-toring system can be properly designed and the water-level data can be accurately interpreted.

Sites at which there is a high degree of geologic variation require more exten-sive (and costly) water-level monitoring systems than sites which are compara-tively more homogeneous in nature. The degree of geologic complexity is often not known or appreciated during the early phases of a testing program and it may require several stages of drilling, well installation, water-level measurement, and analysis of hydrogeologic data before the required understanding is achieved.

Piezometers or Wells?

Ground-water level measurements are typically made in piezometers or wells. Most ground-water monitoring systems associated with assessing ground-water quality are composed of wells rather than piezometers.

Piezometers are specialized monitoring installations whose primary purpose is the measurement of hydraulic head. Generally, these installations are relatively small (less than 1 inch in diameter if a well casing is used) or in some applications may not include a well casing and just consist of tubes or electrical wires connected to pressure or electrical transducers. Piezometers are not typically designed to obtain ground-water samples for chemical analysis, although the term piezometer has been applied to pressure measuring devices which have been modified to collect ground-water samples (Maslansky et al., 1987). Piezometers have traditionally had the greatest application in geotechnical engineering for measuring hydraulic heads in dams and embankments.

Wells are commonly the primary devices in which water levels are measured as part of a monitoring system. They differ from piezometers in that they are typically designed so that ground-water samples can be obtained. As such, they are larger than piezometers (larger than 1.5 inches in diameter), although sampling devices have been developed which allow ground-water samples to be obtained from smaller diameter wells.

Approach to System Design

Design of a water-level monitoring system should begin with a thorough review of available existing data. This review should be directed toward developing a conceptual model of the site geologic and hydrologic conditions. The conceptual model of the hydrogeologic system is used to determine the locations of an initial array of wells. Tentative decisions regarding drilling depths and the zone or zones to be screened should also be made using those data. Existing wells may be incorporated into the array if suitable information as to well construction details are available. Boring/well logs, surficial geologic and topographic maps, drainage features, cultural features (e.g., well fields, irrigation, and buried water pipes), and rainfall and recharge patterns (both natural and man-induced) are several of the major factors that need to be assessed as completely as possible.

The available data should be reviewed to identify:

- the depth and characteristics of relatively permeable saturated zones and aquitards which may be present beneath a site;
- depth to the water table and the likelihood of encountering perched or intermittently saturated zones above the water table;
- probable ground-water flow directions;
- presence of vertical hydraulic gradients;
- features which might cause ground-water levels to fluctuate, such as

well-field pumping, fluctuating river stages, unlined ditches or impoundments, or tides;

• probable frequency of fluctuation; and
• existing wells that may be incorporated into the water-level monitoring program.

The practical limitations of where wells can be located on a site should not be overlooked during this phase of the system design. Wells can be located almost anywhere on some sites; however, on other sites, buildings, buried utilities, and other site features can impose limitations on siting wells.

Number and Placement of Wells

The number of wells required to assess ground-water flow directions beneath a site is dependent on the size and complexity of the site conditions. Simple and smaller sites require fewer wells than larger or more hydrogeologically complex sites.

Many sites have more than one saturated zone of interest where ground-water flow directions need to be assessed. Permeable zones may be separated by less permeable zones. In these cases, several wells screened at different depths may be required at several locations to adequately assess flow directions in, and between, each of the saturated zones of interest.

The minimum number of wells required to estimate a ground-water flow direction within a zone is three (Todd, 1980; Driscoll, 1986). However, the use of three wells is only appropriate for relatively small sites where the configuration of the water table or potentiometric surface is planar in nature, as shown on Figure 9.3.

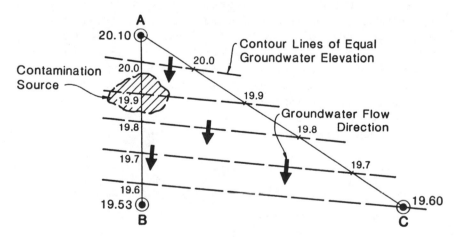

Figure 9.3. Assessing ground-water flow directions at a small site with a planar water table surface.

Generally conditions beneath most sites require more than three wells. Lateral variations in the hydraulic conductivity of subsurface materials, localized recharge patterns, drainage channels, and other factors can cause the potentiometric or water table surface to be nonplanar.

On large sites an initial grid of six to nine wells is usually sufficient to provide a preliminary estimation of ground-water flow directions within a target ground-water zone. Such a configuration will generally allow the complexities in the water table or potentiometric surface to be identified. After an initial set of data is collected and analyzed, the need for and placement of additional monitoring installations can be assessed to fill in data gaps and/or to further refine the assessment of the potentiometric or water-table surface.

Figure 9.4 shows a site where leakage from a buried pipe has caused a ground-water mound to form. In this situation, a three-well array would not provide sufficient data to detect the presence of the mound, and could result in a faulty assessment of the ground-water flow direction beneath the site.

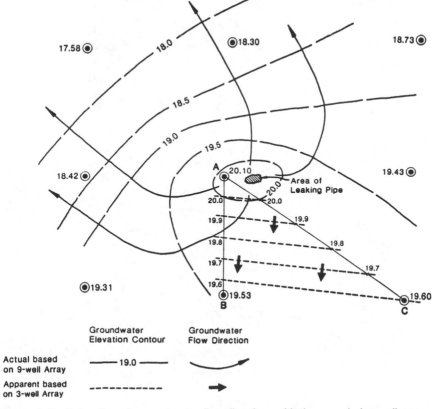

Figure 9.4. Estimation of ground-water flow directions with three- and nine-well array.

Screen Depth and Length

When well locations are established, screen depths should be chosen. Screen depths are generally determined during the drilling operation after a geologic log has been prepared, depending on the amount and quality of the data available prior to drilling.

Wells used to assess flow directions within a zone are usually screened within that zone at similar elevations. Highly layered units may require screens in each depth zone that is isolated by lower hydraulic conductivity layers (Figure 9.5a). Where the units are dipping, it is generally more important to place the screens in the same zone even if the screens are not placed at similar elevations (Figure 9.5b).

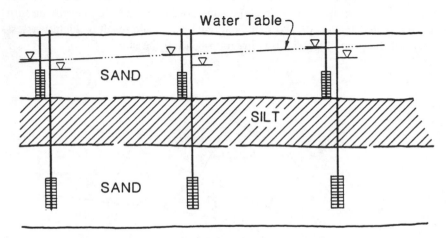

a - Horizontal Strata

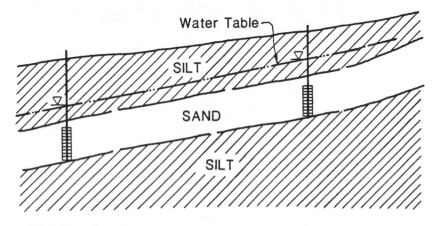

b - Dipping Strata

Figure 9.5. Well screen placement in horizontal and dipping strata.

Similar well-screen lengths should be used and the screen (and sand pack) should be placed entirely within the zone to be monitored. This will result in obtaining a water level which is representative of the zone being monitored and minimize the probability of allowing contaminants, if present, to migrate between zones in the well bore. If the well screen is open to several zones, then a "composite" water level will be measured which may not be representative of any single zone and will add to the difficulty in interpreting the water-level data. Typical well screen lengths are 5 to 10 feet.

If multiple saturated zones are present beneath a site, it is generally necessary to install several wells screened at differing depths at a single location. Such installations allow the assessment of both horizontal and vertical hydraulic gradients. If few reliable data are available for a site, it is desirable that the initial hydrologic characterization start with the uppermost zone of interest. During this initial work a limited number of deeper installations can be installed to provide data to assess the need for additional deeper installations.

Construction Features

Water-level monitoring points can be installed using a variety of methods and configurations (Figure 9.6). Typically, the installations are constructed in drilled boreholes, although driven well points can be used to provide water-level data in shallow, unconfined saturated zones. Drilling and monitoring device installation procedures are discussed in Chapter 7 of this handbook.

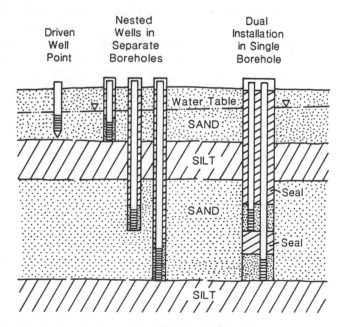

Figure 9.6. Typical monitoring well installation configurations.

At locations where multiple zones are to be monitored, single or multiple installations in the same borehole can be used. If a single well is installed in a borehole, several boreholes will be necessary to monitor multiple zones.

A single installation in a single borehole is preferred because it is easier to install a reliable seal above the well screen when only one well is completed in a borehole. Such a seal is necessary to ensure that the water-level data are representative of the zone being monitored and to assure that contaminants don't move between zones within the borehole. In many situations, especially if a hollow-stem auger is being used to install the well, the cost of installing single installations is only marginally higher than multiple installations in a single borehole.

Multiple installations in a single borehole have been installed successfully as long as an adequate drilling casing diameter is used and care is employed in installing the wells. Installing two 2-inch-diameter wells per borehole should be feasible within 6-inch to 8-inch-diameter well casings. Particular care should be taken to place the sand pack and seal. Seals should be tremied into the borehole as the outer casing is withdrawn. Regardless of whether single or multiple installations are employed, a suitable, protected well head should be installed.

Water-Level Measurement Precision and Intervals

Wells should be accurately located horizontally and vertically, although horizontal surveying is not always required, depending on the size of the site and available base maps. The precision of the horizontal locations is generally not as important as the precision of the elevation survey and water-level measurements.

The top of the well casing (or other convenient measuring point) should be surveyed to a common datum so that water-level measurements can be converted to water-level elevations. The reference point for water-level measurements should be clearly marked at a convenient location on each well casing. This will facilitate reducing measurement error.

The precision of the elevation survey and water-level measurements depends on the slope of the potentiometric or water-table surface and distance between wells. Greater precision is required where the surface is gradual or the wells are close together. Generally reference point elevations should be surveyed and water levels measured with a precision ranging between ± 0.1 foot and ± 0.01 foot.

For example, if water-level fluctuations are occurring over a short period of time, then it may be more important to obtain a set of less precise measurements in a short period of time rather than a very precise set of measurements over a longer period of time. In such cases, measurements made to 0.1 foot may be appropriate. On the other hand, if the potentiometric surface slope is very gradual, more precise elevation control and water-level measurements may be required.

Current environmental regulations generally require that water levels be monitored and reported on a quarterly basis. A quarterly monitoring schedule may be appropriate for sites where water levels fluctuate in response to seasonal conditions such as precipitation or irrigation recharge. However, water levels at many sites respond not only to seasonal factors but also to factors of shorter duration or greater frequency. These factors may include fluctuations caused by tides in coastal areas, daily well pumping, and changes in river stage, among others. Separate zones may also respond differently to the cause of the fluctuations.

During site characterization activities, factors that may cause water levels to fluctuate need to be assessed and their importance evaluated with respect to two issues:

- the time in which a set of water-level measurements need to be obtained; and
- how the flow directions may change as the water levels fluctuate.

With the advent of computer technology, our ability to analyze complex systems at reasonable cost is increasing. Microprocessors connected to transducers allow the collection and analysis of water-level data over extended periods of time. To determine a site-specific monitoring interval, continuous monitoring can be economically accomplished in selected wells screened at differing depths and at varying distances from the cause of the fluctuation. These data can then be used to determine the time frame and intervals in which to obtain water-level measurements and to determine how the various zones beneath the site respond to the cause of the fluctuation.

The period in which the continuous monitoring should be conducted depends on the frequency and duration of the fluctuations. If possible, monitoring should be conducted at times of representative fluctuation. For example, on sites affected by tides, monitoring over several tidal cycles during relatively high and low tides is warranted.

Reporting of Data

Interpretation of water-level data requires that information be available about the monitoring installations and the conditions in which the water level measurements were made.

This information includes:

- Monitoring Installations
  —geologic sequence,
  —well construction features, especially screen and sand pack length, and geologic strata in which the screen is situated,
  —depth and elevation of the top and bottom of the screen and sand pack;
  —measuring point location and elevation; and
  —casing stickup above ground surface.
- Water-Level Data
  —date and time of measurement;
  —method used to obtain the measurement; and
  —other conditions in the area which might be affecting the water level data such as tidal or river stage, well pumping, storm events, etc.

WATER-LEVEL DATA ACQUISITION

For many purposes in ground-water investigations, the accurate determination of water levels in wells or piezometers is paramount. Without accurate measurements, it is not possible to interpret the data to assess ground-water flow directions or, if the data are "apparently" interpretable, a faulty interpretation is made.

Depending upon the ultimate use of the water-level data, the methods and instruments used to collect and record changes in ground-water levels vary substantially. Water-level data acquisition techniques are divided into two major categories for discussion purposes: manual measurements or typically nonrecording methods, and continuous measurements using instruments that provide a record. Although not exhaustive, the following discussion describes techniques most frequently used by the practicing hydrogeologist. These methods are summarized in Table 9.1.

Table 9.1.   Summary of Methods for Manual Measurement of Well Water Levels in Nonflowing and Flowing Wells.

Measurement Method	Measurement Accuracy in Feet	Major Interference or Disadvantage
Nonflowing Wells		
Wetted-tape	0.01	Cascading water or casing wall water
Air-line	0.25	Air line or fitting leaks; gage inaccuracies
Electrical	0.02 to 0.1	Cable wear; hydrocarbons on water surface
Transducer	0.01 to 0.1	Temperature changes; electronic drift; blocked capillary
Float	0.02 to 0.5	Float or cable drag; float size and lag
Popper	0.1	Well noise; well pipes and pumps; well depth
Acoustic probe	0.02	Cascading water; hydrocarbon on well water surface
Ultrasonics	0.02 to 0.1	Temperature changes; well pipes and pumps; casing joints
Flowing Wells		
Casing extensions	0.1	Limited range; awkward to implement
Manometer/ pressure gage	0.1 to 0.5	Gage inaccuracies; calibration required
Transducers	0.02	Temperature changes; electronic drift

Manual Measurements in Nonflowing Wells

Wetted-Tape Method

One of the most accurate techniques used to manually measure ground-water levels is the wetted-tape method. The equipment needed to make a measurement using this method consists of a standard surveyor's steel tape, a block of carpenter's chalk, and a slender lead weight. Steel tapes and hand reels are commercially available in lengths up to 1,000 feet. It is recommended, however, that shorter standard lengths (100, 200, 300, and 500 feet) be used because of weight and cost. Steel tape markings are usually divided only into tenths of feet. Interpolation to the nearest 0.01 foot is possible.

The lead weight is attached to the steel tape end clip with sufficient wire for support, but not enough to be stronger than the tape. This allows the tape to be pulled free should the weight become snagged. The bottom two or three feet of the tape is coated with carpenter's chalk. A water-level measurement is made by lowering the tape slowly into the well, about one or two feet into the water. It is convenient to lower the tape into the water a sufficient distance to allow the tape to read an even foot mark at the top of the well casing or surface reference point. The water-level measurement is calculated by subtracting the submerged distance, as indicated by the absence of change in chalk color, from the reference point at the top of the well.

The practical limit of measurement precision for this method is ± 0.01 foot (U.S. Geological Survey, 1980). Coefficients of stretch and temperature expansion of the steel tape become a concern when water-level measurements are made in wells of higher temperatures or at depths greater than 1,000 feet (Garber and Koopman, 1968). For most ground-water investigations, corrections for these errors are not necessary.

A disadvantage of using the wetted-tape method is that if the approximate depth to water is unknown, too short or too long a length of chalked tape may be lowered into the well, thereby necessitating a number of attempts. Also, water inside the casing, or cascading water, may wet the tape above the actual water level and result in errors in measurement (Everett, 1980).

Air-Line Submergence Method

The air-line submergence method, although less precise than other manual water-level measurement methods, continues to be a preferred technique in wells that are being pumped. To make an air-line measurement of water level in a well, a small, straight tube of accurately known length is installed in the well. This tube, usually 0.375 inch or less in diameter, can be made of plastic, copper, or steel. The air-line and all connections must be air tight, without bends or kinks, and installed to several feet below the lowest anticipated water level. A pressure gage is attached to this line (preferably calibrated in feet of water), as well as a fitting for an air source. In deep wells or where multiple water level measurements are needed, a small air compressor is useful. In shallow wells, a hand air pump is typically used.

A water-level measurement is made when air is pumped into the small tube and the pressure is monitored. Air pressure will continue to increase until it expels all water from the line. Air pressure, which is determined when the pressure gage stabilizes, is used to calculate the height of the water in the tube. If the pressure gage is calibrated in pounds per square inch (psi), a conversion is made to feet by multiplying the psi reading by 2.31. The actual level in the well is determined by subtracting the calculated distance from the air line's length. According to Driscoll (1986), the dependability of measurements made by the air-line device varies with the accuracy of the pressure gage and the care used in determining the initial pressure reading. Depth to water can usually be determined to within 0.25 foot of the true water level. Garber and Koopman (1968) have also shown that the precision of the measurement is mainly dependent upon the accuracy of the pressure gage. They state that even with gages having gradations as small as 0.1 psi, the maximum possible resolution would be 0.23 foot. Unless the air-line method is used in wells of substantial depth, corrections for thermal expansion, hysteresis, fluid density, and barometric pressure are not necessary.

Electrical Methods

Currently, the most favored technique for manual water level measurement is the use of an electrical probe. The most prevalent instrument of this type is that which operates on the principle that a circuit is completed when two electrodes come in contact with the water surface in the well. Other instruments rely on such physical characteristics as resistance, capacitance, or self-potential to produce a signal. Many of these instruments employ a two-wire conductor which is marked every foot, every 5 feet, or at 10-foot intervals. Some newer instruments use vinyl-, epoxy-, or Teflon-clad steel tapes as an insulated electrode and the well casing or grounding wire as the other.

Water-level probes which use self-potential typically have one electrode made of magnesium and the other of brass or steel. When the probe comes into contact with water, a potential between the two dissimilar metals is measured at the surface on a millivolt meter.

If a battery is added to the circuit, the two electrodes may be of the same material, usually brass, lead, or ferrous alloy. When the electrodes come into contact with the water surface, the borehole fluid conducts the current, and a meter, light, or buzzer is activated at the ground surface.

The principles of capacitance and inductance have been used by the U.S. Geological Survey to detect water surfaces (Garber and Koopman, 1968). These were basically specialty instruments and have not been commercially developed. They would, however, have the same apparent accuracy and precision as other electrical probes because the sensing elements are suspended in the well via multiwire conductors.

Errors in water-level measurements using electrical probes result from changes in the cable length and diameter as a function of use, depth, and temperature. After repeated use, the markings on the drop line often have a tendency to become loose and slide (if banded) or become illegible from wear (if embossed). Shallow measurements made with well-maintained electrical probes are typically reproducible to within ± 0.02 foot. Because of kinks in the cable and less than vertical suspension in a well, Barcelona and others (1985) have stated that the accuracy of electrical probes is about 0.1 foot.

A common disadvantage inherent in most electrical probe instruments is that if substantial amounts of oil or other constituents are floating upon the water surface, contact cannot be reliably made. This is a major concern in ground-water investigations involving hydrocarbon releases. Special sensing probes utilizing an optical liquid sensor in conjunction with electrical conductivity are commercially available to measure the hydrocarbon/water interface. Because this type of sensing probe is also suspended from multiconductor wire, the same errors as previously discussed for electrical probes apply.

Pressure Transducer Methods

With the advent of reliable silicon-based strain gage pressure sensors, a unique type of instrument is being commercially marketed for measuring changes in water levels—submersible differential pressure transducers. These transducers contain a 4–20 mA current transmitter and a strain gage sensor. The current transmitter prevents measurement sensitivity from being affected by cable length. Because all sensitive electronics are in the transducer and submerged in a constant temperature environment (the well water), errors due to temperature fluctuations are negligible (In Situ, Inc., 1983). Most transducers used for measuring ground-water levels have a small capillary tube leading from one side of a differential pressure sensor that is vented to the atmosphere. This allows for automatic compensation of barometric pressure. A signal conditioning unit and a power source are required ancillary equipment to make a water-level measurement.

For a discrete water-level measurement, the transducer is lowered a known distance into the well and allowed to equilibrate to the fluid temperature. The distance of submergence of the transducer is read on the signal conditioning unit and is subtracted from the known cable length referenced at the top of the well.

This technique is easily adaptable to continuous monitoring. It also offers several advantages in ease of accurate measurement in both pumping wells and wells with cascading water. Sources of error in this type of instrument include the electronics (linearity, accuracy, temperature coefficient, etc.), temperature changes and inappropriate application (i.e., range and material of construction) of a transducer in a given medium (Sheingold, 1980). Because of the sensitive electronics, care must be taken to avoid any rough handling in the field or in storage.

The accuracy of water-level transducers is dependent upon the type and range (sensitivity) of the device used. Most transducers are rated in terms of a percent of their full-scale capability. For example, a 0 to 5 psi transducer rated at 0.01% will provide measurements to the nearest 0.01 foot. In contrast, a 0 to 25 psi transducer rated at 0.01% will provide measurements to the nearest 0.05 foot (Barcelona et al., 1985).

Float Method

As the name implies, a float is attached to a length of steel tape and suspended over a pulley into the well. At the opposite end of the steel tape, a counterweight is attached. The depth to water is read directly from the steel tape at a known reference point at the top of the casing.

To obtain an accurate measurement using this technique, the absolute length of the float assembly must be measured and subtracted from the steel tape measurement. For greater accuracy, the total amount of float submergence should be calculated and a correction applied. This becomes more critical with smaller diameter floats (Leupold & Stevens, Inc., 1978). This method is used principally to obtain continuous water-level measurements. The accuracy and errors in float-operated devices will be discussed in greater detail in a following section.

Sonic or Audible Methods

Virtually every practicing hydrogeologist has (but should not have) dropped a rock down a well, at one time or another, to determine if water is present. Stewart (1970) investigated and developed a technique to determine the depth to water by timing the fall of a BB (air rifle shot) or a glass marble and recording the time of the return sound of impact. We will not discuss this sonic technique in detail but refer interested readers to the cited reference because of the rather large range of error in measurement (± 5 feet).

Popper

The most simplistic device used to audibly determine the depth to water in a well casing is a popper. This is a metal cylinder 1 to 1.5 inches in diameter and 2 to 3 inches long with a concave bottom. The popper is attached to a steel tape and lowered to within a few inches of the water surface in the well. By repeatedly dropping the popper onto the water surface and noting the tape reading at which a distinctive "pop" is heard, the depth to water is determined (Bureau of Reclamation, 1981).

Because of noise and the lack of clearance, the use of poppers in pumping wells is limited. The accuracy of water-level measurements made by this technique are highly dependent upon the skill of the measurer and the depth of the well. Determination of the water level to within 0.1 foot is usually the detection limit of this procedure.

Acoustic Probe

A unique adaptation of the popper principle was developed by Schrale and Brandwyk (1979), with the construction of an acoustic probe. This electronic device is attached to a steel tape and lowered into the well until an audible sound is emitted from a battery powered transducer contained in the probe. The electric circuit is completed when the two electrodes placed in the bottom of the probe come in contact with the water. As with the previously discussed electrical methods, problems with measurements can occur when hydrocarbons are present or if the well has cascading water. According to the developers of this instrument, a water-level determination is possible to within ± 0.02 foot accuracy.

Ultrasonic Methods

Instruments that measure the arrival time of a reflected transmitted sonic or ultrasonic wave pulse are becoming more common in the measurement of water levels. These instruments electronically determine the amount of time it takes a sound wave to travel down the well casing, reflect off the water surface, and return to the surface. Because the electronic circuitry typically uses microprocessors, this signal is transmitted, received, and averaged many times a second. The microprocessor also calculates the depth to water and displays it in various units. Several of the commercially available instruments simply rest on top of the well casing with nothing being lowered into the well. Rapid determination of water depths in deep wells is a distinct advantage of this technique.

The presence of hydrocarbons on the water surface usually has no effect upon the measurement. Accuracy can be limited by change of temperature in the path of the sound wave and other reflective surfaces in the well (i.e., pipes, casing burrs, pumps, samplers, etc.). Most commercially available hand-held units can measure the depth to water within 0.1 foot if the well's temperature gradient is uniform. Usually, the greater the depth to water from the sensor, the less accurate the measurement. Specialized installations, however, have repeatedly provided water-level measurements accurate to within ± 0.02 foot (Alderman, 1986).

Manual Measurements in Flowing Wells

Casing Extension

When the pressure of a flowing well is sufficiently low, a simple extension of the well casing allows the water level to stabilize so that a depth to water can be directly measured. The direct measurement of the piezometric level by casing extension is practical when the additional height requirement is only several feet. A water level measurement using this technique should be accurate to within ± 0.1 foot because flowing well water levels tend to fluctuate.

Manometers/Pressure Gages

If the pressure of the flowing well is sufficiently high, the use of a casing extension is usually not practical. To measure the piezometric level in such circumstances, the well is sealed or "shut-in" and the resulting pressure of the water in the well casing is measured. Two commonly used instruments to monitor the well pressure are manometers and pressure gages.

A mercury manometer, when properly installed and maintained, has a sensitivity of 0.005 foot of water, and these devices have been constructed to measure ranges in water levels in excess of 120 feet (Rantz et al., 1982). When used to monitor shut-in pressures of wells, an accuracy of ±0.1 foot is typical (U.S.G.S., 1980).

Pressure gages are typically less sensitive to head pressure changes than mercury manometers and, therefore, have only a routine accuracy of ±0.2 foot under ideal conditions when calibrated to the nearest tenth of a foot of water. According to the U.S. Geological Survey (1980), probable accuracy of measuring the pressure of a shut-in well with pressure gages is about 0.5 foot.

When using either of these instruments to measure well pressure, care should be taken to avoid rapid pressure change caused by opening or closing valving used in well sealing. This could create a water hammer and cause subsequent damage to the manometer or pressure gage. Also, field instruments used to monitor pressure should be checked periodically against master gages and standards.

Pressure Transducers

As previously described, pressure transducers can accurately monitor changes in pressure over a wide range. Transducers have been installed in place of pressure gages to determine the piezometric level. If the pressure transducer range is carefully matched with the shut-in well pressure, measurements to ±0.02 foot are obtained. A source of error in these measurements results from changes in temperature in the transducer. Either a transducer unit which has some form of temperature compensation or a unit totally submerged in the well should be used.

Applications and Limitations of Manual Methods

No single method for determining water levels in wells is applicable to all monitoring situations, nor do all monitoring situations require the accuracy and precision of the most sensitive manual measurement technique. The practicing hydrogeologist should become familiar with the various techniques by using two or more of these methods to obtain water levels on the same well. By doing so, the strengths and weaknesses of the monitoring methods will quickly become evident.

Table 9.1 is a summary of the manual measurement techniques previously discussed, with their reported accuracies. Also presented in this summary are several of the principal sources of error or interference accompanying each technique.

This table should be used only as a guide because each monitoring application and the skill of the measurer can result in greater or less measurement accuracy than stated.

Continuous Measurements of Ground-Water Levels

The collection of long-term water-level data is a necessary component of many hydrologic and hydrogeologic investigations. A commonly employed technique is the use of mechanical float recording systems. These devices typically produce a continuous analog record, usually a strip chart, which is directly proportional to the water-level change.

Electromechanical instruments which use a conductance probe with a feedback circuit to drive a strip chart or a punched-tape can successfully monitor rapid changes in water levels. These are used where float-operated systems fail to follow water-level fluctuations.

With the development of field-operable solid-state data loggers and computers, long-term monitoring systems using pressure transducers are finding favor among those conducting hydrogeologic investigations. As with manual water-level measurements, the type of long-term monitoring system employed is dependent upon the investigator's data needs.

Methods of Continuous Measurement

Mechanical—Float Recorder Systems

Instruments which use a float to operate a chart recorder (a drum covered with chart paper and containing a time-driven marking pen) have been used to measure water levels since the beginning of the century. These devices produce a continuous analog record of water-level change, usually as a graph. Depending upon the gage scale and time-scale gearing, a single chart may record many months of water level fluctuations.

Float-operated devices are subject to several sources of error which include float lag, line shift, submergence of counterweight, temperature, and humidity. Leupold and Stevens (1978) detail these errors and suggest methods to correct them. The reader should consult this reference for additional details. For purposes of this discussion, it is noted that when smaller floats are used, the magnitude of error is greatest. For example, float lag, or the lag of the indicated water level behind the true water level due to the mechanical work required by the float to move the instrument gears, can be as much as 0.5 foot for a 1.5-inch float if the force to move the instrument is 3 ounces. This is contrasted to a 0.07 foot error for a 4-inch float and 0.03 foot error for a 6-inch float on an instrument requiring the same 3 ounces of force (Leupold and Stevens, 1978). This error is magnified if the float or float cable is allowed to drag against the well casing. Shuter and Johnson (1961) discuss these problems in measuring water levels in small diameter wells and offer several devices to improve recorder performance.

Because many of the wells constructed in today's ground-water monitoring programs are 2 inches in diameter, caution should be used if a float recording system is installed to obtain continuous water-level measurements.

According to Rantz et al. (1982), if properly installed and operated, long-term water-level measurements in wells are obtainable to about ± 0.01 foot using mechanical float recording systems. This accuracy is based on measurements made in stilling wells used for long-term monitoring of stage height. Because the piezometers and wells typically utilized in monitoring well networks are smaller in diameter, the accuracy for float recording systems used to measure ground-water fluctuations will usually be greater than ± 0.01 foot.

Electromechanical—Iterative Conductance Probes (Dippers)

Iterative conductance probes, commonly referred to as dipping probes or dippers, are electromechanical devices which use an electronic feedback circuit to measure the water level in a well. A probe is lowered on a wire by a stepping motor until a sensor in the probe makes electrical contact with the water. This generates a signal which causes the motor to reverse and retract the probe slightly. After a set time period, the probe is lowered again until it makes contact with the surface, retracts, etc., thus repeating the iterative cycle. The wire cable is connected to either a drum used for chart recording or a potentiometer whose output signal is proportional to the water level (Grant, 1978).

Dipping probes have several advantages over float recording systems. The well can be of smaller diameter and the system can accommodate some tortuosity in the well casing. Because the sensing probe is electromechanical, greater depths to water can be monitored without the mechanical losses associated with float systems. When water level fluctuations are cyclic or change moderately rapidly, the dipping probe better reflects the oscillations in the water levels of smaller wells.

Data Loggers

Data loggers consist of microprocessors connected to transducers which are installed in the well or wells being monitored. The microprocessors consist of hardware and software which allow the automated collection of water-level data over various time periods. The data can be easily manipulated by transfer to a computer database. The use of this equipment is becoming more common and a variety of equipment is commercially available.

ANALYSIS, INTERPRETATION, AND PRESENTATION OF WATER-LEVEL DATA

The primary use of ground-water level data is to assess in which direction ground water is flowing beneath a site. The usual procedure is to plot the location of wells on a base map, convert the depth-to-water measurements to elevations, plot the water-level elevations on the base map, and then construct a ground-water elevation contour map. The direction of ground-water flow is estimated by drawing ground-water flow lines perpendicular to the ground-water elevation contours (Figure 9.4).

The relatively simple approach to estimating ground-water flow directions described above is suitable where wells are screened in the same zone and the flow of ground water is predominantly horizontal. However, with the increased emphasis on detecting the subsurface position of contaminant plumes or in predicting possible contaminant migration pathways, it is becoming evident that the assumption of horizontal flow beneath a site is not always valid. Increasingly, flow lines shown on vertical sections are required to complement the planar maps showing horizontal flow directions (Figure 9.7) to illustrate how ground water is flowing either upward or downward beneath a site (Figure 9.8).

Ground water flows in three dimensions, and as such can have both horizontal and vertical (either upward or downward) flow components. The magnitude of either the horizontal or the vertical flow component and the direction of ground-water flow is dependent on several factors.

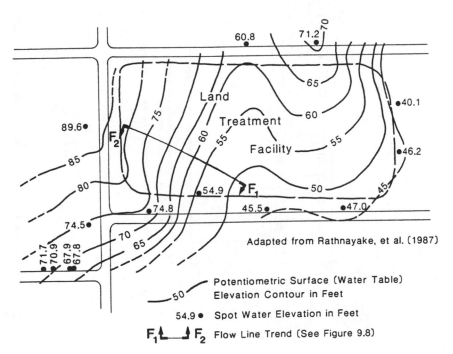

Figure 9.7. Potentiometric surface elevation contour map.

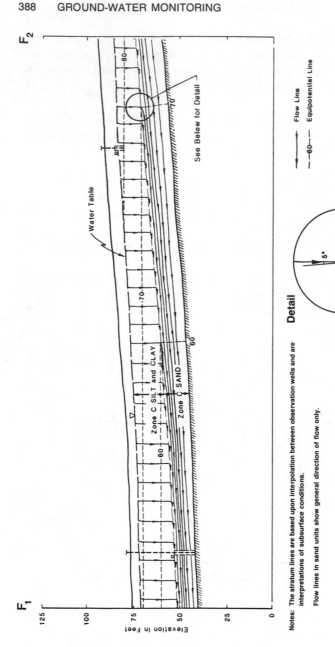

Detail

Detail of the refraction of a typical flow line at the geologic contact.

Notes: The stratum lines are based upon interpolation between observation wells and are interpretations of subsurface conditions.

Flow lines in sand units show general direction of flow only.

Figure 9.8. Section showing vertical flow directions.

Recharge and Discharge Conditions

In recharge areas, ground water flows downward (or away from the water table) while in discharge areas ground water flows upward (or toward the water table). Ground water migrates nearly horizontally in areas between where recharge or discharge conditions prevail. For example, on Figure 9.9 well cluster A is located in a recharge area, well cluster B is located in an area where flow is predominantly lateral, and well cluster C is located in a discharge area. Note on Figure 9.9 that wells located adjacent to one another, but finished at different depths, display different water-level elevations.

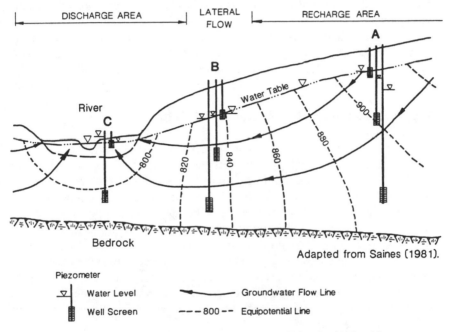

Figure 9.9. Ideal flow system showing recharge and discharge relationships.

Aquifer heterogeneity. This refers to a condition where aquifer properties are dependent on *position* within a geologic formation (Freeze and Cherry, 1979) and needs to be considered in evaluating water-level data. While recharge or discharge may cause vertical gradients to be present within a discrete geologic zone, vertical gradients may be caused by the contrast in hydraulic conductivity between aquifer zones. This is especially evident where a deposit of low hydraulic conductivity overlies a deposit of relatively higher hydraulic conductivity, as shown in Figure 9.8.

Aquifer anisotropy. This refers to an aquifer condition in which aquifer properties vary with *direction* at a point within a geologic formation (Freeze and Cherry, 1979). For example, many aquifer zones were deposited in more or less horizontal layers, causing the horizontal hydraulic conductivity to be greater than the vertical hydraulic conductivity. This condition tends to create more pronounced vertical gradients (Fetter, 1980) which are not indicative of the actual direction of ground-water flow. In anisotropic zones, flow lines do not cross potential lines at right angles and flow will be restricted to higher elevations than in isotropic zones showing the same water-level conditions.

Detailed discussions of each of these factors are beyond the scope of this section. The reader is referred to Fetter (1980) and Freeze and Cherry (1979) for more detailed discussions of the effects of these aquifer conditions on ground-water flow.

The practical significance of the three factors discussed above is that ground-water levels can be a function of either well-screen depth or of well position along a ground-water flow line or, more commonly, a combination of the two. For these reasons, considerable care needs to be taken in evaluating water-level data.

Approach to Interpreting Water Level Data

The first step in interpreting ground-water level data is to make a thorough assessment of the site geology. The vertical and horizontal extent and relative positions of aquifer zones and the hydrologic properties of each zone should be determined to the extent possible. It is difficult to overemphasize how important it is to have as detailed an understanding of the site geology as possible. Detailed surficial geologic maps and geologic sections should be constructed to provide the framework to interpret ground-water level data.

The next step in interpreting ground-water level data is to review monitoring well installation features with respect to screen elevations and the various zones in which the screens are situated. The objective of this review is to identify whether vertical hydraulic gradients are present beneath the site and to determine the probable cause of the gradients.

One method that can be used to assess the distribution of hydraulic head beneath a site is to plot water level elevations versus screen midpoint elevations. An example of such a plot is shown in Figure 9.10 for wells completed within a layered geologic sequence. The figure indicates that a steep downward hydraulic gradient, on the order of 0.85, exists within the sandy silt to silty clay layer. However, in the lower layers, the vertical component of flow is substantially less both within and between the layers.

Once the presence and magnitude of vertical gradients and the distribution of data with respect to each zone are established, the direction of ground-water flow can be assessed. If the geologic system is relatively simple and substantial vertical gradients are not present, a planar ground-water elevation contour map can be prepared which shows the direction of ground-water flow. However, if multiple zones of differing hydraulic conductivity are present beneath the site, several

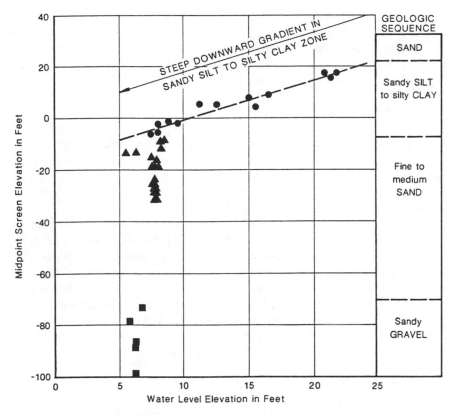

Figure 9.10. Water level elevation vs. midpoint screen elevation for well screened in stratified geologic sequence.

planar maps may be required to show the horizontal component of flow within
each zone (typically the zones of relatively higher hydraulic conductivity) and vertical sections are required to illustrate how ground water flows between each zone.

For the example presented in Figure 9.10, the data indicate that flow is
predominantly downward within the upper silt/clay zone. Flow within the lower
zone appears to be largely horizontal, although a vertical component of flow is
indicated between the sand and the underlying gravel layer.

The examples presented above show downward vertical gradients which are
indicative of recharge areas. Sites can also be situated within discharge areas where
the vertical components of flow are in an upward direction.

The presence of vertical gradients can be anticipated in areas where sites are:

- underlain by a layered (heterogeneous) geologic sequence, especially
  where deposits of lower hydraulic conductivity overlie deposits of substantially higher hydraulic conductivity; or
- located within recharge or discharge areas.

It should be noted that site activities can locally modify site conditions to such an extent that ground water flows in directions contrary to what would be expected for "natural" conditions. Drainage ditches can modify flow within near-surface deposits and facility-induced recharge can create local downward gradients in regional discharge areas among others.

Transient Effects

Ground-water flow directions and water levels are not static and can change in response to a variety of factors such as seasonal precipitation, irrigation, well pumping, changing river stage, and fluctuations caused by tides. Fluctuations caused by these factors can modify, or even reverse, horizontal and vertical flow gradients and thus alter ground-water flow directions.

Time series water-level data are required to assess how ground-water flow directions change in response to these factors. Figure 9.11 shows data for several wells finished at differing depths in an area influenced by changing river stage. The data indicate that river stage affects water levels but that the direction of horizontal and vertical flow gradients do not substantially change with river fluctuation. However, the fluctuations do affect the length of time over which each set of ground-water level measurements should be made. In this case, measurements were made in less than one hour to minimize the effects of the fluctuations on the interpretation of ground-water flow directions.

Contouring of Water-Level Elevation Data

Typically, ground-water flow directions are assessed by preparing ground-water elevation contour maps. Water-level elevations are plotted on base maps and linear interpolations of data between measuring points are made to construct contours of equal elevation (Figure 9.7). These maps should be prepared using data from measuring points screened in the same zone where the horizontal component of the ground-water flow gradient is greater than the vertical gradient. The greatest amount of interpretation is typically required at the periphery of the data set. A reliable interpretation requires that at least a conceptual analysis of the hydrogeologic system has been made. The probable effects of aquifer boundaries, such as valley walls or drainage features, need to be considered.

In areas where substantial vertical gradients are present, the area ground-water flow maps need to be supplemented with vertical sections which show how ground water flows vertically within and between zones (Figure 9.8). These sections should be oriented parallel to the general direction of ground-water flow and should account for the effects of anisotropy.

Computer contouring and statistical analysis (such as Kriging) of water-level elevation data are becoming more popular (McKown and others, 1987). These

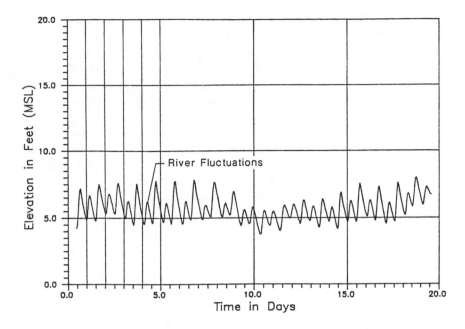

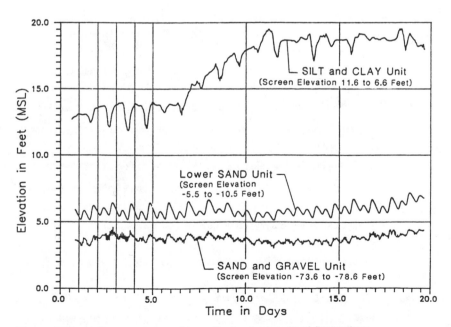

Figure 9.11.   Influence observed in wells due to river level fluctuations.

tools offer several advantages, especially for large data sets. However, the approach and assumptions which underlie these methods should be thoroughly understood before they are applied and the output from the computer should be critically reviewed. The most desirable approach would be to interpret the water-level data using both manual and computer techniques. If different interpretations result, then the discrepancy between the interpretations should be resolved by further analysis of the geologic and water-level data.

The final evaluation of water-level data should encompass a review of geologic and water-quality data to confirm that a consistent interpretation is being made. For example, at a site where contamination has occurred, wells that are contaminated should be downgradient of the site (based on the water-level data). If this is the case, then a consistent interpretation is indicated. However, if wells that are contaminated are not downgradient of the site, based on water-level data, then further evaluation is required.

REFERENCES

Alderman, J. W., 1986, FM Radiotelemetry Coupled with Sonic Transducers for Remote Monitoring of Water Levels in Deep Aquifers, Ground Water Monitoring Review, Vol. 6, No. 2, pp. 114–116.

Barcelona, M. J., J. P. Gibb, J. A. Helfrich, and E. E. Garske, 1985, Practical Guide for Ground-Water Sampling, U.S. EPA, Robert S. Kerr Environmental Res. Lab. EPA-600/2-85-104, Ada, OK, pp. 78–80.

Bureau of Reclamation, 1981, Ground Water Manual, U.S. Dept. of Interior, Bureau of Reclamation, U.S. Govt. Printing Office, Denver, CO, pp. 195–199.

Driscoll, F. G., 1986, Ground Water and Wells, Johnson Division, St. Paul, MN.

Everett, L. G., 1980, Ground Water Monitoring, General Electric Company, Schenectady, NY, pp. 196–198.

Fetter, C. W., 1980, Applied Hydrogeology, C. E. Merrill Publishing Co., Columbus, OH.

Freeze, R. A., and J. A. Cherry, 1979, Groundwater, Prentice Hall, Englewood Cliffs, NJ.

Garber, M. S., and F. C. Koopman, 1968, Methods of Measuring Water Levels in Deep Wells, U.S. Geological Survey Survey Techniques of Water Resources-Inv. Bk. 8, Chap. A1.

Grant, D. M., 1978, Open Channel Flow Measurement Handbook, Instrumentation Specialties Company, Lincoln, NB, pp. 6–7.

Heath, R. C., 1983, Basic Ground-Water Hydrology, U.S.G.S. Water Supply Paper 2220.

In-Situ, Inc., Owner's Manual: Hydrologic Analysis System, Model SE200, In-Situ, Inc., Laramie, WY, pp. 7–11.

Leupold & Stevens, Inc., 1978, Stevens Water Resources Data Book, Leupold & Stevens, Inc., Beaverton, OR.

Maslansky, S. P., C. A. Kraemer, and J. C. Henningson, An Evaluation of Nested Monitoring Well Systems, in Ground-Water Monitoring Seminar Series—Technical Papers, USEPA CERI-87-7, 1987.

McKown, G. L., G. W. Dawson, and C. J. English, 1987, Critical Elements in Site Characterization in Ground-Water Monitoring Seminar Series-Technical papers, USEPA CERI-87-7.

Rantz, S. E., et al., 1982, Measurement and Computation of Streamflow: Volume 1. Measurement of Stage and Discharge, U.S. Geological Survey Water-Supply Paper 2175, U.S. Govt. Printing Office, Washington, DC, pp. 63–64.

Rathnayake, D., C. D. Stanley, and D. H. Fujita, 1987, Groundwater Flow and Contaminant Transport Analysis in Glacially Deposited Fine Grained Soils: A Case Study in Proceedings of the FOCUS Conference on Northwestern Groundwater Issues, NWWA, May 1987, Portland, OR.

Saines, M., 1981, Errors in Interpretation of Ground-Water Level Data, Ground Water Monitoring Review, Vol. 1, No. 1, pp. 56–61.

Schrale, G., and J. F. Brandwyk, 1979, An Acoustic Probe for Precise Determination of Deep Water Levels in Boreholes, Ground Water, Vol. 17, No. 1, pp. 110–111.

Sheingold, D. H., 1980, Transducer Interfacing Handbook, Analog Devices, Inc., Norwood, MA.

Shuter, E., and A. I. Johnson, 1961, Evaluation of Equipment for Measurement of Water Levels in Wells of Small Diameter, U.S. Geological Survey Cir. 453.

Stewart, D. M., 1970, The Rock and Bong Techniques of Measuring Water Levels in Wells, Ground Water, Vol. 8, No. 6, pp. 14–18.

Todd, K. D., 1980, Groundwater Hydrology, 2nd ed., John Wiley and Sons, NY.

U.S. Geological Survey, 1980, National Handbook of Recommended Methods for Water-data Acquisition; Chapter 2—Ground Water, U.S. Department of Interior, Geological Survey, Reston, VA.

10

Methods and Procedures for Defining Aquifer Parameters

John Sevee

INTRODUCTION

The storage and movement of ground water through soil and rock obey certain physical laws. These laws are represented mathematically and are used to quantitatively describe the behavior of ground water within a particular hydrogeologic setting. Certain physical parameters, such as hydraulic conductivity, storativity, and aquifer thickness, must be determined in order to solve the mathematical relationships describing ground-water behavior. Determination or measurement of these parameters is a primary purpose of many field investigations. Once defined, the parameters can be utilized with the appropriate mathematical relationships or equations to calculate such items as ground-water flow rate and direction, aquifer yield, or the behavior of chemicals transported in ground water.

Relative to ground-water monitoring, a quantitative description of an aquifer and its properties can be utilized: (1) to optimize well placement for describing a chemical plume; (2) to determine the appropriate depth to install a monitoring well to encounter a chemical plume in a layered aquifer system; or (3) to decide the minimum number of wells required to detect a leak from a buried tank for a given probability of detection. For instance, to properly locate monitoring wells around a lined landfill for the purpose of detecting liner leakage requires an understanding of the direction of ground-water flow. However, this is not only dependent on the configuration of the water table beneath the landfill but also, in anisotropic conditions, on the variation in hydraulic conductivity of the geo-logic materials. Certain aquifer properties can be used to calculate monitoring well recharge rates and zones of pumping influence created during well purging or ground-water sampling. Computer modeling efforts have shown that the zone of pumping influence around a monitoring well varies with the ratio of hydraulic conductivities of the artificial filto-pack material to that of the natural geologic material (Cohen and Rabold, 1987). The relative contribution of water to the well from each of these materials, and therefore the degree of "representativeness" of the water sample could not be estimated without a quantitative understanding of the aquifer's physical properties. A quantitative description of ground water and aquifer behavior, therefore, becomes important in establishing a reliable ground-water monitoring program.

This chapter defines the terminology associated with hydrogeologic parameters and presents typical laboratory and field methods for measuring or estimating these parameters. Bulk density, water content, porosity, hydraulic conductivity and permeability, ground-water velocity, specific storage and specific yield, trans-missivity, and aquifer compressibility are defined herein. Test methods for defining the aquifer parameter known as dispersivity, which is related to spreading of a chemical plume moving with ground water are not presented, although methods have been presented elsewhere in the literature (Gelhar et al., 1985; Moltz et al., 1986; Sudicky and Cherry, 1979; Bentley and Walter, 1983; Fried and Ungemach, 1971). Dispersivity is both a time- and scale-dependent quantity that describes a process that is presently poorly understood. Determination of dispersivity at typical project scales is extremely costly and requires periods of several years to measure. Because of these constraints, the reader is directed to the literature for descriptions of dispersivity testing methods.

Ideally, all aquifer parameters should be measured in the field (i.e., in situ) under the anticipated ground-water conditions. However, some of the parameters can be measured reasonably well in the laboratory on representative samples of unconsolidated geologic material and then be applied to the field situation. Representative fractured, jointed, or solution-channeled rock samples are cur-rently impractical to obtain for routine laboratory analysis and the results are difficult to apply to in situ conditions. Some of the parameters can be estimated from physical characteristics of the aquifer material. For instance, the hydraulic conductivity of sandy materials is often estimated from the results of grain-size distribution analyses of the material. Commonly used laboratory and field methods, which can be used to measure or estimate these parameters, are described herein.

From a practical standpoint, all points within a particular hydrogeologic set-ting cannot be sampled and tested. Therefore, collected measurements, however many there may be, are typically extrapolated across a study area as a matter of practicality. The more area that is sampled and tested, the greater the level of confidence in the predictions from any quantitative analyses. Because unlimited testing is not practical, selective testing is used to define average or representative properties for a particular parameter. Sometimes the values used in an analysis are a combination of different types of testing, although this approach must be used with caution. Different test methods may use different size samples or may average values over a greater or lesser volume. Sample volume variation can result in different parameter estimates (Bear, 1972; Parker and Albrecht, 1987). Understanding the geology can help in selecting the sample size or the in situ test methods to be used. Cross-checks between test methods (e.g., comparing slug test results with aquifer grain-size analyses), can be of value in verifying or evaluating test results or extrapolating parameter estimates over a broader area using the less-costly test method (Delhomme, 1974). Geostatistics may be a valuable method for interpolating or extrapolating field or laboratory results and concurrently determining the estimation error (Delhomme, 1978).

The limitations of a particular test method should be incorporated in the selection process of any test method. A parameter test method may be appropriate for one application and not another.

BULK DENSITY

Hydraulic conductivity of geologic materials is a function of several factors, bulk density being one (Lambe and Whitman, 1969). Hydraulic conductivity typically decreases with increasing bulk density. Bulk density also affects aquifer compressibility and, therefore, aquifer storativity (Lambe and Whitman, 1969). Bulk density, consequently, plays an important role in understanding aquifer properties and behavior.

Total or wet bulk density of a soil or rock is defined as the total weight or mass of soil or rock, including any water, in a unit volume of material. The dry bulk density of a soil or rock is the weight of the dry solids per unit volume of material, and is related to the total or wet bulk density by:

$$\gamma_d = \gamma_w/(1+w) \tag{10.1}$$

where:

w = water content of the soil or rock calculated on a dry weight basis, expressed as a decimal,

γ_d = dry bulk density, weight or mass per unit volume, and

γ_w = wet bulk density, weight or mass per unit volume.

Typical units of density are grams per cubic centimeter (gm/cm^3) or pounds per cubic foot (lb/ft^3). Methods for determining water content are presented below.

The dry bulk density is related to the specific gravity of the rock or soil solids by:

$$\gamma_d = \frac{G_s \gamma_w}{1 + w\ G_s/S} \tag{10.2}$$

where:

G_s = specific gravity of the solids (dimensionless),
S = degree of saturation expressed as a decimal,
γ_w = density of water, weight or mass per unit volume,
w = water content of the soil or rock calculated on a dry-weight basis, expressed as a decimal.

The degree of saturation represents the fraction of the pore space filled with water. The specific gravities of some typical soil and rock constituents are presented in Table 10.1. A comparison of Equations 10.1 and 10.2 indicates that the dry and wet bulk densities must be less than the particle density ($G_s \times \gamma_w$) of the soil or rock solids (except for solids with a specific gravity less than 1). This is due to the incorporation of void space (i.e., pores or fractures) in the bulk material. Density of soil particles or rock matrix (particle density) typically ranges from 160 to 180 pounds per cubic foot, whereas dry bulk densities of soils typically range from 90 to 130 pounds per cubic foot. Unweathered nonporous rock dry bulk densities approach the specific gravity of the constituent minerals in the rock. Weathered or porous rock dry bulk densities may approach that of soils. Typical bulk densities of various geologic materials are given in Table 10.2.

Table 10.1.   Typical Specific Gravities of Soil and Rock Constituents.

Gypsum	2.32	Dolomite	2.87
Montmorillonite	2.78	Aragonite	2.94
Orthoclase	2.56	Biotite	3.0–3.1
Kaolinite	2.6	Augite	3.2–3.4
Illite	2.6–2.86	Hornblende	3.2–3.5
Chlorite	2.6–3.0	Limonite	3.8
Quartz	2.66	Hematite, hydrous	4.3±
Talc	2.7	Magnetite	5.2
Calcite	2.72	Hematite	5.17
Muscovite	2.8–2.9		

Table 10.2.   Natural Bulk Densities of Typical Soils and Rocks.

Description	Bulk Density Dry	(lb/ft³) Wet
Uniform sand, loose	90	118
Uniform sand, dense	109	130
Nonuniform sand, loose	99	124
Nonuniform sand, dense	116	135
Glacial till	132	145
Soft glacial clay	80	110
Stiff glacial clay	100	129
Soft, slightly organic clay	65	98
Soft, very organic clay	45	89

continued

Table 10.2.   Continued.

Description	Bulk Density Dry	(lb/ft³) Wet
Rock:		
Granite	160	170
Dolerite	185	190
Gabbro	185	193
Basalt	175	180
Sandstone	125	162
Shale	125	150
Limestone	135	162
Dolomite	155	162
Quartzite	—	165
Gneiss	180	185
Marble	160	170
Slate	160	170

Mechanical Density Testing

Total bulk density of a soil can be measured by obtaining undisturbed cores or block samples. After a sample is obtained, the total bulk density is determined by measuring the volume of the sample and its total weight. The total bulk density is calculated by dividing the sample weight by its volume. Dry bulk density can be obtained by drying the sample after volume measurement but prior to weight measurement. Dry density can be calculated by measuring the average water content (see Water Content section) of the sample and using Equation 10.1 once the total bulk density is known.

Core samples of geologic materials are commonly obtained by pushing a thin-walled tube into the soil while installing a borehole (Figure 10.1). The American Society of Testing and Materials (ASTM) recommends a method (ASTM Test Method D 1587) for obtaining thin-walled tube samples from a borehole. This sampling technique is used principally for soft clays and loose silts. Although the method can be used to sample loose sand, significant sample disturbance often occurs and is difficult to prevent. Some disturbance is inevitable with any sampling method and can affect the density of a tube sample (Hvorslev, 1949a). These effects may increase or decrease the measured density depending on the natural density of the soil. Stress relief caused by removal of a soil sample from some depth below the ground surface may also alter the density of the sample. Various versions of the tube sampler are used for soils of differing consistency. A synopsis of sampler types is presented in Winterkorn and Fang (1975).

A block sample is obtained by cutting an undisturbed block of soil from the base or wall of an open excavation. This type of sample is obtained as illustrated in Figure 10.1. The block of soil is typically surrounded by a section of tubing or a square box without covers, and the space between the sample and the container is filled with tamped fine sand or paraffin. A 10- to 12-in. square box with

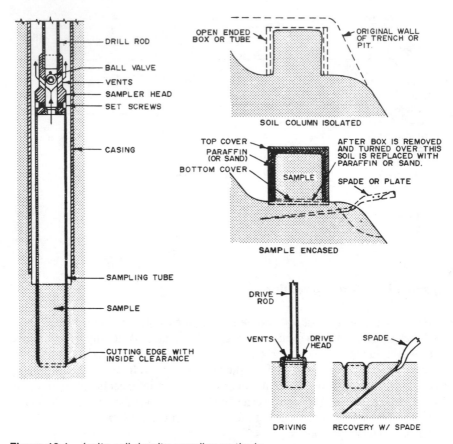

Figure 10.1. Insitu soil density sampling methods.

easily dismantled sides and covers is often used, especially in block sampling of sands. Isolation of the soil block will relieve in-situ stresses and may cause some expansion of the soil, but block sampling is still the best available method for obtaining large undisturbed samples of very stiff and brittle soils, partially cemented soils, and soils containing coarse gravel and stones. The method can be used in all soils except when cohesion is so small that a soil block cannot be isolated.

In order to preserve the moisture content of any soil sample, it should be immediately wrapped in foil or cellophane and covered with paraffin. The wrapped sample should be kept in a cabinet or room which is maintained at a high humidity to prevent desiccation of the sample, until it is ready for analysis.

ASTM describes other methodologies for obtaining in situ bulk density measurements by mechanical means. These methods involve removal of a volume of soil and measuring the weight and volume of the removed soil. For the drive-cylinder method (ASTM Test Method D 2937) a metal cylinder is driven into the ground, as shown in Figure 10.1, and removed with the sample inside. The retrieved sample size and weight are measured and the density calculated. Drive sampling is primarily suited for soft to stiff cohesive soils, silt, and loose to medium-dense fine sand. Compaction of the sample can occur when great force is required to push or drive the sampler into the soil.

The balloon and sand-cone methods involve excavating a hole and placing the excavated soil on a scale for weighing. The volume of the sample is then determined by balancing the scale using water (balloon method) or a clean sand of a known density (sand-cone method). These methods (ASTM Test Methods D 2167 and D 1556) have been used extensively for controlling fill compaction.

ASTM also presents a method for coring rock (ASTM Test Method D 2113). However, if the rock is heavily fractured, the measured bulk density of the core may be severely affected by sampling disturbance. Because coring procedures generally require the use of fluid to cool and lubricate the core barrel, erosion of fracture fillings and remolding of the sample can occur if the rock is highly weathered or poorly indurated.

Determining Density by Gamma-Ray Attenuation

Gamma-ray attenuation can be used to determine soil bulk density by placing two boreholes into the ground—one containing a detector and the other containing a gamma-ray source. The distance between the source and the detector tubes is typically on the order of 1 to 2 feet. The degree of gamma-ray absorption by the soil or rock is a function of the density of material between the source and detector. As the density increases, the degree of gamma-ray absorption also increases. If the boreholes are cased with a metal casing, for example, a correction is required for casing adsorption of the gamma radiation.

Some devices have been developed in which both source and detector are in the same common probe and the entire unit is lowered down a single borehole. When using this method, corrections must be made for the degree of gamma-ray absorption by any casing within the borehole. Each individual site requires calibration of the equipment. The gamma-ray method is also affected by the water content of the soil. Therefore, the water content of the soil must be known in order to utilize this method to calculate bulk density.

ASTM Test Method D-2922 describes the gamma-ray attenuation method used for in situ bulk density determination. This method involves placing a movable gamma-ray source at depths of up to one foot beneath the ground surface. The gamma-ray detector is located within the base of the device, which remains on the ground surface above the source. Therefore, whereas the above-described method measures horizontally between two boreholes, this method measures vertically from the ground surface. The surface may be lowered by excavation and the method repeated.

WATER CONTENT

Water content can be used to estimate the soil porosity and density if the degree of saturation of the soil is known. In partially saturated soils, water content is related to the relative permeability and matric suction (Freeze and Cherry, 1979). Because water content reflects the porosity of the soil, it provides a measure of the water held in storage by a soil or rock. The change in water content during gravity drainage of an initially saturated sample of soil is a measure of the specific yield.

There are two commonly used definitions for water content: volumetric and gravimetric. The volumetric water content is expressed as the volume of water relative to the total sample volume.

$$w_v = V_w/V_t \qquad (10.3)$$

where:

$$w_v = \text{volumetric water content expressed as a decimal, and}$$
$$V_w \text{ and } V_t = \text{volume of water in the sample and the total volume of the sample, respectively.}$$

Water content is typically expressed as a decimal or as a percent (i.e., decimal value multiplied by 100). With this definition, if the material is saturated, the volumetric water content is approximately equal to the total porosity of the soil or rock. Volumetric water content is typically used when examining the behavior of partially saturated soils (Hillel, 1980). Volumetric water content is most easily determined by calculation from the gravimetric water content.

The gravimetric water content (on a dry weight basis), w_w, is determined by dividing the weight of water in a given sample by the dry weight of solids in the sample (i.e., $w_w = W_w/W_s$). The two different water content definitions are related by:

$$w_v/w_w = \gamma_{dry}/\gamma_w \qquad (10.4)$$

where:

$$\gamma_w = \text{density of water in units of weight per unit volume, and}$$
$$\gamma_d = \text{bulk dry density of the soil sample in units of weight per unit volume.}$$

Laboratory Measurement of Water Content

The water content of a soil is commonly determined in the laboratory by drying at a temperature of 105°C to 115°C (ASTM Test Method D2216). The method simply involves weighing the moist soil sample, drying the specimen, and then reweighing the sample to determine the weight loss due to drying. For clay and organic soils, drying at temperatures above 115°C may result in the loss of chemically or physically bound water or, in the case of organic soils, weight loss by burning of the organic fibers. Certain soils and compounds have water of hydration that can be released at relatively low temperatures, thus exhibiting a water content that is temperature dependent. For example, gypsum has several different hydrated states and varying the drying temperature can result in different water contents. The use of a standard temperature range, therefore, provides consistency between measurements.

Field Measurement of Water Content

The water content of soil can also be obtained by various field methods. The most common method utilizes a probe containing a radiation source of fast neutrons (Americium or Radium) and a detector. The radiation source releases fast neutrons that are decelerated when hydrogen atoms are encountered in the soil or rock. The decelerated neutrons are reflected by the hydrogen to a detector that counts the slowed neutrons. Because water is the primary source of hydrogen, the intensity of the slowed neutron radiation reaching the counter probe is proportional to the water content. ASTM describes the neutron deceleration method for measuring the water content of soil (ASTM D3017).

Measurable sources of hydrogen can commonly occur in materials such as clay or organic matter, which contain hydrogen within the soil structure. Furthermore, water may be trapped between clay mineral plates (interstatial) or within unconnected pores in the rock or soil. These factors contribute to errors in measurement of the ''mobile'' water within the media using the radiation method. Mobile water is that water which can move or drain from the soil or rock under a gravitational pressure gradient.

The neutron deceleration method commonly gives the average water content over a 6-inch diameter sphere around the point of measurement. However, this volume may vary due to the soil density or source strength and the actual volume of measurement is uncertain in most cases. In soils that naturally contain hydrogen, such as organic or clay soils, a lengthy calibration procedure may be required to provide reasonably useful results. Despite the calibration required, two advantages to this method are: (1) a large number of measurements can be made in a short time, and (2) the probe can be used in either uncased or cased boreholes. Typically boreholes are cased with aluminum or steel to provide intimate contact with the surrounding soil when water content determinations with the neutron method are used. This avoids cavities that may fill with water or air between the tube and the soil.

Another common field method consists of using gypsum blocks or tensiometers to infer the water content from calibration curves. These methods are used to monitor moisture changes in the partially saturated zone of soil. In the case of the gypsum block, electrodes running to a power supply are connected to the block which is carefully buried in the soil. Backfilling the block with natural soil is critical to maintain the proper moisture-tension characteristics between the block and the undisturbed soil. The electrical resistance of the block is measured and varies with the water content of the surrounding soil. A careful calibration procedure is required to calibrate block resistance to soil water content. A more detailed description of this and other methods is presented in Gardner (1965).

POROSITY

Most rocks and soils are composed of solid mineral particles separated by void spaces. In soil and rock, there may be intergranular pore spaces (primary porosity), as well as fractures (secondary porosity). In the case of most soils, the void spaces between particles usually form a series of interconnected pores. Pores in a rock matrix may not be visible to the naked eye. Primary porosity is that porosity due to voids between the soil or rock grains; root holes, cavities, worm holes, and fractures may cause secondary porosity. Fractures may form as a result of faulting, jointing, foliation, or fissuring.

The volume of the total pore space in a material relative to the overall volume of the rock or soil is termed total porosity, n:

$$n = V_v/V_t \qquad (10.5)$$

where:

V_v = volume of voids in a sample, and
V_t = total sample volume.

Total porosity is expressed either as a dimensionless decimal (which must be less than one) or as a percent (i.e., decimal value times 100). Typical values for total porosity of various geologic materials are presented in Table 10.3.

Table 10.3.   Typical Total Porosities.

Material	Total Porosity (%)
Unaltered granite and gneiss	0–2
Quartzites	0–1
Shales, slates, mica-schists	0–10
Chalk	5–40
Sandstones	5–40
Volcanic tuff	30–40
Gravels	25–40
Sands	15–48
Silt	35–50
Clays	40–70
Fractured basalt	5–50
Karst limestone	5–50
Limestone, dolomite	0–20

Ground water moves and is stored within the pores and fractures of soil or rock. Porosity is therefore an important parameter in describing ground-water behavior and is quantitatively related to various other ground-water parameters, such as hydraulic conductivity, flow velocity, transmissivity, and storativity. Porosity will affect the zone of influence during sampling of a monitoring well by its influence on storativity. Because of its direct influence on flow velocity, the appropriate distance between downgradient monitoring well locations and potential ground-water contaminant sources is related to porosity.

A soil having a broad range of grain sizes generally has a lower porosity than a soil with uniform grain sizes (de Marsily, 1986). This is a result of the finer particles filling in the void spaces between the coarser particles for the soil with a broad range in grain sizes, thus lowering the overall porosity. Clay soils typically have a higher total porosity than sands, silts, and gravels. The porosity of rock is typically much less than that of soils except where the rock is highly weathered or partially dissolved. Karst formations may have very high secondary porosity due to solutioning of the carbonate.

Water content is related to the porosity by:

$$n = 1/(S/G_S w_w + 1) \qquad (10.6)$$

where:

n = the porosity, expressed as a decimal,
w_w = the gravimetric water content, expressed as a decimal fraction,
S = the degree of saturation, if known, typically assumed to be 1 for saturated soils, expressed as a decimal fraction, and
G_S = the specific gravity of the soil solids or rock.

Dry bulk density is related to porosity by:

$$n = 1 - \gamma_d/(\gamma_w G_s) \qquad (10.7)$$

The density of the sample can be determined as discussed earlier. Goodman (1980), describes several methods for laboratory determination of porosity.

A word of caution is warranted relative to porosity. Porosity is used to estimate the saturated flow velocity ($\bar{v}$) of water within pore spaces by the relationship:

$$\bar{v} = Ki/\tilde{n} \qquad (10.8)$$

where:

K = hydraulic conductivity, and
i = hydraulic gradient in the direction of the mean seepage velocity $\bar{v}$, and
$\tilde{n}$ = effective porosity.

The porosity used in Equation 10.8 is the effective porosity, $\tilde{n}$, which always has a lower value than the total porosity. Some water in pore spaces may be held onto soil particles by molecular binding forces (Mitchell, 1976). The soil may contain dead-end pores or unconnected pores which contain water, but through which no water flow is occurring. Therefore, caution should be exercised in selecting effective porosities. Effective porosity is difficult to measure and is typically selected by intuition or experience. Tracer experiments can be used to estimate effective porosities but the procedure is fraught with difficulties. When porosity is very low, laboratory errors may become significant. Field variation in porosity, however, may exceed laboratory errors and suggests multiple in situ test sites.

HYDRAULIC CONDUCTIVITY AND PERMEABILITY

The rate of water movement through a soil was first described mathematically by Darcy (1856). By studying the flow of water through sand columns, Darcy developed a relationship between the filtration velocity, the hydraulic gradient, and a coefficient, K, which has come to be known as the hydraulic conductivity. K is a function of both the medium through which the fluid is moving, and of the fluid itself. In many engineering texts, K is also known as the coefficient of permeability. As a result, the two terms are used interchangeably in hydrogeologic applications. Hydraulic conductivity is expressed in units of length per time, such as meters per second (m/s) or feet per day (ft/day).

Another term, intrinsic permeability, k, is used to describe the part of K that depends only on the medium in which a fluid is flowing. Intrinsic permeability has the units of length squared, such as cm^2 or mm^2, or the darcy ($0.987 \times 10^{-12} m^2 = 1$ darcy).

For granular porous media, Darcy's law can be written as:

$$v_s = Ki \tag{10.9}$$

where:

v_s = specific discharge in units of length per time,
K = the hydraulic conductivity in units of length per time, and
i = dimensionless hydraulic gradient in the direction of v_s.

This definition has been modified to describe flow in a fracture (Louis, 1974). Table 10.4 presents typical hydraulic conductivity values for various geologic materials. Generally, the finer the soil particle size, the lower the hydraulic conductivity value. The difference in K ranges between silts or clays and sands is a result of the smaller effective pore sizes in clays and silts than in sands. Soils which contain a broad range of grain sizes, such as a glacial till, typically have lower K values than a uniformly sized soil, such as a beach sand. Darcy's law

Table 10.4. Typical Hydraulic Conductivities.

Geologic Material	Range of K (m/s)
Coarse gravels	10^{-1}–10^{-2}
Sands and gravels	10^{-2}–10^{-5}
Fine sands, silts, loess	10^{-5}–10^{-9}
Clay, shale, glacial till	10^{-5}–10^{-13}
Dolomitic limestones	10^{-3}–10^{-5}
Weathered chalk	10^{-3}–10^{-5}
Unweathered chalk	10^{-6}–10^{-9}
Limestone	10^{-3}–10^{-9}
Sandstone	10^{-4}–10^{-10}
Unweathered granite, gneiss, compact basalt	10^{-7}–10^{-13}

is the cornerstone for evaluating ground-water flow behavior. However, the relationship is valid only as long as the velocity remains within a particular range of values. As the hydraulic gradient is increased, the water velocity increases and friction loss within the pores or fractures correspondingly increases. This phenomenon is analogous to flow through a pipe. Above a critical velocity, frictional losses are no longer linearly related to i, and Darcy's law must be modified or it becomes invalid. Some authors have suggested that an upper limit to the applicability of Darcy's equation for porous media be established by relating the velocity to a Reynold's Number (Bear, 1972). Reynold's Number for a porous medium can be defined as:

$$R_e = \bar{v}d\rho/\mu \qquad (10.10)$$

where:

$\bar{v}$ = mean velocity of water in pores in units of length per time,
ρ = fluid unit density in units of mass per volume,
μ = the viscosity of the fluid in units of mass per time-length, and
d = mean diameter of the pores as estimated from the effective grain size diameter in units of length.

In porous media, d is typically selected as the particle size for which 10% of the sample is smaller. In fractured media, d becomes the fracture width. Darcy's law is considered valid up to an R_e of between 1 and 10. Between an R_e of 10 to 100, turbulent flow begins, and beyond 100, turbulence predominates and Darcy's law is invalid.

There also appears to be evidence, although some is conflicting, that for clay or other fine-grained materials Darcy's law may be invalid for very low gradients (Jacquin, 1965a, b). Desaulniers et al. (1986) performed a field investigation in a thick clayey glacial till that supports the concept of threshold gradients. A relationship for Darcy's law incorporating a threshold gradient is suggested as shown in Figure 10.2. Below a particular threshold gradient, i_o, the hydraulic conductivity may be essentially zero for certain materials. For this case, Darcy's law is revised to:

$$v_s = K(i - i_2) \text{ for } i > i_1 \tag{10.11a}$$

$$v_s = 0 \qquad \text{for } i < i_o \tag{10.11b}$$

The value of K varies with the type of fluid flowing within the soil or rock and depends on the viscosity and density of the fluid, such that:

$$K = k\rho/\mu = kg/v \tag{10.12}$$

where:

ρ = the fluid density in units of mass per volume,
μ = the viscosity of the fluid in units of mass per time-length,
g = the acceleration of gravity, e.g., 32 feet/sec^2,
v = kinematic viscosity in units of area per time, and
k = intrinsic permeability of the media through which the fluid is flowing in units of square length.

Both density and viscosity are a function of temperature. The effects of temperature in density and viscosity are shown in Table 10.5. Of the two parameters, viscosity is the more sensitive to temperature changes.

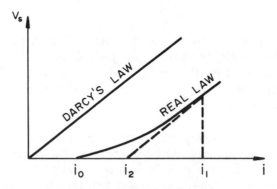

Figure 10.2.   Darcy's law at small hydraulic gradients.

If the hydraulic conductivity of a media is known at one temperature, its value can be calculated at different temperatures by using the above relationship. If laboratory determinations of K using water of a known temperature are available, the laboratory results can then be used to estimate the field hydraulic

Table 10.5.   Variation of Properties of Pure Water with Temperature.

Temperature (°C)	Density (gm/cm³)	Viscosity (X 10^{-2} dyne-sec/cm²)
0	0.99987	1.7921
1	0.99993	1.7313
2	0.99997	1.6728
3	0.99999	1.6191
4	1.00000	1.5674
5	0.99999	1.5188
6	0.99997	1.4728
7	0.99993	1.4284
8	0.99988	1.3860
9	0.99981	1.3462
10	0.99973	1.3077
11	0.99963	1.2713
12	0.99952	1.2363
13	0.99940	1.2028
14	0.99927	1.1709
15	0.99913	1.1404
16	0.99897	1.1111
17	0.99880	1.0828
18	0.99862	1.0559
19	0.99843	1.0299
20	0.99823	1.0050
21	0.99802	0.9810
22	0.99780	0.9579
23	0.99756	0.9358
24	0.99732	0.9142
25	0.99707	0.8937
26	0.99681	0.8737
27	0.99654	0.8545
28	0.99626	0.8360
29	0.99597	0.8180
30	0.99567	0.8007
31	0.99537	0.7840
32	0.99505	0.7679
33	0.99473	0.7523
34	0.99440	0.7371
35	0.99406	0.7225
36	0.99371	0.7085
37	0.99336	0.6947
38	0.99299	0.6814
39	0.99262	0.6685
40	0.99224	0.6560

conductivity at a different ambient temperature by an adjustment for viscosity, ignoring the slight change in density. The expression

$$K_f = (\mu_l/\mu_f)\, K_l \tag{10.13}$$

uses K_l from the laboratory determination; K_f is the field estimation, and μ_f and μ_l are the respective viscosities at the field and laboratory temperatures. In situ hydraulic conductivity testing corrections are rarely necessary because ground-water temperatures at depths of up to 200 feet from the ground surface seldom vary more than about 2°C from the average annual air temperature.

Other factors also affect the magnitude of K; these factors include, but are not limited to, bulk density, grain-size distribution, relative fraction of silt or clay, and, in fractured soil or rock, fracture width and frequency.

Laboratory Determination of K

Two laboratory methods are used for measuring hydraulic conductivity—the falling-head and constant-head permeameter test methods. The apparatuses shown in Figure 10.3 illustrate both the falling-head method and the constant-head method. In both tests, water moves through a test specimen under the influence of gravity alone; in both tests, the specimen is placed in a tube or cylinder and is usually remolded in the process of placing it into the cylinder. If an undisturbed sample is placed into a permeameter, a means of assuring no leakage along the boundary between the sample and the cylinder must be devised. This is difficult to accomplish with certainty in practice. Rubber membranes, silicon, or wax have been used to seal the sides of specimens.

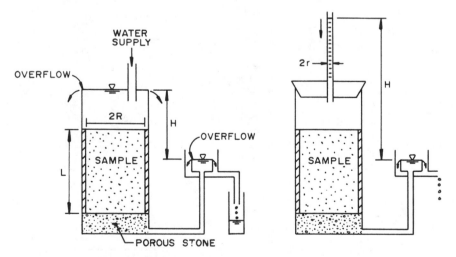

Figure 10.3.   Constant head (left) and falling head (right) permeameters.

In the constant-head test procedure, a known rate of water is allowed to pass through the specimen under a controlled hydraulic head or gradient condition. The hydraulic conductivity can then be computed by using Darcy's law:

$$K = QL/AH \tag{10.14}$$

where:

H = total hydraulic head difference between the ends of the soil specimen in units of length,

L = length of the soil specimen,

A = cross-sectional area of the soil specimen perpendicular to the flow direction, and

Q = flow rate in units of volume per unit time.

In the falling-head procedure, the rate of fall of the water level in a tube elevated above the top of the specimen is monitored. The head across the specimen is measured at two different times and inserted into a modified form of Darcy's law:

$$K = \frac{aL}{A(t_2 - t_1)} \ln (h_1/h_2) \tag{10.15}$$

where:

a = cross-sectional area of the water level monitoring tube,

A = cross-sectional area of the soil sample,

L = length of the sample in the direction of flow, and

h_1 and h_2 = the hydraulic head, in units of length, across the specimen at times t_1 and t_2, respectively.

The falling-head method is generally applicable for materials with hydraulic conductivities ranging from 10^{-7} to 10^{-3} cm/sec. The constant-head method is generally applicable to materials with hydraulic conductivities ranging from 10^{-3} to 10^{-1} cm/sec.

For materials with very low hydraulic conductivities, a larger hydraulic head difference is required in order to move water through the specimen in a reasonable time for laboratory measurement. Low-permeability material testing is typically carried out in a high-pressure permeameter or a triaxial cell apparatus. Figure 10.4 illustrates a schematic diagram of a triaxial permeameter. The sample is placed into the permeameter and a differential fluid pressure, i.e., hydraulic head, is placed across the sample. This apparatus has the advantage of creating a back-pressure condition which dissolves gas bubbles within the specimen (Bishop and Henkel, 1962). The use of extremely high gradients may cause Darcy's law to become invalid due to turbulence and also risk hydraulic piping along the sides of the specimen.

The triaxial cell method has the additional advantage of the specimen being placed inside a rubber membrane which is confined by fluid compression to the walls of the specimen. The rubber membrane helps to minimize the leakage along the sides of the specimen—a problem that is inherent to other permeameter

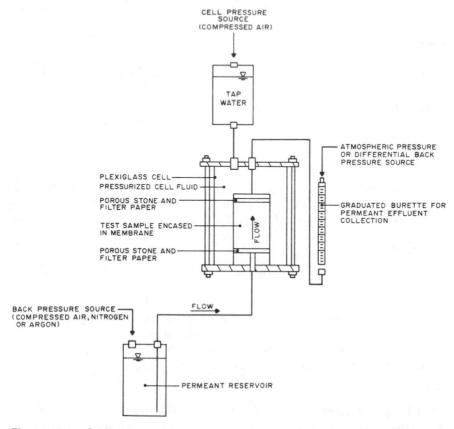

Figure 10.4. Schematic of a constant head triaxial cell permeameter.

methods. Dye can be injected into the test specimen at the end of the test to check for sidewall leakage or piping.

Soil samples are sometimes remolded prior to testing in a triaxial cell using a Harvard miniature mold that has a diameter of about 1.5 inches. The specimen is remolded using a tamping rod attached to a spring so that a constant force can be delivered with each stroke of the rod. Soil samples, remolded using this or other compaction methods (ASTM Test Method D 698 and D 1557), can be trimmed and placed in a triaxial cell. The triaxial or high-pressure permeameters typically are useful on soil or rock with hydraulic conductivities in the range of 10^{-10} to 10^{-4} cm/sec.

Laboratory testing for hydraulic conductivity has several limitations. One major limitation is that a sample with dimensions on the order of a few inches may not be representative of the in situ soil. A correction may have to be applied to stony soils because stones within a soil are usually removed prior to remolding and placement of the specimen into a permeameter. In such cases, soil particles should not exceed one-third the diameter of the test specimen. Another major limitation is that remolding the soil may remove natural structure, such as root holes, bedding structure, and fissures, which may control the conductivity of the soil in its natural setting. Finally, saturation is critical to interpretation of the test results. The degree of saturation has been shown to have a significant effect on the observed hydraulic conductivity (Freeze and Cherry, 1979).

An advantage of laboratory methods for determining hydraulic conductivity, which is difficult to duplicate in the field, is testing of an undisturbed soil sample at various orientations. For example, in a horizontally bedded soil, if flow is forced through a vertically oriented sample, the resulting flow is a measure of K across the bedding. However, if the same sample is trimmed and oriented so that flow during the test is parallel with the bedding planes of a soil, a measurement of the hydraulic conductivity at right angles to the first value can be obtained. If a sample is oriented at an angle other than perpendicular or parallel with the bedding planes, the value must be resolved into the directional components of hydraulic conductivity (Bear, 1972).

When using the falling- or constant-head methods, it is best to prepare remolded test specimens from slightly moist soils. After the specimen is in place, filling of the permeameter cylinder with water should begin from the bottom of the sample to displace entrapped air. Flushing the sample with carbon dioxide prior to flooding with water will improve the rate of saturation. Oven-drying of natural soils which contain silt, clay, or organic matter prior to specimen preparation and testing may limit the ability to saturate the specimen and modify the structure of the soil to such a degree that an accurate measurement of hydraulic conductivity is unlikely. The preferred saturation method is using back pressure (Black and Lee, 1973). This can be easily accomplished in a triaxial cell or specially constructed constant-head test cylinders.

Estimation of K from Soil Grain Size

Hydraulic conductivity can also be inferred from the grain-size distribution of the soil. Hazen (1911) empirically related the effective particle size to hydraulic conductivity, such that:

$$K = Cd_{10}^2 \qquad (10.16)$$

where:

   K = hydraulic conductivity in cm/sec,
   d_{10} = article size (measured in mm) below which ten percent of the cumulative sample has a smaller size, and
   C = constant which ranges from 1 to 1.2.

This method was developed for estimating the hydraulic conductivity of sand filters. Consequently, its use is generally limited to uniformly graded sands, that is sands with a uniformity coefficient of less than 5.0 (Hazen, 1911).

Fair and Hatch (Todd, 1959) proposed another method for estimating hydraulic conductivity which utilizes grain-size data from the entire distribution curve. This method is useful for sandy soils with minor amounts of silt and clay (generally less than 20%). The method assumes that hydraulic conductivity is related to the shape of the grain-size curve and grain characteristics by an empirical mathematical regression:

$$K = \frac{\rho g}{\mu} \left[\frac{n^3}{(1-n)^2} \right] \left[m \left(\frac{\Theta}{100} \sum \frac{P}{d_m} \right)^2 \right]^{-1} \tag{10.17}$$

where:

- K = hydraulic conductivity in units consistent with units of p, g, and d_m,
- μ = viscosity of the fluid in units of mass per time-length,
- ρ = fluid density in units of mass per unit volume,
- m = packing factor, 5,
- Θ = sand grain shape factor; 6.0 for spherical grains and 7.7 for angular grains,
- n = porosity, expressed as a decimal,
- P = percentage of sand held between adjacent sieves,
- d_m = geometric mean of rated sizes of adjacent sieves in units of length, and
- g = acceleration of gravity, e.g., 32 feet/sec².

This equation is dimensionally correct for any consistent set of units.

The Hazen estimation methods do not account for density effects, which typically cause much less variation in hydraulic conductivity than spatial variation of grain size distribution. Powers (1981) describes a method for estimating K developed from grain-size analysis and in situ density of the sand. From the grain-size analysis, the mean grain size, d_{50}, and the uniformity coefficient, C_u, must be determined (Lambe, 1951). The in situ density of the soil can be measured or estimated from standard penetration test results (Gibbs and Holtz, 1957). Using this information, the hydraulic conductivity is estimated from the charts shown in Figure 10.5. This method is useful for sands and gravels.

Slug Tests

In situ slug tests in wells or piezometers are popular for hydraulic conductivity testing in both soil (overburden) and rock. Slug tests involve removing, adding, or displacing a quantity of water in a well or piezometer and monitoring the change in water level with time. In vertically oriented wells or piezometers, this method provides a measure of the horizontal hydraulic conductivity.

The slug test is similar to a falling-head laboratory test in that the rate of water-level decline or increase, as the water level attempts to equilibrate with natural piezometric conditions, is a function of the hydraulic conductivity of the soil and the geometry of the well or screened interval.

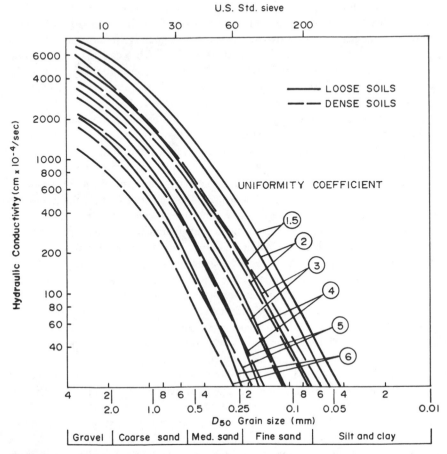

Figure 10.5. Estimate of hydraulic conductivity of sands and gravels.

In very permeable formations, it may be impossible to measure water-level changes with time because the water-level equilibration is almost immediate. The use of an electronic data logger, with a pressure transducer or strip chart recorder, can facilitate the collection of data in this situation. It is important to note that in some cases the hydraulic conductivity of a formation may be so great (as it is in gravels or limestone solution cavities), that head losses caused by construction of the well or piezometer, rather than the formation's actual hydraulic conductivity, may control the rate of water level change. Testing in formations of very low hydraulic conductivity may require long periods of data collection (e.g., days or weeks) or a pressure-test method (see Pressure Tests section).

In situ well or piezometer test methods can be modified so that a constant rate of water is added or extracted from the casing. This is similar to a constant-head laboratory test and requires knowledge of the flow rate, the head differential from the background piezometric condition, and the well geometry.

Slug test results are often analyzed using the method of Hvorslev (1949b). This method allows for various well and aquifer geometries, but is based on a quasi-steady-state solution of the flow equations. Consequently, it is only valuable as an approximation; difference of up to a factor of 0.3 to 0.5 can be observed relative to the Cooper et al. (1967) method. The lack of conceptual rigor limits the accuracy of this method. Solutions for various well and aquifer geometries are also available (Lambe and Whitman, 1969; Hvorslev, 1949b).

The method developed by Cooper et al. (1967) is based on the analytical solution of transient flow equations. This method was developed for wells that fully penetrate an aquifer, and has been adapted to a type-curve matching procedure. Hydraulic conductivity and storage coefficient can be estimated using this method of analysis. The method solves for transmissivity (see Transmissivity section), which, if the aquifer thickness is known, can be used to calculate the hydraulic conductivity. If the well screen partially penetrates the aquifer, the screen length can be substituted for the aquifer thickness and an approximation of the hydraulic conductivity can be obtained. Additional type curves have been developed by Papadopulos et al. (1973) to cover a broader range of storage coefficients and well sizes. This method was updated by Bredehoeft and Papadopulos (1980) for estimating hydraulic conductivity in low-permeability formations.

The Cooper et al. (1967) method utilizes type curves, as shown in Figure 10.6, to match field data of water level above or below static, h, at any time, t, relative to the initial water level above or below static, h_o, at t = 0, to the computed curves presented in dimensionless parameters, $r_w^2 S/r_c^2$ and $2Tt/r_w^2$. The well screen radius, r_w, and the inside well casing radius, r_c, are needed to solve for storativity, S, and transmissivity, T. These tests can have a zone of influence of up to several hundred feet (Sageev, 1986) depending on the properties of the soil or rock and the slug volume. Typically, the greater the slug volume and/or the lower the storativity, the greater the zone of influence. Herzog and Morse (1986) point out that this method may require very long data collection periods for low-conductivity materials. However, for formation hydraulic conductivities of 10^{-6} cm/sec or greater, testing can typically be completed in one day.

The method of Nguyen and Pindar (1984) also allows estimation of both the storage coefficient and the hydraulic conductivity. The procedure is slightly more complicated than the Cooper et al. (1967) method. However, the procedure incorporates constant well discharge or recharge, as well as transient water-level changes. Herzog and Morse (1986) point out that this method has advantages over the Cooper et al. (1967) method. The method was designed for wells screened over only a portion of the aquifer thickness (i.e., partially penetrating wells), which is the more frequent case in practice; the method typically requires less than one day of field time to complete. This method can also be used for angled boreholes and, therefore, can be used to evaluate directional effects on the observed hydraulic conductivity. However, Herzog and Morse (1986) caution about careful measurement of water levels, suggesting that accurate recordings of the hydraulic head be taken by using, for instance, pressure transducers with an

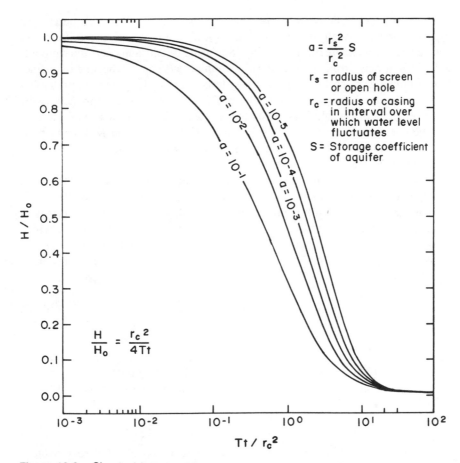

$$a = \frac{r_s^2}{r_c^2} S$$

r_s = radius of screen or open hole

r_c = radius of casing in interval over which water level fluctuates

S = Storage coefficient of aquifer

$$\frac{H}{H_o} = \frac{r_c^2}{4Tt}$$

$a = 10^2$

$a = 10^{-1}$

$a = 10^{-5}$

$a = 10^{-4}$

$a = 10^{-3}$

Figure 10.6. Slug test type curves.

electronic data logger or strip chart recorder. If discrete water-level measurements are made using a hand-held meter, erratic results can lead to interpretation difficulty.

Several investigators have analyzed slug-test type curves for fractured media (Gringarten and Ramey, 1974; Karasaki, 1986). These curves often do not have unique solutions unless information on fracture characteristics is known. Alternately, the fractured media can be assumed to be equivalent to a porous medium and the methods described above are applicable. Karasaki et al. (1988) examined various fracture conditions local to the test well.

Skin effects involve locally increasing the conductivity near the well by opening fractures (positive skin) or decreasing the conductivity near the well (negative skin) by filling natural fractures with drilling mud or drill cuttings. Negative skin effects can make the apparent measured hydraulic conductivity less than the actual in situ hydraulic conductivity. Negative skin effects can also be created by disturbing a naturally layered soil and forming a more or less uniform soil zone along the wellbore during drilling. Smearing of silts and clays during borehole drilling may create a negative skin. Skin effects can be checked in unfractured formations by performing laboratory tests of the aquifer material, preferably using undisturbed samples. One of the problems with most skin models is the nonuniqueness of the solutions. Knowledge of the skin properties and/or aquifer storage coefficient can make the solution possible. Faust and Mercer (1984), Ramey and Agarwal (1972), Moench and Hsieh (1985), and Sageev (1986) have examined skin effects.

Either rising-head or falling-head tests can be performed. However, the values for the two types of tests in a single well can vary by up to a factor of 100; typically the falling-head result is greater than the rising-head result. The errors in measurement are believed to be associated with well installation effects, which cause a disturbance of the aquifer material around the borehole. Milligan (1975) suggests that the "best" hydraulic conductivity estimate, based on the two values, is obtained by:

$$K = (K_{RH} K_{FH})^{1/2} \qquad (10.18)$$

where:

K_{RH} = hydraulic conductivity as determined by rising-head method, and
K_{FH} = hydraulic conductivity as determined by falling-head method.

Packer Testing

Hydraulic conductivity is commonly measured in consolidated rock using a packer test. This method typically gives a measure of the horizontal conductivity. The arrangement for performing packer tests in boreholes is illustrated in Figure 10.7. This method is used for testing rock in which the walls of the borehole are stable. Inflatable packers are generally used to isolate the interval of the borehole to be tested. A single packer is used to test a section of borehole between the bottom of the boring and the packer location. Typically, single packer testing is performed as the hole is advanced. After drilling to a desired depth, the packer is inserted at a selected depth above the bottom of the borehole. The packer is then inflated using water or a gas, and water is injected in the borehole for a given length of time to test the "packed-off" portion of the hole. After the test, the packer is removed and the hole advanced in depth. This procedure can be repeated as many times as desired.

With the use of two packers, any position or interval along the hole can be selected for testing. The interval may be preselected, or it may be selected based on core descriptions or observed fractures. The two-packer system is usually inserted into the borehole after it has been completed. The portion of the fill tube between the two packers contains openings through which water can flow once

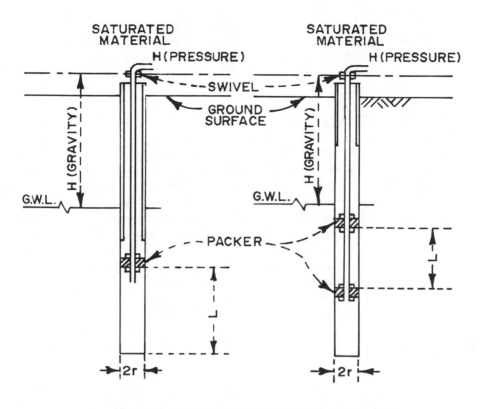

TESTS MADE DURING DRILLING   **TESTS MADE AFTER HOLE IS COMPLETED**

SATURATED MATERIAL

H (PRESSURE)

SATURATED MATERIAL

H (PRESSURE)

SWIVEL

GROUND SURFACE

H (GRAVITY)

H (GRAVITY)

G.W.L.

G.W.L.

PACKER

L

2r

2r

L

TOTAL H = H(GRAVITY) + H(PRESSURE)

Figure 10.7. Insitu packer testing.

the packers have been inflated. Water is forced under pressure into the fill tube and the test is run for the desired length of time. When testing with two packers, the usual procedure is to begin testing at the bottom of the hole and then proceed upward. This practice reduces the likelihood that testing may reduce the stability of the hole walls, which could trap the packer test equipment in the borehole. Depending on the flow rate, head losses within the piping system may be critical to the interpretation of the test results. Skin effects due to drilling, particularly when drilling with mud, can also critically influence the results of the packer testing (Faust and Mercer, 1984). Skin effects can also be caused by core removal, which causes stress changes that can close fractures (Neuman, 1987).

Other types of packers, including compression packers and leather cups, are also used. However, these packers are prone to leakage, which may cause an erroneous interpretation of test findings and result in an overestimate of hydraulic conductivity. Inflatable packers generally yield the best results because they can form a tight seal against the borehole wall, even if the hole is rough-walled or out-of-round.

Before performing the test, the borehole should be cleaned of any cuttings or drill fluids. This involves surging or bailing the hole. The presence of drillers mud or failure to clean the borehole may result in a lower measured hydraulic conductivity than the actual hydraulic conductivity of the rock.

The length of the packer test section is generally governed by the character of the formation. Typically, a 5- or 10-foot length is used. However, variations from these standard lengths can be accommodated by modifying the interval between packers. Depending on the hydraulic conductivity of the rock, the test interval may have to be lengthened or shortened in order to obtain a value for the given conditions and the test duration and pressure. The duration of the test should be sufficient to provide measurable flow volumes. The test section should never be shortened to the point at which the ratio of interval length to hole diameter is less than 5, if the standard horizontal flow equations are to be used for analysis. Test results may become invalid as vertical flow becomes an important component.

It is important to measure the piezometric conditions in the packed-off section of the hole prior to beginning the test. If not available prior to the test, the piezometric conditions must be estimated to set a test pressure. Test pressures should not exceed 0.5 psi per foot of depth as measured from the ground surface to the top of the test section. The purpose for not exceeding this pressure is that excessive hydraulic pressure can induce hydraulic fracturing, thus causing an increase in the measured hydraulic conductivity relative to the undisturbed hydraulic conductivity. The test pressure should be measured by a gauge located as near the well head as possible. Thus, the pressure at the well head is observed without losses due to meters or pumps or turbulence close to the pump. When multiple tests are being conducted in the same borehole at various depths, residual pressures may occur from previous testing and may have to be accounted for in the testing. Preferably, residual pressures should be allowed to come to equilibrium prior to any subsequent testing.

Typically, packer tests are run for a period of from 15 minutes to two hours. Generally, the lower the hydraulic conductivity of the formation, the longer the test. During the first 5 minutes of the test, readings of the flow meter monitoring the amount of injected water should be taken every thirty seconds; thereafter, readings should be made at 5-minute intervals for the remainder of the test period. During the initial portion of the test, an expansion of the packer device may cause a flow of water that is not indicative of formation hydraulic conductivity. Normally the test is run in two parts at two different pressures. For example, during the first part of the test, the borehole could be pressurized at 15 psi for a period of 15 to 30 minutes. After the flow rate has stabilized, the pressure may be increased to 30 psi during the second part of the test.

Packer tests are often performed using pumps supplied on the drill rig. Depending on the type of pump, water flows of 25 to 250 gallons per minute can be achieved. For very low flow rates, a meter calibrated in fractions of a gallon is necessary. However, where high flow rates are expected, a meter calibrated in 5 to 10 gallon-per-minute increments may be satisfactory. Typically, totalizer-type meters are used rather than instantaneous flow meters. However, both meter types have been used in practice. Occasionally, in very low hydraulic conductivity formations, no change in volume will occur other than that caused during the initial pressurization of the packer system. In these cases, the length of the test can be increased to allow a measurable quantity of water to flow into the rock. An estimate of the upper limit of the hydraulic conductivity can be calculated if, even after extending the test period, no flow is observed.

The data required for computing the hydraulic conductivity include the borehole radius, the pressure at the well head, the depth of the borehole and the packer(s), the flow rate, and the height of the well head above the ground surface. The diagram shown in Figure 10.7 illustrates the information necessary to calculate the hydraulic conductivity. The formulas for calculating hydraulic conductivity are:

$$K = \frac{Q}{2\pi LH} \ln \frac{L}{r} \; ; L > 10r \tag{10.19}$$

$$K = \frac{Q}{2\pi LH} \sinh^{-1} \frac{L}{2r} \; , \; 10r > L > r \tag{10.20}$$

where:

   K = hydraulic conductivity in units of length per unit time,
   Q = constant rate of flow into the hole in units of volume per unit time,
   L = length of the portion of the hole tested,
   H = total differential head of water in units of length,
   r = radius of hole tested in units of length,
   ln = natural logarithm, and
sinh^{-1} = inverse hyperbolic sine.

These formulas are valid for calculating hydraulic conductivity when the thickness of the stratum tested is at least 5L, and they are considered to be more accurate for tests below the water table than above it. Multiple pressure tests can be utilized to evaluate potential problems, such as leakage, with the packer testing. Tests conducted at three or four different pressures, increasing from zero and decreasing back to zero, are often done.

Pressure Tests

In formations of very low hydraulic conductivity (i.e., less than 1×10^{-7} cm/sec), pressure tests, sometimes called pulse tests, are more appropriate.

Various investigators have examined pressure tests in low-permeability formations (Wang et al., 1978; Neuzil, 1982; Forster and Gale, 1981; Neuman, 1987). These tests are often analyzed using the type-curve procedures.

In a pressure test, a packer system is placed into the borehole and the packer(s) inflated. An increment of pressure is applied to the zone between the packers. The decay of pressure is monitored and plotted versus time. The rate of pressure decay is related to the storage coefficient and the hydraulic conductivity of the formation. Pressure response data are typically collected using pressure transducers with electronic data loggers or strip-chart recorders.

This test is generally used in low hydraulic conductivity rock formations and must compensate for skin effects and packer adjustment during the application of pressure. An understanding of the presence and orientation of fractures is necessary to select an appropriate type curve to analyze test data. The skin effect within the borehole may be critical in evaluating in situ hydraulic conductivity test data. The presence of a drilling mud filtercake or the smearing of fine-grained material along the borehole walls, created by the drilling operation, may result in test data that reveal an apparent low hydraulic conductivity. Drill cuttings in an unmudded hole may also create skin effects by washing into joints or fissures or coating the sides of the hole. This may be particularly true with air-rotary drilling in fractured rock. The skin effect may be impossible to detect, as indicated by simulation studies (Faust and Mercer, 1984).

Tracer Tests

Single-well and multi-well tracer tests have been performed to determine horizontal hydraulic conductivity in both unconsolidated deposits and rock.

In the single-well test, a tracer is injected into a well, the tracer moves radially away from the well, and then the well is pumped to recover the tracer. The tracer can consist of an easily-measured, nonreactive, nondecaying solute. Tritium, visually identifiable dyes, or electrolyte (e.g., chlorides) have been used. The test can involve multilevel injection points to measure formation response at various depths. Alternately, in a multiwell test, sampling wells are placed radially away from a continuously screened injection well and monitored to detect the arrival of tracer. Multi-well tests involve injection of a tracer into one well and extraction by sampling from other wells. Similar to the single-well test, the wells may have multilevel ports for injection and sampling.

These tests require careful planning and experience in interpretation of the results. They can be time-consuming and expensive and, therefore, often are used only on projects in which budgets can accommodate the effort.

Vertical Hydraulic Conductivity Using Packer Tests

Burns (1969) has proposed a method for estimating vertical and horizontal hydraulic conductivity in homogeneous granular rock using a procedure similar to a standard packer test. The method involves grouting a well casing into the borehole and then perforating the casing at two depths separated by several feet (see Figure 10.8). A packer is inflated between the two sets of perforations to hydraulically separate them. Greater assurance of a seal is obtained through the use of two packers at the end of the interval containing the perforations. Using pressure transducers above and below the packers, the pressures in the borehole are monitored as a known flow volume is injected through the perforations above the packer. The test is stopped when the lower transducer indicates a pressure response at least 10 times the sensitivity of the gauge. The results are calculated using graphical techniques (Burns, 1969).

Packer tests in cased wells rely upon the quality of grouting between the casing and borehole wall. Leaks within the cavity between the perforations will result in an overestimation of the vertical hydraulic conductivity. In low hydraulic conductivity formations, the time required to reach background conditions, which is necessary prior to conducting the test, may be excessive unless a pressure test

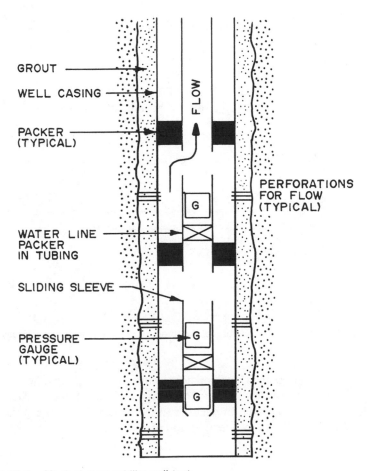

Figure 10.8. Vertical permeability well test.

is performed. Interpretation of the calculated conductivity may be difficult in a fractured medium. Tests similar to the one described above have been developed by Prats (1970) and Hirasaki (1974).

Vertical Hydraulic Conductivity from Seepage Pits

Vertical hydraulic conductivity has also been measured where the water table is very deep using seepage pits (see Figure 10.9). One method relies on the development of a steady-state or near steady-state vertical seepage pattern. The method utilizes an equation describing the theoretical seepage for a given pit geometry:

$$Q = K_v(B + AH)L \qquad (10.21)$$

where:

Q = steady-state flow rate in units of volume per time,
L = length of trench in units of length,
B = pit base width in units of length,
H = pond depth in pit in units of length,
K_v = vertical hydraulic conductivity in units of length per unit time, and
A = varies from 2 to 4 depending on geometry of trench.

Theoretical development of this equation can be found in Polubarinova-Kochina (1962) and Harr (1962).

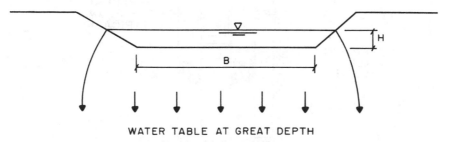

WATER TABLE AT GREAT DEPTH

Figure 10.9. Seepage pit for insitu measurement of vertical hydraulic conductivity.

During use of the method, precipitation and evaporation must be accounted for because the test may last several days. Also, capillary forces in the unsaturated zone may be important for small basins in which the capillary rise is on the same order as the basin depth. Capillarity at the front of the downward moving seepage face may also affect the test results during the period near the start of the test.

The test method involves excavating a shallow pit. During excavation, smearing of the soil on the pit walls must be avoided, as this may reduce the apparent hydraulic conductivity value. Excavation by shovel is often required. A major advantage of this method is that the soil or rock is being tested in an essentially undisturbed condition. Therefore, if the soil has secondary permeability due to fissures or rootholes, the test will include their effects on the measured hydraulic conductivity.

Typical pit sizes range from a few square meters up to 100 square meters. The pit is filled with water, but water must be added so as not to erode the pit base and sidewalls, thus creating a suspension of fine soil particles that may settle and form a flow barrier on the base of the pit. The soil below the pond base can be instrumented with piezometers to assure that steady-state seepage has occurred. However, often the rate of seepage is used to determine if steady-state conditions are attained. The test must be monitored continually to maintain the desired water level in the pit. Minor fluctuations in the pit water level may not be important, but large deviations should be avoided. These tests may last several days or up to a week, depending on the soil type.

Vertical Hydraulic Conductivity from Pumping Tests

Vertical hydraulic conductivity of an aquitard can be obtained from pumping test data (see Transmissivity section). Vertical leakage, L, of an aquitard overlying an aquifer can be an important parameter in evaluating the direction of ground-water flow and in estimating the area of influence of a pumping well. This value is related to the vertical hydraulic conductivity and the thickness of the aquitard:

$$L = K_v M \qquad (10.22)$$

where:

K_v = average vertical hydraulic conductivity in units of length per time, and
M = vertical thickness of the aquitard in units of length.

Different methods have evolved for calculating L directly from pumping test data. The first analytical solution to a leaky aquitard problem, which ignored storage in the aquitard and held the piezometric surface constant in the aquitard, was by Hantush (1956). Hantush (1960) developed a method in which storage of the aquitard can be accounted for. Cooley and Case (1973) developed a method for a water table aquitard. Figure 10.10 illustrates the definition of leakage.

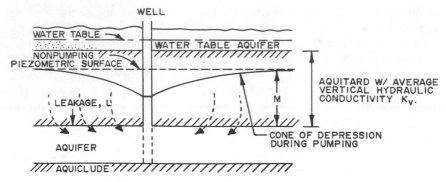

Figure 10.10.   Definition of vertical leakage.

DIRECT MEASUREMENT OF GROUND-WATER VELOCITY

The need to calculate the velocity of ground water is one reason that hydraulic conductivity is such an important parameter in hydrogeologic investigations. Ground-water velocity is used to determine how rapidly ground water or dissolved constituents in ground water are migrating in the ground-water system.

Tracer tests have been used to estimate in situ flow velocities. The method determines flow velocity, which is related to hydraulic conductivity by the relationship

$$\bar{v} = KI/\bar{n} \tag{10.23}$$

where:

$\bar{v}$ = mean ground-water particle velocity in units of length per time,
I = natural or induced hydraulic gradient expressed as a unitless decimal,
$\bar{n}$ = effective porosity of the formation expressed as a unitless decimal, and
K = hydraulic conductivity in units of length per time.

This equation can be used for both unconsolidated materials and highly fractured rock. The effective porosity is a measure of only that porosity which contributes to movement of water within the unconsolidated material or rock. It is commonly less than the total porosity due to closed-end pores and double-layer effects.

If the direction of ground-water movement is known, a nonreactive dye, electrolyte, or radionuclide tracer can be injected into the formation at one point, through a well, and its presence monitored at a downgradient point. The amount of fluid that is injected should be controlled to avoid altering the natural hydraulic gradient. Knowing the distance between the two wells and the breakthrough time of the tracer, the ground-water velocity can be calculated by dividing the distance by the breakthrough time. Furthermore, by knowing the head difference between the two wells and the effective porosity, the average hydraulic conductivity between the two wells can be estimated using Equation 10.23. The assumptions for a natural gradient tracer slug are that natural gradient is not disturbed, the tracer does not react with the geologic formation, and the conditions between the injection and monitoring wells are uniform so that the flow is directly from one well to the other. The use of dyes or tracers is less useful and more risky in fractured or karst terrains, where flow directions are controlled by unobservable fractures or solution channels and the actual direction of flow is difficult to estimate.

A recently developed electronic velocity meter, which is lowered down a well, can measure both flow direction and rate (Kerfoot, 1982). The meter has a radial array of temperature sensors which extend from the base of the downhole portion of the meter. A prong located in the center of the sensor array is heated by an internal battery source when the test begins. The surrounding sensors are then used to electronically sense temperature changes and calculate the direction and rate at which the heat is being carried by the moving ground water. Testing at various depths in a screened well may indicate a variety of flow directions. These may be due to aquifer heterogeneities or ambient currents within the well bore. The measured rate of ground-water flow, when the apparatus is used in a well, may also be affected by the converging flow at the well, due to the greater hydraulic capacity of the well relative to the geologic formation.

The borehole dilution method provides another means of measuring the ground-water velocity local to a well. This method involves isolating a section of borehole using packers. Between the packers an electrolyte or radioisotope solution is injected, but maintained at the same piezometric conditions as the surrounding formation. As groundwater flows through the well section it removes some of the solution causing a dilution with time. The rate of dilution has the relationship to groundwater velocity of:

$$\bar{v} = - \frac{W}{n\bar{\alpha}At} \ln \left(\frac{\leftarrow C}{C_0} \right) \qquad (10.24)$$

where:

$\bar{v}$ = estimated mean groundwater particle velocity near the well in units of length per time,

W = volume of isolated well segment,

C_0 = initial concentration of solute in units of mass per volume,

t = elapsed time from start of test,

C = solute concentration at t, in units of mass per volume,

A = vertical cross-sectional area of isolated well segment,

$\bar{n}$ = effective porosity of formation, expressed as unitless decimal, and,

$\bar{\alpha}$ = correction factor ranging from 1.5 to 3.

This procedure is limited to soils and rock where the seepage velocity is significant relative the natural diffusion potential of the solute. $\bar{\alpha}$ ranges depending on well design, but is typically about 2.0 for wells without a sand pack.

During the procedure, uniform mixing of the solute must be maintained so that sampling or measurement of the remaining concentration represents an average mix. This method is useful in identifying zones of higher seepage velocity where contaminants would migrate at a greater rate. The method, therefore, can be used to select wellscreen depths for monitoring well placement. The method is described in detail in Drost, et al., 1968.

SPECIFIC STORAGE AND SPECIFIC YIELD

Specific storage and specific yield are measures of the water released or gained by a respective decrease or increase in the piezometric level in an aquifer. They are dependent on the compressibility of the aquifer, the compressibility of water, and the porosity of the formation, whether rock or unconsolidated material. These two terms become important when evaluating transient aquifer behavior, such as determining what portion of an aquifer is contributing to a monitoring well screen during sampling.

Specific Yield

If an initially saturated soil or rock is allowed to drain under gravity alone, the water content will decrease to a certain value which is dependent on the grain size or fracture characteristics of the material. Small pores may not drain because the water retention forces are greater in small pores than in large pores. A clay or peat may drain very little, whereas a sand or gravel may drain almost completely. The amount of water remaining in the soil is a measure of its retention capacity. The amount of water that escapes by gravity drainage is defined as the specific yield or drainable porosity (see Figure 10.11). The specific yield is a measure of the volume of water gained or released from a unit surface area of aquifer due to a respective unit rise or decrease in water table level. Specific yield is expressed as a decimal or as a percent. Typical values of specific yield are presented in Table 10.6.

The specific yield (also known as drainable porosity) of a soil may be estimated in the laboratory by allowing the soil specimen to free-drain under gravity conditions in a moisture cabinet where the relative humidity is maintained at 100%. The difference between the water contents before and after drainage is an estimate of the drainable porosity. Specific yield or drainable porosity is also a function of time. For example, in a well-sorted sand, one investigation found that 40% of the drainage occurred after the first few hours, but drainage continued for a period of up to 2.5 years (de Marsily, 1986). Therefore, specific yield can vary in value for a soil depending on the period of interest.

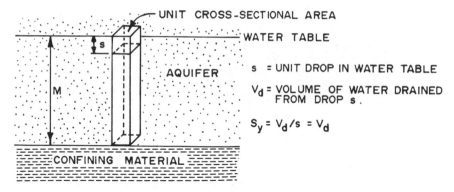

UNIT CROSS-SECTIONAL AREA

WATER TABLE

AQUIFER

s = UNIT DROP IN WATER TABLE

V_d = VOLUME OF WATER DRAINED FROM DROP s.

$S_y = V_d / s = V_d$

CONFINING MATERIAL

Figure 10.11. Definition of specific yield in unconfined aquifer.

Table 10.6. Representative Specific Yields.

Rocks	Specific Yield, %
Clay	1–10
Sand	10–30
Gravel	15–30
Sand and gravel	15–25
Sandstone	5–15
Shale	0.5–5
Limestone	0.5–5

Specific Storage

In confined aquifers, water is released by expansion of the water and compression of the aquifer voids under a decrease in piezometric pressure. The specific storage is a measure of the volume of water released from a unit volume of aquifer per unit decrease in piezometric head. Figure 10.12 illustrates the definition of specific storage. This is a result of: (1) an increase in effective stress between the aquifer particles or blocks when the water level is lowered; and (2) an expansion in the volume of water due to the decrease in pressure (see Walton, 1970 or Freeze and Cherry, 1979).

Storativity is equal to the specific storage divided by the saturated aquifer thickness. Specific storage is "unitless," whereas storativity has the units of reciprocal length, such as 1/ft or 1/m. The amount of water released by a confined aquifer for a unit decrease in head is usually much less than that for an unconfined aquifer. This is because the amount of water released by aquifer compression and water expansion is small compared to that which drains from a soil under the influence of gravity. Values for the specific storage typically range from 10^{-5} to 10^{-3}, which is typically a fraction of the specific yield. In situ determinations of specific storage and specific yield are obtained from pumping tests (see below). Specific

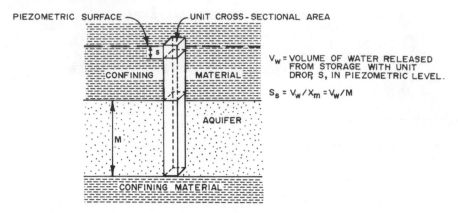

PIEZOMETRIC SURFACE   UNIT CROSS-SECTIONAL AREA

V_W = VOLUME OF WATER RELEASED FROM STORAGE WITH UNIT DROP S, IN PIEZOMETRIC LEVEL.

$S_S = V_W / X_m = V_W / M$

CONFINING   MATERIAL

AQUIFER

M

CONFINING MATERIAL

Figure 10.12. Definition of specific storage in confined aquifer.

storage can also be estimated from type-curve analysis of slug-test data (see Slug Tests).

TRANSMISSIVITY

A term related to hydraulic conductivity which is commonly used in aquifer evaluations, is transmissivity. Transmissivity, like hydraulic conductivity, is useful for calculating ground-water flow rates and recharge capacities of wells. This term represents the average water transmission characteristics over the entire aquifer thickness. Transmissivity was originally introduced by Theis (1935) as the product of hydraulic conductivity and saturated thickness for horizontally confined aquifer flow. Transmissivity, T, is defined as the average of all horizontal hydraulic conductivities at various depths multiplied by the vertical saturated thickness of the aquifer.

$$T = \bar{K}M \qquad (10.25)$$

where:

   T = transmissivity in units of length squared per unit time,
   M = saturated thickness in units of length, and
   $\bar{K}$ = average horiztonal hydraulic conductivity.

$\bar{K}$ represents an average horizontal hydraulic conductivity which may vary with horizontal orientation. Relating this definition to Darcy's law, transmissivity becomes the rate of flow under a unit horizontal hydraulic gradient through the entire thickness of an aquifer of unit width perpendicular to the direction of ground-water flow. Accordingly, units for transmissivity are in length squared per unit time, such as ft^2/day or m^2/sec. For convenience, transmissivity is sometimes expressed in units of gallons per day per foot.

Transmissivity can be determined from slug-test data (see Slug Tests) or aquifer pumping tests, or it can be estimated from laboratory data. In a uniform confined aquifer of constant thickness, transmissivity remains constant provided the piezo-metric level does not decline below the top of the aquifer. In an unconfined aquifer, transmissivity varies with the saturated thickness, M. As the water table raises or lowers, the transmissivity correspondingly increases or decreases. Further-more, the term loses its value for three-dimensional situations. Because trans-missivity carries with it the assumption of horizontal flow, if vertical seepage components exist, for example, near a partially penetrating well or in an uncon-fined aquifer with significant drawdown, the term losses its value in describing flow behavior.

Determination of Transmissivity by Pumping Tests

A pumping test is the most commonly used procedure for determining aquifer transmissivity, specific storage, and/or specific yield. Pumping tests are also useful in examining boundary effects from recharge zones (lakes and rivers) and low conductivity boundaries (rock walls, clay). But because a pumping test may in-fluence a relatively large volume of the aquifer some distance from the pumping well (as opposed to a retrieved sample specimen), the resulting aquifer proper-ties that are measured represent an ''average'' over that volume. The portion of the aquifer influenced by a pumping test can be varied by modifying the pumping rate and/or the length of the test. In fractured media, pumping tests can be used to evaluate the collective effect of many fractures.

A pumping test typically consists of a central pumping well and one or more nearby observation wells. The decline in the ground-water levels in observation wells is monitored as the central well is pumped at a constant or variable rate. The location and configuration of the pumping well and observation wells are dependent on aquifer properties. Common types of idealized aquifers include un-confined or water-table aquifers, confined aquifers, leaky aquifers, or multilayered aquifers. The local hydrogeology will determine what portion of the aquifer is screened by the pumping well and how the results should be interpreted.

Boundary conditions, such as an impervious zone or a recharging river, can influence the results of the testing and must be incorporated into the design of the pumping test. If useful estimates of transmissivity and storativity are desired, the pumping well should be located far enough away from boundaries to permit recognition of drawdown trends before boundary conditions influence the draw-down measurements. When more than one boundary is suspected to be present, it is desirable to locate the observation wells so that the effects of encountering the first boundary on well drawdowns are stabilized prior to encountering the second boundary condition. In the case of a recharge boundary, such as a lake or river, if the intent is to induce recharge from the river, then the pumping well should be located as close to the boundary as possible.

To assist in designing a pumping test, preliminary transmissivity (T) and storativity (S) values can be estimated using reasonable ranges of aquifer properties or laboratory test data. Estimates of these values will assist in selecting a pumping rate, observation well locations, and the pumping well location. Because the pumping well's cone of influence expands with time after the beginning of pumping, the distance to observation wells relative to the duration of the test should be considered. The cone of influence depends on the aquifer storativity and transmissivity and the discharge rate of the pumping well. Assuming typical T and S values, estimates can be made before the pumping test to determine the magnitude of drawdown at a particular observation well. The ability to measure the drawdowns and any outside influences, such as barometric pressure or tidal effects, will determine the desired amount of drawdown at a particular observation well.

Observation Well Positioning

Most pumping test analyses assume that the flow within the aquifer system is predominantly horizontal (Walton, 1970). This condition is not necessarily satisfied in close vicinity to a partially penetrating pumping well or in layered aquifers. The degree of vertical ground-water movement also depends on the length of the screen; if the well does not fully penetrate the aquifer, significant head losses due to vertical velocities or turbulence associated with converging flow may exist. In the case of partially penetrating wells, some investigators recommend a minimum distance between the nearest observation well and the pumping well of $1.5M$ $(K_H/K_V)^{1/2}$, where M is the thickness of the aquifer in units of length, and K_H and K_V are the horizontal and vertical hydraulic conductivities of aquifer, respectively. Other investigators suggest a minimum distance of at least two times the aquifer thickness between the pumping well and the nearest observation well to avoid the zone of vertically converging flow for a partially penetrating well. For a fully penetrating well, the convergence in the vertical direction is generally not a concern, and closer distances can be utilized. If the above conditions cannot be met, piezometers should be placed in both the upper and lower 15% of the aquifer, and the drawdowns in the piezometers averaged for computational purposes. If the aquifer is stratified, the observation wells can be screened over the same interval as the pumping well, assuming that flow is predominantly horizontal throughout the pumping depth. Multilayered aquifer analysis of pumping test data are available (Neuman and Witherspoon, 1969; Boulton and Streltsova, 1975).

The type of aquifer being tested also influences the distance from the pumping well to the observation wells. In a confined aquifer, the decline in the piezometric surface spreads rapidly because water is coming from storage which is a function of the aquifer compressibility. However, in an unconfined aquifer, the water moving to the pumping well is principally draining from the pore spaces of the aquifer material at the water-table position and is coming only in part from aquifer compressibility. This results in a slower expansion of the zone influenced by the pumping well. In general, observation wells in a confined aquifer can be spaced further from the pumping well than wells in an unconfined aquifer.

A common well array for a pumping test consists of four observation wells in two sets (2 wells per set), with one set placed orthogonally from the other. If anisotropic conditions are suspected, a single row of observation wells cannot be used to estimate the directional dependence of transmissivity. A minimum of three observation wells, none of which are on the same radial, are required to separate out the anisotropic behavior (Neuman et al., 1984).

The distance of the observation wells from the pump well depends on the considerations discussed previously and the anticipated boundary conditions. In an area in which complex boundary conditions are anticipated, additional observation wells may be required for the proper interpretation of the test results. The observation wells should be as small in diameter as feasible to minimize the response time. Pneumatic or electronic transducers placed in a sealed borehole are very useful because these instruments respond to pressure changes within the aquifer with virtually no flow equalization time.

When setting up the pumping test, the discharge from the pumping well should be located as far as possible from the pumping and observation wells. This will minimize the potential for the discharge water to infiltrate back into the aquifer and affect test results. Preferably, the discharge should be positioned in a stream, river, pond, or some other surface water body.

Pretest Data Collection

Water levels in the pumping well and observation wells should be measured several days in advance of the pumping test to determine the static water levels within the wells. Because the water levels in the observation wells will be compared with the respective prepumping levels, it is important that the initial water levels and any variability be incorporated into the analysis of the test. If the test is in an area affected by tides, it may be necessary to monitor the daily tidal cycle at each of the observation wells. Each observation well may react differently to the tidal effect. In a shallow water table aquifer, frozen conditions may influence the results of the pumping test by causing conditions similar to a leaky aquifer or delayed storage effects. Precipitation recharge can cause increasing water levels in observation wells and thus cause an over-estimate of transmissivity. It is usually desirable not to conduct a pumping test in a water-table aquifer during a period of heavy rainfall. Evapotranspiration may cause declining water levels in a water-table aquifer for extended-period pumping tests. Water-level measurements made several days prior to and after the pumping test are useful in evaluating evapotranspiration and recharge effects in a shallow water-table aquifer.

If the aquifer is confined, barometric changes may affect the water levels in the wells. An increase in barometric pressure may cause an increase in groundwater levels and, conversely, a decrease in barometric pressure may cause a decrease in water levels. A means for correcting this effect is to monitor the barometric changes at the site of the pumping test and the ground-water level changes in observation wells at the pumping test site several days prior to the test. The barometric pressure changes and the associated water-level changes can be correlated by plotting both values on a graph (Walton, 1970). These graphs can be used to correct water-level data to account for barometric pressure effects. Alternately, the ratio of the water level change for a respective change in barometric pressure can be used directly to correct water-level data. This ratio is called the barometric efficiency of the well. For relatively short tests, that is, tests of less than 12 hours, barometric pressure variations are not of concern unless a significant weather front passes through the area during the test.

In addition to changes in atmospheric conditions, tectonic events or cyclic loading can cause increases or decreases in pore pressures and water levels within an aquifer. The effects of passing railroad trains, automobiles, or earthquakes on water levels in the observation wells have been described (Walton, 1970). Stepped-rate pumping tests are useful prior to running the actual pumping test. The step test allows drawdown data to be collected at the pumping well and a pumping rate selected based on drawdown response. A step test is performed by selecting 3 or more pumping rates that bracket the proposed pumping rate for the test. Each rate is pumped for one hour or until the water levels in the pumping well have stabilized. The pumping is then stopped and the water levels in the well allowed to recover for a period of one hour. Alternately, the pumping rates are increased sequentially after the water level or drawdown has stabilized without the recovery period.

Water Level Measurement During Pumping Tests

During the early part of the pumping test, frequent water-level measurements are required and sufficient manpower or automatic recording devices should be available to obtain the measurements. Defining the time-drawdown curves with accuracy in the early part of the test, when water levels in the wells are changing most rapidly, may require one person to measure water levels at each observation well. Data acquisition systems consisting of pressure transducers in each observation well, attached to a data logger, can accomplish the same goal and provide an increased number of water-level measurements throughout the test. Even if an electronic data logger is used, occasional manual measurements are suggested in the event that the transducers or electronics fail.

The length of the test depends on the aquifer setting and boundary conditions. In practice, economic factors and time constraints also may play a role in determining the length of pumping tests. In most cases, the time-versus-drawdown plots for each well should be prepared during the pumping test to assist in determining if the test is successfully achieving its design goals. Graphs constructed on semi-log paper with drawdown being plotted on the arithmetic scale and time plotted on the log scale are typically used. Once a straight line portion of the suggested graph is obtained, estimates of the transmissivity and storativity can be made. This generally occurs within the early part of the test in a confined aquifer. However, in a water-table aquifer, the straight line condition may require longer pumping periods.

Analysis of Pumping Test Data

Analysis of field pumping test data generally results in determining transmissivity and storativity for the aquifer. Depending on the geologic setting and the design of the pumping test, leakage from a confining layer can also be determined. The reliability of the analysis depends on the design of the pumping test and its principal features such as duration of the test, number of observation wells, and method of analysis. The analysis of the pumping test can utilize steady-state data, near-steady-state data, or transient data.

Steady-state equations can be used when well drawdowns have reached equilibrium and no longer change with time as a result of recharge from a river or lake. In order to apply a steady-state method of analysis, the pumping test must be run long enough so that additional drawdown nearly ceases. Thus, the observed drawdowns in nearby observation wells and in the pumping well itself are in equilibrium with the pumping rate. Analytical and graphical solutions can be applied to the well response data (Walton, 1970; Kruseman and DeRidder, 1970). If the recharge is occurring diffusely from a zone either above or below the pumping formation through a leaky aquitard, the appropriate steady-state solution must be used (Hantush, 1956).

The solution of the ground-water flow equation for transient conditions in a confined aquifer was first solved by Theis (1935). Since then, additional solutions have been developed for varying assumptions and boundary conditions. Cooper and Jacob (1946) utilized the series expansion of the well function to develop an alternative method for solving the transient problem. The transmissivity and storage coefficients can also be calculated from the solution of the well function using a type curve or grapical method (Theis, 1935; Jacob, 1940).

The solution of the transient flow equation for a uniform, confined aquifer where flow is horizontal and radial toward the well is often used to analyze test data, particularly data from the early part of the test. This solution assumes a homogeneous isotropic and infinite aquifer with a uniform thickness throughout. Furthermore, it assumes that the transmissivity and storage coefficients do not vary with time or distance from the pumping well, no leakage of water is assumed to occur from either above or below the aquifer, and the aquifer is in equilibrium. However, the rate of decline decreases as the pumping period extends. The source of water to the well is only that which is released from elastic storage as the aquifer compresses and the water expands due to the decreasing piezometric head. The analysis assumes that the water is released instantaneously as the head declines. The analysis also assumes that the well is of infinitesimal diameter relative to the area influenced by the pumping. Walton (1970) describes this solution as the well function for non-leaking, isotropic, artesian aquifers with fully penetrating wells in constant discharge conditions.

In a water-table aquifer, which is underlain by an impervious geologic unit, with a fully penetrating well, water drains from the aquifer as the water table is lowered and enters the well. However, drainage of water from the soil pores may not occur instantaneously with drawdown. Water is assumed to be released instantaneously as the pressure is lowered due to expansion of the water and compression of the aquifer (Boulton, 1963).

It is possible to interpret the results of a pumping test in a medium that is anisotropic in the horizontal plane (Neuman et al., 1984). In this case, the anisotropic medium is transformed to an equivalent isotropic medium for analysis. The analysis requires drawdown data from a minimum of 3 observation wells.

When the pumping test is conducted next to a recharge or impervious boundary, the time-drawdown data may not fit the predicted theoretical curves. In the case of a recharge boundary, the time-drawdown curve will begin to flatten out to a point at which drawdown no longer occurs with time. This is due to recharge to the aquifer for the particular piezometric conditions created by the pumping. If an impervious boundary is encountered by the drawdown, the rate of drawdown will hasten relative to a condition without a boundary. In either case, a more detailed analysis, possibly using image well theory, will be required. The use of image wells is discussed in a number of texts, including Walton (1970).

If the medium is fractured, interpretation of pumping test data becomes very difficult unless the medium is highly fractured. The degree of drawdown in any particular fracture intercepted by the observation wells is a function of the degree of interconnection with the fractures at the pumping well. The degree of interconnection is a function of the permeability of the individual fractures, their areal extent, and the frequency of fracture intersections. It would not be uncommon in poorly fractured materials, where individual fractures are not well connected, for drawdown in one observation well some distance from the pumping well to exceed that observed at an observation well closer to the pumping well. Methods have been developed for analyzing pumping tests in fractured media idealized cases (Gringarten, 1982; Karasaki, 1986; Gringarten et al., 1974; Ramey, 1970).

Estimation of Transmissivity

Because transmissivity is the product of hydraulic conductivity of an aquifer and the thickness of that aquifer, individual test data can be used to estimate the transmissivity of an aquifer. For instance, if an aquifer is fifty feet thick and samples are taken at 10-foot intervals from a boring through the aquifer, these samples can be tested in the laboratory for hydraulic conductivity. The results can then be multiplied by the sample interval and the products summed for the aquifer. The sum provides an estimate of the entire aquifer transmissivity. Alternately, if the samples represent specific layers within the aquifer, the hydraulic conductivity for each layer can be multiplied by the respective layer thickness and these products summed for all the layers in the aquifer.

Grain-size analysis, with the use of Hazen's approximation is an economic means of estimating the transmissivity in uniform sand aquifers. Many grain-size tests can be performed quickly, thereby providing a more complete representation of the entire aquifer.

COMPRESSIBILITY

The storage coefficient for a confined aquifer is a function of the elasticity of the water and the aquifer material. The vertical deformation, i.e., subsidence or heave, of an aquifer can be calculated using the formation compressibility and changes in the intergranular pressure due to such events as ground-water extraction or fluid injection into an aquifer. Therefore, as part of a ground-water monitoring plan, vertical aquifer deformation may be critical to observing the impact of, for example, ground-water pumping on land surface activities. Excessive land surface deformations can lead to building cracking, pipeline breaks, landslides, or unacceptable grade changes.

Documentation of subsidence due to pumping of an aquifer is abundant in the literature (Gambolati and Freeze, 1973; Riley, 1969; Gambolati et al., 1974; Davis and Rollo, 1969; Bull, 1975; Yerkes and Castle, 1969; Poland and Davis, 1969). The amount of subsidence can be calculated if the change in intergranular pressure is known or can be estimated. The aquifer compressibility relates the change in vertical thickness of the aquifer to a corresponding change in intergranular pressure. A complete review of the theory and methods of calculation can be found in the geotechnical literature (Taylor, 1958; Terzaghi and Peck, 1948; Lambe and Whitman, 1969). An introduction to estimating land subsidence from ground-water level changes is given in Bouwer (1978). Land subsidence can be measured by using settlement platforms, compaction recorders, and electrical strain gauges (Lofgren, 1961; U.S. Department of the Interior, 1985). A typical settlement device is shown in Figure 10.13. Such settlement monitoring devices can provide a continuous and/or long-term record of the rate and magnitude of vertical deformation of an aquifer and any overlying formations.

Vertical deformation can be estimated using undisturbed samples of the formation. The samples are tested for vertical height change under different vertical stress levels in a one-dimensional consolidation test. This test is typically limited to application in silts and clays, because samples of these soils preserve their shape for trimming. However, if representative samples can be obtained, the method is applicable in any geologic material.

The consolidation test has a long history of use in the geotechnical field and a standard method for conducting the test has been specified by the American Society of Testing and Materials (ASTM Test Method D 2435). This test involves obtaining a representative sample of the soil and placing it in a rigid cylinder as shown in Figure 10.14. A piston is placed upon the surface of the confined sample which is loaded with weights. The amount of deformation for each load is used to calculate the elasticity of the soil. The rate at which deformation occurs is a function of the compressibility of the mineral matrix and the hydraulic conductivity of the material. The test is typically run on a saturated sample and water must be allowed to discharge from the sample in order for deformation to occur. The simplified theory of consolidation assumes that the hydraulic conductivity

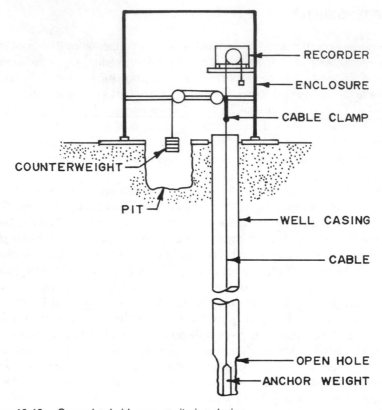

Figure 10.13. Ground subsidence monitoring device.

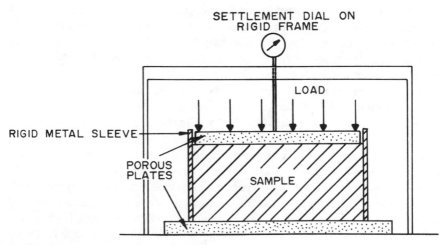

Figure 10.14. Schematic of a one-dimensional consolidometer.

does not vary with the amount of consolidation, and the water and solid soil elements are incompressible. For limited deformations, these assumptions are usually reasonable. The modulus of elasticity of the formation is calculated by the following equation from the test results:

$$dV/V_o = d\sigma/E \qquad (10.26)$$

where:

dV and $d\sigma$ = changes in the volume of the sample for a given change in pressure,
$\quad$ V_o = initial volume of the sample, and
$\quad$ E = bulk modulus of elasticity in units of force per unit area.

Modulus of elasticity has the units of area per unit force. Typical value of E for different materials are given in Table 10.7. This equation assumes that the load due to the consolidation process is entirely transformed to the soil or rock skeleton rather than the water. Furthermore, because the behavior of soil and rock is not totally elastic, the test should be performed over the stress range of interest. A detailed discussion of the test procedure and data presentation are given by Lambe (1951).

Table 10.7. Bulk Moduli of Elasticity for Various Soils and Rocks.

Material	E, kg/cm^2
Dense gravel and sand	1,000–10,000
Dense sands	500–2,000
Loose sands	100–200
Dense clays and silts	100–1,000
Medium clays and silts	50–100
Loose clays	10–50
Peat	1–5
Granite	2–6×10^{-5}
Microgranite	3–8×10^{-5}
Syenite	6–8×10^{-5}
Diorite	7–10×10^{-5}
Dolerite	8–11×10^{-5}
Gabbro	7–11×10^{-5}
Basalt	6–10×10^{-5}
Sandstone	0.5–8×10^{-5}
Shale	1–3.5×10^{-5}
Mudstone	2–5×10^{-5}
Limestone	1–8×10^{-5}
Dolomite	4–8.4×10^{-5}
Coal	1–2×10^{-5}

The modulus of elasticity of the formation is also related to the specific storage of the aquifer. Walton (1970) shows that:

$$S = \gamma_w \, n \, \beta m \, (1 + \alpha/n \, \beta), \qquad (10.27)$$

where:

S = specific storage, unitless decimal,
γ_w = bulk density of water in weight percent volume,
n = average porosity of the aquifer, unitless decimal
β = reciprocal of the bulk modulus elasticity of the water,
α = reciprocal of the bulk modulus of elasticity E, of the soil or rock comprising the aquifer, and
m = aquifer thickness in units of length.

α can be obtained for some materials using a standard laboratory one-dimensional consolidation test as described above. The reciprocal of the bulk modulus for water, β, equals 4.7×10^{-7} cm^2/kg. Thus, from the soil compressibility the specific storage of the confined aquifer can be estimated.

REFERENCES

American Society of Testing and Materials, Section 4, Volume 04.08, Philadelphia, PA.

Bear, J., 1972. Dynamics of Fluids in Porous Media, Amer. Elsevier, New York.

Bentley, H. W. and G. R. Walter, 1983. Two-Well Recirculating Tracer Test at H-2; Waste Isolation Pilot Plant (WIPP), Southeast New Mexico, Hydro-Geochem. Inc., Tuscon, AZ.

Bishop, A. W. and D. J. Henkel, 1962. The Measurement of Soil Properties in the Triaxial Test, St. Martins Press, New York.

Black, D. K. and K. L. Lee, 1973. Saturating Laboratory Samples by Back Pressure. J. Soil Mech. and Found. Div., Vol. 99, SMI, ASCE.

Boulton, N. S. and T. D. Streltsova, 1975. New Equations for Determining the Formation Constants of an Aquifer from Pumping Test Data, Water Resources, 11 (1).

Boulton, N. S., 1963. Analysis of Data from Non-equilibrium Pumping Tests Allowing for Delayed Yield from Storage, Proc. Inst. Civ. Eng., Vol. 26.

Bouwer, H., 1978. Groundwater Hydrology, McGraw Hill, New York.

Bredehoeft, J. D. and I.S. Papadopulos, 1980. A Method for Determining the Hydraulic Properties of Tight Formations, Water Resour. Res., 16(1), pp. 233–238.

Bull, W. B., 1975. Land Subsidence Due to Groundwater Withdrawal in the Los Banos-Kettleman City Area, California. Part 2. Subsidence and Compaction of Deposits, USGS Prof. Paper 437–F.

Burns, W. A., Jr., 1969. New Single-Well Test for Determining Vertical Permeability, Trans AIME, Vol. 246, pp. 743–752.

Cohen, R. M. and R. R. Rabold, 1987. Numerical Evaluation of Monitoring Well Design, Proc. First Nat. Outdoor Action Conf. on Aquifer Restoration, Ground Water Monitoring and Geophysical Methods, NWWA, 267–284.

Cooley, R. L. and C. M. Case, 1973. Effect of a Water Table Aquitard on Drawdown in an Underlying Pumped Aquifers, Water Resour. Res., 9(2), pp. 434–447.

Cooper, H. H., Jr., J. D. Bredehoeft and I. S. Papadopulos, 1967. Response of a Finite Diameter Well to an Instantaneous Charge of Water, Water Resource Res., 3(1), pp. 263–269.

Cooper, H. H., Jr. and C. E. Jacob, 1946. A generalized Graphical Method for Evaluating Formation Constants and Summarizing Well Field History, Trans. Am. Geophys. Union, Vol. 27, pp. 526–534.

Darcy, H., 1856. Les Fontains Publiques de la Ville de Dijon, Dalmont, Paris.

Davis, G. H. and J. R. Rollo, 1969. Land Subsidence Related to Decline of Artesian Head at Baton Rouge, Lower Mississippi Valley, U.S.A., Proc. Tokyo Symp. on Land Subsidence, IASH-UNESCO.

Delhomme, J. P., 1978. Kriging in the Hydrosciences, Adv. Water Resour., 1(5), pp. 251–266.

Delhomme, J. P. 1974. La Cartographie d'une Grandeur Physique à partir de données de Differentes Qualités. Proc. Meet., Int. Assoc. Hydrogeol., Mem. 10(1), pp. 185–194.

de Marsily, G., 1986. Quantitative Hydrogeology, Academic Press, New York.

Desaulniers, D. E., R. S. Kaufmann, J. A. Cherry, and H. W. Bentley, 1986. ^{37}Cl–^{35}Cl variations in a diffusion-controlled groundwater system, Crochimica et Cosmochimica Acta, Vol. 50.

Drost, W., D. Klotz, A. Koch, H. Moser, F. Neumaier, and W. Rauert, 1968. Point Dilution Methods of Investigating Groundwater Flow by Means of Radioisotopes, Water Resour. Res., 4.

Faust, C. R. and J. W. Mercer, 1984. Evaluation of Slug Tests in Wells Containing a Finite-Thickness Skin, Water Resour. Res., 20(4), pp. 504–506.

Forster, C. B. and J. E. Gale, 1981. A Field Assessment of the Use of Borehole Pressure Transients to Measure the Permeability of Fractured Rock Masses, Lawrence Berkeley Lab., Pub. LBL 11829.

Freeze, R. A. and J. A. Cherry, 1979. Groundwater, Prentice Hall, New Jersey.

Fried, J. J. and P. Ungemach, 1971. Determination In-situ du Coefficient de Dispersion Longitudinale d'un Milieu Poreaux Naturel. Cr. Acad. Sci. Paris, 272, Serie A.

Gambolati, G. and R. A. Freeze, 1973. Mathematical Simulation of the Subsidence of Venice. 1. Theory, Water Resour. Res. 9; pp. 721–733.

Gambolati, G., P. Gatto and R. A. Freeze, 1974. Mathematical Simulation of the Subsidence of Venice. 2. Results, Water Resour. Res. 10, pp. 563–577.

Gardner, W. H., 1965. Methods of Soils Analysis, Agronomy Monograph No. 9, Am. Soc. Agron., Madison, Wisconsin, pp. 82–127.

Gelhar, L. W., A. Mantoglou, C. Welty and K. R. Rehfeldt, 1985. A Review of Field-Scale Physical Solute Transport Processes in Saturated and Unsaturated Porous Media, EPR EA-4190, Elec. Power Res. Instit. Palo Alto, CA.

Gibbs, H. J. and W. G. Holtz, 1957. Research on Determining the Relative Density of Sands by Spoon Penetration Testing, Proc. 4th Inter. Conf. Soil Mech. Found. Eng., Vol. I.

Gillham, R. W. and J. A. Cherry, 1982. Contaminant Migration in Saturated Unconsolidated Geologic Deposits, Recent Trends in Hydrogeology, Special Paper 189, Geological Soc. of Am., Boulder, CO.

Goodman, R. E., 1980. Introduction to Rock Mechanics. John Wiley and Sons, New York, N.Y.

Gringarten, A. C., and H. J. Ramey, Jr., 1974. Unsteady-State Pressure Distributions Created by a Well with a Single Horizontal Fracture, Partial Penetration, or Restricted Entry. Soc. Pet. Eng. J., Vol. 14, No. 4, AIME.

Gringarten, A. C., H. J. Ramey, Jr., and R. Raghavan, 1974. Unsteady-State Pressure Distributions Created by a Well with a Single Infinite-Conductivity Vertical Fracture, Soc. Pet. Eng. J., Vol. 14, No. 4, AIME.

Gringarten, A. C., 1982. Flow-Test Evaluation of Fractured Reservoirs, Geologic Society of America, Special Paper 189.

Hantush, M. S., 1964. Hydraulics of Wells in "Advances in Hydroscience", Academic Press, Inc., New York, N.Y.

Hantush, M. S., 1960. Modification of the Theory of Leaky Aquifers, J. Geophys. Res. 65(11), pp. 3713–3725.

Hantush, M. S., 1956. Analysis of Data from Pumping Tests in Leaky Aquifers, Trans. Am. Geophys. Union, Vol. 37, No. 6.

Harr, M. E., 1962. Groundwater and Seepage, McGraw Hill, New York, N.Y.

Hazen, A., 1911. Discussion of "Dams on Sand Foundations" by A. C. Koenig, Trans. ASCE Vol. 73.

Herzog, B. L. and W. J. Morse, 1986. Hydraulic Conductivity at a Hazardous Waste Disposal Site: Comparison of Laboratory and Field-Determined Values, Waste Mgmt. and Res., Vol. 4, Academic Press, London.

Hillel, D., 1980. Applications of Soil Physics, Academic Press, New York, N.Y.

Hirasaki, G. J., 1974. Pulse Tests and Other Early Transient Pressure Analyses for In-situ Estimation of Vertical Permeability, Trans. AIME, Vol. 257, pp. 75–90.

Hvorslev, M. J., 1949a. Subsurface Exploration and Sampling of Soils for Civil Engineering Purposes, Waterways Exp. Station, Vicksburg, Miss.

Hvorslev, M. J., 1949b. Time Lag in the Observation of Ground-Water Levels and Pressures, U.S. Army Waterways Experiment Station, Vicksburg, MS.

Jacob, C. E. and S. W. Lohman, 1952. Nonsteady Flow to a Well of Constant Drawdown in an Extensive Aquifer. Trans. AGU, 33(4).

Jacob, C. E., 1940. On the Flow of Water in an Elastic Artesian Aquifer, Trans. Am. Geophys. Union, Vol. 21, Part 2, pp. 574–586.

Jacquin, C., 1965a. Etude des Écoulements et des Équilibres de Fluides Dans les Sables Arqileux. Rev. Inst. Fr. Pet. 20(4).

Jacquin, C., 1965b. Interactions entre l'argile et les Fluides Ecoulement a' Travers les Argiles Compactes. Rev. Inst. Fr. Pet. 20(10).

Karasaki, K., J. C. S. Long and P. A. Witherspoon, 1988. Analytical Models of Slug Tests, Water Resour. Res., 24(1), pp. 115–126.

Karasaki, K., 1986. Well Test Analysis in Fractured Media, Ph.D. Thesis, Lawrence Berkeley Laboratory, Univ. of California, Berkeley, CA.

Kerfoot, W. B., 1982. Comparison of 2-D and 3-D Ground-Water Flowmeter Probes in Fully-Penetrating Monitoring Wells, Proc. of the Second Nat. Symp. on Aq. Restoration and Ground Water Monitoring, NWWA.

Knutson, G., 1966. Tracers for Groundwater Investigations, Proc. Intern. Symp., Groundwater Problems, Stockholm, Sweden, Pergamon Press, Oxford, England.

Krausman, G. P. and N. A. De Ridder, 1970. Analysis and Evaluation of Pumping Test Data, International Inst. for Land Reclamation and Improvement, Netherlands.

Lambe, T. W., 1951. Soil Testing for Engineers, John Wiley & Sons, Inc., New York, N.Y.

Lambe, T. W. and R. V. Whitman, 1969. Soil Mechanics, John Wiley & Sons, New York.

Loftgren, B. E., 1961. Measurement of Compaction of Aquifer Systems in Areas of Land Subsidence, USGS Prof. Paper 424-B.

Louis, C., 1974. Introduction a l'Hydraulique des Roches. Bull. Bur. Rech. Geol. Min., Ser. 2, Sect. III, No. 4.

Milligan, V., 1975. Field Measurement of Permeability in Soil and Rock, Proceeding of Conf. on In-situ Measurement of Soil Properties, ASCE.

Mitchell, J. K., 1976. Fundamentals of Soil Behavior, John Wiley & Sons, New York.

Moench, A. F. And P. A. Hsieh, 1985. Slug Testing in Wells with a Finite-Thickness Skin, Proc. 10th Workshop Geoth. Reserv. Eng., Stanford, CA.

Moltz, F. J., O. Guven, J. G. Melville and J. F. Keely, 1986. Performance and Analysis of Aquifer Tests with Implications for Contaminant Transport Modeling, EPA/600/2-86/062, U.S. EPA, Robert S. Kerr Env. Res. Lab., Ada, OK.

Neuman, S. P. and P. A. Witherspoon, 1969. Theory of Flow in a Confined Two-Layer Aquifer System. Water Resources Research, 5 (4), pp. 803–816.

Neuman, S. P., G. R. Walter, H. W. Bentley, J. J. Ward, and D. D. Gonzalez, 1984. Determination of Horizontal Aquifer Anistropy with Three Wells, Groundwater, 22 (1), pp. 66–72.

Neuman, S. P., 1987. Theoretical and Practical Considerations of Flow in Fractured Rocks, The 1987 Distinguished Seminar Series on Ground Water Science, NWWA, December 1987.

Neuzil, C. E., 1982. On Conducting the Modified "Slug" Test in Tight Formations, Water Resour. Res. 18(2), pp. 439-441.

Nguyen, V. and G. F. Pinder, 1984. Direct Calculation of Aquifer Parameters in Slug Test Analysis, Groundwater Hydraulics, AGU, Water Resource Mono. 9, Washington, D.C.

Papadopulos, I. S., J. D. Bredehoeft and H. H. Cooper, 1973. On the Analysis of "Slug Test" Data, Water Resour. Res., 9(4), pp. 1087-1089.

Parker, J. C. and K. A. Albrecht, 1987. Sample Volume Effects on Solute Transport Predictions. Water Resour. Res., 23(12), 2293-2301.

Pickens, J. F., and G. E. Grisak, 1981. Scale-Dependent Dispersion in a Stratified Aquifer, Water Resour. Res., 17 (4), pp. 1191-1211.

Poland, J. F. and G. H. Davis, 1969. Land Subsidence Due to the Withdrawal of Fluids, Reviews in Engineering Geology II, Geol. Soc. Am., Boulder, Colorado.

Polubarinova-Kochina, 1962. Theory of Ground Water Movement, Translated by R. de Wiest, Princeton University Press, Princeton, N.J.

Powers, J. P., 1981. Construction Dewatering, John Wiley and Sons, New York, N.Y.

Prats, M., 1970. A Method for Determining the Net Vertical Permeability Near a Well from In-situ Measurements, Trans. AIME, Vol. 249, pp. 637-643.

Ramey, H. J., Jr., 1970. Approximate Solutions for Unsteady Liquid Flow in Composite Reservoirs, J. of Can. Pet. Tech.

Ramey, H. J., Jr., and R. G. Agarwal, 1972. Annulus Unloading Rates as Influenced by Wellbore Storage and Skin Effect, Trans. Vol. 253, AIME.

Riley, F. S., 1969. Analysis of Borehole Extensometer Data From Central California. Proc. Tokyo Symp. on Land Subsidence, IASH-UNESCO.

Sageev, A., 1986. Slug Test Analysis, Wat. Resour. Res. 22(8), pp. 1323-1333.

Sudicky, E. A. and J. A. Cherry, 1979. Field Observations of Tracer Dispersion Under Natural Flow Conditions in an Unconfined Sandy Aquifer, Water Poll. Research Canada, 14, pp. 1-17.

Taylor, D. W., 1958. Fundamentals of Soil Mechanics, John Wiley & Sons, New York, N.Y.

Terzaghi, K. and R. B. Peck, 1948. Soil Mechanics in Engineering Practice, John Wiley & Sons, New York.

Theis, C. V., 1935. The Relation Between the Lowering of the Piezometric Surface and the Rate and Duration of Discharge of a Well Using Groundwater Storage, Am. Geophys. Union Trans., Vol. 16, pp. 519-524.

Todd, D. K., 1959. Ground Water Hydrology, John Wiley & Sons, Inc., New York.

U.S. Department of the Interior, 1985. Earth Manual, Bureau of Reclamation, U.S. Government Printing Office, Washington, D.C.

Walton, W. C., 1970. Ground Water Resource Evaluation, McGraw-Hill, New York, N.Y.

Wang, J. S. Y., T. N. Narasimhan, C. F. Tsang, and P. A. Witherspoon, 1977. Transient Flow in Tight Fractures, Proc. of Invitational Well Testing Symposium, Lawrence Berkeley Lab., Pub. LBL 7027.

Winterkorn, H. F. and H. Y. Fang, 1975. Foundation Engineering Handbook. Van Nostrand Reinhold, New York, N.Y.

Yerkes, R. F. and R. O. Castle, 1969. Surface Reformation Associated with Soil and Gas-Field Operations in the United States, Proc. Tokyo Symp. on Land Subsidence, IASH-UNESCO.

11

Ground-Water Sampling

Beverly Herzog, James Pennino, and Gillian Nielsen

INTRODUCTION

Objectives of ground-water monitoring programs include meeting regulatory requirements, industrial or municipal waste disposal site monitoring, ambient ground-water quality monitoring, research, and general bacteriological and chemical quality monitoring. Each objective may result in development of a different ground-water sampling protocol. However, the ultimate objective of all monitoring programs is to obtain a sample of the water that is as representative of the actual ground-water quality as possible. This objective should be kept in mind throughout the ground-water monitoring program, because everything that is done, from the installation of the wells through the sampling and analysis of the ground water, will affect the reliability and interpretation of the data collected. This chapter, which focuses on factors affecting ground-water samples, explains in detail how actions prior to, during, and subsequent to sampling can have a profound effect on the analytical results. No strong recommendations for any particular sampling protocol are made in this chapter. However, a critical review of the current thinking on sampling ground water and technical criteria which will allow the reader to select a sampling protocol suitable for meeting the objectives of a given sampling program are provided. To put this chapter into its proper framework, a discussion of the various types of sampling objectives follows.

REGULATORY OBJECTIVES OF GROUND-WATER SAMPLING

Resource Conservation and Recovery Act (RCRA)

Hazardous Waste Disposal

The rules developed by the U.S. Environmental Protection Agency (EPA) under RCRA pertain mainly to releases of hazardous wastes from controlled waste-management facilities. The general design of a ground-water sampling program under RCRA requirements is essentially the same as those for all other programs described in this chapter, with the significant exception of the special statistical analysis requirements. Under RCRA, ground-water quality data are collected at specified intervals, and evaluated using stipulated statistical procedures strictly defined in the regulations. The general requirements of this regulation are discussed in greater detail elsewhere in this book.

In order to meet the statistical analysis requirements of this regulation, the monitoring well network and sampling program must be carefully planned. Specifically, the sampling program has to be designed to include one year of quarterly sampling of wells that are either upgradient of the waste disposal area or are capable of yielding samples representative of ground water that has not been affected by the regulated facility. Each time the downgradient wells are sampled, each sample is divided into four portions. The results of analyses of the portions are averaged, and this average is compared to the background concentration of the given chemical constituent or physical property established for the upgradient or background well(s). This comparison is performed using the Cochran's Approximation to the Behrens-Fisher Student's t-test.

The main impact of RCRA is on the frequency of sampling and the special equipment and handling required to properly split the samples. The equipment and handling procedures are discussed in greater detail in the various procedural sections of this chapter. A discussion on splitting samples for replication can be found in the section on "Preservation Guidelines." In addition, the Contract Laboratory Program quality assurance/quality control (QA/QC) requirements are used increasingly in RCRA monitoring.

Solid Waste Disposal

Subtitle D of RCRA applies to solid waste disposal facilities (landfills, mining waste piles) that are not specifically designed for hazardous waste disposal or sites known to contain hazardous wastes. While relatively little attention has been given to this section of RCRA (compared to the hazardous waste provisions), Subtitle D requires states to develop plans for solid waste management. Indirectly, Subtitle D has resulted in more states requiring monitoring of nonhazardous waste disposal sites, including sanitary landfills (Gordon, 1984).

Ground-water monitoring at nonhazardous waste sites typically involves monitoring of indicator parameters such as specific conductance, total dissolved solids, and chlorides. Occasionally, ground-water samples are screened for heavy metal and organic contaminants.

Underground Storage Tanks (UST)

Subtitle I of RCRA provides for the regulation of underground storage tanks. This regulation has recently stimulated a considerable monitoring effort at service stations and at industrial sites where chemicals are stored in buried tanks. Monitoring at these facilities involves both detection monitoring, where there is no known leak, as well as assessment monitoring, where there is evidence or suspicion of a leak. Monitoring at service stations may include sampling ground water for both dissolved hydrocarbons and floating-free phase hydrocarbon layers. Sampling techniques are complicated by the difficulty of obtaining a sample of the water without contaminating the sampling device as it passes through a floating layer of hydrocarbon.

Sampling of wells designed to monitor leaks from buried industrial chemical tanks may require sampling of heavier-than-water liquids which may accumulate at the bottom of an underlying aquifer. Purging procedures for these wells, and for those wells in which floating hydrocarbons are present, must be carefully performed to avoid mixing and spreading of immiscible layers through the saturated zone.

Comprehensive Environmental Response, Compensation, and Liability Act (CERCLA)

There are no specific ground-water monitoring requirements under CERCLA regulations similar to those under RCRA regulations. Ground-water monitoring is performed as part of the ground-water investigation portion of the remedial investigation phase of the cleanup program. The ground-water sampling protocols discussed in this chapter will be applicable to the performance of any CERCLA-required ground-water investigation.

Investigations at sites on the National Priority List often require analytical support from laboratories certified by the U.S. EPA Contract Laboratory Program. In order for the analytical data to be meaningful, adherence to a well-defined quality control and documentation program is necessary for both sampling and transportation procedures. These procedures include extreme care in every phase of field sampling from special preparation of sample bottles prior to sampling to packaging of the samples prior to transportation to the laboratory. Because of the value of these field quality control measures, it is recommended herein that many of these procedures be included for all types of ground-water monitoring situations.

Clean Water Act (CWA)

The Clean Water Act affects ground-water monitoring indirectly. The CWA applies to various federal programs to control pollution of surface water. As a result of these efforts, many municipalities and industries have constructed sewage and industrial waste treatment facilities that utilize sludge drying beds, impoundments, land farming, and landfilling for storage and disposal of waste. While the CWA has provided mechanisms for the classification, monitoring, and cleanup of surface water, it has virtually no provisions for ground-water pollution control. Although the CWA did not require ground-water monitoring, many of these storage and disposal facilities have caused ground-water contamination that has been detected in private and public water supply systems. Therefore, many states are requiring monitoring of these facilities for contamination that may not be covered under RCRA or CERCLA.

Section 208 of the Clean Water Act provided funds for developing plans for dealing with water-quality problems within each of the 50 states. This section of the CWA does not require ground-water monitoring, but it does encourage the incorporation of regional ground-water planning in any water-quality management plan. As a result, some states have developed regional monitoring programs to determine ambient ground-water quality and to detect regional changes in ground-water quality as a result of human activities.

Ground-water sampling may be required by state or local officials at facilities that were constructed as a result of CWA requirements. Sampling in these cases is not defined in the regulations and typically involves detection monitoring for parameters commonly found in leachate from sewage, sewage sludge, or industrial waste impoundments. Leachate from waste impoundments, drying beds, and landfarms is usually monitored by detection of chlorides and nitrates in the ground water. Because these chemicals are also found in septic tank discharges, agricultural runoff, and road deicing salt, interpretation of ground-water sampling results is often difficult. Because the regulations governing these facilities apply to surface-water monitoring and because no ground-water monitoring protocols are defined in the regulations, the ground-water sampling that is performed is often not well planned, further complicating the interpretation of the data. The ground-water sampling plans for such facilities should include adequate leachate and ground-water samples as well as background data to allow for proper detection of contaminant discharges from the facility.

Ground-water quality monitoring as part of a regional water quality plan will involve monitoring for a wide range of contaminants that could occur in the region, as well as chemical parameters of general interest such as pH, carbonate, hardness, and others. These sampling plans should also include the various drinking water parameters, as defined by the National Primary Drinking Water Regulations, even though they may not be suspected in the activities of industries within the region, because they may be present as natural substances in the environment.

Safe Drinking Water Act (SDWA)

Under the Safe Drinking Water Act, ground-water sampling may be done for compliance with one or more of four programs: the Underground Injection Control (UIC) program; the Sole Source Aquifer program; the National Primary

Drinking Water Regulations; and indirectly, the Wellhead Protection program. The general importance of all of these aspects of the SDWA to ground-water monitoring is discussed in detail elsewhere in this book.

Under the UIC program, many states are requiring ground-water monitoring at injection wells. The type of monitoring required is detection monitoring, designed to detect leaks from the injection wells at various depths. The sampling protocol employed may involve use of sampling devices capable of sampling at great depths, because many injection wells inject wastes thousands of feet below potable drinking water zones. The sampling protocol may include special parameters designed to detect water-quality changes in the injection zone miles from the point of injection. A comprehensive sampling protocol will be sensitive to water-quality changes in the potable water aquifers that may be affected both miles from the point of injection and along the vertical length of the injection well.

The Sole Source Aquifer and Well Head Protection programs have similar objectives in terms of monitoring—both involve detection monitoring protocols. Such monitoring protocols involve a wide range of sampling tools appropriate to the various types of special monitoring wells, industrial wells, and private wells available within the protected area. The parameter list will have to be extensive if the aquifer or protection zone is near numerous ground-water contamination sources.

Under the National Primary Drinking Water Regulations, many states are involved in extensive monitoring of public water systems, including many ground-water supplies. Ground-water supplies have different water-quality problems than surface-water supplies because they typically are higher in dissolved solids. Ground-water supply treatment has usually focused on iron removal and hardness adjustment. More recently, the detection of organic chemicals in these ground-water supplies has necessitated more extensive sampling at the wellhead and from monitoring wells placed between the well field and potential contaminant sources. Sampling protocols usually involve monitoring for specific contaminants at the well field and water distribution system and comparing them to federal- or state-mandated standards.

Toxic Substances Control Act (TSCA)

TSCA gives the U.S. EPA the authority to regulate individual toxic substances and mixtures of toxic substances. Normally, this means controlling toxic substances prior to and during manufacture. However, EPA could use this law to control toxic substances discharged to ground water via household waste-disposal systems, and it could provide a mechanism for funding monitoring of the occurrence, migration, and transformation of toxic substances in ground water (Gordon, 1984).

NONREGULATORY OBJECTIVES OF GROUND-WATER SAMPLING

Many state agencies and some federal agencies (e.g., U.S. Geological Survey) are involved in ambient ground-water quality monitoring. These agencies use existing water-supply wells and new monitoring wells to determine ambient ground-water quality and monitor changes in water quality over time on a local as well as a regional basis. Many of these sampling programs are designed to detect changes in ground-water quality as a result of nonregulated, nonpoint pollution sources such as agricultural activity and urban runoff. Special sampling considerations for such programs include difficulties in documenting field conditions, availability of records on construction of existing monitoring wells, and problems with proper preservation and handling of samples collected and shipped by untrained personnel.

Numerous public and private institutions are involved in ground-water research projects that include installation and sampling of monitoring wells. It is not practical to discuss all of the possible sampling protocols that may be used in various research projects. In fact, many research projects are attempting to evaluate existing practices or develop new sampling procedures. However, sampling protocols and equipment described in the following sections of this chapter will provide a source of comparison and baseline information for the researchers.

Another objective of ground-water sampling is the determination of physical and chemical properties of ground water that may affect the efficiency of wells, plumbing, and industrial process equipment. Wells in a municipal water system may be sampled for various chemical parameters such as calcium, magnesium, hardness, and iron, which determine the capacity of the ground water to plug or encrust well screens, casings, and other plumbing. Bacteriological analyses for organisms such as iron bacteria and sulfate-reducing bacteria are performed to determine the potential for biofouling in wells. These analyses sometimes require special sampling devices to physically remove accumulations from inside the well.

Depending on the intended use of the ground water, many other properties (color, temperature) and constituents may be tested to determine the suitability of the water for bottled water and other beverages, industrial processes such as quenching water, or irrigation. Because many of these uses require careful control of water quality, special sampling devices may be necessary to provide continuous monitoring of the color, turbidity, or pH of the water. Some parameters, such as dissolved gases and alkalinity, which affect corrosion potential as well as taste, are very time-dependent and must be determined in situ (in the well or in the water as it is pumped from the well) or within minutes after collection. The special sampling requirements of these latter objectives, as well as others discussed above, are addressed in the section titled "Ground-Water Sampling Plan."

DOCUMENTATION

The purpose of any sampling program is to produce ground-water samples that are representative of in situ ground-water conditions and suitable for chemical analysis. To assure that these goals are met, the sampling effort should be thoroughly documented to cover three basic elements: accountability, controllability, and traceability (Seanor, 1984). Accountability assures that the program design meets its goals by answering the questions of who, what, when, where, and why. Controllability refers to the checks in the system to assure that the procedures actually used are those identified in the sampling plan. Traceability is documentation of what was done, when it was done, how it was done, and by whom it was done. This is often covered in the chain of custody. Such documentation is an important element of any quality assurance program. It is perhaps more critical in sampling than in any other phase of a monitoring program because the sampling element is repeated and occurs in the field, creating a great opportunity for deviation from the established sampling protocol.

Data generated in the field may be used by many people over a period of years and may have to be presented in a court of law. Proper documentation is necessary for data interpretation and for the data to hold up under legal scrutiny. Many data from older monitoring programs have come into question due to lack of documentation, with the end result being that some old monitoring systems have had to be replaced solely for lack of documentation. While many aspects of a monitoring program can only be documented once (e.g., well drilling techniques and construction details), sampling procedures can and should be documented for every sampling trip and for every well.

Several authors (Barcelona et al., 1985; Cheremisinoff et al., 1984; deVera et al., 1980; Scalf et al., 1981; Fenn et al., 1977; Wilson and Dworkin, 1984) have provided lists of data that should be recorded in a field notebook on every trip. Table 11.1 is a compilation of the items suggested by these authors. While

Table 11.1. Items to Be Documented in Field Notebook.

Identities and responsibilities of sampling team members
Purpose of sampling (e.g., surveillance, compliance, etc.)
Name and location/address of sampling project
Location references (maps, photographs, etc.)
Type of site (landfill, underground storage tanks, etc.)
Name and address of field contact
Declared waste components and concentrations (if available)
Date and time of sample collection
Weather conditions
Well data (well depth, depth to water, well construction material, etc.)
Purging data (type of equipment, volume removed)
Purging parameters (changes in temperature, Eh, pH, and specific conductance during purging)
Sampling device
Field-measured parameters for sample (temperature, Eh, pH, dissolved oxygen,
    specific conductance, etc.)
Sample appearance (color, turbidity, oils)
Sample odors, if noticed
Sample containers (type, size, and number; field indentification numbers)
Sample filtration, if any (method, filter pore size)
Preservatives, if any (type and in which sample bottles)
Thermal preservation (e.g., transportation in an ice chest)
Pertinent field observations (e.g., bent well, odor in the air, etc.)
Deviations from established sampling protocol (including problems encountered, such
    as malfunctioning equipment)
Other quality control measures such as decontamination of equipment and collection of field blanks

not all items are necessary for every well at every location, field personnel should follow the basic rule that it is impossible to overdocument one's field work. Because sampling situations vary widely, the records that must be entered in the field log book also vary. As a general rule of thumb, if you are not sure whether the information is necessary, you should record it. The field book should contain sufficient information so that someone else can later reconstruct the sampling event solely from the data contained in the log book.

In order to avoid confusion among projects, a single, separate bound log book should be kept for each project or sampling site. The log book should be kept in a safe place when it is not in the field, and a recent copy should be stored separately in case the original is destroyed. The importance of good documentation cannot be overemphasized because the lack of it can render expensive data suspect or useless.

FACTORS AFFECTING GROUND-WATER SAMPLES

Introduction

Obtaining a representative sample of ground water for water-quality analysis can be difficult. Some workers believe that a representative sample is obtained after a certain number of well casing volumes have been purged from the well. Others believe that the sample should be obtained after the temperature, conductance, and pH of the water being removed from the well have become constant or provide nearly constant readings. Still others have proposed using a flow-through well design (Schmidt, 1986) in which the ground water is sampled as it stands in the well without purging prior to sampling.

The problem of obtaining a representative sample is further compounded by spatial variations in ground-water quality, which can occur over distances as small as 3 cm (Ronen et al., 1987). This heterogeneity problem is discussed in more detail elsewhere in this book.

Any approach to purging and sampling of wells must account for possible changes in ground water and contaminant chemistry during purging and sampling. The importance of these chemical changes has not been sufficiently emphasized in previous literature because the chemical reactions are not well understood in natural water systems and because any additional analyses for special geochemical parameters are costly. However, these chemical changes have a profound effect on the concentration and form of the contaminants to be measured. These measurements form the foundation for site investigations, evaluations, and remediation efforts—in short, they are the basis for everything done in the field of contaminant hydrogeology.

The chemical changes mentioned previously involve various physical, chemical, and bacteriological changes that occur in the aquifer when a well is installed and sampled. A detailed discussion of the chemistry involved is beyond the scope of this book. The following sections briefly describe some of these chemical changes and how they affect water-quality samples.

Monitoring Well Installation

Well Installation Trauma

The methods and procedures used for installing a monitoring well can affect the quality of the sample drawn from the well. According to Walker (1983), "the magnitude of installation trauma is related to physicochemical differences between the natural ground water and substances introduced during well installation (drilling fluids, seals, and backfills)." It is possible to introduce contaminants into the subsurface using any of the available drilling techniques for monitoring well installation. Some contaminant sources such as gasoline, diesel fuel, hydraulic fluid, lubricants, and paint are inherent to drill rigs used in monitoring well installation (Fetter, 1983).

Even at monitoring well installation sites at which no expense was spared to avoid contamination of the well, well installation trauma occurs. Possible outside influences include: tools and screens that were steam cleaned but still had visible bits of mud, topsoil, or grass on them where they touched the ground while being maneuvered into the hole; acetone mist that drifted over the open borehole as screens and casings were being carefully degreased near the hole; bits of contaminated (or uncontaminated) topsoil inadvertently kicked down the borehole during well installation; lubricants and hydraulic fluids that dripped off the rig as it stood over the borehole; gasoline on drillers' gloves from filling the rig tanks from a field service truck—gloves which were later used to handle drilling tools and casings; and exhaust fumes from generators and vehicles only feet from the hole. Considering the myriad of potential sources of well installation trauma, it is impossible to install a sterile monitoring well. Even if all of the above problems could be eliminated, we cannot eliminate bacteria that may not already exist in the aquifer, but which exist in the atmosphere and on everything and everyone involved with the monitoring well installation.

Although the casing and screens installed in the hole may be made of relatively inert substances, the welding rod, welding preparation compounds, pipe dope (a chemical compound used to prepare pipe surfaces prior to joining and welding), bentonite, filter-pack, centralizers, packers, and cement are all potential sources of contaminants. Even if these materials contained no contaminants, they represent foreign bodies in an "open wound" (the borehole) that has been created in the subsurface. These foreign substances generally attempt to come into equilibrium with the chemistry of the ground water and the geologic materials that surround them. For instance, if the oxidation state of iron in a steel screen is different than the oxidation state of iron in the ground water, iron may either dissolve into the ground water or precipitate onto the screen. The precipitation/solution of iron may have an important effect on the concentration of other metals, such as arsenic and lead, in the ground-water sample (Hem, 1985).

Bentonite is commonly used in drilling fluids and as annular seal material in wells. The cation exchange capacity of clays, including bentonite, could have an effect on the presence and concentration of various cations (calcium, magnesium, sodium) in the ground water. Because these elements are often major components of the dissolved solids in ground water often due to aquifer lithology (e.g., limestone, calcareous sands, gravels), any change in their concentrations in the vicinity of the well will change the chemical equilibrium of the ground water. Changes in this equilibrium will ultimately affect the concentration of any contaminants which form complexes with these ions or which may be adsorbed onto the aquifer materials. Bentonite may also provide a source of contamination to a well. Analyses of lead in samples of bentonite used in well sealing and in drilling fluids revealed concentrations of lead ranging from 0.2 to 0.4 mg/L (Table 11.2).

Table 11.2.   Total Lead Concentrations in Samples of Bentonite Well Sealant and Drilling Additive.[a]

	Sealant 1	Sealant 2	Additive	Control
Lead	0.2	0.4	0.4	<0.05

All concentrations are in milligrams per liter (mg/L).
< This symbol means that the concentration was below the detection limits of the laboratory instrument used in the analysis.

[a]All samples were obtained from bulk samples of bentonite used in selected monitoring well installations in Minnesota. Sealant 1 and Sealant 2 were different samples of bentonite used for two different phases of a monitoring well installation program at one site. The identities of the manufacturer(s) of the sealants were unknown to the analysts. The drilling additive (a substance used to hold the borehole open during rotary drilling) used at this site was manufactured by Wyo-Gel. (Use of product name is for clarity only and does not constitute an endorsement.) Each sample was prepared by mixing approximately 6 mg of bentonite with 10 mL of concentrated nitric acid, diluting with distilled water, and then decanting the clear portion of the solution into sample bottles. pH in the sample was maintained at 1 to 2 pH units. Lead analyses were performed by atomic absorption. The control was prepared by adding some of the nitric acid to distilled water.

As discussed in Chapter 8, a monitoring well must be properly developed before it is sampled. One reason for well development is to physically remove some of the potential contaminants, described above, that may have been introduced during the installation of the monitoring well. Vigorous development by jetting or flushing increases the likelihood of removing drilling mud residues, grease and oil, and other potential contaminants. Water introduced to the formation during well development is usually much lower in dissolved solids than the ground water. Hence, aquifer materials may dissolve into this intoduced water to restore the solution equilibrium.

Even if all of the water that is introduced into the well is removed (this is doubtful because the water may not be pumped or air-lifted from the same areas of the aquifer that it went into during jetting or flushing), water remaining in the aquifer materials surrounding the well may be stripped of its more soluble minerals. The force of the development effort may physically remove minerals and contaminants adsorbed onto the aquifer materials, as well as the finer aquifer materials themselves. These problems can be alleviated by using well development methods that do not introduce outside water or air into the well.

Although state-of-the-art technology may have been employed in installing a new monitoring well, it should be remembered that the well represents a scar in the aquifer that may take months or years to heal. Furthermore, it is unlikely that samples of water from this well will yield information that is representative of aquifer water quality until equilibrium has been restored. Even when the aquifer does return to equilibrium, this equilibrium is upset every time a sample is collected. While these problems cannot be eliminated, an understanding of well installation trauma allows the selection of a well design and construction program that will minimize the problem.

Well Installation Method

Drilling methods are covered in detail in Chapter 6. The three most popular drilling methods discussed were fluid rotary drilling, auger drilling, and percussion drilling. Here we are only concerned about the possible effects that a specific drilling method can have on water samples taken from wells installed by that drilling method. A major problem arises from the use of drilling additives. Based on this, if geology were not a concern, the general order of preference for drilling methods would be auger drilling first, then percussion drilling, with rotary drilling last. In addition to additives, the contact between the drilling apparatus and the geologic formations can cause cross-contamination between water-bearing units. Table 11.3 presents a list of advantages and disadvantages of various drilling methods compiled from a large selection of publications (Barcelona et al., 1985; Fetter, 1983; Gillham et al., 1983; Keely and Boateng, 1987a, 1987b; Kill, 1986; Scalf et al., 1981).

As Table 11.3 illustrates, the use of drilling fluids is necessary for some drilling methods. Drilling fluids are used to remove drill cuttings from the borehole, to prevent the borehole from caving, to seal the borehole to prevent fluid loss, and to cool, lubricate, and clean the drill bit. However, these fluids can be a problem when drilling monitoring wells because they can interfere with later water-quality analyses. The major types of water-based drilling fluids are fresh water, water with clay (bentonite) additives, water with polymeric additives, and water with both clay and polymeric additives. Air-based fluids include dry air, mist, foam (with surfactant), and stiff foam (containing strengtheners such as polymers and bentonite) (Driscoll, 1986). A few manufacturers add isopropyl alcohol to their foaming agents.

Bentonite, which is used both in drilling fluids and as a well seal, can have several effects on water chemistry (Gillham et al., 1983). Due to a high cation exchange capacity, bentonite may mask the presence of metals in the ground water by absorption of the metals. Additionally, most bentonites contain about 4% to 6% organic matter and can therefore contribute organic constituents to the sample.

Table 11.3.   Effects of Drilling Methods on Water Samples.

Method	Advantages	Disadvantages
Air rotary	•Drilling fluid is not always used, minimizing contamination and dilution problems	•When more than one water-bearing zone is encountered and hydrostatic pressures are different, flow between zones occurs after drilling is completed but before the hole is cased and grouted •Oil from compressor may be introduced to geologic system •Use of foam additives which contain organic materials can interfere with both organic and inorganic analyses
Mud rotary		•Drilling fluid which mixes with formation is difficult to remove and can cause contamination •Fluid circulation can cause vertical mixing of contaminants •Drilling fluids and additives can interfere with subsequent water quality analyses •Lubricants may cause contamination
Bucket auger	•No drilling fluid is used, minimizing contamination and dilution problems	•Large diameter hole makes it difficult to assure adequate grouting •Must continuously add water in soft formations
Solid stem auger	•No drilling fluid is used, minimizing contamination and dilution problems •Can avoid use of lubricants	•Because auger must be removed before well can be set, vertical mixing can occur between water-bearing zones •Can cause vertical mixing of both formation water and geologic materials
Hollow stem auger	•No drilling fluid is used, minimizing contamination and dilution problems •Can avoid use of lubricants •Formation waters can be sampled during drilling	•Can cause vertical mixing of geologic materials •Can cause vertical mixing of formation waters if augers are removed before well is installed

continued

Table 11.3.   Continued.

Method	Advantages	Disadvantages
Hollow stem auger (continued)	•Well can be installed as augers are removed, decreasing interaction with water from higher water-bearing zones	
Cable tool	•Little or no drilling fluid required	•Contamination is possible if drilling fluid is used •Slight potential for vertical mixing as casing is driven
Jetting	•May be only alternative where rig cannot get in	•Large quantities of water or drilling fluid are introduced into and above sampled formation •Cannot isolate zone with a grout seal

Diagnostics for a bentonite problem include the presence of particulate bentonite or high concentrations of sodium in water samples.

Organic/biodegradable additives can interfere with bacterial analyses and organic-related parameters (Scalf et al., 1981). Revert®, one such organic additive, has been blamed for bacterial contamination of water-supply wells (Richard, 1979). Few scientific studies documenting the effects of additives on water quality have been published. Brobst and Buszka (1986) studied the effects of guar bean, guar bean with breakdown additive, and bentonite on chemical oxygen demand (COD), chloride and sulfate concentrations in ground-water samples. COD was found to be the most effective indicator of contamination due to drilling fluids. Despite repeated development, bentonite wells required approximately 140 days to return to background COD levels. Background COD levels were achieved in 320 and 50 days with guar bean and guar bean plus breakdown additive, respectively.

Because of the effects of organic drilling fluids on microbial activity, Barcelona et al. (1983) concluded that predominantly inorganic clay drilling fluids were preferable to organic drilling fluids for drilling monitoring wells. Fresh water is preferable to either type of drilling fluid. In any case, because *any* drilling fluid can affect water samples, extra care must be taken to assure the integrity of water samples from fluid-drilled wells. Development may take longer than for wells drilled without fluids. Water quality should be monitored during well development for COD and other fluid-specific parameters to determine when background levels are achieved. Samples of any fluid that enters the hole, including water, should be analyzed by a laboratory to determine possible effects on water quality. Without such added precautions, the quality of data from fluid-drilled wells may be questioned.

Well Maintenance Effects

Various well maintenance activities such as replacing or improving components of the well (screen, casing, pump and plumbing, grouting), redevelopment, and disinfection can disturb the equilibrium of the ground-water system just as initial well installation did. Repairs or improvements to the well can allow bacteria, lubricants, and other foreign matter to enter the well. Redevelopment can result in aeration of the ground water, introduction of large amounts of water with a different quality, stripping of mineral coatings from formation materials, filter pack materials and screens, and changes in density of the filter pack and formation materials. Redevelopment with acid or other chemicals will have a marked effect on the ground water and formation materials present in the vicinity of the well.

Disinfection with chlorine, iodine, or other disinfection agents will introduce constituents to the ground water that, if not removed by redevelopment or pumping, can profoundly affect ground-water quality.

Because well maintenance activities can alter water quality, these activities should be clearly documented. Possible chemical effects of well maintenance activities must be carefully examined and reviewed during data interpretation.

Ground-Water Sampling Plan

The objective of most ground-water quality monitoring programs is to obtain samples that are representative of existing ground-water conditions, or samples that retain the physical and chemical properties of the ground water within an aquifer. Earlier segments of this chapter have discussed at length the variety of factors associated with the design, installation, and development of monitoring wells that can influence the capability of ground-water monitoring wells to yield "representative" samples. All too often, however, attention is focused primarily on this area of concern while the human influences that can arise during collection, handling, preservation, and analysis of samples are ignored. These human influences can result in potentially serious random errors in data being generated in the field and/or the laboratory, from which erroneous conclusions can be drawn during the course of an investigation.

The key to minimizing this human error element is to develop and implement a comprehensive, site-specific ground-water sampling and analysis program. Ideally, a ground-water sampling plan should be developed even before the monitoring well network is installed. A good sampling and analysis program must address several key issues, as indicated in Table 11.4. Once the objectives of a ground-water monitoring program have been defined, the next most important issues to be decided are the locations to be sampled, the frequency of sampling events, and the parameters to be monitored. These factors must be defined prior to the development of a ground-water sampling and analysis program because they are the key criteria around which the entire program is developed.

Table 11.4. Sampling Plan Design Elements.

1. Objectives of the groundwater sampling and analysis program;
2. Site-specific parameters of concern to be sampled and analyzed;
3. Location, condition, and access to sampling points (e.g., wells, discharge points, surface water, etc.) to be included in the program;
4. Number and frequency of samples to be collected;
5. Sampling protocol—well purging procedure and equipment needs
   —field parameter monitoring/sample screening
   —sample collection: parameter specific techniques and equipment needs
   —field QA/QC controls
6. Field sample pretreatment requirements—filtration
   —preservation
7. Sample shipment to the analytical laboratory—sample handling
   —delivery method
   —transport time
8. Sample documentation, Chain-of-Custody requirements;
9. Sample chemical analysis—identification of analytical methods
   —storage/holding times
   —laboratory QA/QC

Purging

General

After sample locations and chemical parameters have been selected, the sample collection method must be determined. One of the first issues in sample collection is the determination of well purging protocol. Purging is necessary because a ground-water sample must be representative of formation water. Water that has been standing in a well is typically not representative of formation water because water in the well above the well screen is not free to interact with formation water, is in contact with well construction materials (i.e., casing) for extended periods of time, is in direct contact with the atmosphere, and is subject to different chemical equilibria. This stagnant water often has a different temperature, pH, oxidation-reduction potential, and total dissolved solids content than the formation water (Seanor and Brannaka, 1983). Rust and scale from the monitoring well can interfere with laboratory analysis (Wilson and Dworkin, 1984), as can bacterial activity (Scalf et al., 1981). Volatile organic compounds and dissolved gases in the stagnant water column in the well may volatilize or effervesce in as few as two hours. A field study by Barcelona and Helfrich (1986) concluded that purging of standing water from the well was the dominant factor affecting precision of sampling results. They found that errors caused by improper purging were greater than those associated with sampling mechanisms, tubing, and well construction materials. The goal of purging is therefore to create a situation that will allow the well to provide a sample which is representative of formation water while creating a minimal disturbance to the ground-water flow regime.

Despite the value of purging in minimizing the chemical differences inside and outside of the well, it can create problems in some hydrogeologic situations. The most common problem with purging is overpumping of the formation, which itself can result in the acquisition of unrepresentative samples. If purging is conducted at a higher rate than that used for well development, fine particles may be drawn into the well increasing the turbidity of the sample and providing sites for adsorption of metals and organic chemicals. In addition, in fine-grained materials, overpumping may lower the water table sufficiently to dewater a portion of the saturated zone, exposing a portion of the formation materials to air and other gases and, if they are present, substances floating on top of the water table (hydrocarbons). When purging ends, the water table rebounds and the ground water is exposed to the air and other gases or substances present in the formation materials. Overpumping may also cause dilution, which could mask the presence of contamination. In situations in which the well does not intersect a contaminant plume, overpumping can draw contamination toward the well, causing the contamination to become more widespread.

At least one researcher (Keely, 1982) has advocated overpumping as a means of creating a gradient toward the well, and collecting time series samples to intersect increasingly larger portions of the aquifer. Changes in samples collected during overpumping could then be used to determine whether the well is located at the edge or in the center of a plume.

Besides influencing the migration of contaminants in ground water, well purging procedures can have a significant influence on ground-water sample chemistry. Pumping causes pressure changes that can affect dissolved gas equilibria (U.S. EPA, 1986) in both confined and unconfined aquifers. A reduction in pressure caused by pumping can cause dissolved gases to come out of solution. This causes a shift in chemical equilibrium that may result in precipitation of some cations. This shift can cause calcium carbonate encrustation of well screens and the precipitation of contaminants such as lead and other heavy metals.

Because of the significant influence that well purging can have on contaminant hydrogeology, some scientists advocate no purging for certain situations. Gillham et al. (1983) contended that purging wells in fine-grained sediments may strip the sample of volatile organic compounds. They also contend that purging can cause bias from mixing stagnant and formation waters.

Giddings (1983) also perceived problems when purging low-yielding wells, especially in open boreholes penetrating fractured bedrock or in sand, which may be dewatered by purging. In these cases, turbulent flow would result, causing excessive aeration of samples. As an alternative to purging, Gillham (1982) suggested using a depth-specific sampling device, such as a syringe. This could be lowered to the well intake or to the producing zone of an open hole and provide a representative sample without purging.

Purge Volumes

Three philosophies for determining the volume of water that should be purged from a monitoring well prior to sampling appear in the literature. The first specifies that a given number of bore volumes be purged, the second advocates purging until certain field-measured parameters have stabilized, and the third specifies that the purge volume should be based on the hydraulic performance of the individual well. Proponents of the latter method usually recommend combining the last two methods.

Criteria based on numbers of bore volumes. Suggested numbers of bore volumes to be purged from a well prior to sampling range from less than 1 to more than 20 volumes. Regulations commonly specify the removal of 3 to 5 or 4 to 6 bore volumes. In the context in which the term is used in this chapter, 1 bore volume is defined as the volume of water standing in the well above the top of the well screen. The screened area and sandpack are not included in this volume because water in these areas is free to interact with formation water. Humenick et al. (1980) found that representative samples could be obtained after removing less than 1 bore volume from wells that were situated in a confined sandstone monitoring solution uranium mining activities. Fenn et al. (1977) suggested purging a minimum of 1 bore volume, but preferred 3 to 5. They stated that purging was less critical in high-yielding wells. However, they stipulated that for high-yielding wells, if purging is performed with a pump placed near the static water surface in the well, 3 to 5 bore volumes should be removed. If the pump intake were placed at the bottom of the well screen and the sample collected from the pump discharge line, only 1 bore volume had to be removed. For slowly recharging wells, they suggested that the well be pumped dry and allowed to recover before sampling.

Gillham et al. (1983) suggested a range of 1 to 10 bore volumes. They recommended placing the pump intake close to the static water surface so that stagnant water would move up the well. This would keep the volume to be removed in the range of 5 to 10 bore volumes. They also warned against pumping more than 10 bore volumes because of the problems associated with overpumping.

Several other authors present different numbers of bore volumes to be purged. Scalf et al. (1981) presented 4 to 10 bore volumes as a common range. They warned that more purging may be necessary to produce aquifer water if the pump is set at the screen and suggested lowering the pump during purging. Wilson and Dworkin (1984) suggested a minimum of 5 to 6 bore volumes when sampling for volatile organics. Pettyjohn et al. (1981) also looked at sampling for organic contaminants and advocated the removal of at least 10 bore volumes at a rate of at least 500 ml/min. Unwin and Huis (1983) stated that purging of up to 20 bore volumes was common.

Gibb et al. (1981) and Schuller et al. (1981) collected data on inorganic constitutents from wells at six sites after increasing numbers of bore volumes had been purged. They have been widely quoted for their conclusion that removing 4 to 6 well volumes is adequate for most situations. However, they also concluded that the best method to ensure the acquisition of aquifer-quality water was to determine the purge volume with a pumping test and confirm the volume by measuring field parameters.

It is obvious that it is not possible to recommend that a specific number of bore volumes be removed from monitoring wells during purging. The range of suggested volumes is too large and the cost of improper purging is too great to permit such a recommendation.

Criteria based on stabilization of indicator parameters. The second school of thought regarding purging volumes advocates that the well should be purged until certain field-measured parameters have stabilized. The most common field-measured parameters include temperature, pH, Eh, and specific conductance. Summers and Brandvold (1967) presented one of the earliest studies of water-quality changes with discharge time. They found that flowing wells were sensitive to pH, temperature, and specific conductance. Most variations in ion concentration occurred at low concentration levels, indicating that these ions are most affected by pH, temperature, and pressure changes. Citing this work, Wood (1976) recommended that temperature, pH, and specific conductance be used to determine purge volumes for U.S. Geological Survey studies. He went on to say that pH was the most sensitive of these parameters.

Later researchers looked at the sensitivity of these parameters in a variety of ground-water monitoring situations. Marsh and Lloyd (1980) studied flowing wells in fissured limestone and found temperature and specific conductance to be the most sensitive parameters. Humenick et al. (1980) found monitoring wells in a confined sandstone to be most sensitive to temperature and pH.

Gibb et al. (1981) and Schuller et al. (1981) measured a series of inorganic parameters in the laboratory in addition to the field parameters. They found that calcium, chloride, potassium, magnesium, and sodium were unaffected by purging, whereas iron and zinc were the most significantly affected. Because different wells are sensitive to different parameters, these researchers recommend measuring four field parameters—temperature, pH, Eh, and specific conductance. Gibs and Imbrigiotta (1990) performed a similar experiment, concentrating on volatile organic compounds. They concluded that stabilization of field parameters may not accurately signify stabilization of volatile organic chemical concentrations.

Criteria based on hydraulic and chemical parameters. The third school of thought advocates that information about formation hydraulic parameters be combined with field-measured parameters to determine purge volumes. Transmissivity can be determined by either a pumping test or a slug/bail test. For monitoring wells, pumping test data should be analyzed by a method that accounts for storage in the well, such as the method by Papadopulos and Cooper (1967).

Slug/bail tests are preferred for slowly recharging wells because it is difficult to maintain a pumping rate that is low enough to avoid pumping these wells dry. Methods of determining formation transmissivity are covered in detail elsewhere in this book and are not repeated here.

Given formation transmissivity, casing diameter, and purging rate, it is possible to calculate the length of purging time required to obtain a respresentative sample. Figures 11.1 and 11.2 (from Gibb et al., 1981) can be used for this determination. Figure 11.1 presents the percent of formation water versus time for different transmissivities, while Figure 11.2 shows percent of formation water versus time for various well casing diameters. Both figures assume a discharge rate of 500 mL/min, which was chosen because it was high enough to be practical without causing excessive turbulence in the well.

Because hydraulic parameters can change over time, this calculation only provides an estimate of the proper purging volume. Field-measured parameters can be monitored when the actual purging volume approaches the estimated volume to more accurately determine that the well has been adequately purged.

Criteria based on time series sampling. The most rigorous method of determining the volume of water to be purged was presented by Lee and Jones (1983).

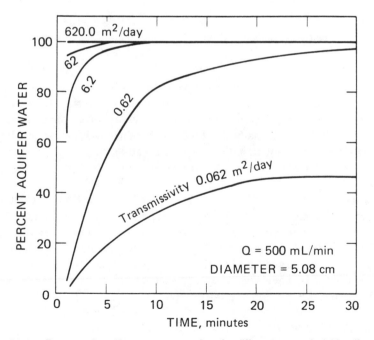

Figure 11.1.   Percent of aquifer water versus time for different transmissivities (from Gibb et al., 1981).

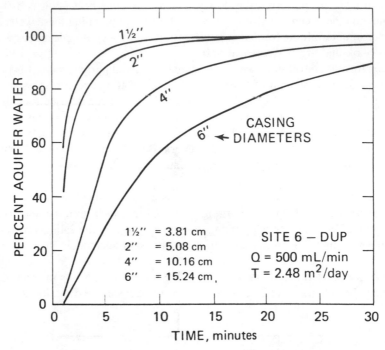

Figure 11.2.   Percent of aquifer water versus time for different well casing diameters (from Gibb et al., 1981).

They recommend collecting samples at one, two, four, six, and ten bore volumes and analyzing for the constituents of concern in addition to such bulk chemical properties as specific conductance, temperature, and dissolved oxygen. Because water quality changes seasonally, they believe that this procedure should be performed twice a year for the first three years. Repeating the procedure once every three to four years thereafter will confirm that the required purge volume has not changed.

The work of Gibb et al. (1981) and Gibs and Imbrigiotta (1990) indicates that the method of Lee and Jones (1983) should be applied, whenever funding allows, to determine the number of bore volumes to be purged from rapidly recovering wells. When this method is impractical, purge volumes should be determined considering both hydraulic and field parameters. A specified number of bore volumes should be used only when regulatory agencies do not allow the use of a more accurate method for determining purge volume.

Purging Considerations for Slowly Recovering Wells

While the problem of obtaining a representative sample from rapidly recovering wells has received much attention, the problem of slowly recovering wells has been virtually ignored. Gillham et al. (1983) contended that wells in fine-grained sediments should not be purged because purging may strip the sample of volatile organic compounds by exposing the formation water to air. They further argued that purging can cause bias resulting from mixing stagnant and formation waters. Giddings (1983) perceived a similar problem with purging low-yielding wells. Fenn et al. (1977) suggested waiting until the well had recovered to its original static water level before collecting the sample. Other researchers (Unwin and Huis, 1983; Barcelona et al., 1985) recommended that the sample be collected during well recovery. They asserted that care must be taken to assure that the well is not emptied to below the top of the screen because to do so would cause aeration of the formation water and, thus, the sample. For very slowly recovering wells, Barcelona et al. (1985) proposed that the sample be collected in small aliquots at two-hour intervals. Both publications further advocated that the sample be collected at a lower flow rate than used for purging to minimize sample disturbance. None of these authors presented data to justify their recommendations on sampling in fine-grained materials. In practice, samples are commonly collected the day after the well is purged.

Data on chemical changes during the recovery of slowing recharging wells are sparse. Griffin et al. (1985) observed changes in volatile organic concentrations in water from three monitoring wells. They found that most concentration changes occurred within the first 48 hours. Purging the well a second time, after allowing it to recover for 24 hours, did not increase concentrations of volatile organic compounds. However, their data set was too small to yield conclusive recommendations on optimal sampling time.

McAlary and Barker (1987) conducted a laboratory test of volatilization losses of organic compounds during ground-water sampling from fine-grained sand. They found volatilization losses of up to 70% within 1 hour after purging if the water level was lowered below the top of the screen. They also found volatilization losses to be less than 10% when water was standing less than about 6 hours.

Herzog et al. (1988) expanded on the work of Griffin et al. (1985) by including more wells and more volatile organic compounds. Samples were collected before purging and at increments up to 48 hours after purging from wells finished in geologic materials that have hydraulic conductivities of less than approximately 1×10^{-5} cm/sec. Some of the data generated by their study are presented in Figure 11.3. They found no statistical difference between samples collected at from 2 to 48 hours after purging. Samples collected before purging had concentrations of volatile organics that were either statistically lower or not different than those collected after purging, depending on the compound.

Purging Equipment

Once the required purge volume has been determined, the questions of how to purge the well remain. The choices for equipment to be used for purging the well are essentially the same as those for sampling, with the exception of discrete

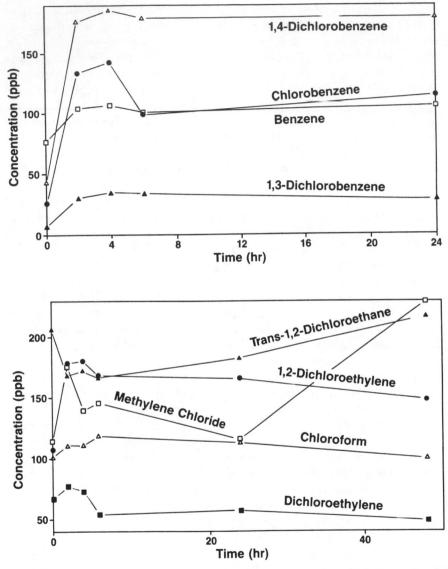

Figure 11.3. Changes in volatile organic chemical concentration with recharge time in a slowly recovering well. Zero time is for sampling before purging (from Herzog et al., 1988).

samplers. These devices are discussed later in this chapter. It is desirable to be able to use the same device for both purging and sampling to minimize the possibility of cross-contamination by minimizing the number of items that enter the well. However, if sampling equipment is not dedicated to individual wells and must be cleaned between each well, it may be impossible to adequately clean the inside of pumping equipment or tubing used for sampling. In this case, purging with a pump is commonly combined with sampling with a separate device, such as a bailer, syringe, or other discrete sampler. For this reason, a dedicated bailer may be attractive as a purging device for low-yielding wells.

Purging Methods

The final question related to purging regards the depth within the well at which the purging device should be placed. Two obvious choices are: (1) the top of the column of water; and (2) the top of the well screen. Each option is widely applied.

A quantifiable answer was determined in a laboratory study by Unwin and Huis (1983). In their study, which ignored the effects of drawdown, they found that pumping near the static water surface was most effective in providing representative ground-water samples. Approximately seven well volumes, depending on the pumping rate, were required to achieve a representative sample. This number, of course, was also dependent on the transmissivity of the laboratory aquifer material. Pumping from below the static water surface introduces uncertainty that all stagnant water has been removed due to the migration of water from above the pump intake. The recommendation that can be made based on these results is that if a device used for purging is immediately used afterward for sampling, purging from the level of the well screen is acceptable. However, if different devices are used for purging and sampling, purging should occur at the static water surface or, as suggested by Keely and Boateng (1987a), in a scheme whereby the purging device is lowered down the well as successive bore volumes are removed.

Sampling Devices

Once a well has been purged, the next task is to collect a ground-water sample for field and/or laboratory analysis. One of the greatest factors influencing this and subsequent portions of the sampling and analysis program is the selection of an appropriate ground-water sampling device that may or may not be the same device used for well purging. There are several factors that must be carefully evaluated, with site-specific program elements in mind, when selecting a ground-water sampling device.

The initial consideration in selecting a sampling device is whether or not the wells will accommodate the device. While this seems intuitively obvious, it is important to consider that the wells may not be plumb, may have constrictions in the casing (i.e., at joints), or may contain other obstructions that make the effective inside diameter of the well smaller than the inside diameter of the casing. Alternately, if the monitoring wells are not in place, it may be more prudent to first select a sampling device that meets the requirements of the monitoring program and then select the size of the casing used in the wells. Whatever the case, the smaller the inside diameter of the well, the more limited the selection of sampling devices becomes. All of the devices described herein will fit into a 3-inch inside diameter well, most can be installed in a 2-inch inside diameter well, several can be used in wells of 1.5 inches inside diameter or less.

It is increasingly important that both the sampling device and the sampling techniques used in a monitoring program be compatible with the analytical power of modern laboratories. The high cost of chemical analyses, particularly where determinations in the parts-per-billion range are required, mandate that groundwater samples be of the highest integrity. Thus, the sampling device chosen for a monitoring program must be evaluated to ensure that any physical or chemical alteration of the sample, caused either by the device's construction (i.e., materials from which it is made) or its method of delivering the sample to the surface, is minimized. It is important, for example, that: (1) the materials in the sampling device neither sorb contaminants from or leach contaminants into the sample; (2) neither the pH state nor the redox potential (Eh) of the sample is altered; and (3) no volatile organic constituents present in the ground water are stripped from the sample as a result of sample aeration, degassing, or pressure changes.

The materials used in the construction of sampling equipment must be considered with respect to the sampled parameters. Where warranted, inert materials should be specified to limit the possibility of leaching constituents from the materials into the water sample, or adsorption of contaminants from the water sample onto sampler materials (Barcelona et al., 1983). A number of studies have been performed for the purpose of determining the suitability of a variety of materials for use in ground-water monitoring programs (Barcelona et al., 1983, 1984, 1985; Curran and Tomson, 1983; Miller, 1982; Sosebee et al., 1982; Pettyjohn et al., 1981; Scalf et al., 1981).

Because the subsurface environment is generally under different temperature, pressure, gas content, and oxidation-reduction potential conditions than those at the surface, precautions must be taken to ensure that the sampling device consistently transports a representative sample to an appropriate storage vessel. Devices that introduce air or noninert gas into a sample, or that cause a sample to undergo significant pressure changes from the sampling depth to the surface are less desirable from the standpoint of preserving the chemical quality of the sample. Sampling systems in which there exist constrictions in the water flow path (i.e., intricate valving, which may produce an "orifice effect") can readily change the pH of the sampled water by changing the carbon dioxide (CO_2) partial pressure of the water. Further, systems that introduce dissolved oxygen could cause oxidation of ferrous iron to ferric iron, which has a marked impact on the speciation and concentration of many chemical constituents in a sample. The oxidation of iron can also affect the pH of the water through iron hydrolysis reactions, which can have a significant impact on both organic and inorganic chemical constituents (Lee and Jones, 1983). Turbulence and depressurization can result in significant changes in a sample's original content of dissolved oxygen and carbon dioxide and volatile organic compounds (Barcelona et al., 1983). All of these effects must be minimized in order to preserve the integrity of the representative sample.

Ideally, the same device is used to purge and sample the ground-water monitoring well. This reduces time and labor expenses, and potential errors associated with equipment decontamination are avoided. In addition, using one device reduces the amount of equipment to be transported in the field. However, despite these obvious advantages, a device selected for both purging and sampling must be designed with a mechanism to permit regulation of the pumping rate. While it is desirable to have a device capable of high pumping rates for well purging activities, it is critical to be able to reduce rates to very low discharge rates (i.e., approximately 100 mL/minute) for sample collection to minimize the potential for sample aeration and degassing (Barcelona et al., 1984). The flow-control mechanism, however, should not be one that involves "throttling down" the device by means of valving, because this introduces the potential for partial pressure changes and subsequent chemical alteration of samples. Rather, the flow control mechanism should exert direct control over the action of the sample delivery mechanism, such as varying the power supplied to an electric submersible pump by means of a rheostat. As an alternative it may be more practical in some situations, such as in large diameter wells with substantial volumes of water, to select two separate devices for use in a monitoring program—one capable of a high flow rate for well purging and another capable of drawing a sample from a well without significant agitation.

The depth from which samples must be taken should also be considered. The deeper the sampling interval or pump setting, the more head the device must overcome to deliver a sample to the surface. Thus, the pumping lift capability of the device determines whether or not the device is suitable for individual applications. It should also be noted that the deeper the sampling interval, the more time-consuming the purging and sampling operation becomes. Generally, the selection of available sampling devices becomes more limited with greater sampling depth.

Ease of operation, cleaning, and maintenance are important but frequently overlooked considerations in sampling device selection. As Barcelona et al. (1984) pointed out, one of the major sources of poor precision in sampling results is sampling device operational problems. This could be due to any one of several factors—either the device and accessory equipment are too complicated to operate efficiently under field conditions, the operator is not familiar enough with the device to operate it properly, or the operational manual supplied with the device does not clearly outline the procedures for proper sample collection with the device. Thus, it is not only important to select a device that is simple to operate, but also to train the operators of the device thoroughly in its operation. The value of data generated by a monitoring program is directly proportional to the individual operator's skill and familiarity with the operation of the device used for sampling.

Some sampling devices, by virtue of their construction, may be difficult to clean, and thus may create potential problems of cross-contamination between wells if the same device is used to sample for more than one well. This may result in the generation of anomalous monitoring data, which are open to erroneous interpretation. Thus, sampling devices should be easy to disassemble for cleaning and should be able to survive the rigors of decontamination (i.e., steam cleaning, solvent rinsing, etc.). The concept of dedicating sampling devices to individual wells essentially removes the potential problem of cross-contamination, but to be cost-effective requires the employment of relatively inexpensive devices.

Field operation efficiency requires that it must be possible to solve equipment maintenance problems in the field. The more mechanically simple the device, the easier it is to repair in the field, if repair should become necessary. Operational problems should be easily recognized and diagnosed, replacement parts should be readily available and inexpensive, and a minimum number of tools should be required to perform necessary maintenance. The device should require a minimum amount of routine maintenance to ensure its reliable performance. Some of the devices described herein are too complex for field maintenance, requiring repairs by trained technicians. This may necessitate the return of the device to the manufacturer for servicing, which can result in unacceptable delays in sampling schedules.

Reliability and durability are two additional factors related to maintenance that warrant attention. Devices used in some monitoring programs must be capable of operating for extended periods of time in ground-water environments that may contain a variety of chemical constituents. These may pose hazards such as corrosion of metallic parts and degradation of synthetic materials (i.e., polyvinylchloride, polyethylene, silicone, etc.). This is especially true where devices are dedicated to wells and thus are continually exposed to a potentially chemically hostile environment.

Remote locations of some monitoring wells require that the sampling device and accessory equipment selected for a monitoring program (i.e., tubing or tubing bundles, hose reels, battery packs, generators, compressed air sources, controlling devices, etc.) be highly portable. While some devices and accessory equipment can be hand-carried to remote sites, some manufacturers have mounted their equipment on backpack frames and some have provided small wheeled carts for their equipment in an effort to increase portability. Other equipment is too bulky and heavy to be transported in the field without being vehicle-mounted.

Both the initial capital cost and the operational cost (including maintenance cost) of the sampling device and accessory equipment are important considerations. However, while cost is commonly one of the overriding factors in the selection process, it is probably much less important than other factors already discussed. It is pointless to use cost as the only criteria for selecting a sampling device when sampling for highly sensitive chemical parameters, especially considering the cost of chemical analyses and the potential cost of litigation. Conversely, the idea that the most expensive device is best is not always valid. For some devices, the initial purchase price may pale in comparison to the cost of operation and maintenance. Some initially inexpensive devices may require, for example, the use of considerable amounts of compressed gas (i.e., nitrogen), which can prove to be expensive, thus making the device less cost-efficient in the long run.

A wide variety of sampling equipment is available for use in small-diameter ground-water monitoring wells. The devices available can be put into three broad categories: grab mechanisms (inluding bailers and syringe devices), suction-lift mechanisms (including centrifugal and peristaltic pumps), and positive-displacement mechanisms (including gas-drive devices, gas-operated bladder pumps, electric submersible pumps, and gas-driven piston pumps). Table 11.5 presents a summary of the characteristics of these sampling devices. The advantages and disadvantages of each type of device for the collection of ground-water samples are presented in Table 11.6.

More recent studies have compared devices for specific types of sampling problems. Stolzenburg and Nichols (1985) found that aeration, caused by the sampling procedure, produced significant chemical changes in iron and trace metal concentrations found in water samples. They found that the bladder pump, peristaltic pump, and conventional bailer introduced the least amount of aeration when fitted with in-line filtration. Schalla et al. (1988) compared the abilities of a stainless steel and Teflon piston pump, a Teflon bailer, a Teflon bladder pump, and a PVC air-lift pump to provide representative samples of volatile chlorinated hydrocarbons, and found no statistical difference among the four devices. This substantiated the work of Muska et al. (1986), which found that overall sampling variability is low regardless of the method used for sampling volatile organic compounds, provided that careful and reproducible procedures are followed.

Filtration

General

The two most prevalent forms of sample pretreatment done in the field at the time of sampling are sample filtration and sample preservation.

"To filter or not to filter"—that is the question and commonly a topic of heated debate when developing sampling protocols. There are two opposing schools of thought regarding the filtration of ground-water samples. One contends that filtration of ground-water samples results in a significant physical and chemical alteration of the sample, thereby destroying any "representativeness" preserved by earlier well installation and sampling efforts. The opposing argument is that when conducting hydrogeologic/geochemical investigations at hazardous and non-hazardous waste sites, investigators are often concerned with only those constituents truly dissolved in ground water, excluding any and all constituents which may be adsorbed onto particulate matter in suspension. Following this argument, sample filtration ensures representative samples. In terms of field practicality, sample chemistry, and laboratory considerations, both viewpoints have their merits. There are legitimate situations in which a ground-water sample should not be filtered prior to analysis, and circumstances demanding filtration to ensure accurate chemical analysis.

Table 11.5. Summary of Characteristics of Sampling Devices Available for Small Diameter Monitoring Wells.

Device	Minimum Well Diameter	Approximate Maximum Sampling Depth	Typical Sample Delivery @ Maximum Depth	Flow Controllability	Materials[a] (Sampling Device Only)	Potential for Chemical Alteration	Ease of Operating, Cleaning and Maintenance	Approximate Cost for Complete System[b]
Bailers	½"	Unlimited	Variable	Not applicable	Any	Slight-moderate	Easy	<$8,100–$8,200
Syringe samplers	1½"	Unlimited	0.2 gal.	Not applicable	Stainless 316, Teflon or polyethylene/glass	Minimum-slight	Easy	<$8,100 (50 mL homemade) $81,500 (850 mL commercially available)
Suction-lift (vacuum) pumps	½"	26 ft.	Highly variable	Good	Highly variable	High-moderate	Easy	$8,100–$8,550
Gas-drive samplers	1"	300 ft.	0.2 gpm	Fair	Teflon, PVC, polyethylene	Moderate-high	Easy	$8,300–$8,700
Bladder pumps	1½"	400 ft.	0.5 gpm	Good	Stainless 316, Teflon/Viton, PVC, silicone	Minimum-slight	Easy	$81,500–$84,000
Gear-drive submersible pumps	2"	200 ft.	0.5 gpm	Poor	Stainless 304 Teflon, Viton	Minimum-slight	Easy	$81,200–$82,000
Helical rotor submersible pumps	2"	125 ft.	0.3 gpm	Poor	Stainless 304, EPDM, Teflon	Slight-moderate	Moderately difficult	$83,500
Gas-driven piston pumps	1½"	500 ft.	0.25 gpm	Good	Stainless 304, Teflon, Delrin	Slight-moderate	Easy to moderately difficult	$83,400–$83,800

Source: Nielsen and Yeates, 1985.
[a]Materials dependent on manufacturer and specification of optional materials.
[b]Costs highly dependent on devices and selection of accessory equipment.

Table 11.6. Advantages and Disadvantages of Several Types of Sampling Devices.

Advantages	Disadvantages

Bailers

Advantages	Disadvantages
•Bailers can be constructed of virtually any rigid or flexible material, including those materials that are inert to chemical contaminants.	•In deep wells, well purging can be difficult and therefore labor- and time-consuming.
•Bailers are mechanically very simple, and thus are easily operated and disassembled for cleaning and repair.	•If the line used with the bailer is not a "noncontaminating" line and is not dedicated to a single well or is not adequately cleaned after each sampling event, cross-contamination between wells can result.
•Bailers are, in comparison with other sampling devices, very inexpensive, making them feasible for dedicated installation in monitoring wells.	
•Bailers can be used to sample water from wells of virtually any depth.	•Aeration, degassing, and turbulence can occur while lowering the bailer through the water column or while transferring the sample from the bailers to the sample container.
•Bailers require no external power source, and are lightweight and highly portable.	•The person sampling the well is susceptible to exposure to any contaminants in the sample.
•Bailers made of flexible material will pass through nonplumb wells.	•Bailing does not supply a continuous flow of water to the surface.
•Bailers can be made to fit any diameter well, and can be made virtually any length to accommodate any sample volume.	•It may be difficult to determine the point within the water column that the sample represents.
•Bailers can provide a "cut" of immiscible contaminants (i.e., petroleum hydrocarbons) from the top of the water column in a well. Transparent bailers are usually used for this purpose.	•Bailer check valves may not operate properly under certain conditions (e.g., high suspended solids content and freezing temperatures).
	•The "swabbing" effect of bailers that fit tightly into a well casing may induce fines from the formation to enter the well, especially if the well has been poorly developed.

Syringe Devices

Advantages	Disadvantages
•The sample taken with a syringe device does not come into contact with any atmospheric gases and is subjected to a very slight negative pressure, thus neither aeration nor degassing of the sample should occur.	•Syringes are inefficient for collecting large volume samples.
	•Syringes cannot be used to purge a well.
•Samples can be collected at discrete intervals and at any depth.	•Syringes are relatively new in the application, therefore they may not be as readily available as other sampling devices.
•Syringes can be made of inert or nearly inert materials.	
•Syringes are not restricted to the limits of suction lift.	•Sample contamination by components of "homemade" sampling devices is possible unless materials are carefully selected.
•The syringe can be utilized as the sample container, thus removing the	•The use of syringes is limited to water with a low suspended solids content.

continued

Table 11.6. Continued.

Advantages	Disadvantages

Syringe Devices (continued)

possibility of cross-contamination between wells.
- Syringes are inexpensive, highly portable, and simple to operate.
- Syringe devices can be used in wells as small as 1¼ in. inside diameter.

- Some leakage has been found to occur around the plunger when syringes are used to sample water containing high levels of suspended solids.

Suction-Lift Pumps

- The flow rate of most suction lift pumps is easily controlled.
- Suction-lift pumps are highly portable and readily available.
- Most suction-lift pumps are inexpensive in comparison to other sampling devices.
- The pump does not contact the sample—only the tubing must be cleaned (peristaltic pump only).
- Suction-lift pumps can be used in wells of any diameter, and can be used in nonplumb wells.

- Sampling is limited to situations in which the potentiometric level is less than 25 feet below the surface.
- A drop in pressure due to the application of a strong negative pressure (suction) causes degassing of the sample and loss of volatiles.
- An electric power source is required for peristaltic pumps.
- The gasoline motor power source used for most centrifugal pumps provides potential for hydrocarbon contamination of samples.
- Pumping with centrifugal pumps results in aeration and turbulence.
- Centrifugal pumps may have to be primed, providing a possible source of sample contaminations.
- Low pumping rates of peristaltic pumps make it difficult to purge the wellbore in a reasonable amount of time.
- Where the sample comes in contact with the pump mechanism of tubing, the choice of appropriate materials for impellers (centrifugal pump) or flexible pump-head tubing (peristaltic pump) may be restrictive.

Gas-Drive Devices

- Gas-drive devices can be utilized in wells of 1¼ in. inside diameter.
- Gas-drive devices are highly portable for most sampling applications, and are inexpensive.
- Discrete depth sampling is possible.
- Gas-drive devices can provide delivery of sample at a controlled, nearly continuous rate.

- If air or oxygen are used as the driving gas, oxidation may occur (causing the precipitation of metals), gas-stripping of volatiles may occur, or CO_2 may be driven from the sample (causing a pH shift). Consequently, air-lift sampling may not be appropriate for many chemically sensitive parameters.

continued

Table 11.6.   Continued.

Advantages	Disadvantages

Gas-Drive Devices (continued)

- The use of an inert driving gas (i.e., nitrogen) minimizes sample oxidation and other chemical alteration.
- Devices can be installed permanently in boreholes without casing.
- Multiple installations can be achieved in a single well or borehole (either temporarily or permanently installed).
- Gas-drive devices can be constructed entirely of inert materials.
- The depth from which samples can be taken with gas-drive devices is limited only by the burst strength of the materials from which the device and tubing are made.

- An air compressor or large compressed-air tanks must be transported to deep sampling locations, reducing portability.
- Application of excessive air pressure can rupture the gas entry or discharge tubing.
- Devices installed permanently in boreholes without casing are difficult or impossible to retrieve for repair; proper installation and operation may be difficult to ensure.

Gas-Operated Bladder Pumps

- Most of these pumps have been designed specifically to sample for low levels of contaminants, so most are or can be made of inert or nearly inert materials.
- The driving gas does not contact the sample directly, thus problems of sample aeration or gas stripping are minimized.
- Bladder pumps are portable, though the accessory equipment may be cumbersome.
- The relatively high pumping rate (in comparison with other sampling devices) allows well purging and large sample volumes to be collected.
- The pumping rate of most of these pumps can be controlled rather easily to allow for both well purging at high flow rates and collection of volatile samples at low flow rates.
- Most models of these pumps are capable of pumping lifts in excess of 200 feet.
- These pumps are easy to disassemble for cleaning and repair.
- Most models of bladder pumps are designed for used in wells of 2-in. inside diameter; some are available for smaller diameter wells.
- Large diameter bladder pumps (i.e., 3¼ outside diameter) are available for large diameter monitoring wells.

- Deep sampling requires large volumes of gas and longer cycles, thus increasing operating time and expense, and reducing portability.
- Check valves in some pump models are subject to failure in water with high suspended solids content.
- Most currently available pump models are expensive, though prices are highly variable.
- Minimum rate of sample discharge of some models may be higher than ideal for the sampling of volatile compounds.

continued

Table 11.6. Continued.

Advantages	Disadvantages

Gear-Drive Electric Submersible Pumps

Advantages	Disadvantages
•These pumps are constructed of inert or nearly inert materials; therefore, they are suitable for sampling organics when optionally available Teflon discharge line is employed. •These pumps are highly portable and totally self-contained, except when auxiliary power sources are employed. •These pumps provide a continuous sample over extended periods of time. •Models are available for both 2-in. and 3-in. (or larger) inside diameter wells. •High pumping rates are possible, making it feasible to use the pump for both well purging and sampling. •Reasonably high pumping rates can be achieved to depths of 150 feet, and depth range can be extended through the use of an auxiliary power source. •These pumps are easy to operate, clean, and maintain in the field, and replacement parts are inexpensive. •In comparison to other pumps offering the same performance, these pumps are inexpensive to purchase and operate.	•There is no control over flow rates; therefore, it is not possible to adjust from a high pumping rate for well purging to a lower rate required for sampling of volatiles. •Sampling in wells with high levels of suspended solids may necessitate frequent replacement of gears. •The potential for pressure changes (cavitation) exists at the drive mechanism; however, this has not been adequately evaluated.

Helical Rotor Electric Submersible Pump

Advantages	Disadvantages
•This pump is portable and relatively easy to transport in the field to remote locations. •This pump is well-suited for use in wells of 2-in. inside diameter. •Relatively high pumping rates are possible with currently available units, thus well purging is possible. •This pump has been specifically designed for monitoring groundwater contamination; therefore, it is constructed of inert or nearly inert materials.	•The currently available pump unit is limited to 125 feet of pumping lift. •High pumping rates with this pump lead to creation of turbulence, which may cause alteration of sample chemistry. •Thorough cleaning and repair in the field may be difficult because the pump is moderately difficult to dissemble. •Water with high suspended solids content can cause aeration problems. •The currently available model is expensive in comparison to other devices offering comparable performance. •The pump must be cycled on/off approximately every 20 minutes to avoid overheating of the motor. •The flow rate cannot be controlled, so the pump may not be suitable for taking samples for analysis of chemically sensitive parameters.

continued

Table 11.6.   Continued.

Advantages	Disadvantages
Gas-Operated Double-Acing Piston Pump	
•Because the sample is isolated from the driving gas, no aeration of the sample occurs.	•Piston pumps are relatively expensive in comparison to other sampling devices.
•The pump provides a continuous sample over extended periods of time.	•The pump is not highly portable—it must be vehicle mounted.
•This pump is relatively easy to operate and is easy to disassemble for cleaning and maintenance, though some maintenance problems (i.e., with the pump motor or valving mechanism) cannot generally be solved in the field.	•Unless the pump intake is filtered, particulate matter may damage the pump's intricate valving mechanism.
•Models of this pump are available for wells of 1¼ in. inside diameter and for well of 2 in. or greater inside diameter.	•The pump's intricate valving mechanism may cause a series of pressure drops in the sample, leading to sample degassing and pH changes.
•The pump uses gas economically.	•Fixed-length tubing bundles may be inconvenient for shallow, low-yield monitoring wells.
•Pumping lifts of more than 500 feet can be overcome with this pump.	•The tubing bundles may be difficult to clean adequately to avoid cross-contamination.
•Flow rate of the pump can be easily controlled by varying the driving gas pressure on the pump.	
•The pump can be made of inert or nearly inert materials.	
•The moderately high pumping rate at great depths allows for collection of large volumes of sample in a relatively short time.	

Source: Nielsen and Yeates, 1985.

Reasons for filtration of ground-water samples include: removal of suspended solids to permit analyses of only the "dissolved" fraction of chemical constituents; determining the percent of suspended solids; separate analyses of constituents attenuated to suspended solids; analysis of "clear" samples using delicate laboratory instrumentation easily clogged by sediment-laden samples; removal of any interference caused by suspended particles when ultraviolet spectrophotometric screening techniques are used (as in analysis for NO_3^-); and determining the concentrations of specific parameters of interest.

Problems with Filtering

While the list of advantages could appear to be convincing, there are as many disadvantages to ground-water sample filtration. Significant changes in groundwater sample chemistry can result from sample filtration (Braids et al., 1987; Stolzenburg and Nichols, 1986). Sample filtration typically involves drawing or pushing a volume of ground water through a filter medium of known pore size by applying either a vacuum or positive pressure to the sample and filtration apparatus. This results in a change in the partial pressure of dissolved gases in the water sample which in turn can result in chemistry changes. Radioactive gases such as radon are also sensitive to changes in pressure across filtration media (Braids et al., 1987).

Volatile organic compounds may be lost in the process of sample filtration if water samples are exposed to the atmosphere or if filtration occurs with pressure changes caused by either a positive pressure or a vacumm applied across the filter. Most volatile organic compounds listed in the volatile category of priority pollutants have low to moderate affinity for solid substrates. Therefore, water samples for volatile analyses are frequently not filtered.

In addition to atmospheric contact, sample oxygenation resulting from sample transfer from the sampling devices to the filtration apparatus frequently occurs. Sample aeration can cause metals, such as iron, to precipitate or can cause co-precipitation or adsorption of metals that were in solution. Sample aeration can also result in the loss of any volatile organic constituents—a particularly serious problem when analyzing samples at the parts-per-billion level.

Sample filtration may also remove chemical constituents that are only slightly soluble in water, such as PCBs or polynuclear aromatic hydrocarbons. If samples are filtered, the data generated by sample analysis may be inaccurate, showing lower concentrations of some constituents than are actually present in the ground water. In these instances, unfiltered samples are typically extracted with an organic solvent in the laboratory prior to analysis. Chemical extraction causes these families of compounds to desorb from suspended solids and appear as if they were in solution.

Filtration membranes and filter papers have also been demonstrated to introduce contaminants, such as phosphorus, to water samples (APHA et al., 1985). Pretreatment of filters in the laboratory can help prevent this problem.

Most ground-water monitoring programs do not require filtration of all samples submitted for analysis. The basis for this decision is often one of practicality. Field filtration can be an arduous task requiring a substantial amount of time and effort when working with turbid or sediment-laden samples. Filtration can also be virtually impossible to perform in the field during the winter months when ambient air temperatures may be below freezing. Some ground-water investigations incorporate private well sampling. Whenever a drinking water source is being studied, samples are not filtered before analysis. Because water taken from private wells is generally not filtered before use, it is desirable to analyze water as it would be consumed, regardless of its physical state. This conservative approach would yield "worst case" concentrations of parameters of concern.

When to Filter

The most common exceptions to a "no filtration" policy include: samples requiring analysis for dissolved metals; samples requiring analysis for parameters, such as PCBs, that are strongly attenuated to suspended particles; long-term monitoring programs that are sensitive to fluctuations in chemical data resulting from variations in the amounts of suspended particles between sampling events; and use of data in hydrogeochemical models that require exacting techniques to distinguish between dissolved and particulate fractions of a chemical species so that parameters such as partition coefficients, adsorption isotherms, and chemical equilibrium can be determined.

The U.S. EPA (1986) recommends that ground-water samples that are to be analyzed for metals be collected in two portions. One sample should be unfiltered for a "total" metals determination and a second sample should be field-filtered using 0.45 μm membrane filters to determine the "dissolved" fraction of metals. By using this approach, the difference in concentration between the total and dissolved fractions may be attributed to the stripping of metallic ions from the suspended particles and any ions sorbed to the particles. While this information may be of some significance at sites where sediment-laden samples are a problem, analytical costs for metals analyses are doubled.

Types of Filtering Devices

Once the decision to filter samples has been made, the filtration apparatus must be selected. There are three main types of filtration apparatus available for use in the field: vacuum filtration, pressure filtration, and in-line filtration devices. Table 11.7 summarizes the selection criteria which must be evaluated with site-specific conditions in mind when choosing filtration equipment. It is beyond the scope of this chapter to evaluate each of the varieties of filtration equipment available on the market today, but it has been addressed by other authors such as Stolzenburg and Nichols (1986). It is important, however, to understand how these three types of filtration equipment can affect ground-water sample chemistry.

Table 11.7.   Filtration Device Selection Criteria.

Appropriate filter pore size
Speed of filtration
Sample volume capacity
Sediment loading rate
Compatibility of filter media with contaminants expected in sample
Field portability
Ease of operation
Ease of decontamination
Reliability of operation under field conditions
Cost to purchase and operate equipment

Both vacuum filtration and pressure filtration involve the transfer of ground water from a sample collection device to the field filtration device. Water is passed through a porous filter, typically made of glass microfibre or cellulose membrane, with a typical pore size of 0.45 μm. In the case of vacuum filtration, the sample is "pulled" through the filter, while in pressure filtration, the ground-water sample is "pushed" through the filter using compressed air or nitrogen as the driving force. In both cases, the filtrate that is generated flows either directly into the sample container, or more often, into a transfer vessel (usually made of glass) and then into the sample container.

While there are some significant differences in terms of use of vacuum and pressure filtration equipment, such as portability, relative filtration rates and ease of decontamination, both vacuum and pressure filtration have the same potential to alter the chemistry of the ground-water sample being filtered. Problems include sample aeration/oxygenation, degassing of volatile constituents, and imparting partial-pressure changes. In addition, both types of apparatus are typically composed of several parts, all of which must be disassembled for thorough decontamination between samples. Thus, the potential for cross-contamination of samples exists. To overcome this particular problem, some manufacturers have developed disposable filtration devices.

To overcome many of the other major inconveniences of use and potential for chemical alteration associated with vacuum and pressure filtration equipment, disposable, in-line filtration devices have been developed. These devices consist of a holder/filter system, typically in cartridge form, in which inlet and outlet connections can be made to enable pressure filtration. The filter cartridge is connected directly to the discharge tubing of the ground-water sampling device. Therefore, the use of in-line filters with any sampling devices which do not incorporate a discharge tube (e.g., bailers) is not possible. In-line filters do, however, reduce problems of sample aeration, exposure of the sample to atmospheric conditions, potential cross-contamination of samples caused by improper equipment decontamination, and degassing of volatile constituents from samples.

Sample Preservation

Introduction

The second most common in-field sample pretreatment process that must be addressed by the sampling and analytical program is sample preservation. It is critical to note that sample filtration, if performed, must be completed before sample preservation to avoid mobilizing constituents that may be attenuated to suspended particles (i.e., metals, PCBs) in the ground-water sample. This would result in the detection of erroneously high "dissolved" concentrations of those constituents when the samples are analyzed in the laboratory.

If a sample of ground water cannot be analyzed immediately upon collection, then it must be stabilized until analysis can be performed. The chemical quality of a ground-water sample begins to change as soon as the sample is extracted from the formation. Indeed, it may be changing as purging and sampling are progressing. If representative data are to be obtained from the analysis of samples, chemical and biological activity in the sample must be stopped or slowed as much as possible. The chemical and biological processes that occur in a water sample can include one or more of the following: (1) formation of complexes; (2) adsorption/desorption; (3) acid-base reactions (hydrolysis); (4) oxidation-reduction reactions and other precipitation-dissolution reactions; and (5) microbiological activity which affects the disposition of many metals, anions, and organic molecules.

These chemical and biological processes cannot be completely stopped, but they can be retarded by available preservation methods. Preservation techniques are limited to pH control, chemical addition, temperature control, and protection from light. Temperature control usually involves keeping the sample cool by refrigeration or packing in ice. This method of sample preservation is usually combined with other preservation techniques. The preservation techniques for various chemical analyses are listed in Table 11.8 (U.S. EPA, 1982).

Reasons for Preservation

When a sample of ground water is exposed to the atmosphere, the partial pressures of major gases in the water sample, such as carbon dioxide, oxygen, and volatile organics may change (U.S. EPA, 1987), resulting in other chemical changes in the sample. For example, a loss of dissolved carbon dioxide from the sample (degassing) will result in an increase in pH. The following chemical equation shows the chemical species involved (Gibb et al., 1981):

$$H^+ + HCO_3^- \gtrless H_2O + CO_2 \text{ (gas)} \tag{11.1}$$

The H^+ ion activity on the left side of the equation represents acid. As the carbon dioxide leaves the water in the form of carbon dioxide gas, more acid (H^+) is consumed to replace the lost carbon dioxide. As the acid is converted to water and carbon dioxide (CO_2 [gas]), the acidity decreases (pH is increased). The loss of carbon dioxide can also result in the precipitation of metals such as calcium (Ca), as illustrated in the following equation:

$$Ca^{+2} + 2\ HCO_3^- \gtrless CaCO_3 \text{ (ppt.)} + H_2O + CO_2 \text{ (gas)} \tag{11.2}$$

The calcium combines with the bicarbonate ion on the left side of the equation to form the calcium carbonate precipitate which appears on the right side of the equation.

Similarly, exposure of the sample to air can introduce oxygen to the water sample and result in the precipitation of any iron cations in solution (as iron hydroxide):

$$O_2 \text{ (gas)} + 4\ Fe^{+2} + 10\ H_2O \gtrless 4\ Fe(OH)_3 \text{ (ppt.)} + 8H^+ \tag{11.3}$$

Other metals of concern in ground-water sampling, such as lead and chromium, may precipitate out of solution by the same mechanism. Furthermore, there may

Table 11.8.   Required Containers, Preservation Techniques, and Holding Times.

Parameter	Container[1]	Preservative[2, 12]	Maximum Holding Time[3]
Bacterial Tests			
1–4. Coliform, fecal and total	P,G	Cool, 4°C 0.008% $Na_2S_2O_3$[6]	6 hours
5. Fecal streptococci	P,G	Cool, 4°C 0.008% $Na_2S_2O_3$[6]	6 hours
Inorganic Tests			
1. Acidity	P,G	Cool, 4°C	14 days
2. Alkalinity	P,G	Cool, 4°C	14 days
4. Ammonia	P,G	Cool, 4°C H_2SO_4 to pH <2	28 days
9. Biochemical oxygen demand	P,G	Cool, 4°C	48 hours
10. Biochemical oxygen demand, carbonaceous	P,G	Cool, 4°C	48 hours
12. Bromide	P,G	None required	28 days
15. Chemical oxygen demand	P,G	Cool, 4°C H_2SO_4 to pH <2	28 days
Inorganic Tests			
16. Chloride	P,G	None required	28 days
17. Chlorine, total residual	P,G	None required	Analyze immediately
21. Color	P,G	Cool, 4°C	48 hours
23–24. Cyanide, total and amenable to chlorination	P,G	Cool, 4°C NaOH to pH >12 0.6g ascorbic acid[6]	14 days[9]
25. Fluoride	P	None required	28 days
27. Hardness	P,G	HNO_3 to pH <2	6 months
28. Hydrogen ion (pH)	P,G	None required	Analyze immediately
31. Kjeldahl and organic 43. Nitrogen	P,G	Cool, 4°C H_2SO_4 to pH <2	28 days
Metals[4]			
18. Chromium VI	P,G	Cool, 4°C	24 hours
35. Mercury	P,G	HNO_3 to pH <2	28 days
Metals, except above	P,G	HNO_3 to pH <2	6 months

continued

Table 11.8. Continued.

Parameter	Container[1]	Preservative[2,12]	Maximum Holding Time[3]
Metals[4] (continued)			
38. Nitrate	P,G	Cool, 4°C	48 hours
39. Nitrate-nitrite	P,G	Cool, 4°C H_2SO_4 to ph < 2	28 days
40. Nitrite	P,G	Cool, 4°C	48 hours
41. Oil and grease	G	Cool, 4°C H_2SO_4 to pH < 2	28 days
42. Organic carbon	P,G	Cool, 4°C HCl or H_2SO_4 to pH < 2	28 days
44. Orthophosphate	P,G	Filter immediately, cool, 4°C	48 hours
46. Oxygen, dissolved probe	G bottle and top	None required	Analyze immediately
Winkler	G bottle and top	Fix on site and store in dark	8 hours
48. Phenols	G	Cool, 4°C H_2SO_4 to pH < 2	28 days
49. Phosphorus (elemental)	G	Cool, 4°C	48 hours
50. Phosphorus, total	P,G	Cool, 4°C H_2SO_4 to pH < 2	28 days
53. Residue, total	P,G	Cool, 4°C	7 days
54. Residue, filterable	P,G	Cool, 4°C	7 days
55. Residue, non- filterable (TSS)	P,G	Cool, 4°C	7 days
56. Residue, settleable	P,G	Cool, 4°C	48 hours
57. Residue, volatile	P,G	Cool, 4°C	7 days
61. Silica	P	Cool, 4°C	28 days
64. Specific conductance	P,G	Cool, 4°C	28 days
65. Sulfate	P,G	Cool, 4°C, add zinc acetate plus sodium hydroxide to pH > 9	7 days
67. Sulfite	P,G	Cool, 4°C	Analyze immediately
68. Surfactants	P,G	Cool, 4°C	48 hours
69. Temperature	P,G	None required	Analyze immediately
73. Turbidity	P,G	Cool, 4°C	48 hours

continued

Table 11.8. Continued.

Parameter	Container[1]	Preservative[2, 12]	Maximum Holding Time[3]
Organic Tests[5]			
Purgeable halocarbons	G, Teflon-lined septum	Cool, 4°C 0.008% $Na_2S_2O_3$[6]	14 days
Purgeable aromatics	G, Teflon-lined septum	Cool, 4°C 0.008% $Na_2S_2O_3$[6] HCl to pH<2[10]	14 days
3,4. Acrolein and acrylonitrile	G, Teflon-lined septum	Cool, 4°C 0.008% $Na_2S_2O_3$[6]	14 days
Phenols	G, Teflon-lined cap	Cool, 4°C 0.008% $Na_2S_2O_3$[6]	7 days until extraction, 40 days after extraction
Benzidines	G, Teflon-lined cap	Cool, 4°C 0.008% $Na_2S_2O_3$[6]	7 days until extraction, 40 days after extraction
Phthalate esters	G, Teflon-lined cap	Cool, 4°C	7 days until extraction, 40 days after extraction
Nitrosamines[7]	G, Teflon-lined cap	Cool, 4°C, store in dark 0.008% $Na_2S_2O_3$[6]	7 days until extraction, 40 days after extraction
PCBs	G, Teflon-lined cap	Cool, 4°C[8] pH 5–9	7 days until extraction, 40 days after extraction
Nitroaromatics and isophorone	G, Teflon-lined cap	Cool, 4°C	7 days until extraction, 40 days after extraction
Polynuclear aromatic hydrocarbons	G, Teflon-lined cap	Cool, 4°C, store in dark 0.008% $Na_2S_2O_3$[6]	7 days until extraction, 40 days after extraction
Haloethers	G, Teflon-lined cap	Cool, 4°C 0.008% $Na_2S_2O_3$[6]	7 days until extraction, 40 days after extraction
Chlorinated hydrocarbons	G, Teflon-lined cap	Cool, 4°C	7 days until extraction, 40 days after extraction

continued

Table 11.8. Continued.

Parameter	Container[1]	Preservative[2, 12]	Maximum Holding Time[3]
Organic Tests[5]			
87. TCDD	G, Teflon-lined cap	Cool, 4°C 0.008% $Na_2S_2O_3$[6]	7 days until extraction, 40 days after extraction
Pesticides Tests			
1–70. Pesticides	G, Teflon-lined cap	Cool, 4°C pH 5–9[8]	7 days until extraction, 40 days after extraction
Radiological Tests			
1–5. Alpha, beta, and radium	P,G	HNO_3 to pH<2	6 months

Source: U.S. EPA, 1982.

1. Polyethylene (P) or Glass (G).
2. Sample preservation should be performed immediately upon sample collection. For composite samples, each aliquot should be preserved at the time of collection.
3. Samples should be analyzed as soon as possible after collection. The times listed are the maximum times that samples may be held before analysis. Samples may be held for longer periods only if data is available and kept on file to show that the specific types of samples under study are stable for the longer time. Some samples may not be stable for the maximum time period given in the table. The sampler/laboratory is obligated to hold the sample for a shorter time if knowledge exists to show that this is necessary to maintain sample stability.
4. Samples should be filtered immediately on-site before adding preservative for dissolved metals.
5. Guidance applies to samples to be analyzed by GC, LC, or GC/MS for specific compounds.
6. Should only be used in the presence of residual chlorine. Use ascorbic acid only in the presence of oxidizing agents. Maximum holding time is 24 hours when sulfide is present. Optionally, all samples may be tested with lead acetate paper before the pH adjustment in order to determine if sulfide is present. If sulfide is present, it can be removed by addition of cadmium nitrate powder until a negative spot test is obtained. The sample is filtered and then NaOH is added to pH 12.
7. For the analysis of diphenylnitrosamine, add 0.008% $Na_2S_2O_3$[6] and adjust pH to 7–10 with NaOH within 24 hours of sampling.
8. The pH adjustment may be performed upon receipt at the laboratory and may be omitted if the samples are extracted within 72 hours of collection. For the analysis of aldrin, add 0.008% $Na_2S_2O_3$.
9. Maximum holding time is 24 hours when sulfide is present.
10. Sample receiving no pH adjustment must be analyzed within seven days of sampling.
11. Samples for acrolein receiving no pH adjustment must be analyzed within 3 days of sampling.
12. When any sample is to be shipped by common carrier or sent through the United States mails, it must comply with the Department of Transportation Hazardous Materials Regulations (49 CFR Part 172). The person offering such material for transportation is responsible for ensuring such compliance. For the preservation requirements of this table, the Office of Hazardous Materials, Materials Transportation Bureau, Department of Transportation, has determined that the Hazardous Materials Regulations do not apply to the following materials: hydrochloric acid (HCl) in water solutions at concentrations of 0.04% by weight or less (pH about 1.96 or greater); nitric acid (HNO_3) in water solutions at concentrations of 0.15% by weight or less (pH about 1.62 or greater); sulfuric acid (H_2SO_4) in water solutions at concentrations of 0.35% by weight or less (pH about 1.15 or greater); and sodium hydroxide (NaOH) in water solutions at concentrations of 0.080% by weight or less (pH about 12.30 or less).

be a loss of organics from solution through adsorption of the organics onto the precipitates (U.S. EPA, 1987). Through the use of preservatives, such as nitric acid, to control the pH, it is possible to reverse this process which often begins as soon as ground water is drawn into the well. Aeration (oxidation) of the water sample is certain to occur during the transfer of a sample from a bailer to a sample bottle. The loss of carbon dioxide (assuming that the ground water is saturated or nearly saturated with carbon dioxide) and the addition of oxygen may even be occurring in the sample as it is withdrawn from the formation by the sampling device. While the introduction of oxygen, as shown by Equation 11.3, would increase hydrogen ion activity (pH decrease), the loss of carbon dioxide with consequent pH increase seems to exert the dominant control on pH. Some researchers have noted an increase in pH in unpreserved samples between field measurement and lab measurement of pH. Therefore, lab measurement of pH in ground-water samples is not considered reliable for characterizing ground water (Hem, 1985).

A complicating factor is the natural presence of iron and, occassionally, heavy metals in some formation materials. If particles of the formation materials move into the well and become part of the sample, then addition of acid to the sample may cause these metals to move into solution, because most metal ions are highly soluble at low pH. To avoid this, the sample should be filtered prior to acidification to remove these formation particles. Other suspended debris (i.e., flakes of rust from oxidized metallic well casing) that may be contributed by the well or other nonformation sources are also removed by filtration prior to acidification. As discussed earlier, field filtration means additional handling of the sample in the field with degassing and possible introduction of oxygen or contaminants. Consequently, dissolved metals of interest may precipitate out of solution and become trapped on the filtration media, making the sample invalid. Samples must be carefully handled and preserved as soon as possible to minimize these chemical changes.

Bacteria that may occur naturally in the ground water or that may be introduced during sampling can also change the chemistry of the sample. Bacteria can convert iron compounds from dissolved to undissolved species and vice versa to obtain energy and nutrients (Hem, 1985). The bacteria may catalyze a reaction (Equation 11.3) to obtain oxygen, resulting in the precipitation of ferric hydroxide [$Fe(OH)_3$]. Conversely, the bacteria may oxidize an organic substance in the water, such as a hydrocarbon, to obtain food and energy. The nitric acid preservative destroys bacteria and redissolves metals that may have precipitated or otherwise been affected by the bacteria.

Bacteria also affect concentrations of nutrients such as nitrate and phosphorus. Samples to be analyzed for these parameters may be preserved with sulfuric acid (which does not contain a source of nitrate as does nitric acid) to destroy the bacteria and prevent increases or decreases in nitrate and phosphorus concentrations that would occur if the bacteria remained active. However, the destruction of the bacterial cells by the acid may increase the concentration of nitrates, phosphorus, and other chemical elements and compounds in the sample which were part of the bacteria and not part of the ground water.

Sampling for bacteria requires careful use of sampling equipment that is as sterile as possible. It may be impossible to completely sterilize sampling equipment in the field. Samples from wells in existing water systems equipped with pumps may be more representative of bacterial conditions in the formation. However, the sample must be obtained prior to any chlorination or filtration apparatus in the water system. If a chlorinated sample must be obtained (for monitoring of water supplies and the effectiveness of chlorination), the sample must be treated with sodium thiosulfate ($Na_2S_2O_3$) to neutralize the residual chlorine in the sample.

Some parameters cannot be stabilized. These include dissolved gases, such as oxygen and carbon dioxide, and parameters such as pH, Eh, and alkalinity. pH (a measure of the degree of acidity or basicity in the water solution) is affected by the presence of carbon dioxide and oxygen. Changes in oxygen and carbon dioxide concentrations already have been discussed. Alkalinity is a measure of the ability of the water solution to neutralize acid. Alkalinity should be determined in the field immediately after collection, ideally under conditions that avoid changes in dissolved gas concentrations and consequent precipitation of compounds such as calcium and magnesium carbonate, which affect the alkalinity measurement. Where possible, alkalinity measurements should be taken from samples filtered by an in-line filter to minimize dissolved gas concentration changes.

Preservation Guidelines

Table 11.8 lists information on sample containers, preservation, and storage for many of the parameters typically analyzed in ground-water samples. Samples should be preserved immediately upon collection. If samples are to be split to allow replicate analyses (as required for RCRA monitoring), then it is ideal to collect a large enough sample in one container, using one preservative dosage, and then distribute the sample to individual containers. Due to sampling equipment and parameter constraints (such as aeration of the sample, precipitation of metal(s), and available well yield), it may be necessary to distribute the split sample directly to individual containers from the sampling pump or bailer. In this case, the water should be distributed to each sample bottle in portions (i.e., if there are three split samples, then each bottle should be filled ⅓, repeating in turn until all three bottles are filled). This is necessary because the ground water pumped or bailed from the well is being derived from an ever-increasing area of the formation. Thus, a sample collected early in the sampling event may not be comprised of the same water as a sample collected later.

Preservatives may be added to empty prepared containers in the laboratory, or they may be added to the sample in the field. It is generally better to have the bottles preserved in the laboratory to minimize handling in the field. However, during sample collection, caution must be taken to avoid overfilling of sample bottles which will cause the preservative to be diluted or lost. Whether samples are preserved in the laboratory or the field, gloves and splashproof goggles should be worn during filling of the sample bottles in case of spillage of the preservatives (especially acids and alkalis). When water is added to acid in the sample bottle, there is always danger of creating a heat-generating reaction which may cause spattering of the acid. Therefore, gloves and goggles should be worn even if sample bottles have been pre-preserved.

Sampling for special parameters used in the determination of encrustation and corrosion potential of ground water (for water well maintenance purposes) requires field analysis of alkalinity, acidity (pH), and total dissolved solids. These parameters should be determined in the field immediately upon sample collection. There is no preservative capable of stabilizing these parameters, especially alkalinity, acidity, and pH. Schock and Schock (1982) recommended pressure filtration prior to determination of alkalinity.

Additional important preservation guidelines are provided in the notes at the end of Table 11.8.

BLANKS

Incorporated into every valid ground-water sampling and analysis program is the use of "blank" samples. Typically, two different types of blanks, field blanks and trip blanks, are used. Field blanks are used whenever a common piece of instrumentation or equipment (e.g., bailers, pumps, water level gauges) is used for more than one well. The purpose of the field blank is to determine whether decontamination procedures are effective at preventing carry-over of contaminants from one sampling location to the next. To produce a field blank, the analytical laboratory must provide the ground-water sampling team with a sufficient amount of deionized or distilled water in clean, unused bottles to permit the samplers to run the water from the bottles through the purging and sampling devices directly back into the original water container. This is done after a well has been sampled and the purging/sampling equipment has been decontaminated. In cases in which high-volume submersible pumps are used to purge large quantities of water from large-diameter wells and a bailer or other device is used to collect samples, field-blank water must be run through both pieces of equipment. This may require that substantial volumes of distilled or deionized water be provided to allow generation of the field blank. Once field blanks have been collected and labeled, the samples are placed in the sample shuttles or coolers for storage and shipment to the laboratory.

Trip blanks are used to determine whether or not sample bottles and/or ground-water sample shipment and/or handling procedures have had an effect on sample chemistry. Trip blanks are prepared by the laboratory. A set of sample bottles is filled with distilled or deonized water from the same source used to prepare field blanks. These filled bottles are then shipped to the ground-water sampling team in the same shuttles or coolers containing the empty sample bottles. These bottles remain in the shuttle or cooler, untouched by the sampling team except for their initial shuttle inspection, for the duration of the sampling event, and are returned to the laboratory unopened, along with the ground-water samples.

The number of field and trip blanks incorporated into a sampling and analysis program is highly dependent upon the level of quality control/quality assurance required. As a general rule of thumb, a 10:1:1 ratio of ground-water samples to field and trip blanks is used. Once all samples and blanks have been collected, labeled, and recorded on Chain-of-Custody forms, sample shuttles or coolers must be transported to the laboratory in the most expedient manner possible so that the chemical integrity of this sample is not altered. As indicated in Table 11.8, every parameter has a unique "holding time," the time that can elapse between sample collection and sample analysis. Samples to be analyzed for some parameters that can be chemically or physically preserved or extracted, such as metals or PCBs, can be stored for longer than a month, while samples to be analyzed for more time-sensitive parameters, such as total coliform or volatile organics, can only be held for a matter of hours or days.

SHIPPING AND STORAGE

Once ground-water samples have been collected and pretreated by filtration and/or chemical preservation and appropriately labeled, they must be immediately transferred into hardcased shuttles or coolers for temporary storage and shipment from the well to the analytical laboratory. These shuttles or coolers, when packed with ice or freezer packs, are instrumental in preserving sample integrity by keeping sample temperatures as close to 4°C as possible, protecting samples from sunlight, and minimizing the risk of sample bottle breakage and loss when transporting samples.

It is important to know the relative storage times of each parameter to be analyzed when planning sample collection and shipment. For example, it would not be wise to collect samples on Friday for Monday delivery to the laboratory if samples are to be analyzed for bacteria. To prevent problems such as this from occurring, it is essential that there be good communication between the sampling team and the laboratory performing the sample analyses.

Shipment method is highly dependent upon the nature of samples to be analyzed. Whenever possible, the ideal method of shipment to the analyzing laboratory is hand delivery by the sampling team. This minimizes the number of people who must handle the samples, reduces the number of people who must sign off on the Chain-of-Custody documentation, and reduces the potential for sample tampering. Commercial overnight couriers are perhaps the most common method of sample shipment. However, problems such as bottle breakage due to rough handling, shuttle/cooler loss, unit weight limitations, and cost can be recurrent when shipping by this means. In addition, some samples cannot be transported due to their "hazard," such as flammability and radiation levels; shipping these samples requires special, more costly transportation arrangements. When sample holding times are sufficiently long, it is possible to ship samples using surface carriers (e.g., bus lines or trucking companies). Although slower, these carriers can take heavier packages and are substantially less expensive.

Regardless of which transporation alternative is selected, shuttles or coolers must be carefully packed, securely sealed, and contents visibly marked "fragile" to minimize the potential for in-transit damage and loss of samples. In cases in which samples are collected from sites under regulated monitoring programs, or where litigation is possible, individual sample bottles as well as sample shuttles or coolers must be sealed with numbered seals or tags (with numbers recorded on the Chain-of-Custody forms) which must be inspected by the receiving laboratory to ensure that the shuttle and sample bottles have not been tampered with.

SUMMARY

Ground-water sample collection is more involved than simply removing a volume of water from a monitoring well, transferring it to a readily available sample container, and shipping it to an analytical laboratory. Every aspect of the sampling program can affect the outcome of the sample analysis, rendering the analysis nonrepresentative of ground-water conditions. A variety of factors can affect the chemical quality of a ground-water sample, including well design, installation, and maintenance; filtration; preservation; and shipping and storage. In addition, humans add a random variability during all of the above phases of the sampling program. Careful planning and development of a site-specific sampling and analysis plan can minimize the adverse effects that these factors can have on sample integrity.

Sampling protocols must be chosen which meet the objectives of the sampling program and which minimize alteration of the sample chemistry. Sampling techniques and sample handling procedures must be documented in the site-specific sampling and analysis plan and relayed by both the laboratory and project management to sampling team members. The sampling team must in turn document all field and sample handling activities. These efforts will ensure that the chemical integrity of the ground-water sample is maintained.

While no one protocol can be recommended for all sampling situations, a thorough understanding of how various factors can affect the sample is necessary to design the sampling program and to interpret the results of sample analysis.

REFERENCES

American Public Health Association, American Water Works Association and Water Pollution Control Federation. 1985. *Standard Methods for the Examation of Water and Wastewater*, Sixteenth Edition.

Barcelona, M. J., J. P. Gibb, J. A. Helfrich and E. E. Garske. 1985. *Practical Guide to Groundwater Sampling*. Illinois State Geological Survey Contract Report 374. 94 pp.

Barcelona, M. J., J. P. Gibb, and R. A. Miller. 1983. *A Guide to the Selection of Materials for Monitoring Well Construction and Groundwater Sampling*. Illinois State Water Survey Contract Report 327. 78 pp.

Barcelona, M. J., and J. A. Helfrich. 1986. Well Construction and Purging Effects on Ground-Water Samples. *Environmental Science and Technology*. Vol. 20, No. 11, pp. 1179–1184.

Barcelona, M. J., J. A. Helfrich, E. E. Garske and J. P. Gibb. 1984. A Laboratory Evaluation of Groundwater Sampling Mechanisms. *Groundwater Monitoring Review.* Vol. 4, No. 2, pp. 32–41.

Braids, O. C., R. M. Burger and J. J. Trela. 1987. Should Ground Water Samples from Monitoring Wells be Filtered Before Laboratory Analysis? *Ground Water Monitoring Review.* Vol. 7, No. 2, pp. 58–67.

Brobst, R. B. and P. M. Buszka. 1986. The Effect of Three Drilling Fluids on Ground Water Sample Chemistry. *Ground Water Monitoring Review.* Vol. 6, No. 1, pp. 62–70.

Cheremisinoff, P. N., K. A. Gigliello, and T. K. O'Niell. 1984. *Groundwater-Leachate: Modeling/Monitoring/Sampling.* Technomic Publishing Company, Inc., 146 pp.

Claassen, H. C. 1982. *Guidelines and Techniques for Obtaining Water Samples That Accurately Represent the Water Chemistry of an Aquifer.* U.S. Geological Survey, Open-File Report 82–1024, Lake, Colorado, 49 pp.

Curran, C. M. and M. B. Tomson. 1983. Leaching of Trace Organics into Water from Five Common Plastics. *Ground Water Monitoring Review.* Vol. 3, No. 3, pp. 68–71.

DeVera, E. R., B. P. Simmons, R. D. Stephens and D. L. Storm. 1980. *Samplers and Sampling Procedures for Hazardous Waste Streams.* U.S. Environmental Protection Agency, Cincinnati, Ohio EPA-600/2-80-18. 69 pp.

Dowd, R. M. 1987. "Review of Studies Concerning Effects of Well Casing Materials on Trace Measurements of Organic Compounds." Handout provided by Waste Management, Inc. at the ASTM Seminar on Standards Development for Ground Water Monitoring, January 20 23, 1987, Tampa, Florida.

Driscoll, F. G., ed. 1986. *Ground Water and Wells.* 2nd Edition. Johnson Division, St. Paul, Minn. 1089 pp.

Dunlap, W. J., J. F. McNabb, M. R. Scalf and R. L. Cosby. 1977. *Sampling for Organic Chemicals and Microorganisms in the Subsurface.* U.S. Environmental Protection Agency, Ada, Oklahoma. EPA-600/2-77-176. 27 pp.

Fenn, D., E. Cocozza, J. Isbiter, O. Braids, B. Yare and P. Roux. 1977. *Procedures Manual for Ground Water Monitoring at Solid Waste Disposal Facilities.* U.S. Environmental Protection Agency. EPA/530/SW-611. 260 pp.

Fetter, C. W. 1983. Potential Sources of Contamination in Groundwater Monitoring. *Ground Water Monitoring Review.* Vol. 3, No. 2, pp. 60–64.

Gibb, J. P., R. M. Schuller, and R. A. Griffin. 1981. *Procedures for the Collection of Representative Water Quality Data from Monitoring Wells.* Cooperative Groundwater Report 7. Illinois State Water and Geological Surveys. 61 pp.

Gibs, J. and T. E. Imbrigiotta. 1990. Well-Purging Criteria for Sampling Purgeable Organic Compounds. *Ground Water.* Vol. 28, No. 1, pp. 68–78.

Giddings, T. 1983. Bore-Volume Purging to Improve Monitoring Well Performance: an Often-Mandated Myth. *Proceedings of the Third National Symposium on Aquifer Restoration and Ground Water Monitoring.* National Water Well Association, Dublin, Ohio. pp. 253–256.

Gillham, R. W. 1982. Syringe Devices for Ground Water Sampling. *Ground Water Monitoring Review.* Vol. 2, No. 2, pp. 36–39.

Gillham, R. W., M. J. L. Robin, J. F. Barker and J. A. Cherry. 1983. *Groundwater Monitoring and Sample Bias.* Environmental Affairs Department, American Petroleum Institute. 206 pp.

Goodenkauf, O. 1983. Ground Water Sampling. *Ground Water Age.* Vol. 17, No. 8, pp. 33–34.

Gordon, W. 1984. *Citizen's Handbook on Ground Water Protection.* Natural Resources Defense Council, Inc. New York, 208p.

Griffin, R. A., B. L. Herzog, T. M. Johnson, W. J. Morse, R. E. Hughes, S. F. J. Chou and L. R. Follmer. 1985. Mechanisms of Contaminant Migration Through a Clay Barrier—Case Study, Wilsonville, Illinois. *Proceedings of the Eleventh Annual Research Symposium of the Solid and Hazardous Research Division.* U.S. EPA, Cincinnati, OH, EPA-600/9-95-13, pp. 27–38.

Hem, D. 1985. "Study and Interpretation of the Chemical Characteristics of Natural Water," U.S.G.S. Water Supply Paper 2254, 263 pp.

Herzog, B. L., S. F. J. Chou, J. R. Valkenburg and R. A. Griffin. 1988. Changes in Volatile Organic Chemical Concentrations After Purging Slowly Recovering Wells. *Ground Water Monitoring Review.* Vol. 8, No. 4, pp. 93–99.

Humenick, M. J., L. J. Turk and M. P. Colchin. 1980. Methodology for Monitoring Ground Water at Uranium Solution Miners. *Ground Water.* Vol. 18, No. 3, pp. 262–273.

Kasper, R. B. and J. A. Serkowski. 1988. Evaluation of Sampling Equipment for RCRA Monitoring in a Deep Unconfined Aquifer. *Proceedings of the Second National Outdoor Action Conference on Aquifer Restoration and Ground Water Monitoring,* National Water Well Association, Worthington, Ohio. pp. 445–455.

Keely, J. F. 1982. Chemical Time-Series Sampling. *Ground Water Monitoring Review.* Vol. 2, No. 4, pp. 29–38.

Keely, J. F. 1986. Chemical Time-Series Sampling. *Ground Water Monitoring Review.* Vol. 2, No. 4, pp. 29–38.

Keely, J. F. and K. Boateng. 1987a. Monitoring Well Installation, Purging, and Sampling Techniques—Part 1: Conceptualization. *Ground Water.* Vol. 25, No. 3, pp. 300–313.

Keely, J. F. and K. Boateng. 1987b. Monitoring Well Installation, Purging, and Sampling Techniques—Part 2: Case Histories. *Ground Water.* Vol. 25, No. 4, pp. 427–439.

Kill, D. 1986. Drilling Practices for Groundwater Monitoring Wells. *Subsurface Monitoring Technology.* University of Wisconsin-Madison. 19 pp.

Lee, G. F. and R. A. Jones. 1983. Guidelines for Sampling Ground Water. *Journal of the Water Pollution Control Federation.* Vol. 55, No. 1, pp. 92–96.

McAlary, T. A. and J. F. Barker. 1987. Volatilization Losses of Organics During Ground Water Sampling from Low Permeability Materials. *Ground Water Monitoring Review.* Vol. 7, No. 4, pp. 63–68.

Marsh, J. M. and J. W. Lloyd. 1980. Details of Hydrochemical Variations in Flowing Wells. *Ground Water.* Vol. 18, No. 4, pp. 366–373.

Miller, G. D. 1982. Uptake and Release of Lead, Chromium and Trace-Level Volatile Organics Exposed to Synthetic Well Casings; *Proceedings of the Second National Symposium on Aquifer Restoration and Ground-Water Monitoring.* National Water Well Association, Dublin, OH. pp. 236–245.

Muska, C. F., W. P. Colven, V. D. Jones, J. T. Scogin, B. B. Looney and V. Price, Jr. 1986. Field Evaluation of Ground Water Sampling Devices for Volatile Organic Compounds. *Procedding of the Sixth National Symposium and Exposition on Aquifer Restoration and Ground Water Monitoring.* National Water Well Association, Worthington, Ohio. pp. 235–246.

Nacht, S. J. 1983. Monitoring Sampling Protocol Considerations. *Ground Water Monitoring Review.* Vol. 3, No. 3, pp. 23–29.

Nielsen, D. M. and G. L. Yeates. 1985. A Comparison of Sampling Mechanisms Available for Small-Diameter Ground Water Monitoring Wells. *Ground Water Monitoring Review.* Vol. 5, No. 2, pp. 83–99.

Papadopulos, I. S. and H. H. Cooper. 1967. Drawdown in a Well of Large Diameter. *Water Resources Research.* Vol. 3, No. 1, pp. 241–244.

PEI Associates, Inc. 1987. Monitoring System Design and Construction in *Groundwater Monitoring Seminar Series-Technical Papers.* U.S. Environmental Protection Agency. 29 pp.

Pennino, J. 1987. "Dissolved Oxygen Changes in Well Water During Purging and Sampling," Unpublished Report.

Pettyjohn, W. A. 1982. Cause and Effect of Cyclic Changes in Ground Water Quality. *Ground Water Monitoring Review.* Vol. 2, No. 1, pp. 43–49.

Pettyjohn, W. A., W. J. Dunlap, R. Cosby and J. W. Keeley. 1981. Sampling Ground Water for Organic Contaminants. *Ground Water.* Vol. 19, No. 2, pp. 180–189.

Rajagopal, R. 1986. The Effect of Sampling Frequency on Ground Water Quality Characterization. *Ground Water Monitoring Review.* Vol. 6, No. 3, pp. 65–73.

Reynolds, G. W. and R. W. Gillham. 1985. Absorption of Halogenated Organic Compounds by Polymer Materials Commonly Used in Ground Water Monitoring. *Proceedings of the Seconds Canadian/American Conference on Hydrogeology.* Nations Water Well Association, Dublin, OH. pp. 125–132.

Richard, M. R. 1979. The Organic Drilling Fluid Controversy: Part I. *Water Well Journal.* Vol. 33, No. 4, pp. 66–74.

Ronen, D., M. Magaritz, H. Gvirtzman and W. Garner. 1987. Microscale Chemical Heterogeneity in Groundwater. *Journal of Hydrology.* Vol. 92, No. 1/2, pp. 173–178.

Scalf, M. R., J. F. McNabb, W. J. Dunlop, R. L. Cosby, and J. S. Fryberger. 1981. *Manual of Groundwater Sampling Procedures.* National Water Well Association. 93 pp.

Schafer, D. C. 1978. Casing Storage Can Affect Pumping Test Data. *The Johnson Drillers Journal.* St. Paul, Minn. Jan.–Feb. 1978, pp. 1–5.

Schalla, R., D. Myers, M. Simmons, J. Thomas and A. Toste. 1988. The Sensitivity of Four Monitoring Well Sampling Systems to Low Concentrations of Three Volatile Organics. *Ground Water Monitoring Review.* Vol. 8, No. 3, pp. 90–96.

Schmidt, K. D. 1977. Water-quality Variations for Pumping Well. *Ground Water.* Vol. 15, No. 2, pp. 130–137.

Schmidt, R. 1986. Personal Communication.

Schock, M. R., and S. C. Schock. 1982. Effect of Container Type on pH and Alkalinity Stability. *Water Research,* Vol. 16, No. 10, pp. 1455-1464.

Schuller, R. M., J. P. Gibb and R. A. Griffin. 1981. Recommended Sampling Procedures for Monitoring Well. *Ground Water Monitoring Review.* Vol. 1, No. 2, pp. 42–46.

Seanor, A. M. 1984. Monitoring Report: Assuring the Quality of Ground Water Samples. *Ground Water Age.* Vol. 18, No. 8, pp. 41–47.

Seanor, A. M. and L. K. Brannaka. 1983. Efficient Sampling Techniques. *Ground Water Age.* Vol. 17, No. 8, pp. 41–46.

Sosebee, J. B., P. C. Geiszler, D. L. Winegardner and C. R. Fisher. 1982. Contamination of Ground Water Samples with PVC Adhesives and PVC Primer from Monitoring Wells. Technical Paper, Environmental Science and Engineering, Englewood, CO. 24p.

Stolzenburg, T. R. and D. G. Nichols. 1985. *Preliminary Results on Chemical Changes in Groundwater Samples Due to Sampling Devices.* EPRI EA-4118, Project 2485-7, Electric Power Research Institute, Palo Alto, CA.

Stolzenburg, T. R., and D. G. Nichols. 1986. Effects of Filtration Method and Sampling on Inorganic Chemistry of Sapled Well Water. *Proceedings of Sixth National Symposium and Exposition on Aquifer Restoration and Ground Water Monitoring.* National Water Well Association, Dublin, OH. pp. 216-234.

Sykes, A. L., R. A. McAllister and J. B. Homolya. 1986. Sorption of Organics by Monitoring Well Construction Materials. *Ground Water Monitoring Review.* Vol. 6, No. 4, pp. 44-47.

Summers, W. K. and L. A. Brandvold. 1967. Physical and Chemical Variations in the Discharge of a Flowing Well. *Ground Water.* Vol. 5, No. 1, pp. 9–10.

Sykes, A. L., R. A. McAllister and J. B. Homolya. 1986. Sorption of Organics by Monitoring Well Construction Materials. *Ground Water Monitoring Review.* Vol. 6, No. 4, pp. 44-47.

Unwin, J. P. and D. Huis. 1983. A Laboratory Investigation of the Purging Behavior of Small-Diameter Monitoring Well. *Proceedings of the Third Annual Symposium on Ground Water Monitoring and Aquifer Restoration.* National Water Well Association, Dublin, Ohio. pp. 257-262.

U.S. Environmental Protection Agency. 1977. *Procedures Manual for Ground Water Monitoring at Solid Waste Disposal Facilities.* EPA/530/SW-611. 269 pp.

U.S. Environmental Protection Agency. 1987. Groundwater Monitoring Seminar Series, CERI-87-7, Chapter IV.

U.S. Environmental Protection Agency Environmental Monitoring and Support Laboratory. 1982. "Handbook for Sampling and Sample Preservation of Water and Wastewater," Cincinnati, Ohio, pp. 386–397.

U.S. Environmental Protection Agency Office of Waste Programs Enforcement, Office of Solid Waste and Emergency Response. 1986. "RCRA Technical Enforcement Guidance Document," OSWER-9950.1, U.S. Government Printing Office, Washington, D.C., 208p., appendices.

U.S. Office of the Federal Register, National Archives and Records Administration. July 1, 1986. "Code of Federal Regulations, 40 CFR Parts 190 to 399," Sections 264.97 to 264.99, pp. 446–453, Appendix IV, pp. 537–538.

U.S. Office of the Federal Register, National Archives and Records Administration. July 1, 1986. "Code of Federal Regulations, 40 CFR Part 300—National Oil and Hazardous Substances Pollution Contingency Plan," pp. 732–860.

U.S. Office of the Federal Register, National Archives and Records Administration. April 17, 1987. "Code of Federal Regulations, 40 CFR Parts 280 and 281, Underground Storage Tanks; Proposed Rules," pp. 12,662–12,864.

Walker, S. E. 1983. Background Ground Water Quality Monitoring: Well Installation Trauma. *Proceedings of the Third National Symposium on Aquifer Restoration and Ground Water Monitoring.* National Water Well Association, Worthington, Ohio. pp. 235–246.

Wilson, L. G. and Dworkin, J. M. 1984. *Development of a Primer on Well Water Sampling for Volatile Organic Substances.* U.S. Geological Survey, Reston, Virginia. 44 pp.

Wood, W. W. 1976. *Guidelines for Collection and Field Analyses of Ground Water Samples for Selected Unstable Constituents.* U.S. Geological Survey Techniques of Water Resources Investigations. Book 1 Chapter D-2. 24 pp.

12

Ground-Water Sample Analysis

Rock J. Vitale, Olin Braids, and Rudolph Schuller

INTRODUCTION

Over the past 15 years, there has been a virtual revolution in the analytical capabilities of laboratories for handling water samples. In the early 1970s, precise and sensitive analysis was available for the common ions and so-called trace metals found in water. At that time, analytical procedures and instrumentation for determining organic substances were limited in sensitivity and scope.

A very general or surrogate analysis to determine the presence of organic compounds was possible by using total organic carbon (TOC), chemical oxygen demand (COD), and biochemical oxygen demand (BOD). These analytical methods either measured the gross amount of carbon in a water sample or measured carbon indirectly by reacting it chemically or biologically to determine how much oxygen would be utilized in carbon oxidation. By today's standards, these methods, although still used for certain legitimate purposes, would be considered only a general indication of the presence of organic materials in water samples.

Determination of specific organic compounds in water was possible in the early 1970s, but was limited in its sensitivity and subject to protocols which were individually developed by each lab or researcher in the field. One of the problems in dealing with organics in water was extracting the organic substances from the water matrix into a matrix that would be amenable to gas chromatographic analysis. The usual approach was to use organic extracting solvents which would be concentrated and then injected into a gas chromatograph. The research of Bellar and Lichtenberg in 1974 resulted in a method for organic volatile compounds that released these compounds from water by purging the sample with air, followed by capturing the released compounds on an exchange resin. This development in dealing with volatile organic compounds enabled the analysis of volatiles to be done rapidly and with high sensitivity. Surveys of public water supplies that were made following this analytical development resulted in the detection of volatile organic compounds (i.e., chloroform) in many public water supplies in the U.S. (*Federal Register,* 1985).

The discovery of volatile organic compound contaminants in public water supplies was accompanied by their detection in ground water which had been contaminated by chemicals associated with industrial processes, wastes, and other anthropogenic sources. The presence of organic compounds, coupled with the ability to determine their presence and concentration, has resulted in a continuing development of analytical scope and sensitivity. As the scope of analysis increased, the selection process for determining desired analytes became more complicated.

SELECTION OF ANALYTICAL PARAMETERS

The selection of analytical parameters for a ground-water investigation is primarily driven by the purpose of the investigation, which is often affected by its regulatory status, existing site conditions, and a knowledge of past site practices.

Ground-water investigations can be grouped into two general categories: those done to determine the natural quality of ground water for academic interest or water supply, and those that are seeking to determine if chemical contaminants are present. In either case, the list of analytes may not be appreciably different. This is because anthropogenic sources of contaminants are so widespread that water unaffected by agricultural or forest management practices, at the very least, is found only in remote locations.

In any given investigation, there may be individual elements or groups of compounds which can be deleted from the analytical list based upon the results of previous sampling events and/or their low probability of detection in ground water. For example, some compounds have extremely low water solubility and their presence in ground water would be considered highly unlikely. Conversely, elements such as aluminum, barium, copper, and vanadium, and compounds such as 2-butanone and 4-methyl-2-pentanone (to name only a few) which were ignored in water quality analysis several years ago, are now of concern and their analysis is required in some ground-water investigations.

A more detailed discussion of the common types of investigations and the typical lists of analytical parameters that are analyzed are presented below. In most cases, the various required parameters consist of a mixture of organic and inorganic constituents in addition to measures of aesthetic quality such as color, turbidity, and odor.

GROUND-WATER INVESTIGATIONS
GOVERNED BY A REGULATORY AGENCY

Existing federal regulations which have established lists of parameters for analysis of ground-water samples include: the Resource Conservation and Recovery Act (RCRA), the Safe Drinking Water Act (SDWA), the Clean Water Act (CWA), and the Superfund Amendments and Reauthorization Act (SARA). Individual states may also have variations on these lists or separate lists and requirements, depending upon the nature of the investigation.

Analytical Requirements Under RCRA

The Resource Conservation and Recovery Act (RCRA) was enacted to regulate activities related to the transport, storage, and disposal of hazardous wastes. As part of the overall regulation, ground water is specifically addressed. Some hazardous-waste disposal facilities and some storage facilities are required to have ground-water monitoring wells. These are specified to include at least one well upgradient of the facility or group of facilities, and at least three wells downgradient.

Monitoring well locations are specified in the configuration mentioned above to provide water-quality information that will indicate if any contamination has occurred in the ground-water regime monitored due to the waste facility and to provide water-level information. The latter is required to establish the direction of ground-water flow and potential direction of contaminant migration.

Water-quality parameters required under RCRA are divided into categories with different requirements for replication and frequency of analysis. The parameters which indicate if ground water is an acceptable drinking water source are included on the U.S. EPA Primary Drinking Water Standards (Table 12.1) which were established under the Safe Drinking Water Act (SDWA) of 1974. Parameters establishing ground-water quality are chloride, iron, manganese, phenols, sodium, and sulfate. Parameters designated as general indicators of ground-water contamination are pH, specific conductance, total organic carbon, and total organic halogen (Table 12.2).

The combination of the above parameters must be used to establish ground-water quality for regulatory compliance, although site-specific parameters may also be required by the regulatory agency. Quarterly monitoring for the site is required for the first year. The indicators of ground-water contamination must also be measured on four replicate samples at each sampling event. This will result in 16 analyses for these parameters in the first year. With 16 analyses performed, statistics can be applied (i.e., Student's t-test) to determine trends and overall data significance.

Following the first year, samples collected to establish ground-water quality must be obtained and analyzed at least annually. The indicators of ground-water contamination must be determined semiannually. Water-level measurements are also required at each sampling event.

Table 12.1. Primary Drinking Water Standards (MCLs).[a]

Metals	
Arsenic	50 μg/L
Barium	1000 μg/L
Cadmium	10 μg/L
Chromium	50 μg/L
Lead[a]	50 μg/L
Mercury	2 μg/L
Selenium	10 μg/L
Volatile Organics	
Vinyl chloride	2 μg/L
Trichloroethene	5 μg/L
Benzene	5 μg/L
Carbon tetrachloride	5 μg/L
1,2-Dichloroethene	5 μg/L
1,1-Dichloroethene	7 μg/L
1,1,1-Trichloroethane	200 μg/L
Semivolatiles	
1,4-Dichlorobenzene	75 μg/L
2,4,5-Trichlorophenol	10 μg/L
Pesticides/Herbicides	
2,4-Dichlorophenoxyacetic acid	100 μg/L
gamma-BHC	4 μg/L
Methoxychlor	100 μg/L
Toxaphene	5 μg/L
Additional Paremeters	
Nitrate	10,000 μg/L
Fluoride	4,000 μg/L

[a]MCL for lead of 5 μg/L has been recently proposed.

Table 12.2. RCRA Hazardous Waste, Subpart F Ground-Water Sampling Frequency and Analysis Requirements.

Sample Frequency	Parameters Analyzed from Upgradient Wells	Parameters Analyzed from Downgradient Wells
First Year After Installation of Monitor Wells:		
Quarterly (every 3 months)	pH[a]	pH
	Specific conductance[a]	Specific conductance
	Total organic carbon[a]	Total organic carbon
	Total organic halogen[a]	Total organic halogen
	Chloride	Chloride
	Iron	Iron
	Manganese	Manganese
	Phenols	Phenols
	Sodium	Sodium
	Sulfate	Sulfate

continued

Table 12.2. Continued.

Sample Frequency	Parameters Analyzed from Upgradient Wells	Parameters Analyzed from Downgradient Wells
First Year After Installation of Monitor Wells (continued):		
	Arsenic	Arsenic
	Barium	Barium
	Cadmium	Cadmium
	Chromium	Chromium
	Fluoride	Fluoride
	Lead	Lead
	Mercury	Mercury
	Nitrate (as N)	Nitrate (as N)
	Selenium	Selenium
	Silver	Silver
	Endrin	Endrin
	Lindane	Lindane
	Methoxychlor	Methoxychlor
	Toxaphene	Toxaphene
	2,4-D	2,4-D
	2,4,5-TP silvex	2,4,5-TP silvex
	Radium	Radium
	Gross alpha	Gross alpha
	Gross beta	Gross beta
	Turbidity	Turbidity
	Coliform bacteria	Coliform bacteria
Subsequent Years (after first year):		
Semiannual (every 6 months)	pH[a]	pH[a]
	Specific conductance[a]	Specific conductance[a]
	Total organic carbon[a]	Total organic carbon[a]
	Total organic halogen[a]	Total organic halogen[a]
Annually	Chloride	Chloride
	Iron	Iron
	Manganese	Manganese
	Phenols	Phenols
	Sodium	Sodium
	Sulfate	Sulfate

[a]Replicate measurements.

Under certain conditions, analytical requirements under RCRA may include the analysis of a very extensive list of parameters included in RCRA Appendix VIII or Appendix IX Constituents.

Many states have adopted the national primary and secondary drinking water regulations or have modified them in part to become more stringent and applied them to ground-water investigations within the state. Although ground water may not meet the drinking water standards in all places, the objective of applying drinking water standards is to provide a goal to which ground water should be treated if, in fact, contaminants have been introduced into them.

Requirements Under Clean Water Act (CWA)

In general, the Clean Water Act focuses upon surface water quality and discharges into the surface waters of the United States. The discharge regulations were established under the National Pollutant Discharge Elimination System (NPDES). This program has resulted in the construction of a number of industrial and municipal wastewater treatment plants that treat wastewater prior to discharge into surface-water bodies.

Under the CWA, a Consent Decree signed by the U.S. EPA required the identification of pollutant chemical classes (or groups of similar types of chemicals) that may be of concern if found in surface water or ground water—the Priority Pollutant List of compounds and elements resulted. This list, shown in Table 12.3, is the most applicable part of the CWA to ground water. Priority pollutant analyses, or parts thereof, have been applied to many ground-water investigations under state and federal regulatory purview. This list of compounds is useful because it includes the organic compounds that are frequently used as raw materials or stored as hazardous waste by industry. The list of inorganic constituents is not all-inclusive, but adds a few toxic elements to the drinking water list of parameters and to a general water-quality analysis.

The national primary and secondary drinking water regulations are also cited under the CWA. Many states have adopted the drinking water regulations or have modified them in part to become more stringent and applied them to ground-water investigations within the state. Although ground water may not meet the drinking water standards in all places, the objective of applying drinking water standards is to provide a goal to which ground water should be treated if, in fact, contaminants have been introduced into them.

Table 12.3.   129 Environmental Protection Agency Priority Pollutants.

Base-Neutral Extractable Organics	Volatile Organics	Pesticides/PCBs
Acenaphthene	Acrolein	Aldrin
Acenaphthylene	Acrylonitrile	alpha-BHC
Anthracene	Benzene	beta-BHC
Benzidine	bis(Chloromethyl) ether	gamma-BHC
Benzo(a)anthracene	Bromoform	delta-BHC
Benzo(a)pyrene	Carbon tetrachloride	Chlordane
3,4-Benzofluoranthene	Chlorobenzene	4,4'-DDT
Benzo(ghi)perylene	Chlorodibromomethane	4,4'-DDE
Benzo(k)fluoranthene	Chloroethane	4,4'-DDD
bis(2-Chloroethoxy) methane	2-Chloroethyl vinyl ether	Dieldrin
bis(2-Chloroethyl) ether	Chloroform	alpha-Endosulfan
bis(2-Chlorisopropyl) ether	Dichlorobromomethane	beta-Endlosulfan
bis(2-Ethylhexyl) phthalate	Dichlorodifluoromethane	Endosulfan sulfate
4-Bromophenyl phenyl ether	1,1-Dichloroethane	Endrin
Butyl benzyl phthalate	1,2-Dichloroethane	Endrin aldehyde

continued

Table 12.3. Continued.

Base-Neutral Extractable Organics (continued)	Volatile Organics (continued)	Pesticides/PCBs (continued)
2-Chloronaphthalene	1,1-Dichloroethylene	Heptachlor
4-Chlorophenyl phenyl ether	1,2-Dichloropropane	Heptachlor epoxide
Chrysene	1,3-Dichloropropylene	PCB-1242
Dibenzo(a,h)anthracene	Ethylbenzene	PCB-1254
1,2-Dichlorobenzene	Methyl bromide	PCB-1221
1,3-Dichlorobenzene	Methyl chloride	PCB-1232
1,4-Dichlorobenzene	Methylene chloride	PCB-1248
3,3′-Dichlorobenzidine	1,1,2,2-Tetrachloroethane	PCB-1260
Diethyl phthalate	Tetrachloroethylene	PCB-1016
Dimethyl phthalate	Toluene	Toxaphene
Di-n-butyl phthalate	1,2-trans-Dichloroethylene	
2,4-Dinitrotoluene	1,1,1-Trichloroethane	**Metals**
2,6-Dinitrotoluene	1,1,2-Trichloroethane	
Di-n-octyl phthalate	Trichloroethylene	Antimony
1,2-Diphenylhydrazine	Trichlorofluoromethane	Arsenic
Fluoranthene	Vinyl chloride	Beryllium
Fluorene		Cadmium
Hexachlorobenzene		Chromium
Hexachlorobutadiene		Copper
Hexachlorocyclopentadiene		Lead
Hexachloroethane	**Acid Extractable Organics**	Mercury
Indeno(1,2,3-c,d)pyrene		Nickel
Isophorone	2-Chlorophenol	Selenium
Naphthalene	2,4-Dichlorophenol	Silver
Nitrobenzene	2,4-Dimethylphenol	Thallium
N-Nitrosodimethylamine	4,6-Dinitro-o-cresol	Zinc
N-Nitrosodi-n-propylamine	2,4-Dinitrophenol	
N-Nitrosodiphenylamine	2-Nitrophenol	
Phenanthrene	4-Nitrophenol	**Miscellaneous**
Pyrene	p-Chloro-m-cresol	
1,2,4-Trichlorobenzene	Pentachlorophenol	Total cyanides
	Phenol	Total phenols
	2,4,6-Trichlorophenol	Asbestos

Requirements Under the Superfund Amendments and Reauthorization Act (SARA)

The SARA legislation has extended the national contingency plan, and some of its provisions have been augmented. Under the subject of analytical parameters, the Priority Pollutant List established under the Clean Water Act has been supplemented with compounds included under the Superfund Contract Laboratory Program (CLP). Complete lists of compounds on the Target Compound List (TCL) and elements on the Target Analyte List (TAL) are provided in Tables 12.4–12.7.

Under a Remedial Investigation/Feasibility Study of a Superfund site, the standard analytical suite is presently referred to as the TCL. Although the TCL includes many parameters, additional parameters could be added if there is

Table 12.4.   Volatile Target Compound List (TCL).[a]

Volatiles	CAS Number
Chloromethane	74-87-3
Bromomethane	74-83-9
Vinyl chloride	75-01-4
Chloroethane	75-00-3
Methylene chloride	75-09-2
Acetone	67-64-1
Carbon disulfide	75-15-0
1,1-Dichloroethene	75-35-4
1,1-Dichloroethane	75-35-3
trans-1,2-Dichloroethene	156-60-5
Chloroform	67-66-3
1,2-Dichloroethane	107-06-2
2-Butanone	78-93-3
1,1,1-Trichloroethane	71-55-6
Carbon tetrachloride	56-23-5
Vinyl acetate	108-05-4
Bromodichloromethane	75-27-4
1,1,2,2-Tetrachloroethane	79-34-5
1,2-Dichloropropane	78-87-5
trans-1,3-Dichloropropene	10061-02-6
Trichloroethene	79-01-6
Dibromochloromethane	124-48-1
1,1,2-Trichloroethane	79-00-5
Benzene	71-43-2
cis-1,3-Dichloropropene	10061-01-5
Bromoform	75-25-2
2-Hexanone	591-78-6
4-Methyl-2-pentanone	108-10-1
Tetrachloroethene	127-18-4
Toluene	108-88-3
Chlorobenzene	108-90-7
Ethyl benzene	100-41-4
Styrene	100-42-5
Total xylenes	100-42-5

[a]U.S. EPA, Aug., 1987.

Table 12.5.   Semivolatile Target Compound List (TCL).[a]

Semivolatiles	CAS Number
Phenol	108-95-2
bis(2-Chloroethyl)ether	111-44-4
2-Chlorophenol	95-57-8
1,3-Dichlorobenzene	541-73-1
1,4-Dichlorobenzene	106-46-7
Benzyl alcohol	100-51-6
1,2-Dichlorobenzene	95-50-1
2-Methylphenol	95-48-7
bis(2-Chloroisopropyl)ether	39638-32-9
4-Methylphenol	106-44-5
n-Nitroso-dipropylamine	621-64-7
Hexachloroethane	67-72-1
Nitrobenzene	98-95-3

continued

Table 12.5.   Continued.

Semivolatiles	CAS Number
Isophorone	78-59-1
2-Nitrophenol	88-75-5
2,4-Dimethylphenol	105-67-9
Benzoic acid	65-85-0
bis(2-Chloroethoxy)methane	111-91-1
2,4-Dichlorophenol	120-83-2
1,2,4-Trichlorobenzene	120-82-1
Naphthalene	91-20-3
4-Chloroaniline	106-47-8
Hexachlorobutadiene	87-68-3
4-Chloro-3-methylphenol	59-50-7
2-Methylnapthalene	91-57-6
Hexachlorocyclopentadiene	77-47-4
2,4,6-Trichlorophenol	88-06-2
2,4,5-Trichlorophenol	95-95-4
2-Chloronapthalene	91-58-7
2-Nitroaniline	88-74-4
Dimethyl phthalate	131-11-3
Acenaphthylene	208-96-8
3-Nitroaniline	99-09-2
Acenaphthene	83-32-9
2,4-Dinitrophenol	51-28-5
4-Nitrophenol	100-02-7
Dibenzofuran	132-64-9
2,4-Dinitrotoluene	121-14-2
2,6-Dinitrotoluene	606-20-2
Diethylphthalate	84-66-2
4-Chlorophenyl phenyl ether	7005-72-3
Fluorene	86-73-7
4-nitroaniline	100-01-6
4,6-Dinitro-2-methylphenol	534-52-1
N-Nitrosodiphenylamine	86-30-6
4-Bromophenyl phenyl ether	101-55-3
Hexachlorobenzene	118-74-1
Pentachlorophenol	87-86-5
Phenanthrene	85-01-8
Anthracene	120-12-7
Di-n-butylphthalate	84-74-2
Fluoranthene	206-44-0
Pyrene	129-00-0
Butyl benzyl phthalate	85-68-7
3,3'-Dichlorobenzidine	91-94-1
Benzo(a)anthracene	56-55-3
bis(2-Ethylhexyl)phthalate	117-81-7
Chrysene	218-01-9
Di-n-octyl phthalate	117-84-0
Benzo(b)fluoranthene	205-99-2
Benzo(k)fluoranthene	207-08-9
Benzo(a)pyrene	50-32-8
Indeno(1,2,3-cd)pyrene	193-39-5
Dibenzo (a,h)anthracene	53-70-3
Benzo(g,h,i)perylene	191-24-2

[a]U.S. EPA, Aug., 1987.

Table 12.6.   Pesticide/PCB Target Compound List (TCL).[a]

Pesticide/PCB	CAS Number
alpha-BHC	319-84-6
beta-BHC	319-87-7
delta-BHC	319-86-8
gamma-BHC(Lindane)	58-89-9
Heptachlor	76-44-8
Aldrin	309-00-2
Heptachlor epoxide	1024-57-3
Endosulfan I	959-98-8
Dieldrin	60-57-1
4,4'-DDE	72-55-9
Endrin	72-20-8
Endosulfan II	33213-65-9
4,4'-DDD	72-54-8
Endosulfan sulfate	1031-07-8
4,4'-DDT	50-29-3
Methoxychlor	72-43-5
Endrin ketone	53494-70-5
Chlordane	57-74-9
Toxaphene	8001-35-2
Arochlor-1016	12674-11-2
Arochlor-1221	11104-28-2
Arochlor-1232	11141-16-5
Arochlor-1242	53469-21-9
Arochlor-1248	12672-29-6
Arochlor-1254	11097-69-1
Arochlor-1260	11096-82-5

[a]U.S. EPA, Aug., 1987.

Table 12.7.   Inorganic Target Analyte List (TAL).

Aluminum
Antimony
Arsenic
Barium
Beryllium
Cadmium
Calcium
Chromium
Cobalt
Copper
Iron
Lead
Magnesium
Manganese
Mercury
Nickel
Potassium
Selenium
Silver
Sodium
Thallium
Vanadium
Zinc
Cyanide

information that indicates the possible presence of specific compounds at the site (i.e., waste products known to be present at the site). As part of the investigation, records of the potentially responsible parties (PRPs) and those of waste handlers are reviewed to determine the composition of materials which could be present at the site. This information should be utilized in making decisions on which parameters should be included (or deleted) in the analytical scheme for the site.

Requirements Under a Site-Specific Administrative Consent Order (ACO)

The preceding sections have dealt with specific requirements for selecting water-quality parameters under several environmental acts. The requirements have been developed to provide a broad-based analytical strategy in order to detect and measure chemical species, particularly contaminants that might be present at a site. In some instances, a regulatory agency will require the facility and/or responsible party to enter into an Administrative Consent Order (ACO). ACOs are legally binding agreements between the regulatory agency and a facility and/or responsible party. A list of compounds and constituents for analysis stems from an ACO and is usually developed on a site-specific basis. Because of this, benefits obtained from an historical data base (i.e., results of previous monitoring programs) can be extremely advantageous. ACOs also reference specific locations and facilities and should take into account the composition of materials being stored, handled, or discharged.

Monitoring programs under an ACO frequently have the same pattern as under other regulations. That is, annual samplings are generally more comprehensive than are quarterly or semiannual samplings. To accommodate this monitoring philosophy, chemical constituents are chosen on the basis of establishing general water quality, scanning for potential contaminants, and assessing the presence of specific compounds and constituents known to be present at the facility.

An example of establishing general water quality would be measuring the major cations and anions for purposes of obtaining a general geochemical characterization. Parts of the SARA TCL volatile or semivolatile analysis may be required to monitor for compounds known or alleged to be present. In this way, the compounds in any given category should be detected (if present). Finally, specific compounds not on the TCL may be required analytical parameters on a site-specific basis. Compounds such as Aldicarb, methyl cellosolve, ethyl butyl ether, butanol, or hundreds of other compounds may be included on the analytical list.

The strategy under an ACO is likely to be similar to that for RCRA or other regulatory programs. Key indicator constituents are chosen for frequent sampling—quarterly, semiannually, or as otherwise specified—and a more comprehensive list of constituents is developed for annual or less frequent measurement.

Analytes Which Are Site-Related

As indicated by the size of the RCRA Appendix IX list, the range of chemicals associated with major manufacturing categories is wide. The Priority Pollutant list was developed from the chemicals most frequently detected in industrial wastewater effluents. However, those waste streams represent only a fraction of the total number of chemicals that are stored, handled, or discharged by industry. A comprehensive guide to industrial waste chemicals is too ambitious for this chapter, but a condensed guide to common contaminants and manufacturing classes is given by SCS Engineers (1985). This guide will indicate the class or classes of chemicals that may be encountered. For other than screening purposes, information from personnel on factory records and previous investigations (if conducted) should be used to create a site-specific list of analytes.

SELECTION OF ANALYTICAL METHOD

Just as important as the selection of the analytical parameters is the selection of the analytical method. The selection of the analytical method is in turn determined by the purpose of the investigation. After establishing the purpose and requirements, an investigator must select the appropriate analytical methods for the parameters of interest. It is the responsibility of the investigator to specify the analytical methodology that he/she wants to utilize for the analysis. Quite often, the investigator may not be aware of the differences between methods. In such a case, it is important to explain to the laboratory *exactly* what the analysis is supposed to accomplish.

As stated earlier, the analytical method will be dependent upon the purpose of the investigation. For example, if the purpose of the investigation is to determine the presence of a specific organic contaminant within certain concentration bounds, the submission of samples for total organic carbon (a nonselective analysis) may not accomplish the objective. Neither would specifying an analytical method that could not obtain the required detection sensitivity.

There may actually be several analytical methods to choose from for the same parameter. Each method should, in the ideal sense, give the same result. However, due to the variables within each method, the results between various methods can vary dramatically. For this reason, some methods may be preferred or even mandated, depending on whether the analytical results are to be prepared for, or in conjunction with, a regulatory agency. For example, EPA may not accept analytical results based upon methods which are not EPA-approved methods.

Specific Requirements for an Analytical Method

Before ground-water samples are submitted to a laboratory, the specific requirements of the analysis, as dictated by the purpose of the investigation, must be communicated to the laboratory so that the investigator does not have to assume that the laboratory understands the requirements of the investigation. Allowing the laboratory to conduct an analysis by its standard procedures could lead to data which is inappropriate (or useless) for the investigation. Perhaps the most important specific requirement for ground-water investigations is the detection limit that will be reported for the requested analysis.

If ground-water samples are to be taken to show that contamination is not present, the level at which contaminants can be detected by current environmental technology must be specified. To say that no contamination exists is correct only to the quantitative extent that the analysis is capable of detecting the contaminant. This level is commonly referred to as a "detection limit." In laypersons' terms, a detection limit is the quantitative point at which the analyte will be detected 99% of the time. Detection limits are usually reported on a weight-by-volume basis (i.e., μg/L or mg/L), or on a statistical basis (i.e., ppb or ppm).

An expensive ground-water sampling and analysis investigation may result in useless information if the detection limits are not low enough for the purpose of the study. An example of this is the analytical detection limit versus a risk determination. Quite often, the primary objective of a ground-water investigation is to assure that human health and the environment are not at risk based upon exposure to contaminants which may be present in the ground water. Accordingly, the ideal detection limits that will meet these objectives are the specific human health-based criteria and environmental-based criteria for the parameters of interest. Obviously the data is of limited usefulness if the analytical detection limits are higher than the most relevant health-based criteria required.

Other specific information that should be discussed with laboratory personnel prior to the sampling and analysis include sample volume requirements, laboratory quality assurance/quality control (QA/QC), chain-of-custody, data package (documentation), sample bottles, preservatives, turnaround time, and reporting requirements.

Description of Analytical Methods

After determining the purpose of the investigation and the specific requirements, the analytical method can be selected. Some of the most popular references for analytical methods are "Methods for Chemical Analysis of Water and Wastes" and "Test Methods for Evaluating Solid Wastes" (SW846) (U.S. EPA 1979, 1986). The following sections will discuss some of the more general methods available for ground-water investigations, some of the more commonly analyzed organic and inorganic parameters, and the potential benefits and problems associated with the various methods.

SCREENING OR DIAGNOSTIC TESTS

Screening or diagnostic tests are procedures that provide an initial indication of the quality of water with an economy of time and dollars. Although they can seldom be used alone because they are screening methods, they can provide valuable information when sampling a large number of wells. These analytical procedures are typically conducted in the laboratory, although some are routinely conducted in the field. Total dissolved solids/specific conductance, total organic carbon, and total organic halogens are examples of screening or diagnostic tests.

Total Dissolved Solids (TDS)/Specific Conductance

The amount of dissolved matter in water is expressed as the concentration of TDS. Usually this measurement represents inorganic matter, yet when organic solvents reach parts-per-million levels, they may also contribute significantly. For purposes of assessing potability, TDS is important because the suggested aesthetic limit is 500 mg/L. In contaminant monitoring programs, this measurement can give an indication of whether the facility in question has released substances into the water by dissolution.

Specific conductance measurement is an indirect way of determining TDS. For inorganic species, the electrical conductivity of the water is directly proportional to the total dissolved salt content. Specific conductance can be measured easily in the field, so a rapid indication of relative TDS can be obtained from existing wells or samples collected during soil boring or well drilling programs. Because specific conductance is an electrical measurement, solutes not contributing ionic species (most organic compounds) will not be reflected in the measurement.

Total Organic Carbon (TOC)

TOC is the measurement of dissolved carbon attributed to organic substances. It may be used as a noncompound-specific measurement of organic matter dissolved in water. The measurement has limitations, as its detection limit is generally about 1 mg/L, putting it in the parts-per-million range. Most organic contaminants are regulated in the low parts-per-billion range. If contaminants occur in this range, a TOC analysis will not be sensitive enough to detect them. It may also be difficult to differentiate between background and downgradient concentrations using TOC measurements, due to the presence of naturally occurring organic compounds. In addition, the TOC analysis will not quantitatively measure volatile organic carbon (i.e., benzene), because these compounds are lost during sample preparation for TOC analysis.

Total Organic Halogen (TOX)

As carbon associated with organic substances can be measured, so also can the halogen elements (i.e., chlorinated, brominated compounds) associated with dissolved organic substances. This analysis is more sensitive than TOC and may be effective in the parts-per-billion range. The results are only indicative of halogenated organics, and cannot in themselves be used to quantify the specific compounds' concentrations. This is because there is no way to determine what gravimetric factor or factors apply to the concentration conversion.

Correlation of TOX with known halogenated compound concentrations as determined by gas chromatography is poor (Plumb and Fitzsimmons, 1984; Stevens, et al., 1984). TOX reflects the gross presence of halogenated compounds. A conversion of equivalent TOX from individual compound determinations does not always match well with the corollary TOX measurement.

Field Screening/Diagnostic Parameters

Analyses for field screening/diagnostic parameters are typically easy to conduct in the field and require only minimum expense, equipment, and training. In addition, these field measurements can result in data that are very representative because multiple measurements, over time, can be easily performed. These parameters include temperature, pH, oxidation/reduction potential, and dissolved oxygen.

Temperature

Although ground-water temperature may appear to be of no value as an indicator of contamination, there are instances when it is useful. Municipal landfills generate considerable heat in the anaerobic decomposition process of the putrescibles. This heat is transferred to the lixiviant water (leachate) which subsequently enters the ground-water system. Because aquifers are good insulators, the warmed water may persist for some distance downgradient. Thus, a zone of warmer-than-ambient water may then be indicative of a leachate plume.

Ground-water temperature for the first hundred or so feet below ground level is approximately that of the local average annual air temperature. Impoundments that leak or are meant to recharge will recharge warmer-than-ambient ground water in the summer and cooler-than-ambient ground water in the winter. These influences may be detected in nearby monitoring wells.

pH

A simple measurement that may be indicative of changes from ambient water-quality conditions is pH. Wastes may be alkaline or acidic to a sufficient extent to similarly affect ground water. On the other hand, a buffered solution, such as municipal landfill leachate, may influence an acidic or alkaline ground water toward neutrality. If the contaminants are solely organic, and do not have acidic or basic functional groups, pH will not be diagnostic of their presence.

Oxidation/Reduction Potential

Ground-water geochemistry is strongly influenced by the oxidation/reduction (redox) potential. Contaminant chemistry can also be influenced by redox potential or may, in turn, influence redox potential. Redox potential can be measured directly with a platinum electrode in millivolt units. This measurement is also known as an Eh measurement. The method of measurement is simple but it requires a flow-through cell for the electrode to avoid atmospheric interferences. If water comes into contact with the atmosphere, the redox potential rapidly changes and trends toward equilibrium with the atmosphere.

Dissolved Oxygen

Related to redox is the dissolved oxygen (DO) content of water. A low DO means that the redox potential will also be low. However, the actual potential is also a function of the oxidation state of dissolved chemical species. These are not measured in a DO measurement. Contaminants that induce an oxygen demand in water will reduce the DO. Thus, a lower-than-ambient DO may be present in a zone of contamination from a municipal landfill or a highly organic waste.

Physical/Chemical Characterizations

Ground-water samples should be scrutinized for visual signs of contamination both in the field and in the laboratory. Color or color colloidal matter may indicate the presence of contaminants. Ground water is less prone to the presence of natural-colored organic substances than surface water. Nevertheless, color could be of a natural origin.

A two-phase (aqueous and nonaqueous) sample is indicative of a substantial introduction of an organic liquid. The organic phase may be either lighter or heavier than water, depending upon its composition. Most organic liquids (fuels, solvents) have an aqueous solubility sufficient to produce a plume of dissolved contaminants in addition to the contamination problems associated with the organic liquid (solvent) itself.

Odors may be present in ground-water samples. Although ground-water samples should never be intentionally smelled, odors may be useful in categorizing the contaminant (gasoline, fuel oil, ester, chlorinated methane, or ethane). Odors may also be misleading as they may be altered by in situ microbial processes or arrive at a well through the vadose zone and not ground water. In the latter case, there will be olfactory evidence of contamination in the well, but laboratory analysis of such ground water may not detect any contaminants.

SPECIFIC ORGANIC COMPOUND ANALYSIS

Organic analyses are typically placed into three fractions: the volatile, base/neutral/acid (BNA) extractables (also referred to as the semivolatile fraction), and the pesticide/PCB fraction. To facilitate discussion of organic parameters, these will be discussed by fraction. Although there are thousands of organic compounds not on the EPA's TCL, it is convenient to discuss organic compounds by referring to EPA's TCL compounds. Many of the aspects discussed below are common to all organic analyses and should provide a basis for selecting an appropriate analytical method. Where applicable, the appropriate EPA method reference will be provided.

Volatile Organic Compounds

The organic fraction analyzed most commonly in ground-water investigations is the volatile fraction. The volatile fraction can be further subdivided into the following classes:

- chlorinated (and/or brominated) hydrocarbons
  (i.e., vinyl chloride, bromomethane, trichloroethylene)
- aromatics
  (i.e., benzene, toluene, chlorobenzene)
- ketones
  (i.e., acetone, 2-butanone)

A list of the TCL volatile organic compounds is provided in Table 12.4 (U.S. EPA, 1987).

Although volatile organics are fairly soluble, the primary fate of volatile organic compounds in surface-water systems is loss to the atmosphere. Despite this, volatile organics can be fairly persistent in ground water. Depending upon the purpose and requirements of the investigation, different methods can be applied to analysis for volatile organic compounds.

Because volatile organic compounds often present health and safety concerns, it is prudent to use field analytical instruments such as screening devices, when sampling for these compounds. An example of such an instrument is the Organic Vapor Analyzer (OVA). The OVA gives an indication of the approximate total volatile organic concentration, but is not capable of identifying specific volatile organic compounds or their individual concentrations without certain modifications to the instrument. The OVA is calibrated to a specific compound, such as methane, and is most sensitive to compounds that are chemically similar to the calibrated compound. Typically, OVAs have detection limits of about 1 ppm.

The OVA is not ordinarily used as a primary analytical method, but is more appropriately used as a screening tool: (1) to measure volatile vapor releases when a well head is opened; (2) to assure that vapors are not present in the samplers' ambient breathing zone; and (3) to provide an estimate of relative contaminant concentrations. While this measurement may give an indication of the presence of volatile contaminants in ground water, it can be deceiving because the measurement is of the air in the well casing and not of the water itself. In-field headspace analysis using an OVA can also be a valuable screening tool (i.e., screening split-spoon samples during well installation).

Another field analytical method for volatile organic compounds is head-space analysis by portable gas chromatography (GC). Figure 12.1 is a schematic showing the major components of a GC. The graphical representation of the compounds as they elute from the GC column are referred to as ''peaks'' and are charted on a gas chromatogram. Figure 12.2 represents a typical gas chromatogram.

Peaks are produced during GC analysis by compounds present in a sample. These peaks elute in the order of their boiling points or melting points and are detected by use of different types of detectors located at the end of the GC column. Detectors are constructed to be sensitive only to a certain type or class of compound. This selectively aids in the identification of specific analytes and is particularly important when analyzing complex samples.

Electron Capture Detectors (ECDs) are very sensitive to chlorinated compounds. Photoionization Detectors (PIDs) are selective to unsaturated compounds such as aromatic compounds and Flame Ionization Detectors (FIDs) are best for most other organic compounds.

Most portable GCs operate by use of short packed GC columns. The field GC is calibrated with a mixture of known concentration standards for those compounds to be analyzed. Once the GC is calibrated, the retention times are established

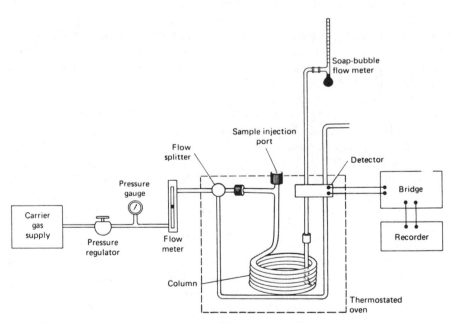

Figure 12.1.   Schematic diagram of a gas chromatograph.
Source: Skoog, D. A., 1985. Principles of Instrumental Analysis, Third Edition.

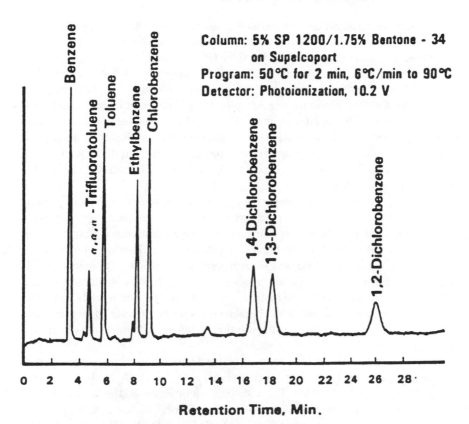

Figure 12.2.   Gas chromatogram of purgeable aromatics.
Source: Federal Register, CFR 40 Part 136.

for each compound. A retention time is the specific point in time that a compound (peak) elutes on the gas chromatogram.

A portable GC is capable of performing compound specific identifications, but cannot be considered absolutely qualitatively correct because it is a single peak response on a GC column. This single peak may represent the compound of interest or perhaps some other compound that happens to elute at the same point in time. Typically, portable GC columns operate under ambient conditions whereas more sophisticated laboratory GC columns are temperature-programmed, although more sophisticated temperature-programmed field GCs are being developed. An advantage of the field GC over the OVA, in addition to its ability to provide compound-specific quantitative results, is that results obtained with a field GC may be acceptable to a regulatory agency as a semiquantitative assessment. This semiquantitative assessment can be used to assist in the selection of which samples should be submitted for quantitative laboratory analysis.

Ground-water samples analyzed by field GC for volatile organic compounds are usually collected by placing 20–25 mL of sample in a 40 mL vial with a Teflon®-lined septum, heating the sample, and allowing the headspace to equilibrate with the ground-water sample. If volatile organic compounds are present in the sample, an equilibrium will be established between the concentration of the compound in the water and the concentration of that compound from the headspace (air). Typically, microliter amounts of air from the headspace are injected into a GC. Liquids are not injected into a GC column that is not temperature-programmed.

One shortcoming of the head-space analysis is the assumption that all ground-water samples will have the same degree of air/water equilibration (partitioning). It is known that this assumption is only semiquantitatively reasonable. One variable that has impacts on the various partitioning coefficients is the presence of salts (Na, Ca, M, K). One way to overcome this problem is to matrix-match all calibration standards with the water samples. This is done by obtaining a sample of ground water that is known to be free of volatile organic compounds, yet similar in other chemical and physical characteristics to the samples to be analyzed. This ground water can be used as the dilution water in the preparation of calibration standards.

A more sophisticated and the most often-used method of volatile organic analysis is the purge-and-trap method using a laboratory GC (EPA Methods 601 and 602) (*Federal Register,* 1984). For volatile organic analysis, the laboratory GC is operated using a temperature-controlled packed column. As such, slightly better separation between compounds (peaks) can be achieved. However, identification by purge-and-trap cannot be considered absolute since it is still only a single peak that matches the retention time of the compound of interest from a calibration standard. Other compounds in nature can also elute at the same time. The real benefit of purge-and-trap analysis is that the analysis will transfer all compounds (some less than others) from the ground water to the sample headspace above the sample. The method detection limit for most volatile organic compounds by purge-and-trap GC is 1 μg/L. Figure 12.3 presents a schematic of the purge-and-trap system.

Because of the volatility of the compounds, the samples for this analysis are collected in 40 mL vials with no headspace (bubbles). The bubbles will act to liberate the analytes of concern the same way the analytical method liberates the compounds from the water sample. Holding times are of particular importance due to analyte losses with time.

The state-of-the-art analytical method for volatile organic compounds which can provide the most confident qualitative and quantitative data is purge-and-trap GC interfaced with a mass spectrometer (MS) as the detector (EPA Method 624) (*Federal Register,* 1984). This is referred to as a GC/MS (Figure 12.4). Simply stated, the MS "listens" for the mass ions unique to the specific compounds that are being analyzed. Typically the MS will listen for the identification ion within

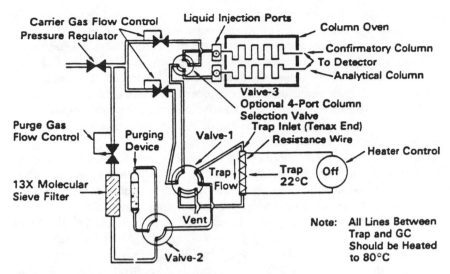

Figure 12.3.   Purge and trap system.
Source: Federal Register, CFR 40 Part 136.

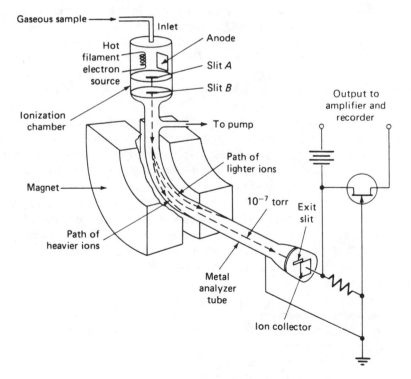

Figure 12.4.   Schematic diagram of a mass spectrometer.
Source: Skoog, D. A., 1985. Principles of Instrumental Analysis, Third Edition.

a certain retention time window, based upon those established during an earlier calibration. The identification of compounds is established confidently because each compound will have a unique mass spectral fingerprint. Because some isomers will result in identical mass spectra, isomer specificity is achieved by GC retention times. An isomer is a compound that is one of several possible orientations for the same organic compound. For example, 1,2-xylenes and 1,4-xylenes are both xylene isomers. Typically, the quantitation limit for volatile organic compounds by GC/MS are 5 μg/L. Recent methodology using capillary columns for volatile organic compounds has shown quantitation limits of 0.01–1 μg/L. A quantitation limit is the point where the analysis will yield results that will have a very high degree of precision. A quantitation limit is always higher than a detection limit. Quantitation limits are used when referring to organic parameters, while detection limits are used when referring to inorganic parameters.

Quite often large peaks may be present on the chromatogram that the MS has not identified as being TCL compounds. Through the use of the data system, these peaks can be compared to a mass-spectral library. Compounds detected during these library searches are referred to as tentatively identified compounds. These identifications should be considered semiqualitative and semiquantitative at best, although in some instances good qualitative identifications are possible.

From a cost standpoint, ground-water samples collected for quantitative volatile organic analysis should be first characterized by GC/MS, with subsequent sampling rounds monitored by the less expensive and more sensitive GC procedures. Volatile organic analysis by GC/MS is typically a little less than twice the cost of analysis by GC alone.

Semivolatile Organic Compounds

Because the vapor pressures of semivolatile compounds (also referred to as extractable compounds) are much lower than those observed for volatile compounds, semivolatile compounds must be removed from ground-water samples via solvent extraction. The primary variable which governs how the TCL semivolatile organic compounds will partition into the solvent is the pH of the sample. As such, the pH of the sample must be varied during the extraction process to assure that all compounds will be extracted. Previously, on EPA's Priority Pollutant List, semivolatile compounds were classified according to the pH at which they were extracted: base, neutral, and acid extractable.

The semivolatile compounds on EPA's TCL can be generally subdivided into the following six classes:

1. Phenolic Compounds
   (phenol, 4-methylphenol)
2. Chlorinated Aromatics
   (pentachlorphenol, 2-chloronaphthalene)
3. Halogenate Ethers
   (4-bromophenylphenylether)

4. Nitrogen Compounds
   (N-nitrosodiphenylamine)
5. Polynuclear Aromatics
   (anthracene, pyrene)
6. Phthalate Esters
   (di-n-butyl phthalate, bis(2-ethylhexyl)phthalate)

A list of the TCL semivolatile compounds is provided in Table 12.5 (U.S. EPA, 1987). Semivolatile compounds have very low solubilities, ranging from 7×10^{-4} mg/L to 31.7 mg/L (U.S. EPA, 1982).

Ground-water samples for semivolatile organic analyses are prepared by taking one liter of ground-water and adjusting the pH at various points during the extraction process. The extraction solvent is usually methylene chloride. The extracts are then combined and concentrated (evaporated) to a final concentration of one milliliter with a gentle flow of nitrogen.

The best analytical method for semivolatile organic compounds is by GC/MS. Due to the great number of extractable organic compounds (both TCL and non-TCL), packed column GC will not provide the chromatographic resolution (peak separation) needed. Therefore, capillary columns are used for extractable GC/MS analysis. Capillary columns are composed of fused silica and can be as long as 100 meters. Capillary column analysis results in sharp peaks, and therefore better separation between peaks and compound sensitivity can be attained.

The semivolatile organic analysis is conducted by injecting microliter amounts of the methylene chloride extract onto the capillary column. The MS analysis then proceeds as per the volatile organic compounds, including library search procedures for non-TCL compounds. The typical quantitation limit for most semi-volatile compounds by low resolution GC/MS is 10 μg/L.

Although GC/MS analysis of semivolatile compounds is the most reliable method from a qualitative and quantitative standpoint, samples can be screened by capillary column GC for the presence of certain classes of compounds. For example, the presence of various hydrocarbons, such as fuel oil, will appear as a cluster of peaks on a capillary column GC analysis. These compounds can then be fingerprinted or can be quantitatively estimated based upon the use of fuel oil standards.

There are some specialized GC/MS methods for semivolatiles that can result in substantially lower quantitation limits than 10 μg/L. Methods such as isotope dilution, single ion monitoring (SIM), and extraordinary concentration techniques can result in parts-per-trillion (ppt) quantitation limits in ground-water samples. Similar quantitation limits can also be achieved by use of a high resolution MS. A good rule of thumb is the lower the quantitation limit, the higher the price for analysis. If the purpose of the investigation demands a lower quantitation limit than achievable by the standard method, this needs to be discussed with the laboratory prior to sample collection. This is important because special sampling procedures and/or larger sample volumes may be necessary in order to achieve the desired quantitation limits.

Pesticides/Herbicides/PCBs

Pesticides/herbicides, and polychlorinated biphenyls (PCBs) are organic compounds that can be particularly toxic at very low concentrations. Because trace concentrations can elicit a toxic response in living organisms, the quantitation limits for these compounds must frequently be in the parts-per-trillion range. These compounds are also semivolatile compounds; however, they are typically analyzed by GC. Most pesticides, herbicides, and all PCBs contain at least one chlorine atom (most contain many chlorines). The ECD is particularly sensitive to chlorinated compounds. Because these compounds are semivolatile in nature, they must also be extracted and concentrated from ground-water samples by solvent extraction. Although methylene chloride is the solvent of choice for the analysis of semivolatiles, it is inappropriate for GC analysis by ECD, because the ECD is sensitive to chlorine. For the analysis of these compounds, the solvent of choice for GC/ECD analysis is hexane.

Typically, one liter of ground-water sample is extracted with hexane. Microliter volumes of hexane are then injected onto the GC column. Quantitation limits for these compounds by GC/ECD are on the order of 10 to 200 ppt in ground-water samples.

As with the analysis of volatile organics by GC, the primary pitfall of pesticide/herbicide analysis by GC is that identification solely by GC cannot be considered qualitatively confident since the identification of most pesticides/herbicides appears as a single peak on a gas chromatogram. Although these compounds can be confirmed further by analysis on a second dissimilar GC column (i.e., GC column with a different type of packing material), false positive results can still be a problem with pesticide/herbicide analysis by GC/ECD.

Some pesticides such as toxaphene and technical chlordane and all PCBs are mixtures of various chlorinated compounds. The GC analysis of these compounds will result in a unique multipeak pattern. The positive results for multipeak pesticides and PCBs can be considered qualitatively confident because the unique multipeak fingerprint is difficult to generate chromatographically unless the analyte is truly present. Although one column will suffice for multipeak identifications, the laboratory should be required to confirm the identification on two dissimilar columns.

Pesticides, herbicides, and PCBs can also be analyzed by GC/MS. These compounds can also be tentatively identified by using data system library search procedures of chromatographic peaks from the semivolatile fraction. However, unless extraordinary methods of MS analysis (i.e., SIM, high resolution MS) are performed, the quantitation limits by GC/MS may not be lower than the health-based criteria for most of these compounds.

SPECIFIC CONSTITUENT INORGANIC ANALYSIS

The inorganic parameters that are typically analyzed in ground-water samples are metals and cyanide. The U.S. EPA has established specific holding times, container types, preservatives, and storage requirements for both organic and inorganic parameters in ground-water (*Federal Register,* 1984). The specific holding times for the TCL organics and TAL inorganics are presented in Table 12.8.

The metals that are commonly analyzed can vary from a list of 8 particularly toxic metals (and cyanide) to a list of 23 on the TAL, which includes more common and "nontoxic" metals (Table 12.7). The detection limits required for inorganic TAL parameters are generally below health-based criteria and secondary standards. Secondary drinking water standards are based upon aesthetic considerations (taste and odor).

Metals and cyanide can exist in either a nonsoluble solid precipitate or in a soluble form. Sampling of ground water can typically result in samples containing large quantities of silts and other solids. To determine in what form various inorganic parameters exist, ground-water samples can be field-filtered through a 0.45 μm filter. Preservation for heavy metal analysis is by addition of nitric acid to a pH of less than 2 and addition of NaOH in a separate bottle to a pH greater than 12 for cyanide analysis.

The two methods that are commonly used for metal analysis are atomic absorption and atomic emission. These instrumental methods of analysis require that the sample be prepared so that any metals that are present will be in an ionic form. Metals are converted to an ionic form by sample digestion. Digestion of a ground-water sample is performed by gentle heating with nitric and/or hydrochloric acid. This digestion will dissolve suspended materials into solution, and therefore the results of the analysis will represent a total analysis. It is important to decide early on whether or not the samples should be filtered and how the resulting analytical data will subsequently be interpreted.

Atomic Absorption

Atomic absorption (AA) is based upon a measured difference in electronic signal between instrumental optics induced by the sample which is present as a gas between these optics (Figure 12.5).

There are three types of atomic absorption methods: flame AA, graphite furnace AA, and cold vapor AA.

For flame AA analysis, the digested sample is aspirated through a very thin tube drawn by suction and introduced into an air-acetylene or nitrous-oxide flame. A beam of light from a lamp specific to the metal being analyzed is focused through the flame. Depending upon the concentration of the analyte in the digest, the optics will measure an electronic difference. Flame AA is used for a wide variety of metals. Some more toxic metals such as arsenic, thallium, lead, and antimony are not sensitive by flame atomic absorption methods at concentrations of environmental concern. These elements must be analyzed by graphite furnace atomic absorption. Detection limits of 5 or less μg/L can be achieved for these elements by graphite furnace AA. Typical flame AA detection limits range from 500–1,000 μg/L for these elements.

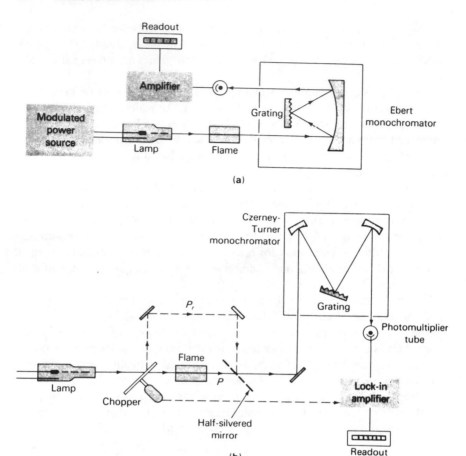

Figure 12.5. Typical flame spectrophotometers: single beam design and double beam design.
Source: Skoog, D. A., 1985. Principles of Instrumental Analysis, Third Edition.

The graphite furnace technique involves placing a staged hollow graphite tube in the path of a beam of light set at the wavelength of the analyte of interest. Microliter amounts of the acid digest are placed (or sprayed) into the entrance port of the graphite tube. The temperature of the graphite tube is increased slowly via electrode circuitry. Initially the liquid is dried within the tube. The temperature is then increased to pyrolysis temperature that will break various complexes in which the analyte of interest may be "tied up." Finally, the temperature is "ramped" to the point at which the analyte will be converted to a gaseous form (atomization). When the gas of the analyte of interest passes through the beam of light set at a wavelength unique to that analyte, an electronic difference (absorbance) is measured which is proportional to the concentration of the analyte.

The cold vapor AA method is used exclusively for the determination of mercury, and can achieve detection limits of 0.1–0.2 μg/L. The theory of cold vapor is similar to that of graphite furnace AA, with one exception. Whereas GFAA generates the gaseous form of the analyte by a temperature increase, the cold vapor technique generates mercury gas by a chemical reaction. The generation of gaseous elemental mercury is done by the rapid addition of a liquid reagent (stannous chloride) after a complex digestion procedure is carried out in a bottle with a shaved glass stopper (i.e., a BOD bottle). The sample (still in the BOD bottle) is then purged with argon, and any gaseous mercury that is liberated passes quickly across the optics; again, the measured electronic difference is proportional to the concentration of mercury present in the sample.

Atomic Emissions

Just as metals can be measured by an absorbance difference in a gaseous form (some better than others), metals can also be measured by a corresponding emission. The determination of metals by emission spectroscopy can result in much lower detection limits for most metals as compared to atomic absorption. The two most critical factors in detecting trace-level analytes are the temperature and stability of the flame. To this end, technology has incorporated both of these factors by use of inductively coupled plasma (ICP) and direct current plasma (DCP) emission spectroscopy.

A plasma is a high-temperature electronic flux that exists at a temperature an order of magnitude higher than conventional flames. In addition, plasmas are extremely stable. The pitfall with ICP (and DCP) is that high concentrations of solids (TDS), salts (Na, Ca), and common elements (iron, aluminum) can result in severe matrix interferences.

Another benefit of the use of ICP in addition to the lower detection limits is simultaneous multielement analysis capability. ICP systems can analyze 15 to 20 metals in a water sample in a two-minute period. The operation of ICP is similar to that of flame AA. A peristaltic pump draws an acid-digested sample into a chamber which sprays the sample into the plasma. The optics measure the difference in emissions as intensity at the wavelengths of interest and record the difference in concentration units.

Cyanide Analysis

The final analyte that is typically examined is cyanide. In basic ground water, cyanide complexes can exist in soluble forms. In acidic ground water, cyanide can exist in a soluble form but can also be liberated as hydrogen cyanide (HCN) gas. Accordingly, ground-water samples to be analyzed for cyanide are preserved with sodium hydroxide to a pH of greater than 12. The analysis of cyanide proceeds by taking advantage of the fact that cyanide is liberated as HCN in a heated, acidic environment. There are two methods commonly used for trace-level cyanide analysis: one is a manual distillation with a subsequent colorimetric determination; and the other is an autoanalyzer method. Both work on the same principles; however, the autoanalyzer performs the analysis with very small volumes of sample. Both methods can result in detection limits of 10 μg/L. This concentration of cyanide is proportional to the intensity of the purple color determined colorimetrically.

Total Phenols

Total phenols can be analyzed by either a distillation/colorimetric method or an autoanalyzer method. The rigorous distillation converts all phenolic compounds to phenol. The phenol is collected as the lighter distillate. The distillate is collected and treated with color-inducing reagents. The concentration of phenol is proportional to the intensity of yellow to orange color, which is determined colorimetrically.

Other Analyses

This discussion has emphasized the parameters most often required in ground-water investigations where the emphasis is determining if there is ground-water contamination. There are, of course, many other analytes which may be of interest, depending upon the purpose of the investigation. These include various nitrogen compounds [i.e., ammonia nitrogen, total Kjeldahl nitrogen (TKN), nitrate-nitrogen, nitrite-nitrogen], sulfur compounds (i.e., sulfate, sulfite, sulfides), phosphorus (i.e., phosphates), and other soap compounds (i.e., surfactants).

Finally, biological analysis may be of interest in some ground-water investigations, particularly when the ground water is to be used for drinking water. One biological indicator that is commonly used is the analysis of total coliform bacteria. Some coliform bacteria are associated with ground-water contamination by septic systems (i.e., fecal bacteria). Most of the biological parameters are examined by allowing the ground water to incubate with an enriched broth which the specific microbes of interest can use as a food source. After a designated period of time, the samples are examined for any activity (i.e., turbidity, the presence of bacteria colonies, etc.).

The procedures for many of these other types of analysis can be found in the "Standard Methods for the Analysis of Water and Wastewater" (APHA, AWWA, WPCF, 1976).

QUALITY ASSURANCE/QUALITY CONTROL

Obviously, selecting the appropriate parameters and methods for analyzing parameters are critical steps to properly assessing ground-water quality. However, just as critical is the care taken during sample analysis, the submission of check samples to "test" the sampling process, and a review of the appropriateness of the data after it has been generated. This process is referred to as Quality Assurance/Quality Control (QA/QC).

Very often the ability to assess the quality of analytical data is in direct proportion to the degree of confidence that is required in the analytical measurements. For example, taking a pH reading of your garden soil just to satisfy your curiosity does not require stringent QA/QC. However, if you wanted to grow expensive plants that require pH 6 to 6.5 to survive, more elaborate QA/QC requirements, such as multiple samples, may be required because the measurement will be used to answer very specific questions (i.e., is the pH between 6 and 6.5). The degree of QA/QC that is implemented should therefore be proportional to the specific requirements for confidence in the analytical measurements.

The amount of QA/QC that is implemented (bottleware, field notes, duplicate samples, blanks, etc.) is also proportional to the available funds for the investigation. Ultimately, the costs involved with ground-water investigations will increase as the need for confidence in the data increases.

There are various reasons to consider high-confidence analytical data important. Obviously, if very expensive decisions are going to be based on analytical data, a high degree of confidence is required. Other reasons to require a high degree of confidence include instances when human health risks are being assessed or when data are being used for litigation purposes.

The following sections will discuss data quality parameters, selection of a laboratory, preparation of quality assurance project plans, and the various aspects of laboratory QA/QC.

Data Quality Parameters

Quality Assurance is a mechanism by which the qualitative and quantitative reliability of the data can be assessed. Quality of analytical measurements can be assessed by the examination of five data quality parameters: precision, accuracy, representativeness, completeness, and comparability.

Precision

A measure of reproducibility between either one sample analyzed twice or a comparison of the results between duplicates (i.e., the sample solution submitted for laboratory analysis in two complete sets of bottleware). Precision is usually represented in one of three ways: percent difference, relative percent difference, and relative standard deviation.

Accuracy

A measure of how close an analytical result is to a "true" value. A "true" value is one that has a certified concentration based upon many analyses. Accuracy is typically expressed as a percentage of the true value.

Representativeness

The degree to which one sampling result can be compared with many results of a larger sample (i.e., another point in an aquifer). Representativeness is assessed by comparing results for several samples taken from the same aquifer (from various depths within the aquifer).

Comparability

The degree to which one result by one analytical method can be compared to an analytical result for the same sample by a different analytical method. For example, the results for cyanide by manual distillation in 10 samples are the same as those obtained by autoanalyzer in these samples. Based upon this, the methods can be called comparable.

Completeness

A measure of the amount of valid data that is obtained compared to the amount of valid data that is needed to fulfill the purpose of the investigation.

The last parameter mentioned deserves further discussion because completeness can be correlated to several grades of quality with respect to analytical data. Absolute confidence in analytical results begins with: obtaining certified clean bottleware; strict adherence to established installation, well development, purging, and sample collection procedures; preservation, chain-of-custody, sample holding times, quality control samples, field notes; and last but not least, analysis by a reputable laboratory that will follow established analytical methods correctly.

Selection of an Analytical Laboratory

The emphasis of the following discussion is the QA/QC which takes place after the samples are delivered to the laboratory. It is important to note, however, that a critical QA element is the selection of a qualified analytical laboratory. Selection of a laboratory should be conducted by individuals who are familiar with the types of analysis that are going to be performed, specific methodologies to be used, and various other requirements that are dictated by the purpose of the investigation. The most basic requirement (which should not be assumed) is that the laboratory has the capability (instrumentation, experience, etc.) to perform the specific analyses required for the investigation.

The laboratory should have a good reputation for quality and must be willing to cooperate with the investigator, who should set specific requirements for sample analysis. Special requirements such as special holding times, sequence of sample analyses, and frequency of laboratory blanks, duplicates, spikes, calibrations, and data package deliverables are just a few of the requirements that need to be addressed by individuals who are knowledgeable in these areas.

One indication of the laboratory's ability to provide quality analytical data is the various state or health department certifications which the laboratory holds. Certifications do not exclusively make a good laboratory, but they do indicate that the laboratory is capable (if required) of generating quality work. With regard to certifications, there are numerous laboratories across the country that have contracts with the EPA to analyze samples for Superfund projects. These laboratories are referred to as CLP laboratories. A common misconception is that these laboratories are "EPA certified" laboratories. This is far from true; the fact is that this is a contract program. Although an onsite audit is conducted and performance evaluations or "test samples" must be acceptable for the lab to receive further consideration, contracts go to the lowest bidder. In order to gain insight into how good or bad a laboratory is, an onsite audit should be performed by a competent (preferably independent) quality assurance chemist.

A properly conducted audit, which may include sending performance evaluation samples, will identify those laboratories that can generate high-quality data from those which are not worth further consideration. Additional aspects which should be considered when selecting a laboratory are the availability and flexibility of scheduling sample arrival and analyses, the location of the laboratory with regard to the mode of sample shipment, and what impacts, if any, the shipping mode will have on the required sample holding times.

Although the financial aspect of analyses is a consideration, investigators should not select a laboratory solely on a cost basis. Beware of bargain basement prices—more often than not, you get what you pay for.

Preparation of a Quality Assurance Project Plan

For large investigations (greater than 50 samples), the investigator should consider preparing a Quality Assurance Project Plan (QAPP). In many instances, when the investigation is being reviewed by a state or federal agency, the preparation of a QAPP is required before work can begin. The purpose of the QAPP is to state the objectives of the investigation with regard to data quality; who will perform each task in the investigation (project responsibilities), including the designation of the analytical laboratory; what protocols will be used for ground-water sampling; sample custody; specific requirements for quality control samples; the specific analytical methods for each analyte; holding times; sample container and preservative requirements; data package requirements; data validation and reduction protocols; frequency of audits; and reports to management. The ultimate purpose of the QAPP is to describe all the whats, wheres, whens, hows, and whys of the investigation. This is necessary so that all steps in the investigation are understood and nothing is left to the laboratory's interpretation (or imagination). Considering the cost of sample analysis, a QAPP is a worthwhile effort whether or not a regulatory agency is involved. The QAPP does not have to be a massive formal document, but rather one that clearly states the needs and the requirements of the investigation.

Laboratory QA/QC

The laboratory QA/QC process usually begins when the laboratory ships sample containers and preservatives to the investigator. After the sampling event, samples are shipped and arrive at the laboratory. The laboratory must follow proper procedures or requirements for chain-of-custody, sample storage and holding time, sample preparation, use of quality-control samples, instrument calibration, sample analysis, laboratory validation, reporting, documentation, and recordkeeping. Figure 12.6 presents the laboratory QA/QC process, beginning with laboratory receipt of samples. The remainder of this chapter briefly discusses each of these elements.

Chain-of-Custody

At the moment the sample bottles are released from the laboratory, a chain-of-custody begins and must be maintained. When samples are received by the analytical laboratory, an internal laboratory chain-of-custody begins and must be maintained. As with the field custody, laboratory custody and transfer of samples must be documented so that the identity and integrity of samples can be established and maintained. Laboratory (and field) chain-of-custody should be considered a fundamental requirement for all investigations.

Sample Storage and Holding Time Requirements

One of the most important aspects of laboratory QA/QC is assuring sample integrity and strict adherence to holding times. A holding time is the time that has elapsed from the moment the ground-water sample is collected to the moment of sample preparation and/or analysis. Holding time requirements are generally specified by the regulatory agency. Results from samples analyzed beyond established holding times should be viewed as questionable. Table 12.8 presents the recommended holding times for the TCL and TAL parameters.

Sample Preparation

Depending on the analysis that has been requested, sample preparation may be required before analysis can proceed. The most common types of sample

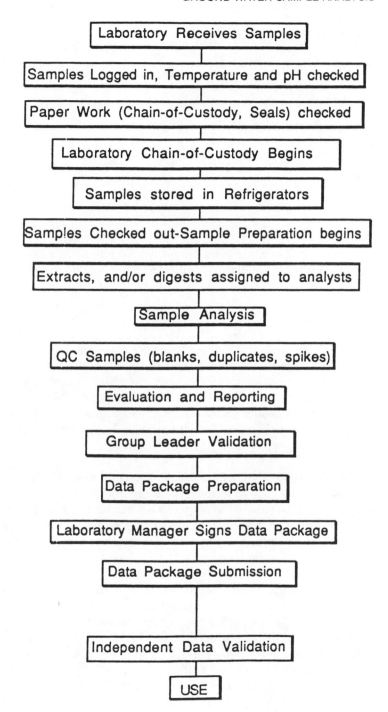

Figure 12.6. Laboratory process.

Table 12.8.   Maximum Allowable Holding Times.[a]

Parameter	Maximum Holding Time	Note
TCL volatiles	14 days	b
TCL semivolatiles	7 days	c,d
TCL pesticides/PCBs	7 days	c,d
Heavy metals (except mercury)	6 months	
Mercury	28 days	
Cyanide	14 days	

[a]*Federal Register,* 1984.
[b]For aromatic compounds, holding time is 7 days if pH is not adjusted to <2 with HCL.
[c]For liquid samples, 7 days until extraction, 40 days for the analysis of extracts.
[d]For solid samples, 14 days until extraction, 40 days for the analysis of extracts.

preparation include extractions, digestions, and distillations. Some holding times are specified to the time of sample preparation and a subsequent time until actual analysis.

Laboratory Quality-Control Sample

The analysis of quality-control samples is another important aspect of laboratory QA/QC which requires input from the investigator and discussion with laboratory personnel because quality-control samples can result in additional costs. Analysis of quality-control samples may also require collection of additional sample volume. The laboratory quality-control samples must be analyzed concurrently with samples for the investigation. These QA samples, which include laboratory blanks, duplicates, and spikes, are described below.

Holding blanks. The analysis of this blank is designed to provide an indication of contaminants that may be introduced during laboratory sample storage. The laboratory should be required to prepare a holding blank when samples arrive at the laboratory. This blank will be stored with the samples until they are analyzed. A holding blank must be prepared by the laboratory to be stored with samples each day samples are received.

Method blanks. A method blank is a portion of deionized water that is carried through the entire analytical scheme, including all sample preparation. It is important that the volume used for method blanks be the same as the samples. Method blanks must be analyzed (purged) every 12 hours for volatile organic compound analysis. Method blanks for volatile organic analysis are typically prepared by the laboratory the same day they are analyzed; therefore, they do not monitor contaminants introduced during sample storage as holding blanks do. Method blanks are also referred to as preparation blanks.

Instrument blanks. An instrument blank is a portion of pure solvent or water that has not gone through sample processing (extraction/digestion). The purpose of this blank is to check for contaminants present in the instrument (and the unprocessed solvent). Instrument blanks should be analyzed immediately after analysis of samples that have high concentrations of a parameter of interest. If contaminants are detected in instrument blanks, the laboratory must institute corrective actions until clean instrument blanks are obtained.

Duplicates. Laboratory duplicates are two separate aliquots of the same sample which have been independently processed and analyzed for the same parameters to determine the precision of the analytical system. The analytical laboratory should perform a duplicate analysis on a minimum of one sample in twenty. Duplicate results are typically compared as relative percent difference (RPD).

$$\text{Relative Percent Difference} = \frac{\text{Sample A} - \text{Sample B}}{\text{Average Sample A} + \text{B}} \times 100$$

Spiked samples. A spike is a sample in which the compound being analyzed is actually added (or spiked) into the sample to determine the accuracy of the analytical system. The question is, can you get back as much as you put in? The results of a spike are expressed in terms of the percent recovery with regard to the amount added. There are two types of spikes that are used by an analytical laboratory: matrix spikes (for both organic and inorganic); and surrogate spikes (organic exclusively).

$$\% \text{Recovery} = \frac{\text{Spike Result} - \text{Unspiked Sample Result}}{\text{Concentration Added}} \times 100$$

Matrix spikes. To determine laboratory accuracy and precision, a sample is analyzed unspiked (to determine a baseline). A second portion (or aliquot) of sample is then typically spiked with five of the TCL volatiles, twelve of the TCL semivolatiles, six of the TCL pesticides, and all of the metals and cyanide (all at predetermined concentrations). The results (percent recoveries) of these spikes are a direct measure of analytical accuracy. A third portion of sample is spiked in the same manner (for organic compounds). The comparison of respective recoveries between the two spikes is a measure of analytical precision. Depending on anticipated levels and types of parameters, the laboratory can be instructed to add additional matrix spike parameters at appropriate concentrations.

Surrogate spikes. Surrogate spikes are added to every sample for organic analysis. A surrogate compound is a special compound synthetically prepared to test the analytical procedure. Surrogate compounds are not found naturally; however, they are similar to several compounds on the TCL. There are three volatile surrogates, six semivolatiles, and one pesticide surrogate which are added at predesignated concentrations.

The surrogate compounds are treated as unknowns, and a percent recovery is calculated concurrently with the analytes of interest. Because the sample characteristics will affect the percent recovery, these percent recoveries indicate the accuracy of the analytical method. More importantly, surrogate recoveries are direct measures of how well the method has worked on each individual sample. If poor recoveries are obtained, the sample should be reextracted and/or reanalyzed a second and possibly third time, using less sample; however, using less sample will result in higher quantitation limits.

The "acceptable" recoveries for surrogate compounds vary depending on the individual compound and the type of compound. The U.S. EPA has generated acceptable ranges for both matrix spike and surrogate recoveries. These ranges were generated from statistical manipulation involving numerous studies based on recoveries reported from many contract laboratories. Some "acceptable" recoveries can be as low as 11%. For some purposes, these recoveries are not acceptable, and the laboratory must be instructed to follow specific criteria for reextraction/reanalysis. One problem which should be noted with multiple reextractions is the required sample volume and holding time constraints. This is why it is a good idea to collect large volumes of sample.

Instrument Calibration

Aspects of instrument calibration include: how the instruments used for analysis will be calibrated, how often the calibration will be checked, what actions will be taken if poor calibration checks are obtained, and who prepared the calibration standard. If instruments are calibrated properly using incorrect concentrations, incorrect data will be generated. The source of standards and spiking solutions must be different so that each can be a check on the other.

The laboratory must perform a multi-point (three or five) calibration to determine instrument sensitivity, stability, and linearity. At a minimum, the following instrument calibration procedures should be conducted:

- a three- or five-point concentration calibration curve
- after every tenth sample (or a time length pertinent to the analysis), a single-concentration calibration check must be performed

If the concentration of the calibration check does not compare well with the previous multipoint calibration, a new five-point calibration must be performed before samples can be analyzed.

Sample Analysis

Once the instrument has been calibrated and blanks have been analyzed verifying that the instrument is free of contaminants, analysis of samples can proceed. An important aspect of sample analysis is the analytical sequence. A highly contaminated sample may severely contaminate the instrument. Carryover or memory effects can generate false positive results for subsequently analyzed samples, particularly during volatile organic analysis.

Laboratory Validation and Reporting

Once the analytical results have been generated, they must be validated. This typically is done by laboratory section heads or the laboratory manager. The Quality Assurance Project Plan (if prepared) should specify how calculations will be verified and document the procedures that will be used to assess how the results of laboratory QA/QC samples impact the data. Analytical results should be reported to *two* significant digits. The laboratory manager should sign the analytical data before results are released and, quite often, the submission of a complete data package is required. The submission of a laboratory data package is a decision that the investigator must make. The importance of the investigation and the implications of the results should be considered when making this decision.

Documentation and Recordkeeping

To assure the integrity of the QA/QC process, full documentation should be provided to account for all laboratory activities. The following are some of the areas requiring laboratory documentation:

- sample log-in procedures, including assignment of laboratory control number, taking pH and temperature, and verifying field paperwork
- internal laboratory chain-of-custody—every transfer must be entered
- verification of refrigerator storage, including daily temperature verification log
- time chronicle to verify holding times
- extraction/digestion instrument logs to verify analyses and analytical sequence
- lab narrative to describe any problems encountered during the analyses
- summary forms allowing brief examination of pertinent QA/QC information
- raw data—every item of data (i.e., standards, blanks, spikes, duplicates, and samples) relating to analysis; this includes *all* instrument printouts and copies of analysts' notebooks
- all information pertaining to the case should be stored in a laboratory file, and all analytical data should be stored on magnetic tape for future reference for a period of at least five years

Independent Laboratory QA Review

Whenever possible, analytical results and laboratory documentation should be independently validated by a qualified, experienced quality-assurance chemist.

Unfortunately, the QA process does not end when the laboratory delivers the data to the investigator. Laboratory chemists and managers may not give special attention to make sure that *your* results are valid. Unless an independent data review is performed to make certain the results are correct, it must be assumed that the laboratory made no mistakes. If erroneous, this may be a costly assumption.

In addition to assessing the validity of the analytical data, an independent data review can also provide the interpretation of analytical bias. Bias is the tendency for the results to be necessarily slanted high or low. For example, if spike recoveries are consistently 130% for benzene and the laboratory reports benzene in a monitoring well at 20 μg/L, intuitively it can be stated that the actual concentration of benzene in the well may be slightly less than the 20 μg/L reported.

Once the data review has been performed, a report should be prepared which qualifies certain areas of the data before the results can be utilized. Another item to consider is data presentation (data reduction). Considering the number of samples and compounds that may be analyzed during an investigation, the number of analytical results can be quite extensive and cumbersome. Typically, a preferred alternative is to reduce the data by use of computer spreadsheets to just those compounds with positive results. Specific codes which provide an indication of data reliability should be placed next to results as appropriate during the independent data validation. Once the validity has been assessed and the positive results have been tabulated, it is highly recommended that individuals familiar with the hydrogeology of the site examine the data tables. By this trouble-shooting process, anomalies and data gaps can be identified.

SUMMARY

The primary thrust of most ground-water investigations is determining ground-water quality. Considering the costs associated with well installation and laboratory analysis, it would seem apparent that all necessary steps should be taken to maximize the results of this effort and the quality of the data collected. This starts with the design of the investigation and this, in turn, translates into the parameters to be analyzed, the analytical methods to be used, and the laboratory QA/QC to be followed.

REFERENCES

APHA, AWWA, WPCF. Standard Methods for the Examination of Water and Wastewater, 14th edition. 1976.

Bellar, T. A., and J. J. Lichtenberg, 1974. Determining Volatile Organics and Microgram-per-Litre Levels by Gas Chromatography. J. Amer. Water Works, December, pp. 739–744.

Federal Register, Wednesday, November 13, 1985. Volume 50. NO219 Part II, Environmental Protection Agency, 40CFR Parts 141–142, National Primary Drinking Water Regulations; Volatile Synthetic Organic Chemical.

Federal Register, Friday, 26 October 1984 Part VIII, Environmental Protection Agency. 40CFR Part 136. Guidelines Establishing Test Procedures under the Clean Water Act.

Plumb, R. H., Jr., and C. K. Fitzsimmons, 1984. Performance Evaluation of RCRA Indicator Parameters. First Canadian/American Conference on Hydrogeology Proceedings, Banff, Alberta, Canada, June 22–26.

SCS Engineers, 1985. Industries and Their Related Hazardous Wastes. Reston, Virginia.

Skoog, D. A. Principles of Instrumental Analysis Third Edition, 1985. Saunders College Publishing.

Stevens, A. A., R. C. Dressman, R. K. Sarrell, and H. J. Brass, 1984. TOX, Is It the Non-Specific Parameter of the Future? U.S. Environmental Protection Agency, Technical Report EPA-600/D-84-169, June.

United States Environmental Protection Agency Aquatic Fate Process Data For Organic Priority Pollutants 1982, Office of Water Regulation and Standards, Washington, DC 20460.

United States Environmental Protection Agency Contract Laboratory Program. Statement of Work for Analysis 8/87.

United States Environmental Protection Agency, Methods for Chemical Analysis of Water and Wastes, 1979. Environmental Monitoring and Support Laboratory, Cinn., Ohio, EPA-600/4-79-020.

United States Environmental Protection Agency, Test Methods for Evaluating Solid Wastes, 1986. Office of Solid Waste and Emergency Response, Washington, DC, SW-846.

13

Organization and Analysis of Water Quality Data

Martin N. Sara and Robert Gibbons

Hydrogeologists and others who use water analyses must make interpretations of individual analyses or of large numbers of analyses at a time. From these interpretations, final decisions are made regarding detection and assessment monitoring programs. Few aspects of hydrogeology have expanded more rapidly than water quality in the last 20 years. The expansion of water quality programs was based on two factors:

- Improvements in analytical methods have greatly increased our ability to determine, with accuracy and precision, a vast number of inorganic elements, and new organic compounds in water. Furthermore, automation of analytical processes has increased the speed of determination, allowing statistically significant studies to be made of constituents which formerly were beyond the analytical capabilities of all but the most sophisticated instrumentation.
- The expansion of water chemistry has been in response to public as well as professional concern about health questions, particularly in relation to radionuclides and trace organic constituents.

As a result, many comprehensive water quality monitoring programs of land disposal facilities have resulted in thousands of individual parameter analyses included in both detection and assessment monitoring programs. Although the details of water chemistry often must play an important part in water analysis interpretation, a fundamental need is for means of correlating analyses with each other and with hydrologic or other kinds of information that are relatively simple, as well as scientifically reasonable and correct. It may be necessary, for example, to make an organized evaluation in a report of the water quality present upgradient from a facility to correlate water quality with environmental influences by comparison. The process of water quality assessment relative to baseline and detection monitoring is shown on Figure 13.1.

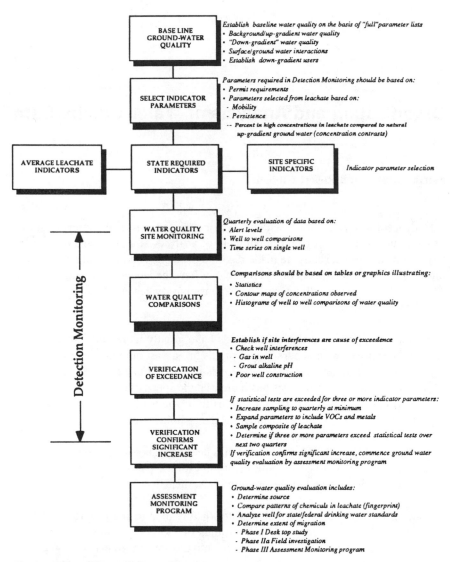

Figure 13.1.   Water quality assessment procedure.

The interpretative techniques and correlation procedures described here do not require extensive application of chemical principles. The procedures range from simple comparisons and inspection of analytical data to more extensive analyses and the preparation of graphs and maps that show significant relationships and allow for extrapolation of available data to an extent sufficient to be most practical and useful in the context of detection and assessment monitoring programs.

The monitoring program name (i.e., baseline or detection), task components, and task descriptions are illustrated on successive columns of Figure 13.1. Each of these task components have individual sections which describe water quality assessment procedures.

BASELINE WATER QUALITY

Characterizing the existing or baseline quality of ground water is an important task for two reasons. First, existing ground-water quality normally defines the baseline conditions for evaluating risks to human health and the environment. Second, existing ground-water quality, in part, determines current uses and affects future uses. In addition, determining ground-water uses is an important initial step in identifying potential downgradient exposure pathways.

In evaluating the background water quality in the area, the investigation must consider not only possible background concentrations of the selected indicator chemicals, but also the background concentrations of other potential leachate constituents. Existing chemical parameters associated with indicator chemicals (i.e., chloride or iron) or other RCRA hazardous constituents may be due to natural conditions in the area, prior releases from the unlined old landfill areas, or prior releases from other upgradient sources in the surrounding area. Assessing ground-water concentrations of water quality constituents is necessary to establish an existing baseline of ground-water quality to which the incremental effects of potential release can be added. For example, existing levels of water quality parameters in an area may be below a level at which risks to human health or the environment would be expected to occur. Adding additional parameter load from a release could, however, result in overall levels of water quality above the significant risk threshold or above applicable environmental standards, even if the facility release is not sufficient by itself to pose a significant risk.

Measuring the ambient concentrations of every RCRA hazardous constituent is not feasible for most baseline studies. To adequately assess background ground-water quality, the investigation should attempt to identify other potential release sources in the area (e.g., CERCLA sites, RCRA facilities, municipal landfills, agricultural areas, or surface water dischargers) and identify which constituents are most likely to have been released by each source. Some of the parameters on the list of background chemicals may also be indicator chemicals, particularly if the facility has experienced a prior release. When determining which chemicals to include on a list of background parameters, the engineer should include all the selected indicator chemicals described as the baseline water quality parameter list in the next section.

In cases where sufficient appropriate historical monitoring data are unavailable, the engineer may need to install a ground-water monitoring system or add to an existing system in order to adequately assess background ground-water quality. The design of this monitoring program should be based on previous chapter guidance on location and flow path monitoring. At a minimum, background water quality should be determined based upon at least two separate samplings of existing or newly installed monitoring wells.

For facilities that have experienced a prior release, the investigation should also establish the results of any sampling or monitoring, or hydrogeological investigations conducted in connection with the release (if available), and should provide references to any reports prepared in connection with the release.

SELECTION OF INDICATOR PARAMETERS

The chemical analysis, with its columns of concentration values reported to two or three significant figures, accompanied by descriptive material related to the source and the sampling and preservation techniques, has an authoritative appearance which, unfortunately, can be misleading.

Low level volatile organics laboratory interferences, monitoring well grout contamination, and sampling bias of the well all provide significant changes to values for constituents and properties that may be different from the values in the original water body. The initial stage of any water quality assessment is the selection of parameters to be used within the monitoring program.

Detection Monitoring Indicator Parameters

Detection monitoring programs require that individual chemical parameters be selected to represent the natural quality of the water and the potential parameters that may be adversely affected or changed through operation of a facility. These parameters, called "indicators," are selected on a basis of a number of criteria:

- required by permit, state or federal regulations, or guidance
- are mobile (i.e., are likely to reach ground water first and be relatively unretarded with respect to ground-water flow), stable, and persistent
- do not exhibit significant natural variability in ground water at the site
- are correlative with constituents of the wastes that are present at the site
- are easy to detect and are not subject to significant sampling and analytical interferences
- are not redundant (i.e., one parameter may sufficiently represent a wider class of potential contaminants)
- do not create data interpretation problems (e.g., caused by common laboratory and field contaminants).

Since most chemical indicators are based on naturally occurring chemical parameters, the following table provides ranges of values occurring in natural aquifers and the persistent and mobile parameters typically seen in leachates from sanitary landfills:

Leachate Indicators	Natural Aquifer Ranges
TOC (filtered)	1–10 ppm
pH	6.5–8.5 pH units
Specific conductance	100–1,000 μm/cm
Manganese (Mn)	0–0.1 ppm
Iron (Fe)	0.01–10 ppm
Ammonium (NH$_4$ as N)	0–2 ppm
Chloride (Cl)	2–200 ppm
Sodium (Na)	1–100 ppm
Volatile organics (EPA Method 624)	<40 ppb

These few indicators present a somewhat limited view of the chemical environment of an aquifer and would especially limit baseline assessment of water quality.

Although individual definitions vary, consider a "complete" analysis of ground water to include those natural constituents which occur commonly in concentrations of 1.0 mg/L or more in potable water. Depending on the hydrogeologic setting, a "complete" analysis would include:

Ammonia (as N)	Total Organic Carbon (TOC)
Bicarbonate (HCO$_3$)	pH
Calcium	Arsenic
Chloride	Barium
Fluoride (F$^-$)	Cadmium
Iron (Fe)	Chromium
Magnesium	Cyanide
Manganese, dissolved (Mg^{2+})	Lead
Nitrate (as N)	Mercury
Potassium	Selenium
Sodium (Na$^+$)	Silver
Sulfate (SO$_4$)	Nitrogen, dissolved (N^2)
Silicon (H$_2$SiO$_4$)	Oxygen, dissolved (O^2)
Chemical Oxygen Demand (COD)	Total Dissolved Solids (TDS)

The volatile organic compounds (VOCs) established in Method 624

In general, the investigator should examine closely the water quality results if these indicators show values above the ranges of natural aquifer waters given above. The volatile organic level of 40 ppb was established from tolerance (prediction) intervals on numerous upgradient wells in 17 facilities (Hurd, 1986) and are due to both collection and analytical interferences.

WATER QUALITY SITE MONITORING

The ground-water data collected during both the site characterization and detection monitoring phases is typically restructured, simplified, and must be presented in a manner that facilitates verification and interpretation. All physical and chemical analytical data are reported via laboratory analysis transmittal sheets. The data are compiled into tables and display formats that facilitate the understanding and correlation of the information.

A list of all monitoring data should be provided for each sampling event and updated as new data become available. The data should include the following: well identification, date of analysis, laboratory, measurement unit, limit of detection, and chemical concentration. The data are then categorized and organized to allow quick reference to specific values. Compilation and evaluation of laboratory data into summary reports must be performed without transcription errors and with careful consideration for the quality of individual data values.

Reporting

Laboratory results for a given analyte generally are presented as a quantified value or as ND (not detected). Chemical data presented in site assessment or detection monitoring programs should present results for each sample according to this protocol. Results are reported either as a quantified concentration or as less than ("<") the method detection limit value (thus, ND results are shown as "<" on the summary report).

To the extent feasible, all laboratory results should be reported in a manner similar to that described above.

Significant Digits

The number of significant digits reported by the laboratories reflects the precision of the analytical method used. Rounding of values is generally inappropriate, as it decreases the number of significant digits and alters the apparent precision of the measurements. Therefore, the investigator retains the number of significant digits in the transcription, evaluation, and compilation of data into secondary reports. Variation in the number of significant digits reported for a given analyte may be unavoidable if there is an order of magnitude change in the concentration of a chemical species from one round of sampling to the next, or if the precision of the analytical methodology differs from one round to the next.

Outliers

Unusually high, low, or otherwise unexpected data values (i.e., outliers) can be attributed to a number of conditions, including:

- sampling errors or field contamination
- analytical errors or laboratory contamination
- recording or transcription errors
- extreme, but accurately detected, environmental conditions (e.g., spills, migration from facility, etc.).

Gross outliers may be identified by informal scanning of the data. This exercise is facilitated by printouts of high and low values of the data. Formal statistical tests also are available for identification of outliers. When feasible, outliers are corrected (e.g., in the case of transcription errors) and documentation and validation of the reasons for outliers are performed (e.g., review of field blank, trip blank, QA duplicate sample results, and laboratory QA/QC data). Field and laboratory QA/QC results, as well as field and laboratory logs of procedures and environmental conditions, are invaluable in assessing the validity of reported suspect concentration values. Outliers that reasonably can be shown not to reflect true or accurate environmental conditions are eliminated from statistical analyses, but continue to be reported with data summaries.

Units of Measure

Units of measure are recorded for each parameter in the laboratory raw data reports. Special care is taken not to confuse $\mu g/L$ measurements with mg/L measurements when compiling, transcribing, or reporting the data.

WATER QUALITY COMPARISONS

The type of water analysis interpretation most commonly required of hydrogeologists is the preparation of a report summarizing the water quality in an aquifer, a drainage basin, or some other area unit which is under study. The writer of such a report is confronted with difficulties of typically large amounts of data from a few sources, and this information must be extrapolated. The finished report must convey water quality information in ways in which it will be understandable by both the regulatory and technical management staff. As an aid to interpreting groups of chemical analyses, several approaches will be discussed that can serve to relate analyses to each other and to provide means of extrapolating data areally and in time. Different types of visual aids which often are useful in reports will be described. The basic methods used in the interpretation are inspection and simple mathematical or statistical treatment to bring out resemblances among chemical analyses; procedures for extrapolation of data in space and time;

and preparation of graphs, maps, and diagrams to show the relationships developed.

Inspection and Comparison

A simple inspection of a group of chemical analyses generally will make possible a separation into obviously interrelated subgroups. For example, it is easy to group together the waters that have dissolved solids or chloride concentrations falling within certain ranges. The consideration of dissolved solids, however, should be accompanied by consideration of the kinds of ions present as well.

Simple inspection of water quality data arrayed in massive sets of numbers probably is the number one analysis technique used by regulatory agencies in deciding if a particular facility is contributing to ground water contamination. These data sets are, however, of limited use in the interpretation of site geologic and hydrogeologic conditions. The simple placement of water quality data, however, on maps and cross sections provides a powerful tool for interpretation of ground water conditions. These data can be arrayed on maps and cross sections in a number of ways to enhance the interpretation of site flow paths and ground water movement.

Figure 13.2 shows the typical tabular array of water quality data. Since these data formats require significant efforts to assimilate, it is recommended that alternative display formats be employed whenever appropriate for detailed understanding of water quality information. Water quality display formats in increasing complexity can be divided into the following categories:

- tabular presentation
- contour maps
- time series displays
- histogram format

- stiff diagrams
- scholler diagrams
- trilinear diagrams
- correlation coefficients

Each of these presentations can have useful application to understanding water quality variations and categories of ground water. Tabular presentations of parameters are a necessary evil that can be eased by use of data summaries and averages. Massive data arrays are always prone to transcription errors, and particular care should be used in proofing of data sets. Computer based spread sheets can ease data reduction time; however, any transcription of data must be carefully checked and rechecked for accuracy.

Contour Maps

Presentation of water quality data into map based formats has been traditionally handled through contouring of monitoring point data. The technique of mapping ground-water quality data by drawing lines of equal concentration of dissolved solids or of single ions, either organic or inorganic parameters, has been used

INORGANIC ANALYSES OF NEWARK MONITORING WELLS
SAMPLING DATE: JANUARY, 1988

WELLS	1C	2C	3B	3C	4C	5C	6C
CALCIUM	1500	2300	370	540	130	75	2300
MAGNESIUM	4300	2200	260	430	53	29	1200
SODIUM	43200	14100	1900	2600	190	150	5100
POTASSIUM	520	53	9.4	12	3.4	73	14
IRON	6.1	3.6	290	7.3	11	20	14
MANGANESE	35	13	2.9	0.35	0.47	0.34	18
NICKEL	1.4	0.45	0.26	<0.05	<0.05	<0.05	0.18
CHLORIDE	72000	30000	3600	5300	390	130	15000
SULFATE	6800	3180	590	760	110	94	1700
TOTAL PHOSPHOROUS (as PO4)	0.08	0.63	0.1	0.41	0.43	0.74	<0.04
NITRATE	<0.1	<0.1	<0.1	<0.1	<0.1	<0.1	<0.1
NITRITE	0.07	0.03	<0.01	<0.01	0.01	0.01	0.01
FLUORIDE	<0.01	<0.01	0.32	0.28	0.43	0.35	0.11
SILICA (as SiO2)	16	24	22	25	27	20	24
CALCIUM HARDNESS	1900	2600	1100	1600	320	15	7100
MAGNESIUM HARDNESS	30100	20400	1600	2100	350	140	5900
TOTAL HARDNESS	32000	23000	2700	3700	670	150	13000
CARBONATE ALKALINITY (as CaCO3)	1	<0.1	<0.1	<0.1	<0.1	250	<0.1
BICARBONATE ALKALINITY (as CaCO3)	550	480	400	350	410	240	380
HYDROXIDE ALKALINITY (as CaCO3)	<1.0	<0.1	<0.1	<0.1	<0.1	<0.1	<0.1
TOTAL ALKALINITY (as CaCO3)	550	480	400	350	410	490	380
TOTAL DISSOLVED SOLIDS	130000	57000	7100	11000	1300	630	30000
pH, FIELD	7.27	7.34	7.69	7.63	8.07	9.02	7.35
pH, LAB	6.6	6.9	7.1	7.6	8.0	10.2	7.2
SPECIFIC CONDUCTANCE, umhos/ca, FIELD	off scale	off scale	15200	17300	1900	200	off scale
SPECIFIC CONDUCTANCE, umhos/ca, LAB	137000	72500	12000	17400	2670	1100	41100
TEMPERATURE, C, FIELD	17	14	15	16	17	19	16
TURBIDITY, HYD	570	72	1400	100	120	350	140
COLOR, APHA UNITS	4	1	1	2	1	2	2

†† NOTE; ALL ANALYSES IN MG/L UNLESS OTHERWISE NOTED

Figure 13.2.   Water quality data display.

in reports published for more than 70 years (Hem, 1970). The applicability of this technique of constructing isogram maps depends on several factors:

- homogeneity of water composition with depth
- gradient of the parameter concentration between measuring points.

Restricted sampling points in either vertical or horizontal directions will limit the usefulness of this technique; however, the typical facility detection or assessment monitoring system should have sufficient isocontour maps. Contour maps can have the form of either closed isopleths, as shown in Figure 13.3, or with open gradient lines, as shown in Figure 13.4. Both of these contour maps show

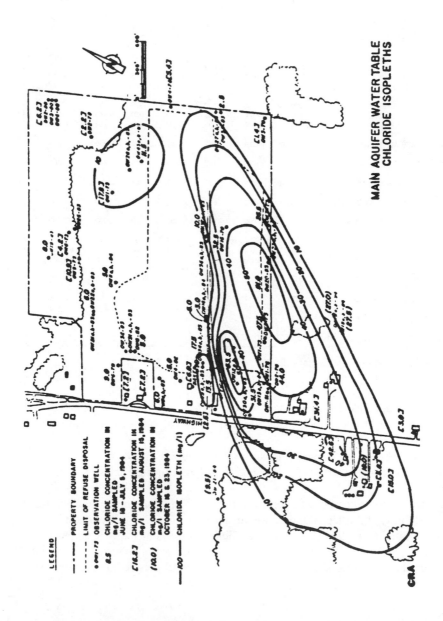

Figure 13.3.   Contour map with closed isopleths.

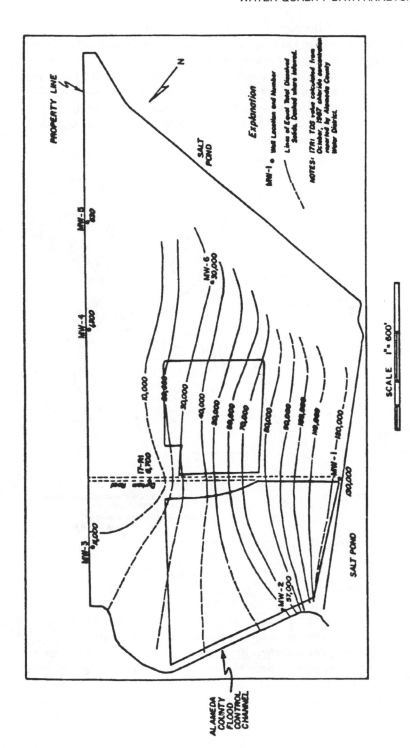

Figure 13.4. Contour map with open gradient lines.

isocontours of chlorides. Because chlorides are typically not affected by precipitation or reactions that would lower concentrations, this parameter serves as one of the best inorganics to use in contour formats.

Additional parameters such as conductivity, temperature, and COD can also serve as conservative parameters for these presentations. On occasion, organic parameters such as chlorinated hydrocarbons can be used for contouring. Figure 13.5 shows such a presentation. As with any data contouring, questionable data should be represented by dashed lines.

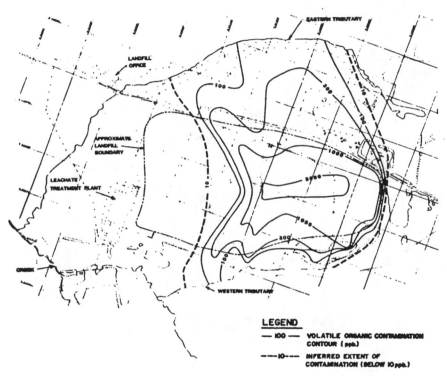

Figure 13.5. Contour map showing organic parameters.

Time Series Formats

Water quality at a single point can change with time. Even with the generally slow movement of ground water, long term detection or assessment monitoring programs can present gradual changes in water quality. These changes can be illustrated by time series presentations that take changes in water quality with time into consideration. Time series can be used to compare individual parameters with time (compare water quality in a well against itself), or multiple parameters in a well with time, or to illustrate changes with time to multiple wells for a common parameter. Figure 13.6 shows a comparison of a number of wells with time for common parameters. Time series presentations can suffer if too much data is presented on one x–y plot. Figure 13.7 shows a time series plot for eight wells for chloride; although only a single parameter is displayed, the variable y scales used in the presentation make interpretation of trends difficult. Time series presentations are most effective when single parameters or functions are compared as shown in Figure 13.8. This illustration compares water level elevations with chloride concentrations. Whether or not the water level elevation truly is related to the chloride concentration decline is a separate question; however, the data are displayed in an easily understood format.

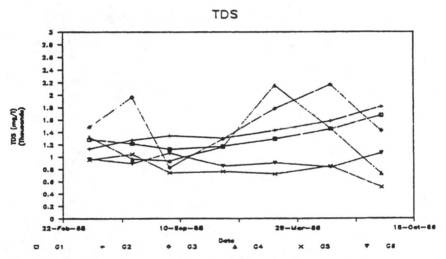

Figure 13.6. Time series comparison of a number of wells.

Histograms

The plotting of ground water data as a series of comparative histograms (or bar graphs) has been a traditional methodology for representation of area water quality variability. Most of the traditional methods of graphing analyses are designed to represent simultaneously the total solute concentration and the proportions assigned to each ionic species for one analysis or group of analyses. The units in which concentrations are expressed in these traditional diagrams are milliequivalents per liter. Hem (1970) provides descriptions as to these bar graph radiating vector plots, circular diagrams, and stiff diagrams so these plotting systems will not be discussed in the typical "whole" analysis interpretations. Water quality data collected in detection or assessment monitoring programs have not

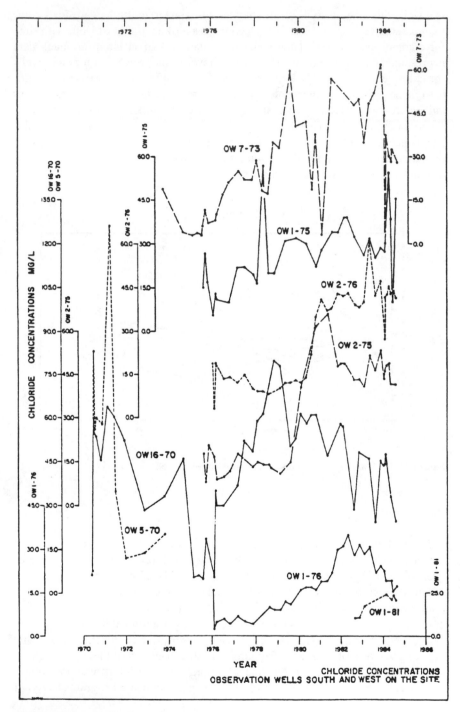

Figure 13.7. Time series comparison of chloride in eight wells.

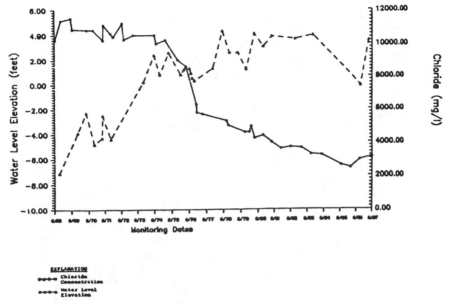

Figure 13.8.   Time series comparison of water level elevation with chloride concentration.

been portrayed in this "whole" analysis format with anions and cations given in the milliequivalents per liter units. Rather, results of water quality analyses are given in milligrams or micrograms per liter and presented in limited display formats of a few parameters. These data, especially volatile organics and hazardous metals, have been displayed as "footprint or fingerprint" histograms illustrating variations in water quality from point to point. Figure 13.9 shows a series of hazardous metal histograms of water quality obtained from individual wells. These data histograms have also been used to track volatile organic plumes and compare the relative proportions of individual organic species from point to point.

Tabular constituent summaries are also a form of comparative histogram formats. Figure 13.10 shows a constituent summary table comparing organic parameters observed in a landfill leachate with organic parameters observed in offsite wells. Many of the constituents do not agree between the landfill leachate "fingerprint" and the offsite monitoring. Care must be taken to use indicator parameters that will not change with time and provide a misinterpretation of the water quality fingerprint.

Trilinear Diagrams

If one considers only the major dissolved ionic constituents in milliequivalents per liter and lumps potassium and sodium together and floride and nitrate with chloride, the composition of most natural water can be closely approximated in terms of three cationic and three anionic species. If the values are expressed as

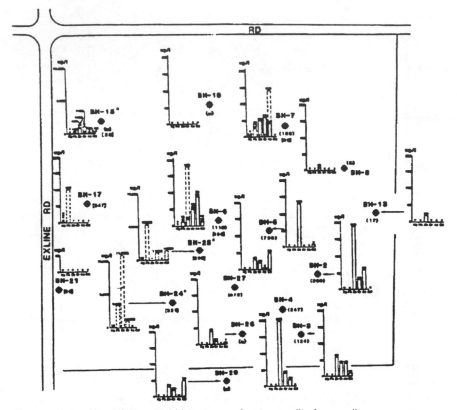

Figure 13.9.   Hazardous metal histograms of water quality from wells.

percentages of the total milliequivalents per liter of cations, and anions, the composition of the water can be represented conveniently by a trilinear plotting technique.

The simplest trilinear plots utilize two triangles, one for anions and one for cations. Each vertex represents 100% of a particular ion or group of ions. The composition of the water with respect to cations is indicated by a point plotted in the cation triangle, and the composition with respect to anions by a point plotted in the anion triangle. The coordinates at each point add to 100%. Most trilinear diagrams are in the form of two triangles bracketing a diamond shaped plotting field first described by Piper (1944).

The trilinear diagram constitutes a useful tool in water-analysis interpretation. Most of the graphical procedures described here are of value in pointing out features of analyses and arrays of data which need closer study. The graphs themselves do not constitute an adequate means of making such studies, however, unless they can demonstrate that certain relationships exist among individual samples. The trilinear diagrams sometimes can be used for this purpose.

CONSTITUENT SUMMARY TABLE

Compound	Landfill GW 1985	EMM-2 UWT 1983	EMM-2 UWT 1984	EMM-2 UWT 1985	EMM-3 UWT 1983	EMM-3 UWT 1984	EMM-3 UWT 1985	EMM-2 LWT 1983	EMM-2 LWT 1984	EMM-2 LWT 1985	EMM-3 LWT 1983	EMM-3 LWT 1984	EMM-3 LWT 1985 (LL)	EMM-2 LST 1983	EMM-2 LST 1984	EMM-2 LST 1985	EMM-3 LST 1983	EMM-3 LST 1984	EMM-3 LST 1985
1,2-Dichlorobenzene	X																		
1,4-Dichlorobenzene	X																		
4-Methylphenol	X																		
Bis(2-ethylhexyl)phthalate	X																		
Methyl naphthalene	X																		
PCB-1242	X																		
PCB-1254	X																		
PCB-1260	X																		
Benzene	X							X			X	X		X	X				
Carbon Disulfide	X								X		X				X				
Chlorobenzene	X								X			X		X					
Ethylbenzene	X							X			X	X		X	X	X			
Methylene Chloride	X											X							
N-Nitrosodiphenylamine (*)	X										X								
Naphthalene	X					X					X								X
Phenol	X										X								
Toluene	X							X			X	X		X	X				
Total Xylenes	X							X	X		X				X				
5-Methylphenol								X			X			X					
1,1,2,2-Tetrachloroethane								X			X				X				
1,1-Dichloroethane										X	X			X	X				
1,2-Dichloroethane										X	X								
1,2-Dichloroethane (**)								X	X		X	X		X	X				
2,4-Dimethylphenol											X								
Acetone											X					X			X
C-3 Alkylbenzene									X		X								
C-4 Alkylbenzene											X								
C-5 Alkylbenzamide									X										
Chloroform								X			X								
Dichlorobenzene								X			X								
Ethyl Ether									X		X								
Ethyl Hexanoic acid											X								
Methylpentanediol									X		X								
Methyl Ethyl Ketone								X			X			X					
Methyl Isobutyl Ketone								X			X								
Methylnonanediol									X										
Oxy Bis Ethoxythane									X		X								
Propanol																			X
Tetrachloroethane								X			X								
Tetrahydrofuran								X			X			X					
Trichloroethane											X			X					
Trimethylbicycloheptanone												X							
Vinyl Chloride								X			X	X							
n-Butylbenzene	X							X			X			X		X			
o-Chlorotoluene								X			X			X					

Figure 13.10. Summary table comparing organic parameters in a landfill leachate with organic parameters in offsite wells.

Figure 13.11 shows a trilinear diagram depicting water chemistry and trends between San Francisco Bay waters and the Newark Aquifer. As with any display of water quality data, the interpretation should be aided by the diagram in a clear-cut comparison or trend.

Trilinear diagrams have become so popular that computer programs have been written to automatically calculate and display the data.

Statistical Treatment of Water Quality Data

Various simple procedures such as averaging, determining frequency distributions, and making simple or multiple correlations are widely used in water analysis interpretation. The more sophisticated applications of statistical methods and, particularly, procedures, which utilize digital computers, are being more and more widely applied. Some potential applications of these techniques will be suggested here. It is essential that proper consideration be given to chemical principles when

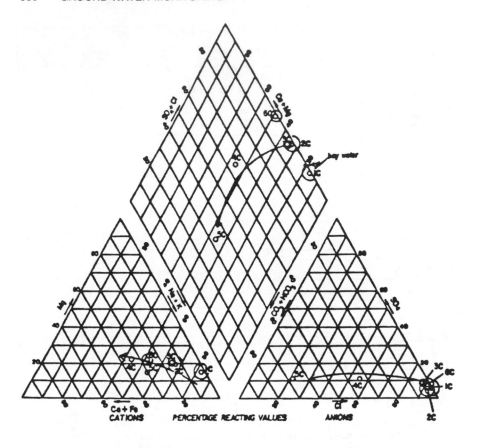

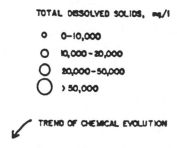

TOTAL DISSOLVED SOLIDS, mg/l

o   0-10,000

O   10,000-20,000

O   20,000-50,000

O   >50,000

TREND OF CHEMICAL EVOLUTION

Figure 13.11. Trilinear diagram showing water chemistry and trends, Newark aquifer.

such applications are made. Most data may require simple averaging and development of basic statistical calculations such as standard deviations and the various test statistics, as shown in Figure 13.12.

The objective of a ground water monitoring program is to determine if the facility is impacting ground water. To accomplish this objective, owner/operators are

SELECTED WATER QUALITY INDICATOR PARAMETERS

Groundwater Level Data

WELL	04/08/85	6/18/85	8/22/85	9/17/85	11/20/85	2/19/86	5/23/86	8/19/86	N	Mean Water Elevation	S.D.
GEI-1	401.43	400.85	399.26	398.64	399.18	400.43	399.10	397.89	8	399.60	1.12
GEI-10	—	—	—	—	—	400.63	399.00	398.00	3	399.21	1.08
GEI-15	—	—	—	—	—	400.34	399.09	397.75	3	399.06	1.06
GEI-2	401.42	400.63	398.63	398.07	398.80	400.12	398.87	397.62	8	399.27	1.23
GEI-3	402.04	401.14	398.80	398.16	400.07	400.65	399.44	398.03	8	399.79	1.35
GEI-4	402.48	401.57	399.34	398.57	400.23	401.06	399.77	398.56	8	400.20	1.33
GEI-5	402.36	401.53	399.57	398.82	399.86	401.01	399.63	398.66	8	400.18	1.23
GEI-6	402.37	401.61	399.61	398.87	399.90	400.96	399.58	398.66	8	400.20	1.24
GEI-7	402.14	401.42	399.81	398.94	399.95	401.05	399.38	398.71	8	400.18	1.15
GEI-8	402.27	401.56	399.90	398.96	400.12	400.98	399.49	398.65	8	400.24	1.19
GEI-9	403.00	402.25	399.81	398.37	400.21	—	—	—	5	400.73	1.68
GEI-10	403.04	402.27	400.19	394.40	396.13	—	—	399.19	6	399.20	3.10
GEI-11	403.04	402.23	400.02	396.10	398.06	—	—	—	5	399.89	2.57
GEI-12	403.08	402.16	399.82	397.72	408.11	401.55	398.80	399.00	8	400.28	1.72
GEI-13	—	—	—	—	—	400.38	399.01	397.63	3	399.01	1.12
GEI-14	—	—	—	—	—	400.29	399.08	399.04	3	399.47	0.58
GEI-15	—	—	—	—	—	400.64	399.10	397.93	3	399.22	1.11
GEI-16	—	—	—	—	—	400.53	399.20	397.95	3	399.23	1.05
GEI-17D	—	—	—	—	—	400.30	398.89	397.39	3	398.86	1.19
GEI-17S	—	—	—	—	—	400.22	399.01	397.59	3	398.94	1.37

Total Dissolved Solids (TSD) (mg/l)

WELL	04/08/85	6/18/85	8/22/85	11/20/85	2/19/86	5/23/86	8/19/86	N	Mean Concent.	S.D.
GEI-1	1200	1210	1120	1170	1290	1450	1670	7	1318	109
GEI-10	—	—	—	—	530	600	539	4	417	242
GEI-15	—	—	—	—	940	1210	1780	3	1323	336
GEI-2	1130	1270	1340	1300	1430	1500	1010	7	1455	189
GEI-3	1400	1960	420	1290	1780	2160	1420	7	1572	449
GEI-4	1320	960	920	1160	2150	1460	726	7	1231	469
GEI-5	950	1040	740	760	720	850	508	7	770	159
GEI-6	960	890	1060	850	900	830	1060	7	932	94
GEI-7	970	990	960	710	670	860	683	7	812	131
GEI-8	960	680	1120	690	660	690	691	7	770	175
GEI-9	970	490	500	—	—	—	—	3	495	5
GEI-10	390	400	420	420	—	—	304	5	406	15
GEI-11	560	530	550	570	—	—	—	4	550	16
GEI-12	650	460	480	480	430	490	441	7	464	22
GEI-13	—	—	—	—	1040	1010	951	3	1000	37
GEI-14	—	—	—	—	2100	2540	2500	3	2407	141
GEI-15	—	—	—	—	1320	—	1390	2	1355	35
GEI-16	—	—	—	—	1400	1330	1350	3	1360	29
GEI-17D	—	—	—	—	1440	1360	1480	3	1427	50
GEI-17S	—	—	—	—	2370	1600	1680	3	1910	325

Chloride Concentrations (mg/l)

WELL	04/08/85	6/18/85	8/22/85	11/20/85	2/19/86	5/23/86	8/19/86	N	Mean Concent.	S.D.
GEI-1	250	250	200	210	360	310	390	7	206.67	72.26
GEI-10	—	—	—	—	19	33	25	4	19.25	12.17
GEI-15	—	—	—	—	140	180	440	3	253.33	333.00
GEI-2	240	260	290	320	330	340	420	7	326.67	49.55
GEI-3	200	280	97	220	330	410	230	7	261.17	97.35
GEI-4	180	91	86	180	540	300	91	7	214.67	164.05
GEI-5	130	160	86	110	110	140	37	7	107.17	39.28
GEI-6	97	110	130	100	130	110	250	7	138.33	51.13
GEI-7	120	96	94	52	56	77	40	7	69.17	21.20
GEI-8	6	9	9	8	8	8	7	7	8.17	0.69
GEI-9	5	7	10	5	—	—	—	4	7.33	2.05
GEI-10	3	4	4	3	—	—	4	5	3.75	0.43
GEI-11	6	8	8	8	—	—	—	4	8.00	0.00
GEI-12	1	12	9	11	10	2	21	7	10.83	5.58
GEI-13	—	—	—	—	190	140	150	3	160.00	21.60
GEI-14	—	—	—	—	550	510	600	3	553.33	36.82
GEI-15	—	—	—	—	160	170	190	3	173.33	12.47
GEI-16	—	—	—	—	140	110	110	3	120.00	14.14
GEI-17D	—	—	—	—	290	260	260	3	270.00	14.14
GEI-17S	—	—	—	—	680	420	360	3	486.67	138.88

Figure 13.12. Selected water quality indicator parameters.

required to place detection monitoring wells in both upgradient and downgradient locations around the facility and to monitor those wells at regular time intervals, typically quarterly, for a series of so-called "indicator parameters," that may vary from facility to facility. The logic of this sampling strategy is that upgradient water quality represents the pristine background conditions for that particular region, and the downgradient water quality represents background regional water quality plus any influence or contamination produced by the facility. Although the logic of this sampling strategy is sometimes questionable, we will proceed assuming that it is appropriate, and describe statistical methods that can be used in detection monitoring programs of this kind. In some cases, particularly when background data are available prior to the installation of the facility, intra-well comparisons may be used (i.e., each well compared to its own history). The major advantage of this approach is that it completely eliminates the spatial component of variability from the comparison. The statistical methods to be described are appropriate for either sampling strategy. The overall methodology is shown in Figure 13.13 and provide the structure for statistical comparisons between wells and intra-well time series statistics.

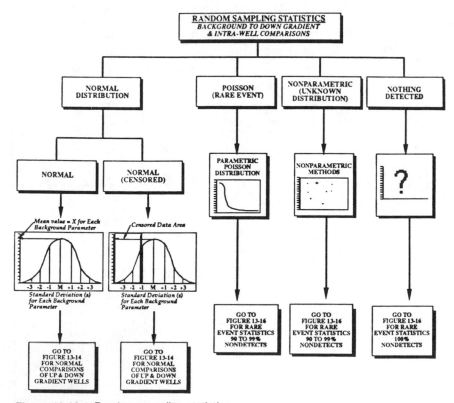

Figure 13.13.   Random sampling statistics.

Assumptions

The statistical methods to be presented are suitable for data that are normally distributed, rare event data that have a Poisson distribution, or data for which the distribution is unknown and/or atypical (i.e., nonparametric). In addition, normally distributed data that have some proportion (less than 90%) below a detection limit (i.e., nondetects) can also be accommodated. We will refer to such data as "censored" and the corresponding distribution as a censored normal distribution. Given these four choices, (i.e., normal, censored normal, Poisson, or nonparametric) the only case that is not covered is when nothing is detected in the background water quality samples (see Figure 13.13). For this case, we will present a method by which practical quantification limits can be obtained based on a small analyte present laboratory calibration study. These limits may in turn be used as detection monitoring decision criteria (i.e., limits) or substituted for nondetects in other analyses.

In addition to selecting the proper sampling distribution, the most critical assumption underlying all of the statistical methods to be presented is that of independence.

These models strictly assume that the observations are the result of some random sampling process and that each observation is an independent random sample from the parent population. In the context of ground water monitoring, this assumption rules out the use of replicate samples, daily samples, and perhaps even monthly samples given how slowly ground water moves. As such, we strongly recommend a quarterly sampling program.

Statistical Overview

In detection monitoring programs, we obtain a sample from a monitoring well and are faced with the decision of whether or not the facility had an impact on the measured concentration of a series of indicator parameters. It is critically important to realize that this new monitoring measurement is not a mean value, but rather, a single new observation. As such, statistical methods for the comparison of mean values (e.g., student's t-test) simply do not apply. From a statistical perspective, the problem is, therefore, to estimate the probability that this new datum was drawn from the population of pristine background ground water for which we only have estimates of its mean and variance obtained from a limited number of upgradient measurements. If we knew that a particular indicator parameter was normally distributed and somehow had the privileged information of knowing the population mean (μ) and variance (σ^2), then we could simply construct the interval $\mu 1.96\sigma$, which would contain 95% of all individual measurements (not means) drawn from that population, and our job would be finished. However, in practice, we never knew the values of μ and σ, but only have the sample based estimates x and s obtained from n independent upgradient measurements. As such, our uncertainty is twofold. First, we have a range of possible values when sampling from a normal distribution with known parameters; and second, we have a range of possible means (x) and standard deviations (s) that could have been obtained from drawing a sample of size n from a normally distributed population with mean μ and variance σ^2. This latter source of uncertainty will require a multiplier that is larger than 1.96 if we are to have reasonable confidence (say 95%) that 95% of the population is contained within the interval. As the number of background water quality measurements approaches infinity, however, the multiplier once again approaches 1.96. When the sample size is small, say n=8, the required multiplier for 95% confidence that 95% of the population is contained is 3.732 (i.e., $\times \pm 3.732$ s). For a sample size of n=30, 95% confidence is achieved using a multiplier of 2.549; and for n=100, the multiplier is 2.233.

These intervals are known as two-sided tolerance intervals in the statistical literature, and are largely due to the work of Wald and Wolfowitz (1946). When we are only concerned with monitoring values that are too large, one-sided tolerance limits can be constructed as x=ks where the multiplier k is somewhat smaller than the previous two-sided tolerance limit factors. Figure 13.14 shows the statistical procedure for evaluation of background and downgradient water quality for parameters that normally have detectable values. For example, for n=8, the one-sided tolerance limit is obtained as x+3.188 s; for n=30, k=2.220; and for n=100, k=1.927. Table 13.1 presents values of the multiplier k for n=4 to 100 that are required to have 95% confidence that 95% of a normally distributed population is contained in the interval (i.e., two-sided) or is below the limit (i.e., one-sided).

Although tolerance intervals are generally quite useful in quality control problems similar to ground-water detection monitoring, even more precise probability statements can be made. For example, in the context of ground-water monitoring, we are generally less interested in what can happen in 95% of all possible samples and more interested in what can happen on the next round of sampling for which measurements are to be obtained from the 2 or 5 or 20 monitoring wells at the facility. Since we know the number of future comparisons (i.e., monitoring wells), we can construct an interval (two-sided) or limit (one-sided) that will contain the next r measurements with 95% confidence. If r, in this case the number of monitoring wells, is reasonably small, it will provide a more conservative test than the corresponding 95% confidence, 95% coverage tolerance interval. For example, for a facility with n=8 background measurements and r=3 monitoring wells, the multiplier for a one-sided 95% confidence, 95% coverage tolerance limit is k=3.188; whereas the corresponding factor for a 95% prediction limit is only k=2.80. However, if the facility had r=10 monitoring wells, the tolerance limit factor is, of course, unchanged; but the prediction limit factor is now k=3.71, which is considerably larger than the corresponding tolerance limit factor. In general, for 95% confidence and 95% coverage, tolerance intervals will be more conservative for facilities with r > 3 monitoring wells and predictive limits will be more conservative for facilities with r ≤ 3 monitoring wells. Given the large number of detection monitoring wells at most modern waste disposal facilities, tolerance intervals may well be the method of choice. Tables 13.2 through 13.4 contain factors for computing one-sided 95% prediction limits based

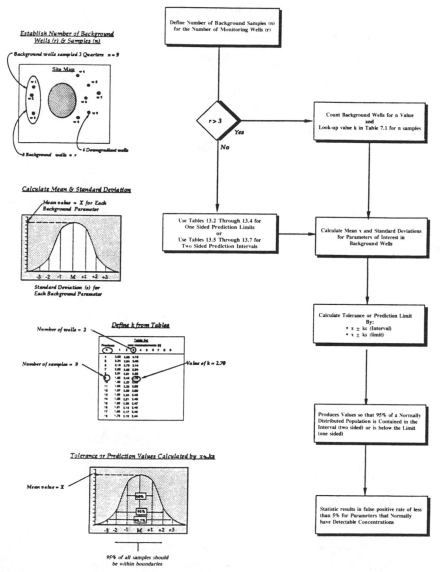

Figure 13.14. Statistical procedure for evaluation of background and downgradient water quality for parameters that have normally detectable values.

on background samples of n=4 to 100, and number of monitoring wells of r=1 to 100. Corresponding two-sided 95% prediction interval factors are provided in Tables 13.5 through 13.7. As in the case of tolerance intervals and limits, these factors are applied as x ± ks (interval) and x+ks (limit).

A detailed description of prediction limits in the context of ground water monitoring problems is provided by Gibbons (1987a).

Table 13.1. Factors (k) for Constructing Two-Sided and One-Sided Normal Tolerance Limits ($\bar{x} \pm ks$ and $\bar{x} + ks$) 95% Confidence That 95% of the Distribution is Covered.

n	Two-Sided	One-Sided
4	6.370	5.144
5	5.079	4.210
6	4.414	3.711
7	4.007	3.401
8	3.732	3.188
9	3.532	3.032
10	3.379	2.911
11	3.259	2.815
12	3.169	2.736
13	3.081	2.670
14	3.012	2.614
15	2.954	2.566
16	2.903	2.523
17	2.858	2.486
18	2.819	2.453
19	2.784	2.423
20	2.752	2.396
21	2.723	2.371
22	2.697	2.350
23	2.673	2.329
24	2.651	2.309
25	2.631	2.292
30	2.549	2.220
35	2.490	2.166
40	2.445	2.126
50	2.379	2.065
60	2.333	2.022
80	2.272	1.965
100	2.233	1.927

Selecting the Number of Background Samples

A question commonly asked of statisticians is: "How many background samples do I need?" This question became even more common when in previous regulations (40 CFR Part 264), owner/operators were required to demonstrate that alternate statistical procedures balanced false positive and false negative results. Fortunately, this requirement is no longer a part of the new RCRA statistical regulation. The reason why this is difficult is that the false negative rate, or one minus statistical power, is dependent on three things: the false positive rate (selected *a priori* by the regulation; that is, 5% for the facility as a whole), the number of background measurements, and the effect size. The effect size describes the smallest difference (typically described in standard deviation units) that is environmentally meaningful.

Since the USEPA is not willing to provide such a minimum effect size, it is completely impossible to derive a sample size that will properly balance false

Table 13.2. Factors For Obtaining One-Sided 95% Prediction Limits For r Additional Samples Given a Background Sample of Size n.

Previous n	Number of New Measurements (r)														
	1	2	3	4	5	6	7	8	9	10	11	12	13	14	15
4	2.63	3.56	4.18	4.67	5.08	5.43	5.74	6.03	6.29	6.53	6.76	6.97	7.17	7.36	7.54
5	2.34	3.04	3.49	3.83	4.10	4.34	4.54	4.73	4.89	5.04	5.18	5.31	5.44	5.55	5.66
6	2.18	2.78	3.14	3.42	3.63	3.82	3.97	4.11	4.24	4.35	4.46	4.56	4.65	4.73	4.81
7	2.08	2.62	2.94	3.17	3.36	3.51	3.65	3.76	3.87	3.96	4.05	4.13	4.20	4.27	4.34
8	2.01	2.51	2.80	3.01	3.18	3.32	3.43	3.54	3.63	3.71	3.79	3.86	3.92	3.98	4.04
9	1.96	2.43	2.70	2.90	3.05	3.18	3.29	3.38	3.46	3.54	3.60	3.67	3.72	3.78	3.83
10	1.92	2.37	2.63	2.82	2.96	3.08	3.18	3.26	3.34	3.41	3.47	3.53	3.58	3.63	3.68
11	1.89	2.33	2.58	2.75	2.89	3.00	3.09	3.17	3.25	3.31	3.37	3.42	3.47	3.52	3.56
12	1.87	2.29	2.53	2.70	2.83	2.93	3.02	3.10	3.17	3.23	3.29	3.34	3.39	3.43	3.47
13	1.85	2.26	2.49	2.66	2.78	2.88	2.97	3.04	3.11	3.17	3.22	3.27	3.32	3.36	3.40
14	1.83	2.24	2.46	2.62	2.74	2.84	2.93	3.00	3.06	3.12	3.17	3.22	3.26	3.30	3.34
15	1.82	2.21	2.44	2.59	2.71	2.81	2.89	2.96	3.02	3.07	3.12	3.17	3.21	3.25	3.28
16	1.81	2.20	2.41	2.57	2.68	2.78	2.85	2.92	2.98	3.04	3.09	3.13	3.17	3.21	3.24
17	1.80	2.18	2.40	2.54	2.66	2.75	2.83	2.89	2.95	3.01	3.05	3.09	3.13	3.17	3.21
18	1.79	2.17	2.38	2.53	2.64	2.73	2.80	2.87	2.93	2.98	3.02	3.07	3.10	3.14	3.17
19	1.78	2.16	2.36	2.51	2.62	2.71	2.78	2.85	2.90	2.95	3.00	3.04	3.08	3.11	3.14
20	1.77	2.14	2.35	2.49	2.60	2.69	2.76	2.83	2.88	2.93	2.98	3.02	3.05	3.09	3.12
21	1.77	2.13	2.34	2.48	2.59	2.67	2.75	2.81	2.86	2.91	2.96	3.00	3.03	3.07	3.10
22	1.76	2.13	2.33	2.47	2.57	2.66	2.73	2.79	2.85	2.89	2.94	2.98	3.01	3.05	3.08
23	1.75	2.12	2.32	2.46	2.56	2.65	2.72	2.78	2.83	2.88	2.92	2.96	3.00	3.03	3.06
24	1.75	2.11	2.31	2.45	2.55	2.63	2.70	2.76	2.82	2.87	2.91	2.95	2.98	3.01	3.04
25	1.74	2.10	2.30	2.44	2.54	2.62	2.69	2.75	2.81	2.85	2.89	2.93	2.97	3.00	3.03
26	1.74	2.10	2.29	2.43	2.53	2.61	2.68	2.74	2.79	2.84	2.88	2.92	2.95	2.99	3.01
27	1.74	2.09	2.29	2.42	2.52	2.61	2.67	2.73	2.78	2.83	2.87	2.91	2.94	2.97	3.00
28	1.73	2.09	2.28	2.42	2.52	2.60	2.66	2.72	2.77	2.82	2.86	2.90	2.93	2.96	2.99
29	1.73	2.08	2.28	2.41	2.51	2.59	2.66	2.71	2.77	2.81	2.85	2.89	2.92	2.95	2.98
30	1.73	2.08	2.27	2.40	2.50	2.58	2.65	2.71	2.76	2.80	2.84	2.88	2.91	2.94	2.97
31	1.72	2.07	2.27	2.40	2.50	2.58	2.64	2.70	2.75	2.79	2.83	2.87	2.90	2.93	2.96
32	1.72	2.07	2.26	2.39	2.49	2.57	2.64	2.69	2.74	2.79	2.83	2.86	2.89	2.92	2.95
33	1.72	2.07	2.26	2.39	2.49	2.56	2.63	2.69	2.74	2.78	2.82	2.85	2.89	2.92	2.94
34	1.72	2.06	2.25	2.38	2.48	2.56	2.62	2.68	2.73	2.77	2.81	2.85	2.88	2.91	2.94
35	1.71	2.06	2.25	2.38	2.48	2.55	2.62	2.67	2.72	2.77	2.81	2.84	2.87	2.90	2.93
36	1.71	2.06	2.25	2.37	2.47	2.55	2.61	2.67	2.72	2.76	2.80	2.83	2.87	2.90	2.92
37	1.71	2.06	2.24	2.37	2.47	2.54	2.61	2.66	2.71	2.76	2.79	2.83	2.86	2.89	2.92
38	1.71	2.05	2.24	2.37	2.46	2.54	2.60	2.66	2.71	2.75	2.79	2.82	2.86	2.88	2.91
39	1.71	2.05	2.24	2.36	2.46	2.54	2.60	2.66	2.70	2.75	2.78	2.82	2.85	2.88	2.91
40	1.71	2.05	2.23	2.36	2.46	2.53	2.60	2.65	2.70	2.74	2.78	2.81	2.85	2.87	2.90
41	1.70	2.05	2.23	2.36	2.45	2.53	2.59	2.65	2.69	2.74	2.77	2.81	2.84	2.87	2.90
42	1.70	2.04	2.23	2.35	2.45	2.53	2.59	2.64	2.69	2.73	2.77	2.80	2.84	2.87	2.89
43	1.70	2.04	2.23	2.35	2.45	2.52	2.59	2.64	2.69	2.73	2.77	2.80	2.83	2.86	2.89
44	1.70	2.04	2.22	2.35	2.44	2.52	2.58	2.64	2.68	2.73	2.76	2.80	2.83	2.86	2.88
45	1.70	2.04	2.22	2.35	2.44	2.52	2.58	2.63	2.68	2.72	2.76	2.79	2.82	2.85	2.88
46	1.70	2.04	2.22	2.34	2.44	2.51	2.58	2.63	2.68	2.72	2.76	2.79	2.82	2.85	2.88
47	1.70	2.03	2.22	2.34	2.44	2.51	2.57	2.63	2.67	2.72	2.75	2.79	2.82	2.85	2.87
48	1.70	2.03	2.22	2.34	2.43	2.51	2.57	2.62	2.67	2.71	2.75	2.78	2.81	2.84	2.87
49	1.69	2.03	2.21	2.34	2.43	2.51	2.57	2.62	2.67	2.71	2.75	2.78	2.81	2.84	2.86
50	1.69	2.03	2.21	2.34	2.43	2.50	2.57	2.62	2.67	2.71	2.74	2.78	2.81	2.84	2.86
60	1.68	2.02	2.20	2.32	2.41	2.48	2.55	2.60	2.64	2.68	2.72	2.75	2.78	2.81	2.84
70	1.68	2.01	2.19	2.31	2.40	2.47	2.53	2.58	2.63	2.67	2.70	2.74	2.77	2.79	2.82
80	1.67	2.00	2.18	2.30	2.39	2.46	2.52	2.57	2.62	2.66	2.69	2.72	2.75	2.78	2.80
90	1.67	2.00	2.17	2.29	2.38	2.45	2.51	2.56	2.61	2.65	2.68	2.71	2.74	2.77	2.79
100	1.67	1.99	2.17	2.29	2.38	2.45	2.51	2.56	2.60	2.64	2.67	2.71	2.73	2.76	2.79

Factor $= t_{(n-1,\ 1-\alpha/r)} \sqrt{1 + 1/n}$

Table 13.3.   Factors For Obtaining One-Sided 95% Prediction Limits For r Additional Samples Given a Background Sample of Size n.

Previous n	Number of New Measurements (r)														
	16	17	18	19	20	21	22	23	24	25	26	27	28	29	30
4	7.71	7.87	8.03	8.19	8.33	8.48	8.61	8.75	8.88	9.00	9.13	9.25	9.36	9.48	9.59
5	5.76	5.86	5.95	6.05	6.13	6.21	6.29	6.37	6.45	6.52	6.59	6.66	6.72	6.79	6.85
6	4.89	4.96	5.03	5.09	5.15	5.22	5.27	5.33	5.38	5.43	5.48	5.53	5.58	5.62	5.67
7	4.40	4.46	4.51	4.56	4.61	4.66	4.71	4.75	4.80	4.84	4.88	4.91	4.95	4.99	5.02
8	4.09	4.14	4.19	4.23	4.27	4.31	4.35	4.39	4.43	4.46	4.50	4.53	4.56	4.59	4.62
9	3.87	3.92	3.96	4.00	4.04	4.08	4.11	4.14	4.18	4.21	4.24	4.26	4.29	4.32	4.34
10	3.72	3.76	3.80	3.83	3.87	3.90	3.93	3.96	3.99	4.02	4.05	4.07	4.10	4.12	4.15
11	3.60	3.64	3.67	3.71	3.74	3.77	3.80	3.83	3.85	3.88	3.91	3.93	3.95	3.98	4.00
12	3.51	3.54	3.58	3.61	3.64	3.67	3.70	3.72	3.75	3.77	3.79	3.82	3.84	3.86	3.88
13	3.43	3.47	3.50	3.53	3.56	3.58	3.61	3.64	3.66	3.68	3.71	3.73	3.75	3.77	3.79
14	3.37	3.40	3.43	3.46	3.49	3.52	3.54	3.57	3.59	3.61	3.63	3.65	3.67	3.69	3.71
15	3.32	3.35	3.38	3.41	3.43	3.46	3.48	3.51	3.53	3.55	3.57	3.59	3.61	3.63	3.64
16	3.27	3.30	3.33	3.36	3.39	3.41	3.43	3.46	3.48	3.50	3.52	3.54	3.56	3.57	3.59
17	3.24	3.27	3.29	3.32	3.35	3.37	3.39	3.41	3.44	3.46	3.47	3.49	3.51	3.53	3.54
18	3.20	3.23	3.26	3.29	3.31	3.33	3.36	3.38	3.40	3.42	3.44	3.45	3.47	3.49	3.50
19	3.18	3.20	3.23	3.26	3.28	3.30	3.32	3.34	3.36	3.38	3.40	3.42	3.44	3.45	3.47
20	3.15	3.18	3.20	3.23	3.25	3.27	3.30	3.32	3.33	3.35	3.37	3.39	3.40	3.42	3.44
21	3.13	3.15	3.18	3.20	3.23	3.25	3.27	3.29	3.31	3.33	3.34	3.36	3.38	3.39	3.41
22	3.11	3.13	3.16	3.18	3.21	3.23	3.25	3.27	3.29	3.30	3.32	3.34	3.35	3.37	3.38
23	3.09	3.11	3.14	3.16	3.19	3.21	3.23	3.25	3.26	3.28	3.30	3.32	3.33	3.35	3.36
24	3.07	3.10	3.12	3.15	3.17	3.19	3.21	3.23	3.25	3.26	3.28	3.30	3.31	3.33	3.34
25	3.06	3.08	3.11	3.13	3.15	3.17	3.19	3.21	3.23	3.25	3.26	3.28	3.29	3.31	3.32
26	3.04	3.07	3.09	3.11	3.14	3.16	3.18	3.19	3.21	3.23	3.25	3.26	3.28	3.29	3.31
27	3.03	3.06	3.08	3.10	3.12	3.14	3.16	3.18	3.20	3.22	3.23	3.25	3.26	3.28	3.29
28	3.02	3.04	3.07	3.09	3.11	3.13	3.15	3.17	3.19	3.20	3.22	3.23	3.25	3.26	3.28
29	3.01	3.03	3.05	3.08	3.10	3.12	3.14	3.16	3.17	3.19	3.21	3.22	3.24	3.25	3.26
30	3.00	3.02	3.04	3.07	3.09	3.11	3.13	3.14	3.16	3.18	3.19	3.21	3.22	3.24	3.25
31	2.99	3.01	3.04	3.06	3.08	3.10	3.12	3.13	3.15	3.17	3.18	3.20	3.21	3.23	3.24
32	2.98	3.00	3.03	3.05	3.07	3.09	3.11	3.12	3.14	3.16	3.17	3.19	3.20	3.22	3.23
33	2.97	2.99	3.02	3.04	3.06	3.08	3.10	3.11	3.13	3.15	3.16	3.18	3.19	3.21	3.22
34	2.96	2.99	3.01	3.03	3.05	3.07	3.09	3.11	3.12	3.14	3.15	3.17	3.18	3.20	3.21
35	2.96	2.98	3.00	3.02	3.04	3.06	3.08	3.10	3.12	3.13	3.15	3.16	3.17	3.19	3.20
36	2.95	2.97	3.00	3.02	3.04	3.06	3.07	3.09	3.11	3.12	3.14	3.15	3.17	3.18	3.19
37	2.94	2.97	2.99	3.01	3.03	3.05	3.07	3.08	3.10	3.12	3.13	3.15	3.16	3.17	3.19
38	2.94	2.96	2.98	3.00	3.02	3.04	3.06	3.08	3.09	3.11	3.12	3.14	3.15	3.17	3.18
39	2.93	2.96	2.98	3.00	3.02	3.04	3.05	3.07	3.09	3.10	3.12	3.13	3.15	3.16	3.17
40	2.93	2.95	2.97	2.99	3.01	3.03	3.05	3.07	3.08	3.10	3.11	3.13	3.14	3.15	3.17
41	2.92	2.94	2.97	2.99	3.01	3.03	3.04	3.06	3.08	3.09	3.11	3.12	3.13	3.15	3.16
42	2.92	2.94	2.96	2.98	3.00	3.02	3.04	3.05	3.07	3.09	3.10	3.11	3.13	3.14	3.15
43	2.91	2.94	2.96	2.98	3.00	3.02	3.03	3.05	3.07	3.08	3.10	3.11	3.12	3.14	3.15
44	2.91	2.93	2.95	2.97	2.99	3.01	3.03	3.04	3.06	3.08	3.09	3.10	3.12	3.13	3.14
45	2.90	2.93	2.95	2.97	2.99	3.01	3.02	3.04	3.06	3.07	3.09	3.10	3.11	3.13	3.14
46	2.90	2.92	2.94	2.96	2.98	3.00	3.02	3.04	3.05	3.07	3.08	3.09	3.11	3.12	3.13
47	2.90	2.92	2.94	2.96	2.98	3.00	3.02	3.03	3.05	3.06	3.08	3.09	3.10	3.12	3.13
48	2.89	2.92	2.94	2.96	2.98	2.99	3.01	3.03	3.04	3.06	3.07	3.09	3.10	3.11	3.12
49	2.89	2.91	2.93	2.95	2.97	2.99	3.01	3.02	3.04	3.05	3.07	3.08	3.10	3.11	3.12
50	2.89	2.91	2.93	2.95	2.97	2.99	3.00	3.02	3.04	3.05	3.06	3.08	3.09	3.10	3.12
60	2.86	2.88	2.90	2.92	2.94	2.96	2.97	2.99	3.01	3.02	3.03	3.05	3.06	3.07	3.08
70	2.84	2.86	2.88	2.90	2.92	2.94	2.95	2.97	2.98	3.00	3.01	3.03	3.04	3.05	3.06
80	2.83	2.85	2.87	2.89	2.91	2.92	2.94	2.95	2.97	2.98	3.00	3.01	3.02	3.03	3.05
90	2.82	2.84	2.86	2.88	2.89	2.91	2.93	2.94	2.96	2.97	2.98	3.00	3.01	3.02	3.03
100	2.81	2.83	2.85	2.87	2.89	2.90	2.92	2.93	2.95	2.96	2.98	2.99	3.00	3.01	3.02

Factor $= t_{(n-1, \; 1-\alpha/r)} \sqrt{1 + 1/n}$

Table 13.4. Factors For Obtaining One-Sided 95% Prediction Limits For r Additional Samples Given a Background Sample of Size n.

Previous n	30	35	40	45	50	55	60	65	70	75	80	85	90	95	100
					Number of New Measurements (r)										
4	9.59	10.11	10.58	11.02	11.42	11.80	12.15	12.49	12.80	13.11	13.40	13.68	13.94	14.20	14.45
5	6.85	7.14	7.40	7.64	7.86	8.06	8.25	8.42	8.59	8.75	8.90	9.04	9.18	9.31	9.43
6	5.67	5.87	6.05	6.22	6.37	6.50	6.63	6.75	6.86	6.97	7.06	7.16	7.25	7.34	7.42
7	5.02	5.18	5.32	5.45	5.57	5.67	5.77	5.86	5.95	6.03	6.10	6.17	6.24	6.31	6.37
8	4.62	4.75	4.87	4.98	5.07	5.16	5.24	5.32	5.39	5.46	5.52	5.58	5.63	5.68	5.74
9	4.34	4.46	4.57	4.66	4.74	4.82	4.89	4.96	5.02	5.07	5.13	5.18	5.23	5.27	5.31
10	4.15	4.25	4.35	4.43	4.51	4.57	4.64	4.69	4.75	4.80	4.85	4.89	4.93	4.98	5.01
11	4.00	4.10	4.18	4.26	4.33	4.39	4.45	4.50	4.55	4.60	4.64	4.68	4.72	4.76	4.79
12	3.88	3.97	4.05	4.12	4.19	4.25	4.30	4.35	4.40	4.44	4.48	4.52	4.55	4.59	4.62
13	3.79	3.87	3.95	4.02	4.08	4.13	4.18	4.23	4.27	4.31	4.35	4.39	4.42	4.45	4.48
14	3.71	3.79	3.86	3.93	3.99	4.04	4.09	4.13	4.17	4.21	4.25	4.28	4.31	4.34	4.37
15	3.64	3.72	3.79	3.86	3.91	3.96	4.01	4.05	4.09	4.12	4.16	4.19	4.22	4.25	4.28
16	3.59	3.67	3.74	3.79	3.85	3.90	3.94	3.98	4.02	4.05	4.08	4.11	4.14	4.17	4.20
17	3.54	3.62	3.68	3.74	3.79	3.84	3.88	3.92	3.96	3.99	4.02	4.05	4.08	4.11	4.13
18	3.50	3.58	3.64	3.70	3.75	3.79	3.83	3.87	3.90	3.94	3.97	4.00	4.02	4.05	4.07
19	3.47	3.54	3.60	3.66	3.70	3.75	3.79	3.82	3.86	3.89	3.92	3.95	3.97	4.00	4.02
20	3.44	3.51	3.57	3.62	3.67	3.71	3.75	3.79	3.82	3.85	3.88	3.91	3.93	3.96	3.98
21	3.41	3.48	3.54	3.59	3.63	.368	3.72	3.75	3.78	3.81	3.84	3.87	3.89	3.92	3.94
22	3.38	3.45	3.51	3.56	3.61	3.65	3.68	3.72	3.75	3.78	3.81	3.83	3.86	3.88	3.90
23	3.36	3.43	3.49	3.54	3.58	3.62	3.66	3.69	3.72	3.75	3.78	3.80	3.83	3.85	3.87
24	3.34	3.41	3.46	3.51	3.56	3.60	3.63	3.67	3.70	3.73	3.75	3.78	3.80	3.82	3.84
25	3.32	3.39	3.44	3.49	3.53	3.57	3.61	3.64	3.67	3.70	3.73	3.75	3.78	3.80	3.82
26	3.31	3.37	3.42	3.47	3.52	3.55	3.59	3.62	3.65	3.68	3.71	3.73	3.75	3.77	3.80
27	3.29	3.35	3.41	3.46	3.50	3.54	3.57	3.60	3.63	3.66	3.69	3.71	3.73	3.75	3.77
28	3.28	3.34	3.39	3.44	3.48	3.52	3.55	3.59	3.61	3.64	3.67	3.69	3.71	3.73	3.75
29	3.26	3.32	3.38	3.42	3.47	3.50	3.54	3.57	3.60	3.62	3.65	3.67	3.70	3.72	3.74
30	3.25	3.31	3.36	3.41	3.45	3.49	3.52	3.55	3.58	3.61	3.63	3.66	3.68	3.70	3.72
31	3.24	3.30	3.35	3.40	3.44	3.48	3.51	3.54	3.57	3.59	3.62	3.64	3.66	3.68	3.70
32	3.23	3.29	3.34	3.39	3.43	3.46	3.50	3.53	3.55	3.58	3.61	3.63	3.65	3.68	3.68
33	3.22	3.28	3.33	3.38	3.42	3.45	3.48	3.51	3.54	3.57	3.59	3.62	3.64	3.66	3.68
34	3.21	3.27	3.32	3.37	3.41	3.44	3.47	3.50	3.53	3.56	3.58	3.60	3.62	3.64	3.66
35	3.20	3.26	3.31	3.35	3.39	3.43	3.46	3.49	3.52	3.55	3.57	3.59	3.61	3.63	3.65
36	3.19	3.25	3.30	3.35	3.39	3.42	3.45	3.48	3.51	3.54	3.56	3.58	3.60	3.62	3.64
37	3.19	3.24	3.29	3.34	3.38	3.41	3.44	3.47	3.50	3.53	3.55	3.57	3.59	3.61	3.63
38	3.18	3.24	3.29	3.33	3.37	3.40	3.44	3.46	3.49	3.52	3.54	3.56	3.58	3.60	3.62
39	3.17	3.23	3.28	3.32	3.36	3.40	3.43	3.46	3.48	3.51	3.53	3.55	3.57	3.59	3.61
40	3.17	3.22	3.27	3.31	3.35	3.39	3.42	3.45	3.47	3.50	3.52	3.54	3.56	3.58	3.60
41	3.16	3.22	3.27	3.31	3.35	3.38	3.41	3.44	3.47	3.49	3.51	3.54	3.56	3.58	3.59
42	3.15	3.21	3.26	3.30	3.34	3.37	3.41	3.43	3.46	3.48	3.51	3.53	3.55	3.57	3.59
43	3.15	3.20	3.25	3.30	3.33	3.37	3.40	3.43	3.45	3.48	3.50	3.52	3.54	3.56	3.58
44	3.14	3.20	3.25	3.29	3.33	3.36	3.39	3.42	3.45	3.47	3.49	3.51	3.53	3.55	3.57
45	3.14	3.19	3.24	3.28	3.32	3.36	3.39	3.41	3.44	3.46	3.49	3.51	3.53	3.55	3.56
46	3.13	3.19	3.24	3.28	3.32	3.35	3.38	3.41	3.43	3.46	3.48	3.50	3.52	3.54	3.56
47	3.13	3.18	3.23	3.27	3.31	3.34	3.38	3.40	3.43	3.45	3.48	3.50	3.52	3.53	3.55
48	3.12	3.18	3.23	3.27	3.31	3.34	3.37	3.40	3.42	3.45	3.47	3.49	3.51	3.53	3.55
49	3.12	3.18	3.22	3.26	3.30	3.34	3.37	3.39	3.42	3.44	3.46	3.48	3.50	3.52	3.54
50	3.12	3.17	3.22	3.26	3.30	3.33	3.36	3.39	3.41	3.44	3.46	3.48	3.50	3.52	3.53
60	3.08	3.14	3.18	3.22	3.26	3.29	3.32	3.35	3.37	3.40	3.42	3.44	3.46	3.47	3.49
70	3.06	3.12	3.16	3.20	3.24	3.27	3.30	3.32	3.35	3.37	3.39	3.41	3.43	3.45	3.46
80	3.05	3.10	3.14	3.18	3.22	3.25	3.28	3.30	3.33	3.35	3.37	3.39	3.41	3.42	3.44
90	3.03	3.08	3.13	3.17	3.20	3.23	3.26	3.29	3.31	3.33	3.35	3.37	3.39	3.41	3.42
10	3.02	3.07	3.12	3.16	3.19	3.22	3.25	3.27	3.30	3.32	3.34	3.36	3.38	3.39	3.41

Factor $= t_{(n-1,\ 1-\alpha/r)} \sqrt{1 + 1/n}$

Table 13.5. Factors For Obtaining Two-Sided 95% Prediction Limits For r Additional Samples Given a Background Sample of Size n.

Previous n	Number of New Measurements (r)														
	1	2	3	4	5	6	7	8	9	10	11	12	13	14	15
4	3.56	4.67	5.43	6.03	6.53	6.97	7.36	7.71	8.03	8.33	8.61	8.88	9.13	9.36	9.59
5	3.04	3.83	4.34	4.73	5.04	5.31	5.55	5.76	5.95	6.13	6.29	6.45	6.59	6.72	6.85
6	2.78	3.42	3.82	4.11	4.35	4.56	4.73	4.89	5.03	5.15	5.27	5.38	5.48	5.58	5.67
7	2.62	3.17	3.51	3.76	3.96	4.13	4.27	4.40	4.51	4.61	4.71	4.80	4.88	4.95	5.02
8	2.51	3.01	3.32	3.54	3.71	3.86	3.98	4.09	4.19	4.27	4.35	4.43	4.50	4.56	4.62
9	2.43	2.90	3.18	3.38	3.54	3.67	3.78	3.87	3.96	4.04	4.11	4.18	4.24	4.29	4.34
10	2.37	2.82	3.08	3.26	3.41	3.53	3.63	3.72	3.80	3.87	3.93	3.99	4.05	4.10	4.15
11	2.33	2.75	3.00	3.17	3.31	3.42	3.52	3.60	3.67	3.74	3.80	3.85	3.91	3.95	4.00
12	2.29	2.70	2.93	3.10	3.23	3.34	3.43	3.51	3.58	3.64	3.70	3.75	3.79	3.84	3.88
13	2.26	2.66	2.88	3.04	3.17	3.27	3.36	3.43	3.50	3.56	3.61	3.66	3.71	3.75	3.79
14	2.24	2.62	2.84	3.00	3.12	3.22	3.30	3.37	3.43	3.49	3.54	3.59	3.63	3.67	3.71
15	2.21	2.59	2.81	2.96	3.07	3.17	3.25	3.32	3.38	3.43	3.48	3.53	3.57	3.61	3.64
16	2.20	2.57	2.78	2.92	3.04	3.13	3.21	3.27	3.33	3.39	3.43	3.48	3.52	3.56	3.59
17	2.18	2.54	2.75	2.89	3.01	3.09	3.17	3.24	3.29	3.35	3.39	3.44	3.47	3.51	3.54
18	2.17	2.53	2.73	2.87	2.98	3.07	3.14	3.20	3.26	3.31	3.36	3.40	3.44	3.47	3.50
19	2.16	2.51	2.71	2.85	2.95	3.04	3.11	3.18	3.23	3.28	3.32	3.36	3.40	3.44	3.47
20	2.14	2.49	2.69	2.83	2.93	3.02	3.09	3.15	3.20	3.25	3.30	3.33	3.37	3.40	3.44
21	2.13	2.48	2.67	2.81	2.91	3.00	3.07	3.13	3.18	3.23	3.27	3.31	3.34	3.38	3.41
22	2.13	2.47	2.66	2.79	2.89	2.98	3.05	3.11	3.16	3.21	3.25	3.29	3.32	3.35	3.38
23	2.12	2.46	2.65	2.78	2.88	2.96	3.03	3.09	3.14	3.19	3.23	3.26	3.30	3.33	3.36
24	2.11	2.45	2.63	2.76	2.87	2.95	3.01	3.07	3.12	3.17	3.21	3.25	3.28	3.31	3.34
25	2.10	2.44	2.62	2.75	2.85	2.93	3.00	3.06	3.11	3.15	3.19	3.23	3.26	3.29	3.32
26	2.10	2.43	2.61	2.74	2.84	2.92	2.99	3.04	3.09	3.14	3.18	3.21	3.25	3.28	3.31
27	2.09	2.42	2.61	2.73	2.83	2.91	2.97	3.03	3.08	3.12	3.16	3.20	3.23	3.26	3.29
28	2.09	2.42	2.60	2.72	2.82	2.90	2.96	3.02	3.07	3.11	3.15	3.19	3.22	3.25	3.28
29	2.08	2.41	2.59	2.71	2.81	2.89	2.95	3.01	3.05	3.10	3.14	3.17	3.21	3.24	3.26
30	2.08	2.40	2.58	2.71	2.80	2.88	2.94	3.00	3.04	3.09	3.13	3.16	3.19	3.22	3.25
31	2.07	2.40	2.58	2.70	2.79	2.87	2.93	2.99	3.04	3.08	3.12	3.15	3.18	3.21	3.24
32	2.07	2.39	2.57	2.69	2.79	2.86	2.92	2.98	3.03	3.07	3.11	3.14	3.17	3.20	3.23
33	2.07	2.39	2.56	2.69	2.78	2.85	2.92	2.97	3.02	3.06	3.10	3.13	3.16	3.19	3.22
34	2.06	2.38	2.56	2.68	2.77	2.85	2.91	2.96	3.01	3.05	3.09	3.12	3.15	3.18	3.21
35	2.06	2.38	2.55	2.67	2.77	2.84	2.90	2.96	3.00	3.04	3.08	3.12	3.15	3.17	3.20
36	2.06	2.37	2.55	2.67	2.76	2.83	2.90	2.95	3.00	3.04	3.07	3.11	3.14	3.17	3.19
37	2.06	2.37	2.54	2.66	2.76	2.83	2.89	2.94	2.99	3.03	3.07	3.10	3.13	3.16	3.19
38	2.05	2.37	2.54	2.66	2.75	2.82	2.88	2.94	2.98	3.02	3.06	3.09	3.12	3.15	3.18
39	2.05	2.36	2.54	2.66	2.75	2.82	2.88	2.93	2.98	3.02	3.05	3.09	3.12	3.15	3.17
40	2.05	2.36	2.53	2.65	2.74	2.81	2.87	2.93	2.97	3.01	3.05	3.08	3.11	3.14	3.17
41	2.05	2.36	2.53	2.65	2.74	2.81	2.87	2.92	2.97	3.01	3.04	3.08	3.11	3.13	3.16
42	2.04	2.35	2.53	2.64	2.73	2.80	2.87	2.92	2.96	3.00	3.04	3.07	3.10	3.13	3.15
43	2.04	2.35	2.52	2.64	2.73	2.80	2.86	2.91	2.96	3.00	3.03	3.07	3.10	3.12	3.15
44	2.04	2.35	2.52	2.64	2.73	2.80	2.86	2.91	2.95	2.99	3.03	3.06	3.09	3.12	3.14
45	2.04	2.35	2.52	2.63	2.72	2.79	2.85	2.90	2.95	2.99	3.02	3.06	3.09	3.11	3.14
46	2.04	2.34	2.51	2.63	2.72	2.79	2.85	2.90	2.94	2.98	3.02	3.05	3.08	3.11	3.13
47	2.03	2.34	2.51	2.63	2.72	2.79	2.85	2.90	2.94	2.98	3.02	3.05	3.08	3.10	3.13
48	2.03	2.34	2.51	2.62	2.71	2.78	2.84	2.89	2.94	2.98	3.01	3.04	3.07	3.10	3.12
49	2.03	2.34	2.51	2.62	2.71	2.78	2.84	2.89	2.93	2.97	3.01	3.04	3.07	3.10	3.12
50	2.03	2.34	2.50	2.62	2.71	2.78	2.84	2.89	2.93	2.97	3.00	3.04	3.06	3.09	3.12
60	2.02	2.32	2.48	2.60	2.68	2.75	2.81	2.86	2.90	2.94	2.97	3.01	3.03	3.06	3.08
70	2.01	2.31	2.47	2.58	2.67	2.74	2.79	2.84	2.88	2.92	2.95	2.98	3.01	3.04	3.06
80	2.00	2.30	2.46	2.57	2.66	2.72	2.78	2.83	2.87	2.91	2.94	2.97	3.00	3.02	3.05
90	2.00	2.29	2.45	2.56	2.65	2.71	2.77	2.82	2.86	2.89	2.93	2.96	2.98	3.01	3.03
100	1.99	2.29	2.45	2.56	2.64	2.71	2.76	2.81	2.85	2.89	2.92	2.95	2.98	3.00	3.02

Factor = $t_{(n-1,\ 1-\alpha/r)}\sqrt{1+1/n}$

Table 13.6. Factors For Obtaining Two-Sided 95% Prediction Limits For r Additional Samples Given a Background Sample of Size n.

Previous n	Number of New Measurements (r)														
	16	17	18	19	20	21	22	23	24	25	26	27	28	29	30
4	9.80	10.01	10.21	10.40	10.58	10.76	10.93	11.10	11.26	11.42	11.57	11.72	11.87	12.01	12.15
5	6.97	7.09	7.20	7.30	7.40	7.50	7.59	7.68	7.77	7.86	7.94	8.02	8.10	8.17	8.25
6	5.75	5.83	5.91	5.98	6.05	6.12	6.19	6.25	6.31	6.37	6.42	6.48	6.53	6.58	6.63
7	5.09	5.15	5.21	5.27	5.32	5.38	5.43	5.48	5.52	5.57	5.61	5.65	5.69	5.73	5.77
8	4.67	4.73	4.78	4.83	4.87	4.92	4.96	5.00	5.04	5.07	5.11	5.15	5.18	5.21	5.24
9	4.39	4.44	4.49	4.53	4.57	4.61	4.64	4.68	4.71	4.74	4.78	4.81	4.83	4.86	4.89
10	4.19	4.23	4.27	4.31	4.35	4.38	4.42	4.45	4.48	4.51	4.53	4.56	4.59	4.61	4.64
11	4.04	4.08	4.11	4.15	4.18	4.21	4.24	4.27	4.30	4.33	4.35	4.38	4.40	4.43	4.45
12	3.92	3.96	3.99	4.02	4.05	4.08	4.11	4.14	4.16	4.19	4.21	4.24	4.26	4.28	4.30
13	3.82	3.86	3.89	3.92	3.95	3.98	4.00	4.03	4.05	4.08	4.10	4.12	4.14	4.16	4.18
14	3.74	3.78	3.81	3.84	3.86	3.89	3.92	3.94	3.96	3.99	4.01	4.03	4.05	4.07	4.09
15	3.68	3.71	3.74	3.77	3.79	3.82	3.84	3.87	3.89	3.91	3.93	3.95	3.97	3.99	4.01
16	3.62	3.65	3.68	3.71	3.74	3.76	3.78	3.81	3.83	3.85	3.87	3.89	3.90	3.92	3.94
17	3.58	3.60	3.63	3.66	3.68	3.71	3.73	3.75	3.77	3.79	3.81	3.83	3.85	3.86	3.88
18	3.53	3.56	3.59	3.61	3.64	3.66	3.68	3.71	3.73	3.75	3.76	3.78	3.80	3.82	3.83
19	3.50	3.53	3.55	3.58	3.60	3.62	3.65	3.67	3.68	3.70	3.72	3.74	3.76	3.77	3.79
20	3.47	3.49	3.52	3.54	3.57	3.59	3.61	3.63	3.65	3.67	3.68	3.70	3.72	3.73	3.75
21	3.44	3.46	3.49	3.51	3.54	3.56	3.58	3.60	3.62	3.63	3.65	3.67	3.69	3.70	3.72
22	3.41	3.44	3.46	3.49	3.51	3.53	3.55	3.57	3.59	3.61	3.62	3.64	3.65	3.67	3.68
23	3.39	3.41	3.44	3.46	3.49	3.51	3.53	3.54	3.56	3.58	3.60	3.61	3.63	3.64	3.66
24	3.37	3.39	3.42	3.44	3.46	3.48	3.50	3.52	3.54	3.56	3.57	3.59	3.60	3.62	3.63
25	3.35	3.38	3.40	3.42	3.44	3.46	3.48	3.50	3.52	3.53	3.55	3.57	3.58	3.60	3.61
26	3.33	3.36	3.38	3.40	3.42	3.44	3.46	3.48	3.50	3.52	3.53	3.55	3.56	3.58	3.59
27	3.32	3.34	3.36	3.39	3.41	3.43	3.45	3.46	3.48	3.50	3.51	3.53	3.54	3.56	3.57
28	3.30	3.33	3.35	3.37	3.39	3.41	3.43	3.45	3.46	3.48	3.50	3.51	3.53	3.54	3.55
29	3.29	3.31	3.34	3.36	3.38	3.40	3.42	3.43	3.45	3.47	3.48	3.50	3.51	3.52	3.54
30	3.28	3.30	3.32	3.34	3.36	3.38	3.40	3.42	3.44	3.45	3.47	3.48	3.50	3.51	3.52
31	3.27	3.29	3.31	3.33	3.35	3.37	3.39	3.41	3.42	3.44	3.45	3.47	3.48	3.50	3.51
32	3.25	3.28	3.30	3.32	3.34	3.36	3.38	3.39	3.41	3.43	3.44	3.46	3.47	3.48	3.50
33	3.24	3.27	3.29	3.31	3.33	3.35	3.37	3.38	3.40	3.42	3.43	3.44	3.46	3.47	3.48
34	3.23	3.26	3.28	3.30	3.32	3.34	3.36	3.37	3.39	3.41	3.42	3.43	3.45	3.46	3.47
35	3.23	3.25	3.27	3.29	3.31	3.33	3.35	3.36	3.38	3.39	3.41	3.42	3.44	3.45	3.46
36	3.22	3.24	3.26	3.28	3.30	3.32	3.34	3.35	3.37	3.39	3.40	3.41	3.43	3.44	3.45
37	3.21	3.23	3.25	3.27	3.29	3.31	3.33	3.35	3.36	3.38	3.39	3.40	3.42	3.43	3.44
38	3.20	3.23	3.25	3.27	3.29	3.30	3.32	3.34	3.35	3.37	3.38	3.40	3.41	3.42	3.44
39	3.20	3.22	3.24	3.26	3.28	3.30	3.31	3.33	3.35	3.36	3.37	3.39	3.40	3.41	3.43
40	3.19	3.21	3.23	3.25	3.27	3.29	3.31	3.32	3.34	3.35	3.37	3.38	3.39	3.41	3.42
41	3.18	3.21	3.23	3.25	3.27	3.28	3.30	3.32	3.33	3.35	3.36	3.37	3.39	3.40	3.41
42	3.18	3.20	3.22	3.24	3.26	3.28	3.29	3.31	3.33	3.34	3.35	3.37	3.38	3.39	3.41
43	3.17	3.19	3.21	3.23	3.25	3.27	3.29	3.30	3.32	3.33	3.35	3.36	3.37	3.39	3.40
44	3.17	3.19	3.21	3.23	3.25	3.27	3.28	3.30	3.31	3.33	3.34	3.36	3.37	3.38	3.39
45	3.16	3.18	3.20	3.22	3.24	3.26	3.28	3.29	3.31	3.32	3.34	3.35	3.36	3.37	3.39
46	3.16	3.18	3.20	3.22	3.24	3.25	3.27	3.29	3.30	3.32	3.33	3.34	3.36	3.37	3.38
47	3.15	3.17	3.19	3.21	3.23	3.25	3.27	3.28	3.30	3.31	3.33	3.34	3.35	3.36	3.38
48	3.15	3.17	3.19	3.21	3.23	3.24	3.26	3.28	3.29	3.31	3.32	3.33	3.35	3.36	3.37
49	3.14	3.17	3.19	3.20	3.22	3.24	3.26	3.27	3.29	3.30	3.32	3.33	3.34	3.35	3.37
50	3.14	3.16	3.18	3.20	3.22	3.24	3.25	3.27	3.28	3.30	3.31	3.32	3.34	3.35	3.36
60	3.11	3.13	3.15	3.17	3.18	3.20	3.22	3.23	3.25	3.26	3.27	3.29	3.30	3.31	3.32
70	3.08	3.11	3.12	3.14	3.16	3.18	3.19	3.21	3.22	3.24	3.25	3.26	3.27	3.28	3.30
80	3.07	3.09	3.11	3.13	3.14	3.16	3.17	3.19	3.20	3.22	3.23	3.24	3.25	3.26	3.28
90	3.05	3.08	3.09	3.11	3.13	3.15	3.16	3.17	3.19	3.20	3.21	3.23	3.24	3.25	3.26
100	3.04	3.06	3.08	3.10	3.12	3.13	3.15	3.16	3.18	3.19	3.20	3.22	3.23	3.24	3.25

Factor $= t_{(n-1,\ 1-\alpha/r)}\sqrt{1+1/n}$

Table 13.7. Factors For Obtaining Two-Sided 95% Prediction Limits For r Additional Samples Given a Background Sample of Size n.

Previous n	Number of New Measurements (r)														
	30	35	40	45	50	55	60	65	70	75	80	85	90	95	100
4	12.15	12.80	13.40	13.94	14.45	14.92	15.37	15.79	16.19	16.57	16.93	17.28	17.62	17.94	18.25
5	8.25	8.59	8.90	9.18	9.43	9.67	9.89	10.10	10.29	10.48	10.66	10.83	10.99	11.14	11.29
6	6.63	6.86	7.06	7.25	7.42	7.57	7.72	7.85	7.98	8.10	8.21	8.32	8.42	8.52	8.61
7	5.77	5.95	6.10	6.24	6.37	6.49	6.59	6.69	6.79	6.88	6.96	7.04	7.12	7.19	7.26
8	5.24	5.39	5.52	5.63	5.74	5.83	5.92	6.00	6.08	6.15	6.21	6.28	6.34	6.40	6.45
9	4.89	5.02	5.13	5.23	5.31	5.40	5.47	5.54	5.60	5.66	5.72	5.78	5.83	5.87	5.92
10	4.64	4.75	4.85	4.93	5.01	5.09	5.15	5.21	5.27	5.32	5.37	5.42	5.47	5.51	5.55
11	4.45	4.55	4.64	4.72	4.79	4.86	4.92	4.97	5.02	5.07	5.11	5.16	5.20	5.24	5.27
12	4.30	4.40	4.48	4.55	4.62	4.68	4.73	4.78	4.83	4.88	4.92	4.96	4.99	5.03	5.06
13	4.18	4.27	4.35	4.42	4.48	4.54	4.59	4.64	4.68	4.72	4.76	4.80	4.83	4.86	4.89
14	4.09	4.17	4.25	4.31	4.37	4.42	4.47	4.52	4.56	4.60	4.63	4.67	4.70	4.73	4.76
15	4.01	4.09	4.16	4.22	4.28	4.33	4.37	4.41	4.45	4.49	4.53	4.56	4.59	4.62	4.65
16	3.94	4.02	4.08	4.14	4.20	4.25	4.29	4.33	4.37	4.40	4.44	4.47	4.50	4.53	4.55
17	3.88	3.96	4.02	4.08	4.13	4.18	4.22	4.26	4.30	4.33	4.36	4.39	4.42	4.45	4.47
18	3.83	3.90	3.97	4.02	4.07	4.12	4.16	4.20	4.23	4.27	4.30	4.33	4.35	4.38	4.40
19	3.79	3.86	3.92	3.97	4.02	4.07	4.11	4.14	4.18	4.21	4.24	4.27	4.29	4.32	4.34
20	3.75	3.82	3.88	3.93	3.98	4.02	4.06	4.10	4.13	4.16	4.19	4.22	4.24	4.27	4.29
21	3.72	3.78	3.84	3.89	3.94	3.98	4.02	4.05	4.09	4.12	4.15	4.17	4.20	4.22	4.24
22	3.68	3.75	3.81	3.86	3.90	3.95	3.98	4.02	4.05	4.08	4.11	4.13	4.16	4.18	4.20
23	3.66	3.72	3.78	3.83	3.87	3.91	3.95	3.98	4.01	4.04	4.07	4.10	4.12	4.14	4.16
24	3.63	3.70	3.75	3.80	3.84	3.88	3.92	3.95	3.98	4.01	4.04	4.06	4.09	4.11	4.13
25	3.61	3.67	3.73	3.78	3.82	3.86	3.89	3.93	3.96	3.98	4.01	4.03	4.06	4.08	4.10
26	3.59	3.65	3.71	3.75	3.80	3.83	3.87	3.90	3.93	3.96	3.98	4.01	4.03	4.05	4.07
27	3.57	3.63	3.69	3.73	3.77	3.81	3.85	3.88	3.91	3.93	3.96	3.98	4.01	4.03	4.05
28	3.55	3.61	3.67	3.71	3.75	3.79	3.83	3.86	3.89	3.91	3.94	3.96	3.98	4.00	4.02
29	3.54	3.60	3.65	3.70	3.74	3.77	3.81	3.84	3.87	3.89	3.92	3.94	3.96	3.98	4.00
30	3.52	3.58	3.63	3.68	3.72	3.76	3.79	3.82	3.85	3.87	3.90	3.92	3.94	3.96	3.98
31	3.51	3.57	3.62	3.66	3.70	3.74	3.77	3.80	3.83	3.86	3.88	3.90	3.92	3.94	3.96
32	3.50	3.55	3.61	3.65	3.69	3.73	3.76	3.79	3.81	3.84	3.86	3.89	3.91	3.93	3.95
33	3.48	3.54	3.59	3.64	3.68	3.71	3.74	3.77	3.80	3.83	3.85	3.87	3.89	3.91	3.93
34	3.47	3.53	3.58	3.62	3.66	3.70	3.73	3.76	3.79	3.81	3.84	3.86	3.88	3.90	3.92
35	3.46	3.52	3.57	3.61	3.65	3.69	3.72	3.75	3.77	3.80	3.82	3.84	3.86	3.88	3.90
36	3.45	3.51	3.56	3.60	3.64	3.67	3.71	3.73	3.76	3.79	3.81	3.83	3.85	3.87	3.89
37	3.44	3.50	3.55	3.59	3.63	3.66	3.70	3.72	3.75	3.77	3.80	3.82	3.84	3.86	3.88
38	3.44	3.49	3.54	3.58	3.62	3.65	3.69	3.71	3.74	3.76	3.79	3.81	3.83	3.85	3.86
39	3.43	3.48	3.53	3.57	3.61	3.64	3.68	3.70	3.73	3.75	3.78	3.80	3.82	3.84	3.85
40	3.42	3.47	3.52	3.56	3.60	3.64	3.67	3.69	3.72	3.74	3.77	3.79	3.81	3.83	3.84
41	3.41	3.47	3.51	3.56	3.59	3.63	3.66	3.69	3.71	3.74	3.76	3.78	3.80	3.82	3.83
42	3.41	3.46	3.51	3.55	3.59	3.62	3.65	3.68	3.70	3.73	3.75	3.77	3.79	3.81	3.82
43	3.40	3.45	3.50	3.54	3.58	3.61	3.64	3.67	3.69	3.72	3.74	3.76	3.78	3.80	3.82
44	3.39	3.45	3.49	3.53	3.57	3.60	3.63	3.66	3.69	3.71	3.73	3.75	3.77	3.79	3.81
45	3.39	3.44	3.49	3.53	3.56	3.60	3.63	3.65	3.68	3.70	3.72	3.74	3.76	3.78	3.80
46	3.38	3.43	3.48	3.52	3.56	3.59	3.62	3.65	3.67	3.70	3.72	3.74	3.76	3.77	3.79
47	3.38	3.43	3.48	3.52	3.55	3.58	3.61	3.64	3.67	3.69	3.71	3.73	3.75	3.77	3.78
48	3.37	3.42	3.47	3.51	3.55	3.58	3.61	3.63	3.66	3.68	3.70	3.72	3.74	3.76	3.78
49	3.37	3.42	3.46	3.50	3.54	3.57	3.60	3.63	3.65	3.68	3.70	3.72	3.74	3.75	3.77
50	3.36	3.41	3.46	3.50	3.53	3.57	3.60	3.62	3.65	3.67	3.69	3.71	3.73	3.75	3.76
60	3.32	3.37	3.42	3.46	3.49	3.52	3.55	3.58	3.60	3.62	3.64	3.66	3.68	3.70	3.71
70	3.30	3.35	3.39	3.43	3.46	3.49	3.52	3.54	3.57	3.59	3.61	3.63	3.65	3.66	3.68
80	3.28	3.33	3.37	3.41	3.44	3.47	3.50	3.52	3.54	3.57	3.59	3.60	3.62	3.64	3.65
90	3.26	3.31	3.35	3.39	3.42	3.45	3.48	3.50	3.53	3.55	3.57	3.58	3.60	3.62	3.63
100	3.25	3.30	3.34	3.38	3.41	3.44	3.46	3.49	3.51	3.53	3.55	3.57	3.59	3.60	3.62

Factor $= t_{(n-1,\ 1-\alpha/r)} \sqrt{1 + 1/n}$

positive and false negative rates of tolerance or prediction limits. As such, we suggest that the number of samples be selected on the basis of the size of the multiplier. For example, a 95% confidence, 95% coverage one-sided tolerance interval multiplier goes from 7.656 to 5.144 for a change in background sample size of 3 to 4; but only 2.566 to 2.523 for a change in background sample size of n=15 to n=16. Even doubling the background sample size to n=30 only decreases the multiplier by approximately ⅓ of a standard deviation unit (i.e., k=2.220). As such; we recommend a background sample size in the range of 16 to 32. If more observations are available, they should, of course, also be included because they will provide even more precise estimates of μ and σ^2.

The previous line of reasoning also suggests how the background sample should be selected. The three choices are: (1) fixed sample of n historical measurements (e.g., 2 years fixed window), (2) the n most recent measurements (e.g., a 2-year moving window), (3) all available historical measurements. When using tolerance or prediction limits or intervals, we recommend the third option as shown on Figure 13.15.

Finally, how shall the n background measurements be selected? Should we obtain n measurements from a single background well, a single measurement from n different background wells, or something in between? In the case of upgradient versus downgradient comparisons, there are two possible strategies. For a facility with eight or more upgradient or background wells, the eight measurements could be used to construct a new tolerance limit specific to each quarterly monitoring event (Figure 13.15). With eight background measurements, the one-sided 95% confidence, 95% coverage tolerance limit multiplier would be 3.188. Although the multiplier is somewhat larger than the value of 2.523 obtained for n=16 background samples, this strategy completely eliminates the temporal component of variability, and will, therefore, yield a smaller standard deviation than if historical measurements are pooled. The result is an effective detection monitoring program that is supported by the new USEPA statistical rule.

For facilities with less than eight background water quality wells, we suggest that at least four upgradient or background wells be monitored quarterly, so that after one year of monitoring, 16 background measurements will be available. Furthermore, four upgradient wells, if widely spaced, will provide a reasonable characterization of the spatial component of variability at the facility. If only a single upgradient well were installed, as required in the previous RCRA regulations, differences between upgradient and downgradient water quality would be completely confounded with spatial variability; that is, is the difference between upgradient and downgradient measurements due to the influence of the facility or simply the difference that one would find by drilling any two holes in the ground? There is, of course, no answer to this question if there is only a single background well.

Nondetects

It is very common in ground water detection monitoring to obtain samples that cannot be properly quantified because their concentrations are less than the limit

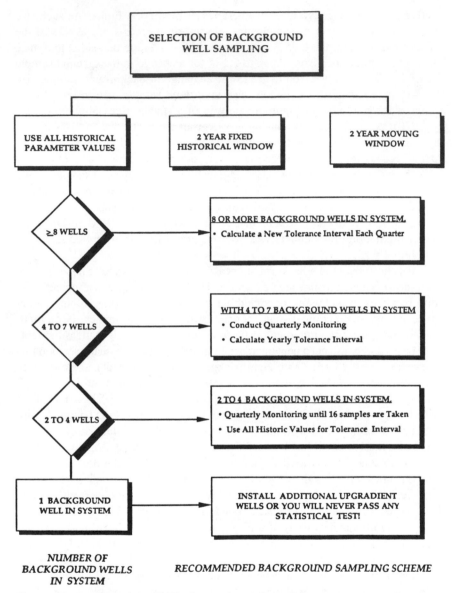

Figure 13.15. Recommended background sampling scheme.

of detection of the analytical instrument. This condition can make the direct application of the previously described statistical prediction and tolerance limits and
intervals problematic because the usual sample statistics x and s are no longer
valid estimates of μ and σ. Statistically, these distributions are termed "censored."
In this section, we consider three different levels of "censoring:" (1) up to 90%

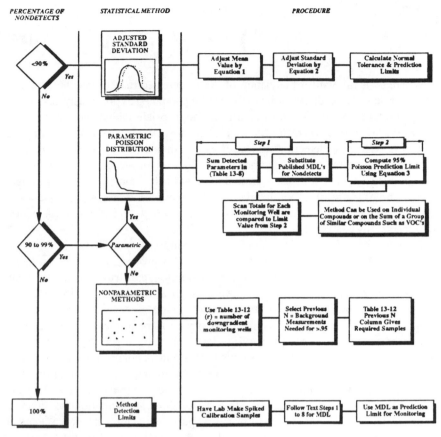

Figure 13.16. Statistical methods for parameters with 90% to 100% nondetects.

nondetects, (2) 91% to 99% nondetects, and (3) nothing detected, as shown in Figure 13.16.

Case 1: *Up to 90% Nondetects.* When at least 10% of the ground water samples have measurable values, the mean and variance of the distribution can be approximated using a method due to Aitchison (1955).

The adjusted mean value is given by:

$$x = \left(1 - \frac{n_o}{n}\right) \bar{x}' \qquad (13.1)$$

Where $\bar{x}'$ is the average of the n, detected values, n_0 is the number of non-detects and $n = n_1 + n_0$ is the total number of samples. The adjusted standard deviation is:

$$s = \sqrt{\left(1-\frac{n_o}{n}\right) s^{2\prime} + \frac{n_o}{n}\left(1-\frac{n_o-1}{n-1}\right) \bar{x}^{2\prime}} \qquad (13.2)$$

Where s' is the standard deviation of the n_i detected measurements. The normal tolerance and prediction limits can then be computed as previously described, using the total sample size n to obtain the appropriate tabled multipler.

Case 2: Compounds detected in 1% to 10% of all background samples (VOCs). When the detection frequency is less than 10%, the previous method of obtaining adjusted mean and variance estimates no longer applies. With limited data, it is difficult to know just what to do. What further complicates this problem is that one of the most important classes of detection monitoring compounds is the Volatile Organic Priority Pollutants (VOCs), which typically have detection frequencies in this range.

To date, the only statistical approach to setting site specific limits for these compounds is described by Gibbons (1987b). This procedure is based on tolerance and prediction limits for the Poisson distribution, a distribution that has been widely used for the analysis of rare events such as suicide, mutation rates, or atomic particle emission. These limits can be applied either to detection frequencies (i.e., number of detected compounds per scan) or to the actual concentrations when recorded in parts per billion (ppb). In the latter case, it is assumed that a measurement of 20 ppb of benzene represents a count of 20 molecules of benzene for every billion molecules of water examined. To the extent that this is an accurate description of the true physical measurement process, Poisson prediction and tolerance limits provide a reasonable approximation that appears to be sufficiently accurate for most practical purposes. This sentiment is echoed in the new 40 CFR 264 statistical regulation:

> Tolerance intervals and prediction intervals have not been widely used by the Agency to evaluate ground water monitoring data. However, the Agency is aware of recent publications that have employed these statistical methods to evaluate ground water monitoring data, especially in evaluating certain classes of chemical compounds (e.g., volatile organic compounds). Several commentors suggested that the Agency incorporate this research into today's final rule, noting that these procedures may be the best way to evaluate data that is below the limit of analytical detection.

In the case of VOCs, the 95% Poisson prediction limit is computed as follows (Figure 13.16):

1. For each USEPA Method 624 volatile organic priority pollutant scan, sum the detected concentrations across the 27 compounds listed in Table 13.8, substituting the published method detection limit (MDL) (see Table 13.8) for those compounds that were not detected. For example, if none of the compounds were detected, the sum for that scan is 154 ppb.

Table 13.8.   Method 624 Volatile Organic Compounds and Published Method Detection Limits.

Compound	Reported MDL
Benzene	4.4
Bromodichloromethane	2.2
Bromoform	4.7
Bromomethane	10.0
Carbon Tetrachloride	2.8
Chlorobenzene	6.0
Chloroethane	10.0
Chloroform	1.6
Chloromethane	10.0
Dibromochloromethane	3.1
1,1-Dichloroethane	4.7
1,2-Dichloroethane	2.8
1,1-Dichloroethene	2.8
trans-1,2-Dichloroethene	10.0
1,2-Dichloropropane	6.0
cis-1,3-Dichloropropene	5.0
trans-1,3-Dichloropropene	10.0
Ethyl Benzene	7.2
Methylene Chloride	2.8
1,1,2,2-Tetrachloroethane	6.9
Tetrachloroethene	4.1
Toluene	6.0
1,1,1-Trichloroethane	3.8
1,1,2-Trichloroethane	5.0
Trichloroethene	1.9
Trichlorofluoromethane	10.0
Vinyl Chloride	10.0

all values reported in μg/L.

2. Compute the 95% Poisson prediction limit as:

$$\frac{y}{n} + \frac{t^2}{2n} + \frac{t}{n} \sqrt{y(1+n) + \frac{t^2}{4}} \qquad (13.3)$$

Where y is the total ppb for all n background scans (i.e., the sum of n individual scan totals), n is the number of background scans, t is the $(1-.05/r)$ 100% point of student's t distribution on $n-1$ degrees of freedom (see Tables 13.9 through 13.11), and r is the number of monitoring wells.

Scan totals for each monitoring well (computed as in Step 1) are then compared to the limit value computed in Step 2.

For other compounds that have detection frequencies in the range of 1% to 10%, the same strategy may be applied, either individually or on the sum of a group of similar compounds, as in case of the VOCs.

Table 13.9.   Values of t For Obtaining One-Sided 95% Poisson Prediction Limits For r Additional Samples Given a Background Sample of Size n.

Previous n	Number of New Measurements (r)														
	1	2	3	4	5	6	7	8	9	10	11	12	13	14	15
4	2.35	3.18	3.74	4.18	4.54	4.86	5.14	5.39	5.62	5.84	6.04	6.23	6.41	6.58	6.74
5	2.13	2.78	3.19	3.50	3.75	3.96	4.15	4.31	4.47	4.60	4.73	4.85	4.96	5.07	5.17
6	2.01	2.57	2.91	3.16	3.36	3.53	3.68	3.81	3.93	4.03	4.13	4.22	4.30	4.38	4.45
7	1.94	2.45	2.75	2.97	3.14	3.29	3.41	3.52	3.62	3.71	3.79	3.86	3.93	4.00	4.06
8	1.89	2.36	2.64	2.84	3.00	3.13	3.24	3.33	3.42	3.50	3.57	3.64	3.70	3.75	3.81
9	1.86	2.31	2.57	2.75	2.90	3.02	3.12	3.21	3.28	3.35	3.42	3.48	3.53	3.58	3.63
10	1.83	2.26	2.51	2.68	2.82	2.93	3.03	3.11	3.18	3.25	3.31	3.36	3.41	3.46	3.50
11	1.81	2.23	2.47	2.63	2.76	2.87	2.96	3.04	3.11	3.17	3.22	3.28	3.32	3.37	3.41
12	1.80	2.20	2.43	2.59	2.72	2.82	2.91	2.98	3.05	3.11	3.16	3.21	3.25	3.29	3.33
13	1.78	2.18	2.40	2.56	2.68	2.78	2.86	2.93	3.00	3.05	3.11	3.15	3.20	3.23	3.27
14	1.77	2.16	2.38	2.53	2.65	2.75	2.83	2.90	2.96	3.01	3.06	3.11	3.15	3.19	3.22
15	1.76	2.14	2.36	2.51	2.62	2.72	2.80	2.86	2.92	2.98	3.02	3.07	3.11	3.15	3.18
16	1.75	2.13	2.34	2.49	2.60	2.69	2.77	2.84	2.89	2.95	2.99	3.04	3.07	3.11	3.15
17	1.75	2.12	2.33	2.47	2.58	2.67	2.75	2.81	2.87	2.92	2.97	3.01	3.05	3.08	3.11
18	1.74	2.11	2.31	2.46	2.57	2.65	2.73	2.79	2.85	2.90	2.94	2.98	3.02	3.06	3.09
19	1.73	2.10	2.30	2.44	2.55	2.64	2.71	2.77	2.83	2.88	2.92	2.96	3.00	3.03	3.07
20	1.73	2.09	2.29	2.43	2.54	2.62	2.70	2.76	2.81	2.86	2.90	2.94	2.98	3.01	3.04
21	1.72	2.09	2.28	2.42	2.53	2.61	2.68	2.74	2.80	2.84	2.89	2.93	2.96	3.00	3.03
22	1.72	2.08	2.28	2.41	2.52	2.60	2.67	2.73	2.78	2.83	2.87	2.91	2.95	2.98	3.01
23	1.72	2.07	2.27	2.40	2.51	2.59	2.66	2.72	2.77	2.82	2.86	2.90	2.93	2.96	2.99
24	1.71	2.07	2.26	2.40	2.50	2.58	2.65	2.71	2.76	2.81	2.85	2.89	2.92	2.95	2.98
25	1.71	2.06	2.26	2.39	2.49	2.57	2.64	2.70	2.75	2.80	2.84	2.88	2.91	2.94	2.97
26	1.71	2.06	2.25	2.38	2.48	2.57	2.63	2.69	2.74	2.79	2.83	2.86	2.90	2.93	2.96
27	1.71	2.06	2.25	2.38	2.48	2.56	2.63	2.68	2.73	2.78	2.82	2.85	2.89	2.92	2.95
28	1.70	2.05	2.24	2.37	2.47	2.55	2.62	2.68	2.73	2.77	2.81	2.85	2.88	2.91	2.94
29	1.70	2.05	2.24	2.37	2.47	2.55	2.61	2.67	2.72	2.76	2.80	2.84	2.87	2.90	2.93
30	1.70	2.05	2.23	2.36	2.46	2.54	2.61	2.66	2.71	2.76	2.80	2.83	2.86	2.89	2.92
31	1.70	2.04	2.23	2.36	2.46	2.54	2.60	2.66	2.71	2.75	2.79	2.82	2.86	2.89	2.91
32	1.70	2.04	2.23	2.36	2.45	2.53	2.60	2.65	2.70	2.74	2.78	2.82	2.85	2.88	2.91
33	1.69	2.04	2.22	.235	2.45	2.53	2.59	2.65	2.70	2.74	2.78	2.81	2.84	2.87	2.90
34	1.69	2.03	2.22	2.35	2.44	2.52	2.59	2.64	2.69	2.73	2.77	2.81	2.84	2.87	2.90
35	1.69	2.03	2.22	2.34	2.44	2.52	2.58	2.64	2.69	2.73	2.77	2.80	2.83	2.86	2.89
36	1.69	2.03	2.22	2.34	2.44	2.51	2.58	2.63	2.68	2.72	2.76	2.80	2.83	2.86	2.88
37	1.69	2.03	2.21	2.34	2.43	2.51	2.57	2.63	2.68	2.72	2.76	2.79	2.82	2.85	2.88
38	1.69	2.03	2.21	2.34	2.43	2.51	2.57	2.63	2.67	2.72	2.75	2.79	2.82	2.85	2.87
39	1.69	2.02	2.21	2.33	2.43	2.50	2.57	2.62	2.67	2.71	2.75	2.78	2.81	2.84	2.87
40	1.68	2.02	2.21	2.33	2.43	2.50	2.56	2.26	2.67	2.71	2.75	2.78	2.81	2.84	2.87
41	1.68	2.02	2.20	2.33	2.42	2.50	2.56	2.62	2.66	2.70	2.74	2.78	2.81	2.84	2.86
42	1.68	2.02	2.20	2.33	2.42	2.50	2.56	2.61	2.66	2.70	2.74	2.77	2.80	2.83	2.86
43	1.68	2.02	2.20	2.32	2.42	2.49	2.56	2.61	2.66	2.70	2.74	2.77	2.80	2.83	2.85
44	1.68	2.02	2.20	2.32	2.42	2.49	2.55	2.61	2.65	2.69	2.73	2.77	2.80	2.82	2.85
45	1.68	2.02	2.20	2.32	2.41	2.49	2.55	2.60	2.65	2.69	2.73	2.76	2.79	2.82	2.85
46	1.68	2.01	2.20	2.32	2.41	2.49	2.55	2.60	2.65	2.69	2.73	2.76	2.79	2.82	2.84
47	1.68	2.01	2.19	2.32	2.41	2.48	2.55	2.60	2.65	2.69	2.72	2.76	2.79	2.82	2.84
48	1.68	2.01	2.19	2.32	2.41	2.48	2.54	2.60	2.64	2.68	2.72	2.75	2.78	2.81	2.84
49	1.68	2.01	2.19	2.31	2.41	2.48	2.54	2.60	2.64	2.68	2.72	2.75	2.78	2.81	2.84
50	1.68	2.01	2.19	2.31	2.40	2.48	2.54	2.59	2.64	2.68	2.72	2.75	2.78	2.81	2.83
60	1.67	2.00	2.18	2.30	2.39	2.46	2.52	2.58	2.62	2.66	2.70	2.73	2.76	2.79	2.81
70	1.67	1.99	2.17	2.29	2.38	2.45	2.51	2.56	2.61	2.65	2.68	2.72	2.75	2.77	2.80
80	1.66	1.99	2.17	2.28	2.37	2.45	2.51	2.56	2.60	2.64	2.67	2.71	2.74	2.76	2.79
90	1.66	1.99	2.16	2.28	2.37	2.44	2.50	2.55	2.59	2.63	2.67	2.70	2.73	2.75	2.78
100	1.66	1.98	2.16	2.28	2.36	2.44	2.49	2.54	2.59	2.63	2.66	2.69	2.72	2.75	2.77

Factor = $t_{(n-1,\ 1-\alpha/r)}$

Table 13.10.   Values of t For Obtaining One-Sided 95% Poisson Prediction Limits For r Additional Samples Given A Background Sample Of Size n.

Previous n	Number of New Measurements (r)														
	16	17	18	19	20	21	22	23	24	25	26	27	28	29	30
4	6.89	7.04	7.18	7.32	7.45	7.58	7.70	7.82	7.94	8.05	8.16	8.27	8.37	8.47	8.57
5	5.26	5.35	5.44	5.52	5.60	5.67	5.75	5.82	5.88	5.95	6.01	6.08	6.14	6.20	6.25
6	4.53	4.59	4.65	4.72	4.77	4.83	4.88	4.93	4.98	5.03	5.08	5.12	5.16	5.21	5.25
7	4.11	4.17	4.22	4.27	4.32	4.36	4.40	4.45	4.49	4.52	4.56	4.60	4.63	4.66	4.70
8	3.85	3.90	3.95	3.99	4.03	4.07	4.10	4.14	4.17	4.21	4.24	4.27	4.30	4.33	4.35
9	3.68	3.72	3.76	3.80	3.83	3.87	3.90	3.93	3.96	3.99	4.02	4.05	4.07	4.10	4.12
10	3.55	3.59	3.62	3.66	3.69	3.72	3.75	3.78	3.81	3.83	3.86	3.88	3.91	3.93	3.95
11	3.45	3.48	3.52	3.55	3.58	3.61	3.64	3.67	3.69	3.72	3.74	3.76	3.78	3.81	3.83
12	3.37	3.40	3.44	3.47	3.50	3.52	3.55	3.58	3.60	3.62	3.65	3.67	3.69	3.71	3.73
13	3.31	3.34	3.37	3.40	3.43	3.45	3.48	3.50	3.53	3.55	3.57	3.59	3.61	3.63	3.65
14	3.26	3.29	3.32	3.35	3.37	3.40	3.42	3.45	3.47	3.49	3.51	3.53	3.55	3.57	3.58
15	3.21	3.24	3.27	3.30	3.33	3.35	3.37	3.40	3.42	3.44	3.46	3.48	3.49	3.51	3.53
16	3.18	3.21	3.23	3.26	3.29	3.31	3.33	3.35	3.37	3.39	3.41	3.43	3.45	3.47	3.48
17	3.15	3.17	3.20	3.23	3.25	3.27	3.30	3.32	3.34	3.36	3.38	3.39	3.41	3.43	3.44
18	3.12	3.15	3.17	3.20	3.22	3.24	3.27	3.29	3.31	3.33	3.34	3.36	3.38	3.39	3.41
19	3.09	3.12	3.15	3.17	3.20	3.22	3.24	3.26	3.28	3.30	3.32	3.33	3.35	3.36	3.38
20	3.07	3.10	3.13	3.15	3.17	3.20	3.22	3.24	3.25	3.27	3.29	3.31	3.32	3.34	3.35
21	3.05	3.08	3.11	3.13	3.15	3.17	3.19	3.21	3.23	3.25	3.27	3.28	3.30	3.32	3.33
22	3.04	3.06	3.09	3.11	3.13	3.16	3.18	3.20	3.21	3.23	3.25	3.26	3.28	3.29	3.31
23	3.02	3.05	3.07	3.10	3.12	3.14	3.16	3.18	3.20	3.21	3.23	3.25	3.26	3.28	3.29
24	3.01	3.03	3.06	3.08	3.10	3.12	3.14	3.16	3.18	3.20	3.21	3.23	3.24	3.26	3.27
25	3.00	3.02	3.05	3.07	3.09	3.11	3.13	3.15	3.17	3.18	3.20	3.21	3.23	3.24	3.26
26	2.99	3.01	3.03	3.06	3.08	3.10	3.12	3.14	3.15	3.17	3.19	3.20	3.22	3.23	3.24
27	2.98	3.00	3.02	3.05	3.07	3.09	3.11	3.12	3.14	3.16	3.17	3.19	3.20	3.22	3.23
28	2.96	2.99	3.01	3.04	3.06	3.08	3.09	3.11	3.13	3.15	3.16	3.18	3.19	3.21	3.22
29	2.96	2.98	3.00	3.03	3.05	3.07	3.09	3.10	3.12	3.14	3.15	3.17	3.18	3.19	3.21
30	2.95	2.97	3.00	3.02	3.04	3.06	3.08	3.09	3.11	3.13	3.14	3.16	3.17	3.18	3.20
31	2.94	2.96	2.99	3.01	3.03	3.05	3.07	3.08	3.10	3.12	3.13	3.15	3.16	3.18	3.19
32	2.93	2.96	2.98	3.00	3.02	3.04	3.06	3.08	3.09	3.11	3.12	3.14	3.15	3.17	3.18
33	2.93	2.95	2.97	2.99	3.01	3.03	3.05	3.07	3.09	3.10	3.12	3.13	3.14	3.16	3.17
34	2.92	2.94	2.97	2.99	3.01	3.03	3.04	3.06	3.08	3.09	3.11	3.12	3.14	3.15	3.16
35	2.91	2.94	2.96	2.98	3.00	3.02	3.04	3.06	3.07	3.09	3.10	3.12	3.13	3.14	3.16
36	2.91	2.93	2.95	2.98	3.00	3.01	3.03	3.05	3.07	3.08	3.10	3.11	3.12	3.14	3.15
37	2.90	2.93	2.95	2.97	2.99	3.01	3.03	3.04	3.06	3.07	3.09	3.10	3.12	3.13	3.14
38	2.90	2.92	2.94	2.97	2.98	3.00	3.02	3.04	3.05	3.07	3.08	3.10	3.11	3.12	3.14
39	2.89	2.92	2.94	2.96	2.98	3.00	3.02	3.03	3.05	3.06	3.08	3.09	3.11	3.12	3.13
40	2.89	2.91	2.94	2.96	2.98	2.99	3.01	3.03	3.04	3.06	3.07	3.09	3.10	3.11	3.13
41	2.89	2.91	2.93	2.95	2.97	2.99	3.01	3.02	3.04	3.05	3.07	3.08	3.10	3.11	3.12
42	2.88	2.91	2.93	2.95	2.97	2.98	3.00	3.02	3.03	3.05	3.06	3.08	3.09	3.10	3.12
43	2.88	2.90	2.92	2.94	2.96	2.98	3.00	3.01	3.03	3.05	3.06	3.07	3.09	3.10	3.11
44	2.88	2.90	2.92	2.94	2.96	2.98	2.99	3.01	3.03	3.04	3.06	3.07	3.08	3.10	3.11
45	2.87	2.89	2.92	2.94	2.96	2.97	2.99	3.01	3.02	3.04	3.05	3.07	3.08	3.09	3.10
46	2.87	2.89	2.91	2.93	2.95	2.97	2.99	3.00	3.02	3.03	3.05	3.06	3.07	3.09	3.10
47	2.87	2.89	2.91	2.93	2.95	2.97	2.98	3.00	3.02	3.03	3.04	3.06	3.07	3.08	3.10
48	2.86	2.89	2.91	2.93	2.95	2.96	2.98	3.00	3.01	3.03	3.04	3.05	3.07	3.08	3.09
49	2.86	2.88	2.90	2.92	2.94	2.96	2.98	2.99	3.01	3.02	3.04	3.05	3.06	3.08	3.09
50	2.86	2.88	2.90	2.92	2.94	2.96	2.97	2.99	3.01	3.02	3.03	3.05	3.06	3.07	3.09
60	2.84	2.86	2.88	2.90	2.92	2.93	2.95	2.97	2.98	3.00	3.01	3.02	3.04	3.05	3.06
70	2.82	2.84	2.86	2.88	2.90	2.92	2.93	2.95	2.96	2.98	2.99	3.00	3.02	3.03	3.04
80	2.81	2.83	2.85	2.87	2.89	2.90	2.92	2.94	2.95	2.97	2.98	2.99	3.00	3.02	3.03
90	2.80	2.82	2.84	2.86	2.88	2.90	2.91	2.93	2.94	2.96	2.97	2.98	2.99	3.01	3.02
100	2.79	2.82	2.84	2.85	2.87	2.89	2.90	2.92	2.93	2.95	2.96	2.97	2.99	3.00	3.01

Factor $= t_{(n-1,\ 1-\alpha/r)}$

Table 13.11.   Values of t For Obtaining One-Sided 95% Poisson Prediction Limits For r Additional Samples Given A Background Sample of Size n.

Previous n	Number of New Measurements (r)														
	30	35	40	45	50	55	60	65	70	75	80	85	90	95	100
4	8.57	9.04	9.46	9.85	10.21	10.55	10.87	11.17	11.45	11.72	11.98	12.23	12.47	12.70	12.92
5	6.25	6.52	6.76	6.97	7.17	7.36	7.53	7.69	7.84	7.98	8.12	8.25	8.38	8.50	8.61
6	5.25	5.44	5.60	5.76	5.89	6.02	6.14	6.25	6.35	6.45	6.54	6.63	6.71	6.79	6.87
7	4.70	4.85	4.98	5.10	5.21	5.31	5.40	5.48	5.56	5.64	5.71	5.78	5.84	5.90	5.96
8	4.35	4.48	4.59	4.69	4.78	4.87	4.94	5.02	5.08	5.14	5.20	5.26	5.31	5.36	5.41
9	4.12	4.23	4.33	4.42	4.50	4.57	4.64	4.70	4.76	4.81	4.86	4.91	4.96	5.00	5.04
10	3.95	4.06	4.15	4.22	4.30	4.36	4.42	4.48	4.53	4.58	4.62	4.67	4.71	4.74	4.78
11	3.83	3.92	4.00	4.08	4.14	4.20	4.26	4.31	4.36	4.40	4.44	4.48	4.52	4.55	4.59
12	3.73	3.82	3.89	3.96	4.02	4.08	4.13	4.18	4.22	4.26	4.30	4.34	4.37	4.41	4.44
13	3.65	3.73	3.81	3.87	3.93	3.98	4.03	4.08	4.12	4.15	4.19	4.23	4.26	4.29	4.32
14	3.58	3.66	3.73	3.80	3.85	3.90	3.95	3.99	4.03	4.07	4.10	4.13	4.16	4.19	4.22
15	3.53	3.61	3.67	3.73	3.79	3.84	3.88	3.92	3.96	3.99	4.03	4.06	4.09	4.11	4.14
16	3.48	3.56	3.62	3.68	3.73	3.78	3.82	3.86	3.90	3.93	3.96	3.99	4.02	4.05	4.07
17	3.44	3.52	3.58	3.64	3.69	3.73	3.77	3.81	3.84	3.88	3.91	3.94	3.96	3.99	4.01
18	3.41	3.48	3.54	3.60	3.65	3.69	3.73	3.77	3.80	3.83	3.86	3.89	3.92	3.94	3.96
19	3.38	3.45	3.51	3.56	3.61	3.65	3.69	3.73	3.76	3.79	3.82	3.85	3.87	3.90	3.92
20	3.35	3.42	3.48	3.53	3.58	3.62	3.66	3.69	3.73	3.76	3.79	3.81	3.84	3.86	3.88
21	3.33	3.40	3.46	3.51	3.55	3.59	3.63	3.66	3.70	3.73	3.75	3.78	3.80	3.83	3.85
22	3.31	3.38	3.43	3.48	3.53	3.57	3.60	3.64	3.67	3.70	3.73	3.75	3.77	3.80	3.82
23	3.29	3.35	3.41	3.46	3.50	3.54	3.58	3.61	3.64	3.67	3.70	3.72	3.75	3.77	3.79
24	3.27	3.34	3.39	3.44	3.48	3.52	3.56	3.59	3.62	3.65	3.68	3.70	3.72	3.75	3.77
25	3.26	3.32	3.38	3.42	3.47	3.50	3.54	3.57	3.60	3.63	3.66	3.68	3.70	3.72	3.75
26	3.24	3.31	3.36	3.41	3.45	3.49	3.52	3.55	3.58	3.61	3.64	3.66	3.68	3.70	3.72
27	3.23	3.29	3.35	3.39	3.43	3.47	3.51	3.54	3.57	3.59	3.62	3.64	3.67	3.69	3.71
28	3.22	3.28	3.33	3.38	3.42	3.46	3.49	3.52	3.55	3.58	3.60	3.63	3.65	3.67	3.69
29	3.21	3.27	3.32	3.37	3.41	3.44	3.48	3.51	3.54	3.56	3.59	3.61	3.63	3.65	3.67
30	3.20	3.26	3.31	3.36	3.40	3.43	3.47	3.50	3.52	3.55	3.58	3.60	3.62	3.64	3.66
31	3.19	3.25	3.30	3.34	3.38	3.42	3.45	3.48	3.51	3.54	3.56	3.58	3.61	3.63	3.65
32	3.18	3.24	3.29	3.33	3.37	3.41	3.44	3.47	3.50	3.53	3.55	3.57	3.59	3.61	3.63
33	3.17	3.23	3.28	3.33	3.37	3.40	3.43	3.46	3.49	3.52	3.54	3.56	3.58	3.60	3.62
34	3.16	3.22	3.27	3.32	3.36	3.39	3.42	3.45	3.48	3.51	3.53	3.55	3.57	3.59	3.61
35	3.16	3.21	3.26	3.31	3.35	3.38	3.42	3.44	3.47	3.50	3.52	3.54	3.56	3.58	3.60
36	3.15	3.21	3.26	3.30	3.34	3.37	3.41	3.44	3.46	3.49	3.51	3.53	3.55	3.57	3.59
37	3.14	3.20	3.25	3.29	3.33	3.37	3.40	3.43	3.45	3.48	3.50	3.52	3.54	3.56	3.58
38	3.14	3.19	3.24	3.29	3.32	3.36	3.39	3.42	3.45	3.47	3.49	3.52	3.54	3.56	3.57
39	3.13	3.19	3.24	3.28	3.32	3.35	3.38	3.41	3.44	3.46	3.49	3.51	3.53	3.55	3.57
40	3.13	3.18	3.23	3.27	3.31	3.35	3.38	3.41	3.43	3.46	3.48	3.50	3.52	3.54	3.56
41	3.12	3.18	3.23	3.27	3.31	3.34	3.37	3.40	3.43	3.45	3.47	3.49	3.51	3.53	3.55
42	3.12	3.17	3.22	3.26	3.30	3.33	3.37	3.39	3.42	3.44	3.47	3.49	3.51	3.53	3.54
43	3.11	3.17	3.22	3.26	3.30	3.33	3.36	3.39	3.41	3.44	3.46	3.48	3.50	3.52	3.54
44	3.11	3.16	3.21	3.25	3.29	3.32	3.35	3.38	3.41	3.43	3.45	3.48	3.49	3.51	3.53
45	3.10	3.16	3.21	3.25	3.29	3.32	3.35	3.38	3.40	3.43	3.45	3.47	3.49	3.51	3.53
46	3.10	3.15	3.20	3.24	3.28	3.31	3.34	3.37	3.40	3.42	3.44	3.46	3.48	3.50	3.52
47	3.10	3.15	3.20	3.24	3.28	3.31	3.34	3.37	3.39	3.42	3.44	3.46	3.48	3.50	3.51
48	3.09	3.15	3.19	3.24	3.27	3.31	3.34	3.36	3.39	3.41	3.43	3.45	3.47	3.49	3.51
49	3.09	3.14	3.19	3.23	3.27	3.30	3.33	3.36	3.38	3.41	3.43	3.45	3.47	3.49	3.50
50	3.09	3.14	3.19	3.23	3.26	3.30	3.33	3.35	3.38	3.40	3.43	3.45	3.46	3.48	3.50
60	3.06	3.11	3.16	3.20	3.23	3.27	3.30	3.32	3.35	3.37	3.39	3.41	3.43	3.45	3.46
70	3.04	3.09	3.14	3.18	3.21	3.24	3.27	3.30	3.32	3.34	3.37	3.39	3.40	3.42	3.44
80	3.03	3.08	3.12	3.16	3.20	3.23	3.26	3.28	3.31	3.33	3.35	3.37	3.38	3.40	3.42
90	3.02	3.07	3.11	3.15	3.18	3.21	3.24	3.27	3.29	3.31	3.33	3.35	3.37	3.39	3.40
00	3.01	3.06	3.10	3.14	3.17	3.20	3.23	3.26	3.28	3.30	3.32	3.34	3.36	3.38	3.39

Factor = $t_{(n-1,\ 1-\alpha/r)}$

A Nonparametric Approach

The previous discussion has been based on the assumption that the distribution of the parameter(s) of interest is known and has a parametric form (i.e., normal, censored normal, or Poisson). In some cases, however, this assumption is unreasonable and a "distribution free" statistical method may be required. In the content of ground-water monitoring, Gibbons (1988a) has adapted the nonparametric prediction limit originally described by Chou and Owen (1986). In contrast to the parametric approach in which we estimate a limit value from the mean and standard deviation of a sample of n previous measurements, the nonparametric approach identifies the required number of samples (n) so that the maximum value of those samples is the 95% prediction limit. Gibbons (1988a) further generalizes the procedure to include the effects of resampling (i.e., taking a vertification sample following a statistically significant ground water monitoring result.)

The 95% nonparametric prediction limit can be easily obtained with the aid of Table 13.12. For example, assume that we have a facility with r=10 downgradient monitoring wells. Furthermore, let us also assume that if we fail a detection monitoring test, we are permitted to resample the well before any further action is taken and if the repeat sample does not fail the test, we return to normal detection monitoring. How many background samples (n) are we required to obtain, so that the maximum observed measurement of those n samples will contain the next r=10 monitoring measurements given the possibility of a single resample of any well that fails the initial test? Inspection of Table 13.12 reveals that a background sample of n=18 measurements provides 94.9% confidence, and a background sample of n=19 provides 95.3% confidence. The answer is, therefore, n=19 background samples must be taken in order to ensure that the maximum of those 19 samples will not be exceeded by the next 10 monitoring measurements (i.e., one at each downgradient well), given that we can resample any well that fails the initial test.

The major advantage of this approach is that it only assumes that the samples are independent and measured on a continuous scale, but no particular distribution is specified. Furthermore, as long as at least one ground water sample has a measurable value, the limit is defined. When nothing is detected in the n background samples, an alternative approach must be taken.

Case 3: What to do when nothing is detected. Statistical methods are of little use without measurable data. Nevertheless, it is surprisingly common to observe a background collection of 10 or so measurements for which nothing was detected. What do we do? Is the tolerance or prediction limit zero? Is it the method detection limit? The answer to this question can only be found by examining the specifics of the analytic measurement process itself (see Gibbons 1988b). Interestingly, the analyst's decision as to whether or not a particular substance is present in a particular sample is completely based on the application of statistical decision rules. Even more remarkable, these decision rules are based on tolerance limits (see Currie 1968) for analyte absent or single concentration detection limit studies and prediction limits for calibration designs (i.e., a series

Table 13.12. Probability That at Least 1 out of 2 Samples Will be Below the Maximum of n Background Measurements at Each of r Monitoring Wells.

Previous n	Number of New Measurements (r)														
	1	2	3	4	5	6	7	8	9	10	11	12	13	14	15
4	.933	.871	.813	.759	.708	.661	.617	.576	.537	.502	.468	4.37	.408	.381	.355
5	.952	.907	.864	.823	.784	.746	.711	.677	.645	.614	.585	.557	.530	.505	.481
6	.964	.930	.897	.065	.834	.804	.775	.748	.721	.695	.670	.646	.623	.601	.580
7	.972	.945	.919	.893	.869	.844	.821	.798	.776	.754	.734	.713	.693	.674	.655
8	.978	.956	.935	.914	.894	.874	.854	.835	.817	.799	.781	.764	.747	.730	.714
9	.982	.964	.946	.929	.912	.896	.879	.863	.848	.832	.817	.802	.788	.773	.759
10	.985	.970	.955	.941	.927	.912	.899	.885	.872	.858	.845	.833	.820	.808	.795
11	.987	.975	.962	.950	.938	.926	.914	.902	.890	.879	.868	.857	.846	.835	.824
12	.989	.978	.967	.957	.946	.936	.926	.915	.905	.895	.886	.876	.866	.857	.847
13	.990	.981	.972	.962	.953	.944	.935	.926	.917	.909	.900	.892	.883	.875	.868
14	.992	.983	.975	.967	.959	.951	.943	.935	.927	.920	.912	.904	.897	.889	.882
15	.993	.985	.978	.971	.964	.957	.950	.943	.936	.929	.922	.915	.909	.902	.895
16	.993	.987	.981	.974	.968	.961	.955	.949	.943	.937	.930	.924	.918	.912	.906
17	.994	.988	.983	.977	.971	.965	.960	.954	.949	.943	.938	.932	.927	.921	.916
18	.995	.990	.984	.979	.974	.969	.964	.959	.954	.949	.944	.939	.934	.929	.924
19	.995	.990	.986	.981	.976	.972	.967	.963	.958	.953	.949	.944	.940	.935	.931
20	.996	.991	.987	.983	.979	.974	.970	.966	.962	.958	.953	.949	.945	.941	.937
25	.997	.994	.991	.989	.986	.983	.980	.977	.975	.972	.969	.966	.964	.961	.958
30	.998	.996	.994	.992	.990	.988	.986	.984	.982	.980	.978	.976	.974	.972	.970
35	.998	.997	.996	.994	.993	.991	.990	.988	.987	.985	.984	.982	.981	.979	.978
40	.999	.998	.997	.995	.994	.993	.992	.991	.990	.988	.987	.986	.985	.984	.983
45	.999	.998	.997	.996	.995	.994	.994	.993	.992	.991	.990	.989	.988	.987	.986
50	.999	.998	.998	.997	.996	.996	.995	.994	.993	.992	.992	.991	.990	.989	.989
60	.999	.999	.998	.998	.997	.997	.996	.996	.995	.995	.994	.994	.993	.993	.992
70	1.00	.999	.999	.998	.998	.998	.997	.997	.996	.996	.996	.995	.995	.995	.994
80	1.00	.999	.999	.999	.998	.998	.998	.998	.997	.997	.997	.996	.996	.996	.995
90	1.00	1.00	.999	.999	.999	.999	.998	.998	.998	.998	.997	.997	.997	.997	.996
10	1.00	1.00	.999	.999	.999	.999	.999	.998	.998	.998	.998	.998	.998	.997	.997

Previous n	Number of New Measurements (r)														
	20	25	30	35	40	45	50	55	60	65	70	75	80	90	100
4	.252	.178	.126	.089	.063	.045	.032	.022	.016	.011	.008	.006	.004	.002	.001
5	.377	.295	.231	.181	.142	.111	.087	.068	.054	.042	.033	.026	.020	.012	.008
6	.483	.403	.336	.280	.233	.195	.162	.135	.113	.094	.078	.065	.055	.038	.026
7	.569	.494	.430	.373	.324	.281	.244	.212	.184	.160	.139	.121	.105	.079	.060
8	.638	.570	.510	.455	.407	.364	.325	.291	.260	.232	.207	.185	.166	.132	.106
9	.693	.632	.577	.326	.480	.438	.400	.365	.333	.303	.277	.253	.230	.192	.160
10	.737	.683	.633	.586	.543	.503	.466	.432	.400	.371	.343	.318	.295	.253	.217
11	.773	.724	.679	.637	.597	.560	.525	.492	.461	.432	.405	.380	.356	.313	.275
12	.802	.759	.718	.679	.643	.608	.576	.545	.515	.488	.461	.437	.413	.370	.331
13	.826	.787	.750	.715	.682	.650	.620	.591	.563	.537	.512	.488	.465	.423	.384
14	.846	.811	.778	.746	.716	.686	.658	.631	.605	.580	.557	.534	.512	.471	.433
15	.863	.832	.801	.772	.744	.717	.691	.666	.642	.619	.597	.575	.554	.515	.478
16	.877	.849	.821	.795	.769	.744	.720	.697	.675	.653	.632	.612	.592	.554	.519
17	.889	.864	.839	.814	.791	.768	.746	.724	.703	.683	.663	.644	.625	.590	.556
18	.900	.876	.854	.831	.810	.789	.768	.748	.729	.710	.691	.673	.656	.622	.590
19	.909	.888	.867	.846	.826	.807	.788	.769	.751	.733	.716	.699	.683	.651	.620
20	.917	.897	.878	.859	.841	.823	.805	.788	.771	.754	.738	.722	.707	.677	.648

continued

Table 13.12.   Continued.

Previous	Number of New Measurements (r)														
n	1	2	3	4	5	6	7	8	9	10	11	12	13	14	15
25	.945	.931	.918	.905	.892	.880	.867	.855	.843	.831	.819	.807	.796	.774	.752
30	.960	.951	.941	.932	.922	.913	.904	.895	.886	.877	.868	.860	.851	.834	.817
35	.970	.963	.956	.949	.942	.935	.928	.921	.914	.907	.900	.893	.887	.874	.860
40	.977	.971	.966	.960	.955	.949	.944	.938	.933	.927	.922	.917	.911	.901	.890
45	.982	.977	.973	.968	.964	.959	.955	.950	.946	.942	.937	.933	.929	.920	.912
50	.985	.981	.978	.974	.970	.967	.963	.959	.956	.952	.949	.945	.941	.934	.927
60	.989	.987	.984	.982	.979	.976	.974	.971	.969	.966	.964	.961	.959	.954	.948
70	.992	.990	.988	.986	.984	.983	.981	.979	.977	.975	.973	.971	.969	.965	.962
80	.994	.992	.991	.990	.988	.987	.985	.984	.982	.981	.979	.978	.976	.973	.970
90	.995	.994	.993	.992	.990	.989	.988	.987	.986	.985	.983	.982	.981	.979	.976
100	.996	.995	.994	.993	.992	.991	.990	.989	.988	.987	.987	.986	.985	.983	.981

of different spiking concentrations in the range at the method detection limit (MPL) (see Hubaux and Vos, 1970; and Clayton et al., 1987).

These method detection limits are defined as the point at which the false positive and false negative rates are both less than 5% for a test of the null hypothesis that the concentration of the analyte in the sample is zero.

To compute method detection limit from "spiked" calibration samples, we can use the method of Clayton et al. (1987) as follows:

1. Select four concentrations in the range of the hypothesized MDL. For example, for Benzene we might select concentrations of 4, 8, 12, and 16 mg/L.
2. Prepare 16 samples; that is 4 replicates at each of the 4 spiking concentrations.
3. Multiple compounds may be examined simultaneously by including them in the same samples; however, the order of their concentrations should be randomized so that one sample does not contain all of the lowest concentrations and another sample all of the highest concentrations.
4. Introduce these 16 samples in the usual daily workload of two or more analysts (e.g., two analysts would receive eight samples each). It is essential that the analysts be *completely* blind to which compounds are present in the samples and their respective spiking concentrations, and that they simply be instructed to perform the standard analytic method in question (e.g., Method 624 VOC Scan).
5. The results of the analysis should be recorded as the square root of the ratio of the compound peak area to the internal standard; that is,

$$\text{response signal} = \sqrt{\frac{\text{peak area count}}{\text{internal standard area count}}} = yi \qquad (13.4a)$$

Transform the spiking concentration

$$xi = \sqrt{x^*_i - 0.1} - \sqrt{0.1} \qquad (13.4b)$$

Where x* is the original spiking concentration

6. For each compound, compute the slope of the regression line of the instrument response signal (y) on the targeted concentration (x) as:

$$b = \frac{\sum_{i=1}^{16}(x_i - \bar{x})(y_i - y)}{\sum_{i=1}^{16}(x_i - \bar{x})^2} \qquad (13.5)$$

Where x is the average of the four target concentrations, and Y is the average of the 16 instrument response signals as defined above.

7. For each compound, compute the variance of deviations from the regression line as:

$$s^2_{y\cdot x} = \sum_{i=1}^{16}(y_i - \hat{y}_i)^2 / (16-2) \qquad (13.6)$$

Where $\hat{y}_i = y_i + b(x_i - \bar{x})$ is the predicted instrument response for target concentration x_i.

8. The method detection limit for n=16 samples is then computed as:

$$MDL^* = (3.46 s_{y\cdot x}/b) \sqrt{1 + \frac{1}{16} + \bar{x}^2 / \sum_{i=1}^{16}(x_i - \bar{x})^2} \qquad (13.7a)$$

Where 3.46 is the $\alpha = \beta = .95$ percentage of the noncentral t distribution on $16-2=14$ degrees of freedom. To express MDL in the original metric (example μ/L), compute:

$$MDL = (MDL^*)^2 + 0.632456\ MDL^* \qquad (13.7b)$$

In the absence of any detected values, this estimated MDL can be used as the corresponding prediction limit for detection monitoring.

We note that there are existing published MDLs for many compounds, includ-

We note that there are existing published MDLs for many compounds, including the Method 624 VOCs. More recently, Practical Quantitation limits have also published in the Method SW-846 regulation. These national values were established under idealized conditions in which both presence and spiking concentration were known to the analyst and very questionable statistical computations were performed (see Gibbons 1988b). As such, it is quite reasonable that such levels will not be reached in routine laboratory practice and that the procedure described here will provide more realistic estimates that are consistent with attainable standards in the routine application of these methods. This view is reiterated in the new 40 CFR 264 statistical regulation:

> The Appendix IX rule (52 FR 25942, July 9, 1987) listed practical quantification limits (PQLs) that were established from "Test Methods for Evaluating Solid Waste" (SW-846). SW-846 is the general RCRA analytical methods manual, currently in its third edition. The PQLs listed were EPA's best estimate of the practical sensitivity of the applicable method for RCRA ground water monitoring purposes. However, some of the PQLs may be unattainable because they are based on general estimates for the specific substance. Furthermore, due to site specific factors, these limits may not be reached. For these reasons the Agency feels that the PQLs listed in Appendix IX are not appropriate for establishing a national baseline value for each constituent for determining whether a release to ground water has occurred. Instead, the PQLs are viewed as target levels that chemical laboratories should try to achieve in their analyses of ground water. In the event that a laboratory cannot achieve the suggested PQL, the owner or operator may submit a justification stating the reasons why these values cannot be achieved (e.g., specific instrument limitations). After reviewing this justification, the Regional Administrator may choose to establish facility specific PQLs based on the technical limitations of the contracting laboratory. Thus, EPA is today clarifying 264.97(h) to allow owners or operators to propose facility specific PQLs. These PQLs may be used with the statistical methods listed in 264.97.

Summary

The statistical methods described here provide a series of general tools by which detection monitoring programs can be designed using indicator parameters that vary from 100% detection to no detection. The methods are completely site specific with the exception of the case of no detection for which they are specific to the monitoring laboratory responsible for the routine analysis of the ground water samples. The statistical procedures are parametric and nonparametric forms of prediction and tolerance intervals and limits, and as such, are consistent with the new RCRA 40 CFR Part 264 statistical regulation. The facility wide false positive rate is restricted to 5%; therefore, quarterly monitoring should result in one false positive decision every five years. Resampling of the well or wells in question should produce even fewer false positive results. Using the suggested sample sizes should produce false negative rates of less than 5% for monitoring requirements in excess of 2 to 3 standard deviation units above the background mean. False positive and false negative results are, therefore, balanced for even modest deviations from background water quality levels.

VERIFICATION OF EXCEEDANCE

When water quality data from a detection monitoring program shows a statistically significant increase, a series of steps should be performed to determine if the parameter increase is due to a release from the facility or from nonfacility based interferences. Three major components of potential exceedance can be defined as:

- Site interference
  - —landfill gas present in well
  - —grout alkaline pH interferences
  - —poor well construction or maintenance
  - —background increase due to upgradient discharge
  - —natural aquifer interferences
- Laboratory interference
  - —transcription errors
  - —laboratory contamination
  - —method or chemical interferences
- A facility release exceedance

Each of the above potential pathways can cause a site to go into an exceedance verification analysis. Some of the interferences are relatively simple to define such as transcription errors and laboratory chemical interferences (e.g., persistent low levels of methalene chloride).

Other interferences are difficult to define and quantify, such as landfill gas cross-contamination or poor well construction. Each of the site interferences will be briefly discussed in the following subsections.

Landfill Gas Cross-Contamination

The presence of landfill gas in monitoring wells is probably one of the most significant cross-contamination problems observed at solid waste sites. The typical gas contaminated well shows persistent low level vinyl chloride. The well may be sampled through bailing or by bladder pumps. Typically, bailed wells show much higher levels of gas cross-contamination due to the bailer passing through the gas on the way out of the well. The cause of this cross-contamination is usually due to monitoring wells screening across the water table. Since part of the screen spans the unsaturated zone, landfill gas can enter the well. The well screens can act as vents for permeable, unsaturated soils so, in effect, wells screened in the unsaturated zone can attract landfill gas.

These gas cross-contamination problems, however, do not relate to true ground water contamination. In a number of cases, hand bailed wells showing vinyl chloride hits produced nondetect VOAs once a bladder pump was installed. To define if gas is causing cross-contamination, the following procedure should be considered:

- If persistent low levels of VOCs are observed from a well, answer the following questions:
  —Is the well screened partially in the unsaturated zone?
  —Is the well bailed to obtain a sample?
  —Can landfill gas be detected in the well with a methane monitor?
  —Is vinyl chloride observed in the water quality results?
  —Is the well casing cracked in the unsaturated zone (observe with down hole TV camera)?

The presence of landfill gas cross-contamination would be suspected if any of the above were observed at a VOC detect well.

Grout Alkaline pH Interferences

Very high pH readings are extremely unusual in ground water quality analysis. Typically, pH levels above 8.5 would be considered very uncommon for natural aquifers (Hem, 1970). However, monitoring wells can produce water with pH levels approaching 14. For wells contaminated with alkaline grout, water pH levels of 11 to 12 are typical. These grout contaminated wells are a result of poor well construction practice by not separating the screened zone far enough from the annular seals of cement/bentonite grout. Since these grouts contain mainly cement, hard set grouts can bleed very alkaline water down into the well screen area and be sampled. This is caused by the following:

- Wells located in low permeability units with strong vertical gradients (i.e., grout bleeds directly downward).
- Grout injected into the screened area of the well.
- Bentonite seals too thin or ineffective.
- Fractured rock providing channels around bentonite seals.

All of the above causes of grout alkaline well contamination can be remedied by proper construction and development of the well using ASTM standard well designs. For those wells already showing high pH levels, the investigation should redevelop the well until reasonable pH levels are obtained. Normal purging of wells during sampling (in low permeability environments) can take many years to reduce pH to background levels. These wells may have to be replaced with ASTM specification wells or continuously pumped at low levels until acceptable pH readings are obtained. Low rate pumping, however, may not be effective since high pH readings may return once purging is stopped.

Poor Well Construction or Maintenance

Interferences due to inter-aquifer connections or surface water entering the screened zone of the well along the annular space can cause exceedance of water quality indicator parameters. Proper well construction based on flow path monitoring can target the uppermost aquifer for monitoring. Well maintenance and inspection can also highlight damaged wells where surface water may be entering the well. Each monitoring well should be visually inspected before sampling to document the condition of the installation.

Background Increase Due to Discharge

Resolving an exceedance of indicator parameters may require inspection of upgradient areas for visible signs of spills or use of chemicals that can cause exceedance of indicator parameters. For example, road salt on upgradient recharge areas can cause a dramatic increase in chlorides in highly permeable uppermost aquifers. Since chloride is a mobile indicator parameter for landfill leachate, road salt can give false positive results in downgradient wells. This interference is especially troublesome with poorly defined monitoring systems that have not adequately defined upgradient and downgradient relationships.

Natural Inter-Aquifer Interferences

Water quality obtained from aquifers can typically exceed drinking water quality standards for many nonorganic and some organic indicators. Interferences with detection monitoring programs have been documented for chlorides (shale bedrock discharge). Arsenic has been reported in bedrock wells from Massachusetts; and even organic indicator parameters such as phenols and TOX have been reported occurring naturally in swamps and near the sea shore, respectively. The best procedure to follow in assessing these potentially natural indicator parameters is to have adequate truly upgradient monitoring wells. Sufficient numbers of background wells should assist in establishing ambient water quality and provide a better statistical basis for water quality comparisons.

Expression of the relationships among ions, or of one constituent to the total concentration in terms of mathematical ratios, is often helpful in making resemblances and differences among waters stand out clearly. These differences can define natural aquifer interferences of indicator parameters. For most comparisons of this type, concentration values expressed in terms of milliequivalents per liter or moles per liter are the most useful.

Ratios are obviously useful to establish chemical similarities among waters; for example, in grouping analyses representing a single geologic terrane, upgradient and downgradient waters in a single aquifer, or a water-bearing zone. Fixed rules regarding selection of the most significant values to compare by ratios cannot be given, but the investigator should consider some thought as to the sources of ions and the chemical behavior of the parameter.

Distinguishing between two contamination sources usually involves comparing different ionic ratios using a variety of graphical methods (i.e., Stiff, Piper, or Schoeller diagrams). A lack of analytical data can impact the use of these approaches exactly as they were intended.

VERIFICATION OF RELEASE

The verification of an exceedance of water quality indicator parameters by ruling out site and laboratory interferences would move the detection monitoring program into a series of steps to confirm the significant increase. These steps can be reduced into a series of bullets:

- increase sampling.
- determine source of exceedance:
  —Sample leachate for fingerprint.
  —Compare leachate fingerprint with observed exceedance.
- expand parameters to include VOCs and metals.
- determine if three or more parameters exceed statistical tests over the next two quarters of sampling.

If confirmation of the significant increase occurs, then the investigation should move toward an assessment monitoring program. Assessment monitoring may include the drilling of individual wells, expansion of analytical parameters, and additional sampling of the detection monitoring wells. Assessment monitoring is not specifically addressed in this manual since the assessment program requires establishment of rate and extent of leachate migration. If no preliminary literature review and detection monitoring design investigations were previously performed, the investigator should use the phased sequence of defining target pathways and basing the assessment monitoring system on knowledge of the site geology and hydrogeology.

REFERENCES

Aitchison, J., 1955. On the Distribution of a Positive Random Variable Having a Discrete Probability Mass at the Origin. J. Amer. Statistical Asso. 50, 901–908.

Chou, Y. M.; and D. B. Owen, 1986. One-Sided Distribution-Free Simultaneous Prediction Limits for Future Samples. Journal of Quality Technology, 18, 96–98.

Clayton, C. A., J. W. Hines and P. D. Elkins, 1987. Detection Limits with Specified Assurance Probabilities. Analytical Chemistry, 59, 2506–2514.

Currie, L. A., 1968. Limits for Qualitative Decision and Quantitative Determination. Analytical Chemistry, 40, 586–593.

Gibbons, R. D., 1987a. Statistical Prediction Intervals for the Evaluation of Ground Water Quality. Ground Water, 25, 455–465.

Gibbons, R. D., 1987b. Statistical Models for the Analysis of Volatile Organic Compounds in Waste Disposal Facilities. Ground Water, 25, 572–580.

Gibbons, R. D., 1988a. A General Statistical Procedure for Ground Water Detection Monitoring at Waste Disposal Facilities. Ground Water, submitted for publication.

Gibbons, R. D., F. H. Jarke, and K. P. Stoub, 1988b. Method Detection Limits. Ground Water, submitted for publication.

Hem, J. D., 1970. Study and Interpretation of the Chemical Characteristic of Natural Water. Geological Survey Water Supply Paper 1473, 2nd ed., U.S. Government Printing Office, Washington, DC, page 363.

Hubaux, A.; and G. Vos, 1970. Decision and Detection Limits for Linear Calibration Curves. Analytical Chemistry, 42, 849–855.

Hurd, M., 1986. Determining the Impact of Land Disposal—The Review of Organic Analytical Data. ETC Internal report, page 28.

Piper, A. M., 1944. A Graphic Procedure in the Geochemical Interpretation of Water Analyses. Air Geophysical Union Trans., Vol. 25, 914–923.

USEPA, 1988. 40 CFR Part 264: Statistical Methods for Evaluating Ground Water Monitoring from Hazardous Waste Facilities; Final Rule. Federal Register, 53, 196 39720–39731.

Wald, A., and J. Wolfowitz, 1946. Tolerance Limits for a Normal Distribution. Ann. Math. Statistics, 17, 208–215.

14

Health and Safety Considerations in Ground-Water Monitoring Investigations

Steven P. Maslansky and Carol J. Maslansky

One of the most important aspects to consider when conducting ground-water investigations is hazard recognition. Look up "hazard" in any dictionary and it will be defined as danger or something that causes danger or difficulty.

Many required tasks performed during ground-water studies can be hazardous. Safety must be emphasized, and all personnel must know how to protect themselves, their co-workers, and the equipment they operate. Publications describing safe operating procedures around heavy equipment utilized in ground-water monitoring operations are available (NWWA, 1980; NDF, 1986) and will not be discussed in detail in this chapter.

Hazard can also be defined, however, as a lack of predictability or uncertainty. This second definition more aptly defines most conditions found in the field. The key to safety in the field is an ability to recognize situations that may produce hazardous conditions, and to plan ahead to avoid or mitigate these conditions.

CONTINGENCY PLANNING

During ground-water investigations, every effort must be made to expect the unexpected. Contingency planning can be thought of as a method to mitigate the impact of Murphy's Law. A careful site history review or previous monitoring and site characterization activities may minimize the possibility of discovering unknown hazardous materials or unstable conditions.

The very nature of hazardous materials work, however, complicates and exacerbates anticipated field conditions and hazards normally associated with clean water work. The possibility of hitting buried utilities, encountering hazardous substances, initiating an equipment fire, or sustaining personal injury must always be considered. At hazardous substance sites, both the possibility of and severity of emergency situations are enhanced. Those directly responsible or peripherally involved with site safety must ensure that personnel on the job are adequately trained, medically qualified, physically fit, and possess a good mental attitude toward safety.

Logistics must be carefully considered. Workers must have adequate protective clothing (amount, appropriate type, proper sizes) and equipment do the job safely and efficiently. Temperature stress and fatigue must be anticipated when employing protective equipment. Appropriate real-time air monitoring devices must be available; they must be working properly and calibrated. Backup instrumentation should be available if needed. Because medical emergencies do occur, the site safety officer or some other designated individual should be trained in standard first aid and CPR.

The site safety plan is a major component of contingency planning. A good site safety plan must address a myriad of topics. If properly conceived and written, the plan will address all aspects of the work plan as well as anticipated emergencies. Typical areas that should be addressed in the safety plan include the following:

1. Safety staff organization, responsibilities of key personnel, and their alternates. This includes identification of the site safety officer, emergency medical officer, and other individuals responsible for implementation and continued enforcement of the on-site health and safety program.
2. Safety and health hazard assessment for site operations. This includes a listing of the known or anticipated site hazards (i.e., chemical, biological, physical), known or anticipated chemical contaminant concentration ranges, and applicable occupational exposure limits (OSHA PELs, NIOSH RELs, or ACGIH TLVs). In addition, site operations should be individually evaluated and safety procedures outlined to mitigate specific hazards.
3. Personal protective equipment requirements. This includes the types of protective clothing and respiratory protection required due to site-specific hazards as well as the development of site-specific action levels to dictate upgrading or downgrading levels of protection.
4. Methods to assess personal and environmental exposure. This includes radiological and meteorological monitoring, real-time or direct-reading air monitoring, measurement of representative worker exposures using personnel air monitoring, perimeter, and time-weighted average

air monitoring. Procedures should be outlined regarding sample collection, instrument use, maintenance, calibration, frequency of sampling, and quality assurance and control.

5. Standard operating safety procedures, work practices, and engineering controls. This section should contain site rules and prohibitions (i.e., buddy system, site hygiene practices, eating/drinking/smoking restrictions), handling procedures for hazardous materials and samples, and sample container handling procedures and precautions. Also included in this section are protocols for activities involving confined spaces, excavations, and welding and cutting. Provisions for night illumination and sanitation facilities are also covered.

6. Site control measures. These measures must include an on-site communication plan, site entry and egress procedures (check-in and checkout), delineation of access points, and site security precautions. A site map should be included that indicates work zones.

7. Personal hygiene and decontamination procedures. Facilities adequate to provide personal hygiene and sanitation for onsite workers must be provided. Decontamination protocols for personnel, vehicles, and equipment should be outlined. The location of decontamination stations should be delineated on the site map.

8. Emergency equipment and medical emergency procedures. The type of emergency equipment should be listed; this equipment should include emergency eyewashes/showers, first-aid supplies including oxygen, fire extinguishers, emergency-use respirators, and spill control equipment and sorbents. First-aid/medical-emergency equipment should be approved by a physician. This portion of the plan may also include provisions for physiological monitoring procedures (heat/cold stress monitoring) and protocols for altering work and rest schedules based on temperature, levels of protection, and field activity.

9. Emergency response plan and contingency procedures. Preplanning and agency contacts should be listed in the event of chemical overexposure, personal injury, fire or explosion, spills or releases, or detection of radioactivity. Instructions for such scenarios should be prepared for posting with a list of all local, municipal, state, and federal emergency contacts. Criteria and procedures for onsite evacuation and initiation of a community alert should be specified. Preplanning with local agencies will ameliorate local fears regarding the potential onsite hazards for emergency responders, as well as potential health hazards from contaminated equipment and personnel. Also required are decontamination procedures for injured personnel and a route map to the nearest medical facility. The identity, roles, chain of command, and methods of communication between all key personnel must be indicated.

10. Logs, reports, and recordkeeping. Examples of all forms, such as air-monitoring data logs, training logs, safety inspection logs, weekly/monthly reports, accident reports, incident reports, employee/visitor

register, and medical certification reports should be included. All exposure and medical monitoring records must be maintained according to OSHA 29 CFR 1910.120. Individuals responsible and recordkeeping methods should be outlined.

The site safety plan should be issued to, reviewed by, and signed by all onsite workers. A statement attached to the plan that the undersigned ". . . has read, been briefed, and will comply with all provisions of the plan . . ." will aid in minimizing potential problems and misunderstandings; the liability of individuals responsible for health and safety is also lessened.

A "tailgate" safety meeting should be held daily to discuss the day's activities; problems encountered the previous day and changes to the formal site safety plan should be discussed. It is advisable to keep written documentation of the topics discussed and the workers present at each tailgate briefing.

HAZARD IDENTIFICATION AND CLASSIFICATION

Electrical Hazards

Electrical hazards include electrical wires, buried cables, and generators; all pose a danger of shock or electrocution if contacted or severed during site operations. Urban or suburban locations often require drilling adjacent to power lines or buried cables. Electrical shock is an overlooked hazard when installing or testing electrical pumps. Capacitors that retain a charge are a common source of electric shock. Lightning is also an electrical hazard, especially when working around metal equipment. The presence of water on the site compounds these hazards.

Electrocution is the most frequent cause of job-related mortality for drillers and helpers (NWWA, 1980). Contact with overhead wires or drilling into buried cables constitute the majority of cases; line arcing has been implicated in a few cases.

To minimize electrical hazards, low-voltage equipment with ground-fault interrupters and water-tight corrosion-resistant connecting cables should always be used. A minimum of 20 feet should be maintained between drilling equipment and overhead wires. This distance may be increased based upon local utility requirements or state and local regulations; distance requirements are often based on line voltage.

Local utilities should be contacted for information regarding buried cables. In many states, an underground utility protective organization, such as One Call or Dig Safe will notify local utilities about proposed digging activities. When working on private property, however, it is the responsibility of the owner to map out specific locations. The correct location of utilities must be determined prior to drilling or excavating; personnel representing the owner of the utility should mark locations and advise personnel on specific safety precautions; the latter information is usually not volunteered and it may be necessary to specifically request safety recommendations. Remote sensing or controlled pit excavations may have to be undertaken at sites where information is limited. It should be remembered that more than one underground line has been found "where it did not belong."

Physical Hazards

Physical hazards are also referred to as general safety hazards. The site itself can be a hazard with unstable slopes, uneven terrain, holes and ditches, steep grades, and slippery, mud-covered surfaces. Sharp debris, such as broken glass and jagged metal, may litter the site. The very act of drilling increases the slip/trip hazard by creating wet working surfaces. Because drilling and excavating operations do not afford clearance on all sides, avenues of egress may be limited; extra care must be taken to avoid being struck by or caught between equipment.

Wearing protective equipment further increases the risk of physical and/or mechanical harm by decreasing hearing, vision, and agility. Constant vigilance is required to avoid injury produced by drilling tools, support equipment, and vehicles.

Good housekeeping around the site under investigation prevents accidents. Likewise, good maintenance of equipment and proper use of hand tools can minimize the potential for personal injury and equipment loss. Personnel should ensure that all emergency shut-offs on heavy equipment are working properly. Care must be exercised around wire line hoists and hoisting hardware, catheads, and rope hoists, as well as moving augers and rotary drill tools. Too many operators, helpers, and inspectors have lost digits, limbs, and lives because they were pulled into moving machinery or struck by objects. The National Drilling Federation's Drilling Safety Guide (NDF, 1986) provides an excellent checklist of safety precautions for working around drilling equipment.

Noise

Noise is a hazard that is usually overlooked. Noise can produce potential hazards because it interferes with normal communication between workers. It may also startle or distract. Noise can also produce physical damage to the ear that may cause pain and temporary or permanent hearing loss. The effect of noise on hearing depends on the amount and characteristics of the noise as well as the duration of exposure (NSC, 1988).

There are three general classes of noise, all of which are found around groundwater monitoring site operations: continuous noise, intermittent noise, and impact-type noise. Continuous noise is noise heard when the drill rig or excavation equipment is running; intermittent noise can be heard over continuous noise, such as when the compressor or pumping equipment is in use; and impact-type noise is produced by hammers or driving tools.

Sounds vary in intensity and are measured in decibels. As the decibel (dB) level rises, the sound increases more rapidly than is perceived. A sound of 90 dB is twice as loud as an 80 dB sound. Low intensity sounds such as quiet conversation measure about 40–50 dB and are quite pleasant; city noise is 60–65 dB; heavy equipment, 85–90 dB; a jackhammer produces 100–120 dB. The 120–130 dB sound from a heavy metal rock group can produce discomfort and temporary hearing loss, while a single exposure to a rifle blast can cause pain and permanent hearing loss.

Noise is a pervasive and insidious cause of hearing loss. In most instances, there is no pain. Prolonged exposure to loud noise (85–90 dB) from heavy equipment can produce hearing loss characterized by an inability to hear sounds, as well as difficulty in understanding or distinguishing various sounds. While loudness depends primarily on sound pressure, it is also affected by frequency, which is perceived as pitch. Most sounds contain a mixture of frequencies; sounds that are composed primarily of high frequency noise are generally more annoying than low frequency sounds. High frequency noise also has a greater potential for causing hearing loss.

Hearing loss can be reduced by using hearing protectors, which act as barriers to reduce sound entering the ear. Protectors include disposable or reusable plugs and ear muffs. Manufacturers supply Noise Reduction Ratings (NRR), based on a system which indicates how much noise reduction is attained with each type of protector. Skeptics should note that with hearing protectors, it is easier to hear co-workers over background noise.

OSHA has established guidelines to prevent occupational hearing loss. Whenever employee noise exposures equal or exceed an average of 85 dB per 8-hour day, employers must implement a hearing conservation program. This program is described in OSHA regulation 29 CFR Part 1910.95.

Temperature Stress

Temperature stress includes heat stress as well as cold injury. Heat stress is caused by overheating of the body and loss of fluids through sweating. Left unrecognized, heat stress may progress to heat stroke, which is a life-threatening condition. Heat-related problems usually occur in individuals who are unaccustomed to heavy workloads and heat or who are in poor physical condition. Obesity, alcohol or drug use, age, and the presence of other complicating factors, such as acute and chronic diseases, also affect the way an individaul responds to hot working conditions.

Heat stress is a major hazard for workers wearing chemical protective clothing in hot environments. Protective clothing limits the dissipation of body heat and prevents evaporation of moisture. Heat rash or "prickly heat" may result from continuous exposure to hot, humid air conditions found inside protective clothing. Reduced work tolerance and increased risk of heat stress is directly related to the ambient temperature and the amount and type of chemical protective clothing worn. The potential to increase heat stress should be assessed when selecting protective clothing. The frequency of rest periods should be based on anticipated workload, ambient temperatures, worker physical fitness, and protective clothing selected (NIOSH, 1985a).

Cold injuries such as frostbite and hypothermia, as well as impaired ability to perform work, are hazards at low temperatures and when there is a significant wind-chill factor. Cold injuries are increased under damp or wet conditions.

Hypothermia may occur in workers wearing protective clothing, especially after episodes of heavy work. Protective clothing offers no insulation and does not retain body heat. Tanks containing breathable air should not be stored outside during cold weather. Breathing cold air can rapidly lead to hypothermia and can also cause lung damage. Symptoms of hypothermia include shivering, followed by numbness, drowsiness, and progressive loss of coordination.

Frostbite occurs when there is local cooling of the body—commonly affected are the ears, nose, hands, and feet. Frostnip is the incipient stage of frostbite, and is characterized by numbness. The affected area, which remains soft to the touch, will initially redden and then become waxy-white. Frostbite occurs when the skin freezes and ice crystals form. The skin becomes hard to the touch and turns mottled white or gray.

Radiation Hazards

Radiation hazards can include radioactive materials from industry, laboratories, and hospital wastes. Three types of harmful radiation can be emitted: alpha particles, beta particles, and gamma waves. Alpha radiation has a very limited capacity for penetration and is usually stopped by clothing and the outer layers of skin. Beta radiation has a greater potential for penetration than alpha particles, and can cause burns to the skin and damage to tissue below the skin. While alpha and beta particles pose only a mild to moderate threat outside the body, they can produce significant damage if materials emitting these particles are inhaled or ingested. Use of protective clothing and respirators, in concert with good personal hygiene and decontamination procedures, offers good protection against alpha and beta particles.

Gamma radiation easily passes through clothing and human tissue and can cause serious damage to the body. Protective clothing affords no protection against gamma radiation. Use of protective clothing and respirators, however, will prevent radiation-emitting materials from entering the body. In the presence of gamma radiation greater than 2 mrem/hr (milliroentgen-equivalent-man per hour) above background, all site activities should cease (NIOSH, 1985a). The EPA has established a "backoff" limit of 1 mrem/hr (Linson, 1988).

Chemical Hazards

Chemical hazards may be encountered during ground-water monitoring investigations. These chemical hazards may include toxic, flammable, explosive, reactive, or corrosive materials. A material may have more than one hazard, such as being flammable and toxic. It must be remembered that not all chemical hazards are found in abandoned drums or buried in the ground. Many chemical hazards are brought onto the site; these include gasoline, diesel, and kerosene fuels, hypochlorites or concentrated bleaching solutions used to kill pathogens in water wells, muriatic acid used in well maintenance work, solutions for decontamination of equipment, explosives used in downhole fracturing and some remote sensing techniques, and compressed gases, such as acetylene, used in cutting and welding.

Flammable liquids and explosive gases or vapors can be encountered at landfills which have received large quantities of organic materials, at hydrocarbon refining or storage areas, and at chemical disposal sites. Of particular concern are methane, a simple asphyxiant which is also explosive, and toxic and explosive hydrogen sulfide gases. Care must be taken especially when drilling through the unsaturated zone; explosive gases may be trapped in areas with natural or artificial deposits of low permeability.

Site history is a valuable aid in determining the type(s) of chemical hazard(s) that may be encountered. It is important to know and understand physical/chemical properties of the chemical hazards expected, including specific gravity, solubility, boiling point, vapor pressure and density, flash point, and flammable or explosive limits.

Flash point is the minimum temperature which produces sufficient flammable vapor to ignite a substance, given a source of ignition. Field sources of ignition include any open flame, exhaust systems, cigarettes, electrical equipment, and static electricity. *Flammable* or *explosive* limits are the percent concentrations in air that will combust given a source of ignition. Gases with narrow limits, such as methane (5% to 15% in air), are lesser hazards than those with wide limits, such as acetylene (3% to 80% in air).

Specific gravity is the weight of a liquid substance relative to water. Materials with a specific gravity greater than water will sink; those with a specific gravity less than water will float. Specific gravity alone is not sufficient for estimating where a chemical will be found in an aqueous environment. *Water solubility* is important in determining how much of the material will mix with water. Although a chemical substance may have a low solubility, high concentrations may be recorded from field samples due to sorption of the material onto fine-grained particles or suspension of the material in water. The presence of other chemical species may enhance water solubility.

Compounds whose *vapor pressures* are greater than one-half of atmospheric pressure (760 mm Hg) at room temperature have the potential for releasing significant vapor concentrations. *Boiling point* is the temperature at which the vapor pressure of a liquid is equal to atmospheric pressure. Chemicals with low boiling points evaporate very quickly.

Vapor density is important in estimating where vapors will occur. Vapors with densities greater than air will collect close to the ground and in low areas, such as trenches. Heavier-than-air gases can be brought out of a drill hole. A pocket of gas under pressure may be suddenly released and forced out of the hole. Drilling fluids can displace gas, and the removal of drilling tools can create a vacuum which literally sucks gas out of the hole. When working with nested monitoring wells, many field personnel have observed that wind across the annulus of one hole can produce a venturi effect and force gas out of an adjacent hole by eduction.

Biological Hazards

Biological agents are living organisms or their products that can cause illness or death to the individual exposed. Biological hazards include hospital, medical office, and laboratory material that may contain infectious wastes. This material may contain microorganisms that cause hepatitis, acquired immune deficiency syndrome, influenza, and other viral and bacterial diseases. Special care should be exercised around landfills in which biological waste from hospitals and medical laboratories may be deposited. Old (pre-RCRA, before 1980) municipal landfills are especially suspect. Fungal spores are often found in landfills. One variety of spore produces histoplasmosis, a respiratory disease which is usually self-limiting, but may in some cases produce severe symptoms and even death.

Many disease-producing microorganisms require a host, or carrier, which transmits the organisms into humans. These carriers include insects or rodents, both common inhabitants of waste-disposal facilities. Diseases that are transmitted by carriers include plague, which is found primarily in New Mexico and is transmitted by rodent-borne fleas; Rocky Mountain Spotted Fever (typhus fever) which is carried by ticks; Lyme Disease, which occurs in the New England and Mid-Atlantic states and is transmitted by the deer tick; and a variety of encephalopathies (inflammation of the brain) transmitted by mosquitoes.

Plants that elicit allergic skin reactions in sensitive individuals, such as poison ivy, oak, and sumac, are biological hazards. Even when not transmitting disease, or producing an allergic response, insects and other invertebrates which produce painful stings or bites should be considered hazardous; bees, wasps, fire ants, and biting flies fall into this category. Biting flies can produce dangerous field situations because they are so distracting.

Many wild animals are attracted to field sites, especially sites that are located in unpopulated areas. Bears, wolves, and wild dogs will investigate equipment left for the night, and may still be present when workers arrive the next morning. Vibrations and water discharge associated with drilling and excavation activities can disturb snakes; snakes are often found in buckets, mud tubs, and other containers left empty overnight.

Common sense can mitigate most biological hazards. Protective clothing should always be worn at landfill locations where waste is exposed. This clothing should include gloves, safety shoes, goggles, coveralls, and a dust mask that covers the nose and mouth when blowing dust is noticed. Insect and invertebrate hazards can be lessened by using repellants; care should be taken to avoid air, soil or water sample contamination with the repellant. Nests can be avoided or removed. If a bee or wasp nest is encountered, a carbon dioxide fire extinguisher can be used to incapacitate the residents until the nest is removed. Taping pant legs and sleeves shut lessens ant and tick bite hazards. Plants which provoke allergic reactions should be identified and removed or avoided.

Toxic Hazards

Exposure to toxic materials may occur through inhalation, skin and eye contact, ingestion, or injection. Inhalation is the most common route of exposure. The water solubility of a vapor or gas is important in determining how much of the inhaled material actually reaches the lungs. Highly soluble gases such as ammonia dissolve readily in the upper respiratory tract, while less soluble gases such as phosgene and nitrogen dioxide readily reach the lung. Chemicals rapidly enter the bloodstream after being inhaled into the lungs.

Many chemicals can be absorbed through the skin. The skin and its film of lipid and sweat often act as an effective barrier; absorption is faster through skin that has been damaged by lacerations or abrasions, inflammation, or sunburn. Organic solvents, such as acetone and toluene, remove lipids from the skin; solvents enhance the permeability of the skin and facilitate skin absorption of other materials. Chemicals can also be absorbed through the eye and enter the bloodstream; the eye can also be easily damaged by chemicals.

Ingesting or swallowing a chemical is an unlikely route of exposure. Chemicals can be ingested, however, if they are left on hands or clothing, or if consumption of food or drink is allowed at the worksite. On dusty sites, ingestion of contaminants adsorbed onto particulates is possible if dust filters are not utilized.

Injection exposure is the least common route of exposure. If an open wound is exposed to a chemical, however, direct contact with the blood is possible.

Factors that influence toxicity of chemicals include the amount and duration of exposure, the route of exposure, and the susceptibility of the individual. Individual susceptibility is determined by many factors including age, sex, diet, inherited traits, overall physical health, and use of alcohol, tobacco products, medications, and drugs.

Acute toxicity of a chemical refers to its ability to produce adverse health effects as a result of a one-time exposure of short duration; such exposure often produces an emergency situation. *Chronic toxicity* is the ability of a chemical to produce systemic damage as a result of repeated exposure. Some chronic effects have a very long latency interval and develop gradually, making it difficult to establish a cause-and-effect relationship.

Exposure to multiple chemicals may result in health effects different from the effect of each alone. *Synergism* is a process in which two or more chemicals produce a toxic effect greater than the sum of their individual effects. Chlorinated solvents and alcohol can each cause liver damage; exposure to chlorinated solvents while drinking large amounts of alcohol can result in excessive liver damage. *Potentiation* occurs when the toxic effect of one chemical is increased by exposure to another chemical, although the second chemical by itself does not cause the effect. Acetone by itself does not damage the liver, but it increases the damage produced by chlorinated solvents.

Systemic toxins are substances that produce damage to specific organs. Many chemicals produce multiple organ effects. Chlorinated solvents, for instance, affect the central and peripheral nervous system, liver, kidney, and heart. *Chemical asphyxiants* are substances which interfere with the transport or use of oxygen by tissues. Carbon monoxide prevents the uptake of oxygen by red blood cells; cyanides poison cellular enzyme systems and prevent tissues from utilizing oxygen.

Irritants produce pain, swelling, and inflammation of exposed tissues. Skin, eyes, lungs, and membranes can be affected by irritants. Severe eye irritants such as acids and alkalis can cause corneal damage and may impair vision. Pulmonary irritants can produce excessive fluid buildup in the lungs, or pulmonary edema. Severe pulmonary edema, which inhibits oxygen exchange, can be life-threatening. Severe pulmonary irritants include chlorine, fluorine, paraquat, sulfur dioxide, and sulfuric acid.

Sensitizers are substances which produce allergic reactions in sensitive individuals. Skin reactions include rashes or blister formation. Itching frequently accompanies allergic skin reactions. Eye irritation is manifested by watery, itching, and reddened eyes. Respiratory sensitizers produce asthma-like reactions or hay fever symptoms. Common sensitizers are organic amine compounds, epoxy resins, and isocyanates.

Carcinogens are chemicals that cause cancer. *Mutagens* are chemicals that cause genetic change by damaging genes or chromosomes. This type of change is called a mutation. Mutuations affect the way cells function and reproduce. Some kinds of mutations can result in cancer; many mutagens are also carcinogens. *Teratogens* are chemicals which cause birth defects by damaging the fetus while it develops in the mother's womb. Some chemicals produce lethal damage and cause the fetus to be aborted. Commonly encountered products such as alcohol, aspirin, and vitamin A can act as teratogens when taken in large quantities.

All unknown chemicals should be considered hazardous until proven otherwise. Care should be taken to prevent contamination and unwarranted exposure. Because even innocuous materials can cause hypersensitivity reactions when in the powdered form, all unknown powders and dusts should be considered hazardous and appropriate protective measures taken for skin, lungs, and eyes. Vapors and many powdered substances can react with perspiration to produce localized skin irritation, and in some cases, severe chemical burns.

Exposure Limits

Exposure limits are used to control employee inhalation exposure to specific chemical substances in the workplace. There are several organizations that recommend exposure levels, including the American Conference of Governmental and Industrial Hygienists (ACGIH), and the National Institute for Occupational Safety and Health (NIOSH), as well as industrial groups. Employers are required by law (29 CFR 1910.1000) to comply with the exposure limits defined by OSHA.

Permissible exposure limits (PELs) were established by OSHA in 1971. PELs are defined for approximately 400 substances; PELs can be changed only by amending the original law by act of Congress. A PEL, as defined in 29 CFR 1910.1000, is the average concentration of a substance to which a typical worker may be exposed to during an 8-hour day of a 40-hour work week.

A *short term exposure limit* (STEL) is defined as the concentration of a substance that a worker may be exposed to for a short interval without experiencing irritation, long-term effects, or acute effects that could interfere with self rescue. Most STEL exposures do not exceed 15 minutes and should not be repeated more than four times during a work day.

The ACGIH recommends *threshold limit values* or TLVs. The most popular TLV is the 8-hour time-weighted average (TWA) value, which is the average concentration that a worker can be exposed to during a typical 8-hour day without suffering any adverse health effects. TLVs are reviewed, updated, and published by ACGIH annually (ACGIH, 1987a).

NIOSH is responsible for assisting OSHA in developing new exposure standards. NIOSH publishes criteria documents that discuss health hazards associated with specific substances or work practices and recommends new, usually lower, exposure levels. These are called *recommended exposure limits* or RELs.

Concentration levels above which a substance is considered to be *immediately dangerous to life and health* (IDLH) are also defined by NIOSH. The IDLH level represents a concentration at which exposure can produce severe health effects. The NIOSH Pocket Guide to Chemical Hazards (NIOSH, 1985b) contains IDLH and REL values.

Confined Space Hazards

A confined space is an environment that has restricted means of access and egress, and is not designed for continuous worker occupancy. In many confined spaces there is limited ventilation and therefore the potential for oxygen deficiency as well as accumulation of toxic or combustible gases. At ground-water monitoring sites, confined spaces include pits and excavations, pipe galleries, tanks, and even the toe of a steep slope if there is a close fence or other obstacle.

Some aspects should be considered prior to entering a confined space. The most important aspect is the actual need to enter the area. Air-monitoring instruments should be readily available; these instruments should be calibrated and appropriate to detect oxygen deficiency, and combustible and toxic gases. Ventilation equipment, protective clothing, or respiratory protection equipment may be necessary and should be on hand for use by trained workers. Rescue or standby personnel should also be ready in case they are needed. It should be noted that over one-half of all workers who die in confined space incidents are attempting to rescue other workers (NIOSH, 1987b).

Risk Versus Hazard

Most sites will pose multiple hazards to workers. The *degree of hazard* refers to the inherent characteristic(s) of a substance that defines it as being hazardous, i.e., flammable, toxic, reactive, radioactive, carcinogenic, corrosive, and so on. The degree of hazard is a function of the specific hazardous properties of the material and most importantly, its concentration. For instance, dioxin (2,3,7,8-TCDD) and saccharin are both laboratory animal carcinogens, but dioxin produces its effects at concentrations that are a million times less than those required by saccharin.

The potential harm that may be exerted upon an exposed worker can be considered the *degree of risk*. It is a function not only of the hazards present but also the likelihood that a worker will encounter these hazards. Risk can be minimized by decreasing contact with the hazard, i.e., using personal protective equipment to protect workers from chemical and biological hazards. Risk can also be reduced by practicing contamination avoidance and good housekeeping, and eliminating working conditions that may produce physical or mechanical hazards.

For instance, installation of monitoring wells adjacent to leaking underground gasoline storage tanks can be hazardous. Other than the hazards associated with drilling itself, gasoline presents additional flammability and toxic hazards. The hazard is dependent on the concentration of gasoline vapors in air, potential sources of ignition present, and the duration of worker exposure. A concentration of approximately 10,000 to 80,000 ppm gasoline vapor in air is flammable in the presence of a source of ignition and sufficient oxygen to support combustion. Exposure of unprotected workers to concentrations of several thousand ppm will produce symptoms of toxicity in a short period of time. The hazard can be identified by the use of real-time air monitoring equipment. The risk can be minimized by purging or venting the borehole, identifying and eliminating potential sources of ignition, and using appropriate personal protective equipment.

SOURCES OF INFORMATION

Many reference texts and online computer systems are available to assist site safety officers and others in assessing potential site hazards and to determine proper control of associated risks. It must be noted that although most reference texts have target audiences (i.e., chemists, toxicologists, firefighters, emergency response personnel, etc.) these texts contain useful information that can be utilized by other disciplines as well.

All texts and computer databases contain errors; some errors are typographical, other errors are derived from incorrect "original" information. For this reason, no less than two sources of information should be consulted, and the latest editions should be utilized, especially when researching industrial hygiene standards or toxicity information. The employment of multiple sources often results in discrepancies between texts regarding chemical/physical parameters and acute toxicity information. In these cases, it is prudent to accept the most conservative number.

Although more than 50 reference texts are available, the following enjoy wide use:

- Chemical Hazard Response Information Systems (CHRIS Manual), U.S. Coast Guard, Vol. 2, Hazardous Substance Data Manual (1985). U.S. Government Printing Office (GPO), Washington, DC.
- Condensed Chemical Dictionary, 11th edition (1987). Gessner G. Hawley, Van Nostrand Reinhold Co., New York, NY.
- Dangerous Properties of Industrial Materials, 7th edition, (1988). Edited by N. Irving Sax, Van Nostrand Reinhold Co., New York, NY.
- Documentation of Threshold Limit Values and Biological Indices, 5th edition, (1986). ACGIH Publications Office, Cincinnati, OH.
- Emergency Action Guides (1989). Association of American Railroads, Washington, DC.
- Fire Protection Guide on Hazardous Materials, 9th edition (1986). National Fire Protection Association, Quincy, MA.
- Guidelines for the Selection of Chemical Protective Clothing (1983). ACGIH Publications Office, Cincinnati, OH.
- Hazardous Material Injuries (1982). Bradford Communications Corporation, Greenbelt, MD.
- The Merck Index, 10th edition (1983). Merck and Co., Inc., Rahway, NJ.
- NIOSH/OSHA Occupational Health Guidelines for Chemical Hazards (1981). NIOSH Publication No. 81-123, GPO, Washington, DC.
- NIOSH Pocket Guide to Chemical Hazards (1985). NIOSH Publication No. 85-114, GPO, Washington, DC.
- Registry of Toxic Effects of Chemical Substances (RTECS, updated periodically). GPO, Washington, DC.
- Threshold Limit Values and Biological Exposure Indices for 1989–1990. (This publication is updated yearly). American Conference of Governmental Industrial Hygienists (ACGIH) Publications Office, Cincinnati, OH.

There are numerous online interactive computer databases that offer information regarding toxicity, chemical/physical properties, and regulatory information on hazardous chemicals. Four of the most useful include:

- *Hazardline,* available through Occupation Health Services, Inc., New York, NY.
- *Toxnet,* available through the National Library of Medicine Toxicology Information Program, Bethesda, MD.
- *CAS Online,* available from Chemical Abstract Services, Inc., Columbus, OH.
- *OHM-TADS,* available from Chemical Information Systems, Inc., Baltimore, MD.

RESPIRATORY PROTECTION

Respiratory hazards may be particulate or gaseous in nature. Inhalation is a major route of exposure for toxic chemicals, biological hazards, and alpha and beta radiation. Respirators provide protection from hazardous contaminants which may be inhaled. Determining the type of respiratory protection is of primary importance when selecting personal protection equipment.

Many workers who have been given respirators to wear (in many cases without fit testing or training) find them uncomfortable and cumbersome. Respirators interfere with smoking and tobacco or gum chewing. If contaminants have been identified or are suspected and a respirator is deemed necessary, it should be worn. Not wearing a respirator under conditions in which one is recommended is short-sighted and stupid. Exposures to high concentrations of toxic substances even for a short period of time can cause serious injury or death. There may be no warning signs or symptoms. Exposures to low concentrations can cause damage to lungs or other internal organs. Exposure to some contaminants may impair vision, affect balance, and produce symptoms of intoxication that could endanger the affected worker as well as co-workers.

Respirators, or respiratory protection devices, are of two basic types: air-purifying and air-supplying. Respirators consist of a facepiece and either an air-purifying device or a source of breathable air.

Air-Purifying Respirators

Air-purifying respirators or APRs selectively remove contaminants from the air by filtration, absorption, adsorption, or chemical reaction. The air-purifying device is typically a particulate filter, or a cartridge or canister containing sorbents for specific gaseous contaminants, or a combination of filter and cartridge/canister. Cartridges are usually attached to the facepiece directly; canisters are attached to the chin of the facepiece or are attached by a breathing hose. APRs usually operate in the negative pressure mode; there are power-assisted APRs, which maintain a positive facepiece pressure during normal breathing conditions.

APRs remove contaminants by passing air through a mechanical filter for particulates, or a cartridge or canister for gases and vapors. These devices are specific for certain types of contaminants, therefore the identity of the hazardous material must be known. The efficiency of the respirator against the contaminant(s) must also be known. Each mask and cartridge or canister is designed for protection against certain contaminant concentrations. Just because a cartridge says it is for use against organic vapors does not mean that it is good for *all* organic vapors. This information is usually available from the manufacturer. Only NIOSH or MSHA-approved equipment should be used. NIOSH periodically publishes a list of all approved respirators and respirator components (NIOSH, 1986).

Specific information should also be available regarding the known or suspected contaminants, including the following:

1. the physical, chemical, and toxicologic properties of each contaminant
2. warning properties, including odor, taste, eye, or respiratory irritation potential
3. the OSHA permissible exposure limit (PEL), NIOSH recommended exposure limit (REL), ACGIH threshold limit value (TLV), or other applicable exposure limits
4. the immediately dangerous to life and health (IDLH) concentration or the lower explosive limit (LEL) for flammable materials

Cartridges or canisters are used against gases or vapors only if the contaminant in question has "adequate warning properties." Warning properties are considered adequate when odor, taste, or irritant effects are noted and persist at concentrations below the OSHA PEL, NIOSH REL, or ACGIH TLVs. These warning properties are essential to the safe use of APRs, because they alert the user to sorbent exhaustion which allows contaminant breakthrough, poor facepiece fit, or other respirator malfunction. Individuals vary in their ability to detect warning properties; it is prudent to verify that respirator users can indeed detect the warning properties of the contaminant in question.

APRs should not be used against identified contaminants that have poor warning properties unless either the respirator is equipped with an approved end-of-service-life indicator (ESLI) or the service life of the sorbent is known.

APRs may not be used in oxygen-deficient atmospheres or in confined spaces. Oxygen deficiency is defined as a concentration of oxygen in the air that is less than 19.5%; normal oxygen concentration in ambient air is 20.9%. APRs are not permitted in any atmosphere with IDLH or LEL concentrations or when the concentration of the contaminant is unknown or exceeds the maximum use concentration stipulated by the manufacturer. NIOSH has published a decision logic for selecting suitable classes of respirators for specific contaminants (NIOSH, 1987c).

Finally, the use of APRs is prohibited when conditions prevent a good facepiece fit. These conditions include beards, large mustaches, long sideburns, scars, and eyeglass temple bars. Wearing of contact lenses has been debated; some agencies allow their use while others do not. Because maintaining a leak-free seal is important to the health and safety of the user, all personnel who wear respirators are required by OSHA to pass a fit-test designed to verify the integrity of the seal.

Cartridges and canisters containing chemical sorbents should not be removed from protective packaging until needed. Once opened, they should be used immediately; efficiency and service life decreases because sorbents begin to absorb humidity and air contaminants even when not in use. Cartridges should be changed regularly to prevent sorbent exhaustion and contaminant breakthrough.

The rule of D's is recommended when changing cartridges. That is, cartridges should be changed *daily,* or more often, if any of the following conditions exist: (1) warning properties are *detected* by the user, (2) it becomes *difficult* to breath, (3) the cartridges are *dirty* or appear *damaged.* Used cartridges should be promptly *discarded.*

Atmosphere-Supplying Respirators

Atmosphere-supplying respirators, or ASRs, supply breathable grade air, not oxygen, to the facepiece via a supply line from a stationary source or from a source carried by the wearer. When air is supplied from a stationary source through a long air line or hose it is called a *supplied-air respirator* (SAR); when the air source is portable and carried by the wearer it is called a *self-contained breathing apparatus* (SCBA).

Atmosphere-supplying respirators, operated in the positive-pressure/pressure-demand mode are recommended for entry into oxygen-deficient IDLH and LEL atmospheres, for known contaminants for which no suitable cartridge exists, and for entry into atmosphere containing unknown concentrations of contaminants.

SCBAs allow workers unhindered access to nearly all areas of the worksite; in confined spaces, however, worker mobility may be impaired. SCBAs are frequently utilized during initial site surveys, during site characterization, for emergency rescue, or for specific site activities that require mobility, such as sampling or working around heavy equipment.

Operating times for SCBAs are either 30 or 60 minutes; these times vary depending on the size and pressure of the air cylinder, the type of work performed, and the fitness of the wearer. A warning alarm sounds when 20% to 25% of the air supply remains.

Some of the disadvantages of using SCBAs include the short operating time and the bulk and weight of the air cylinder. The air tank plus backpack, harness, and regulator may weigh up to 35 lbs.

An alternative to the heavy SCBA is the supplied-air or air-line respirator. Air lines allow longer work intervals than SCBAs, are not heavy, and are less bulky. Workers are tethered to a compressor or air cylinders by a long hose, which can decrease mobility. Workers must be careful not to entangle themselves or equipment on the air line, and frequently must retrace their steps when leaving the area. The air line is vulnerable to punctures, kinking, chemical degradation, and damage from equipment, vehicles, and site debris. All potential on-site hazards to the air line should be removed prior to beginning work. As an alternative, air lines can be placed off the ground or encased in protective sleeves. The length of the line should not exceed 300 feet; experienced workers find that a shorter line is easier to manage.

Air sources for supplied-air respirators may be stationary compressed-air cylinders or a compressor that purifies and delivers breathable grade air to the facepiece. The grade of air supplied from a compressor or a cylinder should be grade D or better. Lesser grades (A–C) of air may contain unacceptably high concentrations of hydrocarbons, carbon dioxide, carbon monoxide, and nitrogen oxides.

Users of supplied-air respirators should also be equipped with an escape pack or egress device. The pack supplies the wearer with 5–15 minutes of breathable air and is designed to allow for emergency exit from the site should the air-line system fail. If exit from the site requires more time than that allowed by the escape pack, SCBA should be utilized.

A combination supplied-air respirator and SCBA combines the features of both. This type of respirator allows entry into and exit from the site using the SCBA, as well as an extended work interval within the contaminated area while attached to the air line. This type of respirator is particularly useful when workers must travel well into the site before reaching the work area, and remain there for a prolonged period of time to perform tasks that do not require excessive mobility. The combination system differs from the escape pack in that it is designed for entry and egress, and provides up to 60 minutes of air.

Respiratory Protection Program

Any employer who provides respiratory protection equipment to his workers is required by law (29 CFR 1910.134) to establish a respiratory protection program. A minimally acceptable program must include the following:

1. written procedures describing the selection and use of respirators
2. training and instruction in the limitations of respirators, and their proper care and use; all such training should be documented
3. selection of respirators based on hazards to which workers are exposed
4. regular cleaning, disinfection, and inspection of respirators after each use or at regular intervals when not in use
5. surveillance of work site conditions and extent of worker exposure to ensure proper selection of respirators
6. regular inspection and evaluation of the program in order to assess its continued effectiveness
7. no individual should be assigned tasks involving respirator use until their medical fitness to use respiratory protection has been evaluated by a physician; a written statement by a physician should stipulate the health and physical conditions considered pertinent to each worker's use or nonuse of respiratory protection equipment

All aspects of the respiratory protection program should be documented in writing and filed for future reference. A record of each employee's medical exam, which should include a pulmonary function test and the physician's fitness statement, should be similarly filed. All fit-test records should be documented and saved. Fit testing and medical examinations should be repeated at least annually.

AIR MONITORING

An essential component of the health and safety program is air monitoring. Air monitoring is an important aspect of the site characterization process which must be performed prior to site entry and then on a regular basis after other activities have been initiated. Site characterization is required by the OSHA rule covering hazardous materials site workers. This information is used to assess the hazards and associated risks to site workers as well as offsite receptors. Identification and quantification of air contaminants is required in order to select appropriate personnel protective equipment and define areas where protective equipment is required. Air monitoring may also be helpful in determining the effectiveness of mitigative activities.

Various monitoring devices can be employed around drilling operations including fixed or portable survey instruments and dosimeters. The devices include instruments for measuring oxygen deficiency, combustible or explosive atmospheres, toxic substances, and radiation. It is very important that workers operating monitoring equipment be thoroughly trained in the use, limitations, and operating characteristics of each piece of equipment.

Instruments selected for use in the field must be capable of generating reliable and useful information. They should be capable of selectively detecting the contaminants of interest, and sensitive at a useful concentration range. Instruments should have a good battery life; they should also be portable, weather resistant, and easy to operate, calibrate, and maintain in the field. It is recommended that intrinsically safe instruments be used when available. Costs of instruments vary greatly, depending on the function of the instrument, its sophistication, desired accessories, and calibration equipment.

There are three different methods used during air monitoring. The method used is dependent upon the type of equipment, the number and training of personnel, and the degree of hazard known or anticipated onsite. Intermittent monitoring involves readings taken when targets of opportunity present themselves or when there is a change in field conditions. Semicontinuous monitoring is utilized when readings are required on a regular basis, i.e., each time drilling tools are removed from the borehole. Continuous monitoring constantly assesses site conditions, i.e., during drilling or excavation activities.

Direct-reading instruments provide information at the time of sampling. Many such instruments can detect contaminants at concentrations as low as 1 ppm. However, quantitative data are difficult to obtain when multiple compounds are present.

Combustible gas indicators (CGIs) measure the risk of fire and explosion from flammable vapors. Readings may be in ppm, percent LEL, or percent combustible gas by volume. It is important to know which type of meter is being used and to understand what the meter readings mean. A CGI may also be used as a toxic meter for flammable materials if the substance is known and the response efficiency of the meter is adequate around a predetermined exposure limit (OSHA PEL, ACGIH TLVs, NIOSH REL, or IDLH). These instruments are internally calibrated for normal oxygen atmospheres, although most will work properly when oxygen levels are somewhat reduced (i.e., >10%). Enriched oxygen concentrations may give false high readings. The manufacturer should be contacted for specific information regarding the minimum and maximum concentrations of oxygen necessary for accurate readings. Acid gases and organic lead, sulfur, and silicon compounds can damage the sensor element.

CGIs measure the total amount of combustible vapor present and cannot differentiate between multiple compounds. Although the instrument is calibrated for a specific flammable gas (the calibrant gas), its relative response to other gases will be different. Temperature also affects response efficiency. Manufacturers supply information on the limitations and relative response efficiency of their instruments to frequently encountered flammable gases.

Oxygen deficiency meters are used to assess the air for oxygen content to determine if respiratory protection is necessary; oxygen content less than 19.5% requires the use of atmosphere-supplying respirators. Oxygen meters are also used to indicate increased oxygen conditions; such conditions may be due to the presence of chemical oxidizers. Oxygen-enriched (>25%) atmospheres increase the risk of combustion. These meters may also be used indirectly to detect the presence of other contaminants. A decrease in oxygen content which is not due to consumption (i.e., combustion or chemical reaction) is generally due to displacement by another substance that may be hazardous.

The oxygen sensor relies on atmospheric oxygen pressure; pressure decreases as elevation increases. To obtain an accurate reading, instrument calibration using clean ambient air should be performed at the altitude at which the instrument is used. Temperature can also affect the response; the normal operating range is usually between 120° and 32°F. Between 0° and 32°F the instrument can be used, but its response time is much slower. High concentrations of carbon dioxide and other acid gases shorten the life span of the sensor. Exhaling into the meter to test its function is therefore not recommended. Some manufacturers offer combination meters that detect combustible gases and oxygen deficiency.

Radiation meters detect the presence of ionizing radiation. Three types of radiation are of major concern: alpha particles, beta particles, and gamma rays. Radiation survey instruments are designed to detect one or more types of radiation. Ion detector tubes are used for measuring high levels of gamma radiation. Proportional detector tubes detect only alpha radiation. Geiger-Mueller tubes are sensitive, and used to detect low levels of beta or gamma radiation. Scintillation detectors are sensitive to low levels of alpha and gamma radiation.

Radiation instruments typically measure exposure rates in milliroentgens/hour (mR/hr), or roentgens/hour. Normal background radiation exposure rates are 0.01 to 0.02 mR/hr.

Detector tubes consist of a glass tube filled with an indicating chemical matrix that changes color in the presence of a specific contaminant or type of contaminant. The length of color change is proportional to the concentration present. The tube is connected to a bellows or piston pump, and air is drawn through the tube by the pump. A long probe or hose can be placed between the tube and pump for sampling remote locations. Tubes may be specific for one contaminant or for a class of contaminants. Some manufacturers produce a "polytube" which is designed to detect the presence of an air contaminant. Additional tubes are used following a decision matrix to aid in the identification of the contaminant.

Detector tubes are inexpensive and easy to use, and can be useful as a screening tool to determine the presence of organic and inorganic contaminants.

Accuracy of detector tubes varies between tubes and manufacturers; some manufacturers report error factors of up to 50% for some tubes. All tubes have expiration dates; shelf life can be reduced by temperature fluctuations. Refrigeration of tubes (at 40°F) will minimize this problem. Temperature, humidity, and interfering substances present at time of use can affect accuracy; manufacturers' instructions usually list limitations for each tube. The presence of interfering substances, i.e., other compounds which also produce a color reaction, often makes interpretation of results difficult. Tubes that produce poor color changes are difficult to use in the field; some tube protocols require multiple pump strokes and an inordinate amount of time to achieve one reading. It is recommended that tube protocols and efficiency be evaluated prior to use under field conditions.

Personal monitors for specific hazards may be worn by individual workers. Monitors are available for combustible gases and oxygen deficiency as well as toxic gases such as carbon monoxide, hydrogen sulfide, cyanides, phosgene, and so on. High-risk workers, i.e., those closest to the potential source of the hazard, are likely candidates for personal monitors. Personal monitors should have an audible alarm, be lightweight and easy to carry, and have good battery life. Dosimeters are also available for organic vapor screening.

Survey instruments are used to detect the presence and total concentration of organic gases or vapors in air. Contaminants are detected at the same time, and there is no identification of individual compounds. Two types of survey instruments are commonly used—the photoionization detector and the flame ionization detector.

Photoionization detectors (PIDs) are capable of monitoring many organic and some inorganic vapors and gases. A PID uses ultraviolet light to ionize gas or vapor molecules. Ions are collected and produce a current; the measured current is proportional to the number of ionized molecules present. The energy required for ionization, measured in electron volts (eV), is called the ionization potential (IP).

A variety of ultraviolet lamps are available, depending on the manufacturer. More commonly used lamps are 9.5, 10.2, 10.6, and 11.7 eV. In order to detect a specific chemical, the energy generated by the ultraviolet lamp of the PID must be equal to or greater than the IP of the chemical. Hexane, with an IP of 10.2 eV, is detected with a 10.2 eV lamp; however, the response efficiency is very poor. A better response is obtained with a higher energy lamp, such as 10.6 or 11.7 eV. Hexane would not be detected by a lamp with an energy less than 10.2 eV. PIDs cannot detect light hydrocarbons, such as methane. Chemical species that are not ionized do not interfere with readings from ionizable substances.

Dust in the air can collect on the ultraviolet lamp, interfere with light transmission, and reduce instrument readings. Most lamps are easily accessible and should be cleaned periodically. Some instruments are equipped with particulate filters; these are useful but must be cleaned or changed frequently. Contaminants may adsorb onto particulates trapped in the filter and interfere with subsequent readings.

High humidity can condense on the lamp and decrease the amount of light reaching the air sample. Humidity also reduces the ionization of chemicals and thereby decreases instrument readings. If water is drawn into the ionization chamber the lamp will short out and must be replaced.

PIDs are normally factory calibrated to one chemical; calibration kits are usually supplied by the manufacturer. The instrument's response to other chemicals varies, depending on the molecular configuration, concentration, and IP. The intensity of the lamp declines slowly with age; however, the ionization energy remains unchanged. The same effect is seen in light bulbs; the intensity of a bulb may decrease with age, but the wattage remains the same. Decreases in ultraviolet light intensity will be perceived during calibration; instrument settings should be adjusted to compensate.

Flame ionization detectors (FIDs) use a hydrogen-fed flame to ionize organic vapors and gases. When the vapors burn, positively charged carbon-containing ions are produced and collected. A current is generated which is proportional to the ions collected. Unlike PIDs, FIDs are capable of detecting virtually all compounds that contain carbon-hydrogen or carbon-carbon bonds. FIDs respond differently to different compounds; however, there is less variability in sensitivity between different substances when compared to a PID. FIDs do not detect inorganic compounds.

Most FIDs can be operated in the survey mode or, with appropriate attachments, in the gas chromatographic (GC) mode. The GC mode is capable of separating components of an air sample using a GC column packed with an inert solid. With proper standards, each constituent can be identified and quantified. The identity of the chemicals of interest must be determined in order to prepare standards. Training and experience are required to successfully operate the GC option.

FIDs are less susceptible than PIDs to high humidity; however, very high humidity conditions will reduce the relative response. A supply of ultra-pure hydrogen is required. FIDs can detect methane; methane is frequently the factory calibrant. At landfills or other field situations where light hydrocarbon gases are found, the FID is not useful for detecting toxic air contaminants. In these cases, the relative responses of both an FID and PID should be assessed in order to determine the source of the readings. The FID will detect methane, while the PID does not. One the other hand, a PID will detect inorganic contaminants while the FID will not.

Portable, programmable GCs are now available. These instruments are not designed for survey work, but they can be valuable for identifying and quantifying onsite contamination. Programmable GCs have also become popular for soil gas and headspace analyses.

Individuals should be trained in the use, maintenance, and calibration of the instruments they operate. Operators should have their own copy of the instrument manual, which should be thoroughly read and understood before attempting to use the instrument. Experience is the best teacher; it is advisable to allow the operator to take the instrument home and experiment with it. Obtaining readings in the garage or kitchen often gives a different perspective to their meaning. That is, not everything that gives a reading is hazardous. Finally, it must be remembered that: (1) no instrument has been designed to detect all possible contaminants;

and (2) a zero reading indicates a lack of instrument response, not zero contamination. In the words of Carl Sagan, ". . . the absence of evidence is not evidence of absence . . ." (Sagan, 1977).

PROTECTIVE CLOTHING

Site workers must be protected against potential hazards. Personal protective equipment (PPE) is utilized to decrease exposure to biological and chemical hazards and to shield against physical hazards. Proper selection and use of PPE should protect the respiratory system, eyes, skin, face, hands, feet, body, and hearing.

The nature of the hazard, based on physical, chemical, or biological properties, and the expected concentrations of contaminants known or anticipated to be present, determine the combination of protective clothing and equipment that will be used. Most organizations dealing with hazardous materials have adopted a four-category system according to the degree of protection afforded.

Level A is worn when the highest level of respiratory, skin, and eye protection is required. A Level A ensemble consists of a pressure-demand atmosphere-supplying respirator, fully encapsulated chemical-resistant suit, inner and outer chemical-resistant gloves, chemical-resistant safety boots (steel toe, shank, and metatarsal protection), and hard hat. Optional equipment might include a cooling system in hot weather, flashover protection, abrasive-resistant gloves, disposable oversuit and boot covers, communication equipment, and safety line. Most drillers will not don Level A protection unless engaged in such practices as installation of investigatory or recovery wells within an area contaminated with highly toxic or corrosive materials.

Level B protection is utilized in areas in which full respiratory protection is warranted, but a lower level of skin and eye protection is adequate (only a small area of head and neck is exposed). Level B consists of a pressure-demand atmosphere-supplying respirator, splash suit (one- or two-piece) or disposable chemical-resistant coveralls, inner and outer chemical-resistant gloves, chemical-resistant safety boots, and hard hat with face shield. Optional items include glove and boot covers, and inner chemical- and flash-resistant fabric coveralls. Many monitoring and recovery wells have been installed in Level B protection, particularly at abandoned hazardous waste sites or at spill sites where the concentration, lack of warning properties, or breakthrough characteristics of the contaminants precluded the use of air-purifying respirators. Level B protection is required by OSHA during characterization activities for sites with unknown hazards.

Level C permits the use of air-purifying respirators. Level B body, foot, and hand protection is normally maintained. Many organizations will permit only the use of approved full-face respirators equipped with a chin- or harness-mounted canister. However, many sites are drilled by personnel wearing half-mask cartridge-equipped respirators. If allowed by the client, the decision of which type to use becomes a trade-off of decreased protection versus increased comfort.

Level D protection consists of a standard work uniform of coveralls, gloves, safety shoes or boots, hard hat, and goggles or safety glasses. Some organizations require personnel who are outfitted in Level D or C protection to carry emergency escape masks which supply 5 to 15 minutes of air, or place these masks around the work site and next to vehicles.

Protective clothing is selected to guard against vapors, splash, flash, and physical contact with chemicals. Clothing should prevent or minimize penetration, permeation, and degradation by the contaminants encountered. Penetration is breakthrough of a chemical through seams, zippers, buttonholes, and the like. Permeation occurs when a chemical soaks into the fabric without altering its physical properties. Degradation occurs when a chemical changes the physical properties of the fabric. One reference, *Guidelines for the Selection of Chemical Protective Clothing* (ACGIH, 1987b), provides a matrix of clothing material recommendations for approximately 300 chemicals, based upon permeation and degradation data. Protective clothing manufacturers also supply chemical compatibility data for their products.

It is advisable to wear disposable clothing over reusable clothing for additional protection and to minimize decontamination procedures. Disposable suits are highly recommended for drilling work. One-piece saran-coated or polythylene-coated Tyveks offer good splash and vapor protection for low-level contamination. Duct tape may be used to seal openings around disposable boots and gloves. Taping increases heat load, however, and can contribute to heat stress. Uncoated Tyveks offer no splash or vapor resistance and should be reserved for training or for sites at which dry particles are the only problem.

Disposable gloves and boot covers may be worn over expensive or leather counterparts that cannot be easily decontaminated. Disposable boots can increase the slip/trip hazard unless they have aggressive soles. Thin boot covers with flat soles are useless in the field and should not be considered. Disposables should be large enough to fit but not so large as to interfere with a worker's ability to perform assigned tasks. Disposable items come in different sizes; the proper sizes should be supplied to all site personnel.

No one PPE ensemble, no matter what combination is used, will be capable of protecting against all hazards. To be effective, PPE must be used in concert with other protective methods as well as good field practices. PPE is not a suit or armor; one should not feel invincible when wearing PPE. Indeed, the use of PPE creates significant worker hazards, such as heat stress, loss of mobility, difficulty in communicating, and impaired vision. It is important, therefore, to select the appropriate level of PPE without over-protecting the worker, because the greater the level of PPE, the greater the associated risk. Health and safety are not mutually inclusive.

SITE OPERATIONS

When working at known or suspected hazardous substance sites, work zones should be established to protect site workers and to minimize the potential risk of injury to workers as well as authorized and unauthorized visitors and untrained support staff. Minimizing the risk of injury also limits potential liability. Site control areas or work zones are established which allow site procedures to be safely conducted while reducing the potential for contacting any contamination present and minimizing the possibility of removing contamination through personnel or equipment leaving the site.

Procedures for minimizing exposure to or transfer of potentially hazardous materials are numerous, and include the following:

1. elimination of unnecessary personnel in the general area and reducing the amount of workers and equipment onsite to that consistent with effective and safe operations
2. establishment of site security and physical barriers to exclude unnecessary personnel and vehicles from the general area
3. establishment of work zones and control points to regulate access to the site
4. implementation of an appropriate contamination avoidance program in order to reduce personnel and equipment exposure to surface and subsurface contamination, and to minimize airborne dispersion of contamination

Various terms have been used to describe work zones at hazardous materials sites (NIOSH, 1985a; HWOER, 1986). The Exclusion Zone (Zone A, Zone 1, Hot Zone, Red Zone) is the area of contamination, and is marked off by the Hot Line. This line is a real or imaginary barrier determined by instrumentation, visual observation, fragmentation distance, or most commonly, by "geopolitical-institutional-topographic boundaries"; these boundaries are usually a fence, road, stream, or some other preexisting natural or human-made boundary.

As a minimum around drilling equipment, the exclusion zone should be defined as the height of the boom, derrick, or mast forming the radius of the zone. For excavation equipment, the minimum radius of the zone is defined as the maximum extension of the boom plus 20 to 30 feet.

The Contamination Reduction Zone (Zone B, Zone 2, Warm Zone, Yellow Zone) acts as a buffer between contaminated areas and clean areas. It is initially established in a noncontaminated or clean area. It is in this zone that personnel are decontaminated at a Personnel Decontamination Station (PDS). Equipment is decontaminated at a separate Equipment Decontamination Station (EDS). The dimensions of the zone are site-specific, not only in terms of logistics, but also in terms of specific site hazards. As a minimum, the zone should be large enough to comfortably contain both the PDS and the EDS.

The Support Zone (Zone C, Zone 3, Cold Zone, Cool Zone, Green Zone, Staging Area) is the outermost portion of the site, and is considered clean or noncontaminated. This zone contains support equipment, trailers, and parking areas. The location of the zone may ideally be established upwind of the prevailing winds, but is usually practically based on site access and location of available utilities and resources.

Level D protection, as a minimum, should be employed around drilling or other heavy equipment. The need for higher levels of protection in each of the work zones is a function of known or suspected hazards and their associated risks. The job function of site workers, the potential length of exposure, weather conditions, and the types of exploration or excavation equipment employed must also be examined before determining the level of protection. For example, a drilling activity may be conducted in Level C with drillers protected against splash and flash. The site geologist, several feet away from the rig, may also be in Level C respiratory protection; however, he may require less skin protection against splash, flash, and vapor contamination.

Although many personnel, equipment, and site restrictions may be employed while conducting ground-water monitoring investigations at a known or suspected hazardous substance site, at a minimum, personnel must adhere to the following:

1. Workers should wear properly selected and fitted protective clothing and respirators at all times when required. Personnel must be given suitable training in the use, limitations, maintenance, cleaning, and storage of protective clothing and equipment.
2. Personnel should not eat, drink, chew gum or tobacco, smoke, take medicines, or perform any other practice onsite that might increase hand to mouth transfer of potentially toxic materials from gloves, unwashed hands, or equipment.
3. Personnel should not have excessive facial hair in the form of heavy mustaches, long sideburns, or beards, which can prevent the proper fit of respirators.
4. Personnel should avoid unnecessary contact with hazardous materials by staying clear of puddles, vapors, mud, discolored surfaces, and containers or site debris.

Many symptoms of temperature stress and toxic overexposure are not readily apparent; site workers should observe each other for any signs of physical or mental abnormalities such as confusion, complexion change, lack of coordination, or changes in speech patterns. Workers should immediately inform each other if they experience headache, dizziness, blurred vision, cramps, nausea, respiratory distress, irritation to eyes, nose, or mouth, or any other signs of distress. It should be noted that many exposures can produce delayed symptoms hours or days after contact.

DECONTAMINATION

Although contamination avoidance is the best procedure at a hazardous materials site, workers' protective clothing is likely to come into contact with contaminated vapors, particulates, drilling mud, and ground water. Walking or driving onsite may contaminate vehicles, equipment, and clothing. During installation and testing of recovery wells, drillers may come into contact with relatively high concentrations of chemical compounds. Drilling tools, as well as development and sampling equipment, may become unavoidably contaminated. These items must be properly decontaminated before being removed from the site, or in the case of sampling equipment, be thoroughly cleaned before the next use. The decontamination procedure will vary greatly depending on the size, condition, and status of the site, the nature of the hazardous materials, and the nature of site acitivies. In general, the more harmful the contaminant, the more extensive or thorough the decontamination should be.

Decontamination is a process by which a hazard is reduced to some predetermined safe level (usually normal background concentrations), by removal, neutralization, absorption, chemical degradation, dilution, covering, or weathering. Decontamination, or "decon," can be broken down into three general categories: environmental decon, safety decon, and health decon.

Environmental decon is performed to protect some aspect of the environment, such as soil, air, or water supply, from low concentrations of a pollutant that may have long-term environmental consequences and, to minimize cross-contamination. This includes contaminants present in the parts-per-million (ppm) or parts-per-billion (ppb) range of contamination.

Safety decon is conducted when a substance is not overly toxic or hazardous, but produces safety problems, such as slippery walking or riding surfaces. Many mild alkalis and diesel fuel fall into this category.

Health decon is performed when the contaminant, by virtue of its toxic, reactive, flammable, or corrosive properties, presents a hazard to site workers or equipment.

When determining proper procedures for decon, both the degree of hazard and the degree of risk to personnel or equipment should be considered. Many sites involve multiple contaminants; information regarding specific chemicals may not be available for uncharacterized sites. In these cases it is necessary to select decon procedures that will cover a broad range of contaminants. The decon protocol initially assumes that personnel or equipment working in the hot or exclusion zone are contaminated until instrumentation reading or visual observation indicate differently. The decon area in the warm zone should be large enough to handle personnel and equipment. Decon procedures can later be downgraded as more information is gathered on the type of contamination and its volume and concentration. The time constraints imposed by adhering to decontamination protocols must always be considered when preparing site work plans.

Contamination avoidance will help to minimize later decon procedures. Equipment, personnel, and the decon area should be kept upwind of the contaminated area if possible. Covering monitoring instruments, tools, and equipment with plastic sheets and tarpaulins can also minimize subsequent cleaning. Care must be taken when placing plastic near hot or moving parts. Disposable clothing, gloves, and boots may be worn over reusables to reduce decon and extend the wear life of more expensive equipment. Porous items such as wooden truck beds or pallets, cloth hoses, hemp ropes, and wooden handles cannot, in many cases, be properly cleaned; these items should be considered expendable and discarded.

Site workers should normally go through a Personnel Decontamination Station (PDS), which consists of several cleaning and rinsing stations as well as clothing and equipment removal stations. Contaminated clothing and equipment should not be taken offsite where others may be exposed to hazardous substances.

In general, wet contamination should be kept wet, and dry contamination should be left dry. Some dry compounds may form solid oxides or other reaction products when wetted; these products may complicate decon procedures or may be more difficult to remove.

Decon operations should start with the simplest methods. For vapors or wet contamination, a general spraying will remove the bulk of contamination, followed by scrubbing of difficult areas if necessary. Dry material should be brushed or scraped off; the contaminated are can then be sprayed and scrubbed if needed. These procedures avoid unnecessary contact with contaminated material by decon personnel.

Trisodium phosphate (TSP) has been used as a general decontamination agent. TSP, however, has a high pH in solution. As a solid or in concentrated solution TSP is a skin and eye irritant; in areas where phosphates have been banned, decon solutions normally considered nonhazardous but containing TSP may have to be stored and disposed of elsewhere. A nonionic, anionic, cationic surfactant solution, otherwise known as liquid detergent, is the best overall decon agent available. Low sudsing liquid laundry detergents are readily available, easily stored and transported in concentrated form, nontoxic, and go into solution easily. Detergent solutions work well against most contaminants encountered. In some cases it may be necessary to utilize special neutralization solutions or solutions containing solvents to effect a thorough decontamination. Whatever decon agent is used, its possible reactivity and suitability for the hazardous materials involved must be evaluated. Not all decontaminants are compatible with each other. It is important that decon personnel understand the potential hazards of the contaminants, as well as those associated with cleaning equipment and decon solutions.

Common sense must be used when determining procedures for decontaminating sampling equipment. Procedures appropriate when working with ppb or ppm concentrations of aqueous phase contaminants will be considerably different than when working with sludges containing high concentrations of toxic materials. Low boiling point, high vapor pressure contaminants found in low concentrations (ppb or ppm), such as chlorinated solvents in ground water, rarely contaminate equipment except by sorption. These substances can be easily removed by a low sudsing detergent solution followed by a water rinse.

Wet decon solutions and water rinses create unwanted runoff that spreads contamination. It is generally a good practice to limit the amount of water utilized. The use of large wash tubs or children's wading pools for decon of personnel can aid in the collection of decon water. Likewise, using small pneumatic garden sprayers as a water source will help minimize the volume of water employed. Remember, it is the pressure rather than the volume that exerts the cleansing action.

The decontamination of vehicles and large pieces of equipment, such as pumps, may be performed on a wash pad constructed so that cleaning solutions and wash water can be recycled or collected for proper disposal. A raised graveled area lined with polyethylene is a lower cost alternative.

It is important that all aspects of equipment, including undercarriage, wheels or tracks, chassis, and cab be thoroughly cleaned. Air filters on equipment operating in the hot zone should be considered highly contaminated and treated as such; contaminated filters should be removed and replaced before equipment leaves the site.

Steam cleaning or high pressure spraying utilizing low volumes of water is the decontamination method of choice for equipment and vehicles. Lower pressure units (90–120 psi) use low sudsing detergent or special neutralizing solutions at a rate generally from 3 to 5 gpm. High pressure spray units (up to 5000 psi) normally do not require decon solutions. Similarly, hot water (120° to 180°F) is usually not necessary when using high pressure sprayers. Field experience has shown that spray units operating at 500 to 800 psi with a flow rate of 3 to 5 gpm are satisfactory; units offering pressures of 1000 psi and flow rates of 1 to 2 gpm are considered ideal.

It is imperative that decon activities be practiced. A quick walk-through of decon procedures at the personnel decon station should be conducted before workers go into the hot zone. For workers utilizing atmosphere-supplying respirators, this will help ensure that sufficient time is allotted to proceed through the decon line. This is especially important on sites where workers will perform self-decon.

MEDICAL MONITORING

The OSHA rule covering hazardous materials workers (29 CFR 1910.120) stipulates that a comprehensive medical surveillance program must be provided to all employees who have been or expected to be exposed to hazardous substances for 30 or more days during a 12-month interval. In addition, workers who wear respirators for any part of 30 days during a one-year period must also be included in a medical monitoring program. Medical monitoring programs are used to establish baseline data which can verify the adequacy of protective methods and determine if exposures have adversely affected the health and well-being of the worker.

The medical tests appropriate to the monitoring program should be determined by a physician, based upon the information provided by the employer regarding potential and actual exposure, respirator and protective clothing use, and job descriptions. Where job duties and exposures are substantially different, several different monitoring protocols may be appropriate. It should be noted that the tests are supposed to be based upon anticipated exposures and not previous exposures. To assist the physician in formulating a medical monitoring program, the OSHA rule recommends that the employer provide the physician with a copy of the OSHA rule and the NIOSH/OSHA/USCG/EPA Occupational Safety and Health Guidance Manual for Hazardous Waste Site Activities (NIOSH, 1985a). The NIOSH/OSHA guidance manual can be obtained from the NIOSH publication office or ordered through the GPO. The OSHA rule for hazardous materials can be obtained from the OSHA consumer affairs office.

All aspects of the medical monitoring program should be conducted at reasonable times and locations so that workers are not discouraged from participating. The employer pays for the program; if given during working hours, the employee must receive normal pay for that time and if given outside normal working hours, the employee must be paid regular wages for the time involved taking and waiting for the examination.

Employees have full access to all medical records concerning medical tests, exposure records, medical opinions, and so on. The examining physician must submit a written report to the employer on each employee certifying the presence or absence of medical conditions which may pose a health risk when working at hazardous materials sites. The physician must also document the fitness of the employee to use different types of respiratory protection and personal protective clothing. Any recommended limitations regarding use of protective equipment or onsite activities should also be documented. All diagnoses and medical conditions that are unrelated to employment will be confidential information between the physician and the employee.

An initial baseline medical exam is required, which includes a full medical history. The exam should be performed prior to any exposures, so that changes in baseline parameters can be more easily associated with hazardous substance exposure. The medical history should include questions concerning family history of specific organ disease (heart, liver, kidney) and cancer, sexual history (onset of menses for women, number of children, fertility problems), dietary habits (food preferences, drinking habits), and over-the-counter and prescription drug usage (birth control pills for women, analgesics, antipyretics, antiseptics, cathartics, stimulants, and vitamins).

It is recommended that a lifestyle section be included to document activities that may augment exposures to hazardous substances, such as making model airplanes (exposure to solvents, epoxy resins, and catalysts), mechanical engine repair (petroleum products, combustion products, asbestos, solvents), painting or woodworking (solvents, thinners, lacquers, wood dust), gardening (pesticides), photography (solvents), glass making (solvents, lead fumes), painting (solvents, thinners, pigments), and sculpting (dusts, wood preservatives, thinners, lacquers).

The physical exam should focus on the pulmonary, cardiovascular, and musculoskeletal systems. Conditions that may predispose an individual to heat stress, such as obesity and lack of physical fitness, should be noted. To assess a worker's capacity to perform while wearing personal protective equipment, a pulmonary function test and EKG should be performed. A stress test may be performed at the discretion of the examining physician. Conditions that may affect performance while wearing protective equipment, such as facial scars, poor eyesight, or orthopedic problems, should be noted.

Medical screening tests frequently employed by examining physicians include blood tests to evaluate liver and kidney function, urinalysis, and complete blood count with differential and platelet evaluation. Specialized tests may be appropriate for workers working with known hazards, i.e., blood tests for lead or other heavy metal contamination, cholinesterase activity (when working with organophosphates), or PCBs.

Periodic exams are required at least yearly; more frequent supplemental exams may be appropriate depending on the type of exposure involved or if symptoms of exposure are noted. An exit exam is also required at the termination of employment. All medical records are to be made available to the worker and maintained by the employer for the period of employment plus 30 years. Specific medical monitoring requirements can be found in the OSHA Hazardous Waste Operations and Emergency Response standard (29 CFR 1910.120) (HWOER, 1989).

TRAINING REQUIREMENTS

The investigation of a site containing known or suspected hazardous substances is a complex task involving a diverse group of engineers, scientists, technicians, and contractors. In the past, however, safety training was often perfunctory or nonexistent for those involved in sample collection, geophysical surveys, or subsurface investigations. Since the promulgation of the OSHA Hazardous Waste Operations and Emergency Response standard, however, employers now have a legal obligation to ensure that workers are trained to work safely at hazardous materials sites.

The OSHA standard requires training for all workers involved in hazardous materials operations at CERCLA (Superfund), RCRA, emergency response sites, and sites designated for cleanup by state or local governments. A minimum of 40 hours offsite training is required for workers at uncontrolled hazardous materials sites; 24 hours is required for workers performing routine operations at RCRA facilities, or sites where hazards have been identified and respiratory protection would not be required. RCRA sites are considered to have more stable working conditions and better identified hazards. An RCRA site with uncharacterized hazards is considered an uncontrolled site; workers involved in the investigation of such a site require 40 hours of training.

During the basic 24 or 40 hours, workers must be trained to recognize hazards and be provided with the skills necessary to minimize those hazards. Workers must be made familiar with the use and limitations of safety equipment and personal protective equipment that may be required onsite.

Workers requiring 40 hours basic training also must receive a minimum of three days field training or on-the-job training under the direct supervision of a trained, experienced supervisor. The 24-hour option requires one day of field experience. All personnel must receive a minimum of 8 hours retraining on an annual basis.

Supervisors and managers must receive an additional 8 hours training on managing hazardous materials operations. A supervisor can be defined as any individual who has responsibility for, or who makes decisions regarding, the health and safety of personnel in the field.

Individuals working at hazardous materials sites have different roles and levels of responsibility. Site hazards and conditions can vary greatly between sites or portions of the same site. Training must therefore be organized to meet the specific needs and levels of comprehension of the individual, the organization, the work assignment, and the specific requirements of a particular site. In order to maximize the needs of the individual and minimize budget and time contraints, a tiered training approach is usually the most effective and efficient.

The first level of a recommended training program is *overview training*. This is equivalent to the basic training that is now required by OSHA for hazardous materials workers. Overview training serves as an introduction to the nature and types of hazards that may be encountered in the field. Although basic in nature, overview training should be designed for a specific audience, and its needs and level of comprehension, in order to be successful. Training is available through the USEPA, numerous consulting firms, and training organizations. Since 1983, the National Water Well Association has sponsored a 40-hour program designed for ground-water personnel.

Certain supervisory, technical support, or field personnel may require intensive training in specific areas of responsibility, such as incident management, respiratory protection, radioactive materials handling, or sample and decontamination procedures. Most organizations do not have the capabilities to develop such *discipline intensive training* programs. Intensive training programs are sponsored by NIOSH, USEPA, and private training organizations.

The final type of training, and probably the most important, is *site-specific training*. During this phase the individual receives detailed training in the actual conditions and hazards which may be encountered while performing specific tasks. A minimum of three days of site-specific or on-the-job training is now required by OSHA. During this training workers should also receive onsite instructions regarding decontamination procedures, emergency escape routes, communications, and the location of emergency equipment.

A record of training for each individual should be maintained to confirm that all workers have received adequate and appropriate training for the tasks assigned, and that each worker's training is up-to-date and in compliance with applicable regulations.

GENERAL SAFETY AND LIABILITY CONSIDERATIONS

The risks associated with ground-water monitoring at hazardous materials sites are many. In addition to the presence of hazardous substances, drillers, equipment operators, field engineers, scientists, and technicians are exposed to a myriad of other hazards and associated risks.

The operation of heavy equipment, with moving parts, electrical systems, and high pressure lines, is dangerous regardless of the nature of the site. Site topography, layout, and the presence of surface and subsurface debris can increase the likelihood of an accident. It is not unusual that the largest single overhead expense is insurance. Workmen's compensation, heavy equipment coverage, general liability, medical, life, unemployment compensation, disability—the list goes on and on.

When an accident occurs, no one is immune to a liability suit. A recent driller's helper fatality that occurred when the rig overturned spawned suits against the driller, the drilling company, the consulting firm, and the contracting state agency. All parties were considered potentially responsible despite the specification which required that ". . . the driller will ensure that the rig is in a stable position prior to its operation." Even if the outcome is favorable to the defendants, they still must pay the costs, in terms of time and money, to defend themselves.

It is difficult sometimes to distinguish between safety and technical specifications when the contract states ". . . the driller will supply all equipment in a safe working condition . . ." Is the inspecting hydrogeologist responsible for ensuring that the driller follow the site safety plan? The inspector is at least responsible to point out major violations of the plan, as well as any condition or activity that jeopardizes the health and safety of those involved. The inspector's supervisor should also be notified, and detailed documentation of each violation and specific actions taken should be noted in his or her daily log or field journal. The inspector should have the authority to shut down a job because of unsafe conditions.

Prequalification of subcontractors with good safety histories and records of adhering to both technical and safety specifications minimizes the potential of a job shutdown. Although not always feasible during the prequalification process, the contractor may find it worthwhile to visit typical job sites and observe the applicant firms' crews at work. Warning flags that suggest a firm may be a safety risk include workers not wearing hard hats and safety shoes while working on or around the rig; not using goggles and gloves while welding, cutting, or grinding; operating equipment with worn cables, pins, sheaves, and air and hydraulic lines; not cleaning up the work site; using improper tools or tools in need of repair; and allowing unauthorized access to the job site. Some practices should immediately disqualify a firm; these include operating equipment with excessively worn parts or with any guard removed, or using improperly assembled, poorly maintained, or unsafe equipment.

Familiarity with all appropriate federal, state, and industry standards is also necessary. Failure to enforce standards increases the risk of injury and enhances liability. Applicable standards should be reviewed periodically. Federal standards include the OSHA General Industry Standard (29 CFR 1910/1926), OSHA Construction Industry Standard (29 CFR 1926); and some state-OSHA standards which augment or supersede federal OSHA regulations. The USEPA and the Army Corps of Engineers have published their own safety and health requirements; contractors to these agencies must comply with these requirements.

There are also industry standards and guidelines that are published on a regular basis. Failure to adhere to these voluntary standards may not only increase the risk of accident or injury, but may increase a firm's liability should a mishap occur. Organizations responsible for industry standards include the American National Standards Institute (ANSI), New York, NY; American Society for Testing and Materials (ASTM), Philadelphia, PA; National Drilling Federation, Columbia, SC; National Fire Protection Association (NFPA), Quincy, MA; and the National Water Well Association (NWWA), Dublin, OH.

SUMMARY

The requirements that must be met before working at hazardous materials sites may seem unwieldly and costly; they have been imposed, however, to safeguard the health and safety of onsite workers. Ground-water monitoring personnel face additional hazards because of the nature of their profession. For additional information the reader is encouraged to obtain the guidance manuals referenced in the text as well as the final OSHA rule (HWOER, 1989) for hazardous materials workers. The EPA Standard Operating Safety Guides (USEPA, 1988) may also be of value.

REFERENCES

(NWWA, 1980). Manual of Recommended Safe Operating Procedures and Guidelines for Water Well Contractors and Pump Installers. National Water Well Association, Dublin, OH, 1980.

(NDF, 1986). Drilling Safety Guides. National Drilling Federation, Columbia, SC, 1986.

(NSC, 1988). Fundamentals of Industrial Hygiene, Third Edition. National Safety Council, Chicago, 1988.

(NIOSHa, 1985). NIOSH/OSHA/USCG/EPA Occupational Safety and Health Guidance for Hazardous Waste Site Activities. DHHS (NIOSH) Publication No. 85-115. Government Printing Office (GPO), Washington, DC, 1985.

Linson, T. Personal communication, USEPA-ERT, HMIRT-NUS, July 28, 1988.

(ACGIHa, 1989). Threshold Limit Values and Biological Exposure Indices for 1989-1990. American Conference of Governmental Industrial Hygienists, Inc., Cincinnati, OH, 1987.

(NIOSHb, 1985). NIOSH: Pocket Guide to Chemical Hazards. DHHS (NIOSH) Publication No. 85-114, GPO, Washington, DC, 1985.

(NIOSHb, 1987). NIOSH: A Guide to Safety in Confined Spaces. DHHS (NIOSH) Publication No. 87-113, GPO, Washington, DC, 1987.

(NIOSH, 1986). NIOSH Certified Equipment List. DHHS (NIOSH) Publication No. 87-102, GPO, Washington, DC, 1986.

(NIOSHc, 1987). NIOSH Respirator Decision Logic. DHHS (NIOSH) Publication No. 87-108, GPO, Washington, DC, 1987.

Sagan, C. The Dragons of Eden: Speculations on the Evolution of Human Intelligence. Random House, Inc., New York, 1977.

(ACGIHb, 1983). Guidelines for the Selection of Chemical Protective Clothing, Vol. I: Field Guide. American Conference of Governmental Industrial Hygienists, Inc., Cincinnati, OH.

(HWOER, 1989). Hazardous Waste Operations and Emergency Response, Final Rule, 29 CFR 1910.120, 54 FR 9294, March 6, 1989.

15

Decontamination Program Design
for Ground-Water Quality Investigations

Gillian L. Nielsen

INTRODUCTION

The ultimate objective of virtually every ground-water investigation conducted under the Comprehensive Environmental Response, Compensation and Liability Act (CERCLA), or "Superfund," and the Resource Conservation and Recovery Act (RCRA), is to determine the presence or absence of subsurface contamination and to assess the extent and environmental significance of that contamination. Throughout these studies, a wide variety of direct and indirect investigative techniques, such as installation of ground-water monitoring wells, soil sampling, soil gas monitoring, and/or geophysical surveys may be implemented to provide qualitative and quantitative physical and chemical information about the site. Using field and laboratory data generated by these investigations, scientists develop an understanding of the chemical nature and subsurface behavior of contaminants, and subsequently design and implement remediation measures in an attempt to "clean up" apparent contamination.

Costs associated with remediating site contamination can be astronomical, and remedial investigations can often culminate in courtroom battles over determining liabilities and environmental damages. Therefore, it is of paramount importance that subsurface samples and field and laboratory data generated by ground-water investigations are truly representative of site conditions.

Equipment decontamination is a very important component of most ground-water contamination investigations. Ironically, this is often the most poorly planned and documented segment of these investigations. Without effective decontamination procedures, any data generated by an investigation are subject to critical scrutiny. Consequently, effective decontamination procedures must be incorporated into every aspect of field investigations, from preliminary soil sampling to ground water sampling. Table 15.1 illustrates examples of various elements of equipment decontamination that should be incorporated into typical ground-water quality monitoring programs.

Table 15.1.   Typical Equipment Decontamination Requirements for Ground-Water Investigations.

Field Activity	Equipment to be Decontaminated
Soil Sampling	—split-spoon samplers —shelby tubes —sample inspection tools —drilling rig —downhole equipment
Installation of Monitoring Wells	—drill rig, augers, drill rod and bits —associated tools —well casing —well screen —well development equipment (e.g., pumps, bailers)

Ground-Water Sample Collection	—water-level gauge(s)
	—well purging device(s)
	—sample collection device(s)
	—sample filtration apparatus
	—instrumentation for field parameter
	  monitoring (i.e., pH meter,
	  conductivity meter, thermometer)

The purposes of decontamination are basically fourfold: (1) to prevent cross-contamination of individual sites and specific sampling locations when using common equipment; (2) to ensure that representative samples are collected for analysis; (3) to ensure proper operation of equipment and instrumentation; and (4) to reduce exposure hazards to workers involved in handling potentially contaminated materials. Chapter 14 of this book discusses personnel decontamination practices in detail. These are necessary to ensure that personnel do not act as a source of contamination and to minimize potential personnel exposure hazards. The primary focus of this chapter will be field equipment decontamination procedures used in ground-water quality investigations.

CURRENT STATUS OF STANDARDIZED
DECONTAMINATION PROCEDURES

While most scientists acknowledge the importance of effective field decontamination, it is ironic that there are currently no standardized procedures to guide these activities. While standard procedures have been established for laboratory equipment decontamination procedures by groups such as the American Society for Testing and Materials (ASTM) and the American Water Works Association (AWWA), no similar set of field equipment decontamination procedures exists. Recognizing the immediate need of such guidance, ASTM is developing (1990) a proposed "Standard Practice for the Decontamination of Field Equipment Used at Waste Sites."

Mickam et al. (1989) conducted an extensive survey of federal and state agencies across the U.S. to evaluate the status of current field equipment decontamination procedures. The need for some form of standardization became readily apparent during this survey. Of the five federal agencies interviewed (including the Department of the Army, Office of the Chief of Engineers, the Nuclear Regulatory Commission, and the National Science Foundation), none had any specific guidance addressing field equipment decontamination protocol. The survey also found that the U.S. Environmental Protection Agency (U.S. EPA, 1986) had no document used nationally to furnish guidance on field decontamination of equipment used in ground-water investigations. The responsibility for recommending and reviewing decontamination procedures has been assigned to regional offices, some of which have independently developed guidance manuals. Many of these documents are significantly different from one another. This lack of continuity between regions makes it impossible for companies with multistate interests to develop corporate decontamination procedures that comply with regulatory agency guidance.

Managers of gound-water investigations are required to either follow state- or region-specific guidance on decontamination procedures (if any) for projects within specific states or regions, or propose site-specific procedures. To do either effectively requires careful evaluation of a number of project-specific variables. Table 15.2 presents a summary of items that must be established on a project-by-project basis before selection of decontamination technique(s).

DESIGN CONSIDERATION FOR DECONTAMINATION PROGRAMS

Of utmost importance in developing a decontamination program is establishing the purpose of the ground-water investigation. This is directly linked to the level of quality assurance/quality control (QA/QC) demanded by the investigation. For example, during the installation of a ground-water monitoring system to act as

Table 15.2. Criteria for Equipment Decontamination Protocol Selection.

- existing federal, state, and/or regional regulatory guidelines
- purpose of the investigation —initial site investigation
  —site assessment
  —remediation
  —litigation-driven
  —leak detection system installation

- media to be sampled —soil
  —ground water
  —surface water
  —waste

- parameters to be monitored —inorganics
  —organics
  —pesticides
  —petroleum hydrocarbons

- anticipated concentrations of contaminants

Logistical Constraints:

- equipment to be decontaminated —size
  —complexity of construction
  —contact with samples submitted for chemical analysis
  —dedicated vs. common devices

- site support —power & water availability
  —site security

- site access

- weather

- management of decontamination wastes —non-hazardous
  —hazardous

- cost

---

a leak detection system around a newly installed underground storage tank system, the required level of QA/QC would be low. No subsurface contamination would be expected at the site unless the tanks were being installed as replacements for old tanks. Therefore, decontamination of equipment used at the site may not be an issue. In contast, however, installation of ground-water monitoring wells at hazardous waste (CERCLA, RCRA) sites would require a higher level of QA/QC throughout every aspect of the investigation to ensure the collection of representative data. The level of QA/QC can be further intensified when an investigation is conducted at a site under litigation. Under these circumstances, not only is sample integrity of concern, but all data must prove to be legally defensible in terms of validity and reproducibility.

Once the purpose of the investigation has been established, the suspected site contaminants must be identified and a representative list of parameters that will be analyzed must be established. This process requires an evaluation of a number of contaminant-specific physiochemical properties to provide information that will be incorporated into the decontamination protocol design. These properties include:

- contaminant physical distribution (i.e., in air, fill, soil, ground water, or surface water);
- chemical matrix (i.e., interfering constituents, or carrier chemicals such as petroleum hydrocarbons or solvents);
- chemical species (i.e., nonvolatile organics, volatile organics, heavy metals, inorganic nonmetals, or others); and
- physical properties (density, volatility, flammability, corrosivity, viscosity, reactivity, natural decomposition or transformation rates).

These are critical factors that direct the selection of decontamination procedures. Decontamination activities must be selected based upon chemical suitability and compatibility with the constituents to be removed during decontamination, and with the concentrations of constituents anticipated. For example, decontamination protocols for an investigation being conducted at a fuel distribution terminal would have to incorporate decontamination procedures such as solvent rinses or special degreasing detergents effective in removing oily substances from all equipment used in the project. However, for an investigation conducted at a metal sludge surface impoundment, metals would be of prime concern and would require different decontamination procedures, such as dilute acid rinses. Method selection can become more complicated if more than one contaminant group is under investigation at a site. For example, if petroleum hydrocarbons and solvents are of concern in an investigation, use of solvents such as acetone or hexane to degrease oily equipment may interfere with efforts to characterize solvent contamination at the site.

Relative concentrations of contaminants influence many components of a ground-water quality investigation, from establishing health and safety guidelines, to sample collection techniques, to decontamination of personnel and equipment. Anticipated concentration levels (i.e., percent-range concentrations vs parts-per-million or parts-per-billion levels), must be considered along with project objectives and QA/QC controls to identify any physical limitations to decontamination.

Matteoli and Noonan (1987) determined through controlled field testing that the time required for effective decontamination was directly related to the construction materials of the equipment being decontaminated. They found that more than three hours of rinsing with clean water were required to lower trichloroethylene (TCE) levels below detection limits for a submersible pump equipped with a rubber discharge hose, compared to 90 minutes for the same pump equipped with a Teflon® hose. They also determined that individual parameters had unique responses to decontamination. For example, they found that decontamination times for Freon 113 were longer than for TCE, while decontamination times for 1,1-dichloroethylene, selenium, and chromium were shorter.

Identifying physical limitations or logistical problems associated with decontamination for a project is essential if the decontamination protocol is to be workable. Many very elaborate decontamination procedures developed and approved for theoretical use may not actually be implemented on a project, due to factors such as time and budget constraints, incompatibility with equipment to be decontaminated, inconvenience, and inability to manage the wastes generated. To avoid this problem, a realistic protocol must be developed which recognizes and accounts for the myriad of project-specific logistical constraints listed on Table 15.2. It is neither logistically reasonable nor safe, for example, to specify that a drilling rig and all necessary support tools and equipment undergo decontamination by solvent or acid washing on a project. Conversely, it is also not reasonable to expect a drilling rig to be "dedicated" to a site unless the project is conducted under the most drastic conditions, such as a high-level radioactive waste site, where it may not be physically possible to decontaminate sufficiently to permit demobilization to another site. Some element of compromise must be incorporated into the decontamination protocol to make the program workable and allow it to meet the objective of the investigation.

As a case in point, Keely and Boateng (1987) found that procedures originally developed for field decontamination of an electric submersible pump were not workable due to the time and patience required on the part of the sampling team to completely disassemble the sampling pump, scrub each individual component, and then reassemble the pump for use in the next monitoring well. The original decontamination procedures were modified to permit recirculation of decontamination solutions, thereby avoiding the need to disassemble the multicomponent submersible pump. The compromise, however, as indicated by the authors, was the potential for carryover of the solvent (acetone) used in the decontamination procedure. The argument used to justify the change in protocol was that the amount of carryover was limited to a few milliliters of water wetting the surface of the pump, and that the device would be immersed in many gallons of water in the casing of the next monitoring well, resulting in potential residual concentrations of acetone in the sub-parts-per-trillion level. This type of compromise may jeopardize the ability to meet the objectives of QA/QC standards of a highly sensitive analytical program. However, it does illustrate the need to anticipate actual field conditions as opposed to ideal field conditions. Under actual field sampling conditions, performance factors such as weather, operator skill, and pressure to complete assigned tasks within a limited time frame can have a significant impact on the quality of actual decontamination techniques.

The most widely applied solution to this problem is to use dedicated equipment, thereby avoiding all but initial equipment decontamination prior to installation.

The second alternative is to reevaluate and modify the program's original decontamination QA/QC specifications.

It is possible to avoid the need to decontaminate well construction materials, such as casing and screen, prior to installation of a well. Many manufacturers can supply precleaned casing and screen which is delivered to a job site in hermetically sealed containers. While the initial cost of these precleaned materials may be somewhat high, the higher cost associated with onsite cleaning of these materials is eliminated, as long as the sealed containers are not opened until use.

It is evident that a variety of factors must be considered when developing a decontamination program for a ground-water investigation. It is also readily apparent that those factors are not only site-specific, but are work task-specific as well. For this reason, it is often necessary to prepare a document which details decontamination procedures on a task-by-task basis. Some states, such as Florida, require that these work plans go one step further and address decontamination procedures for equipment before it is brought to a site for use, as well as onsite field procedures. In a case study of decontamination procedures implemented at a Superfund site in Indiana, Fetter and Griffin (1988) discussed no fewer than eight task-specific components that were incorporated into the field investigation, from well installation to ground-water sample collection.

AVAILABLE DECONTAMINATION PROCEDURES

As previously stated, standard procedures for field equipment decontamination have not yet been established on a national level. There are however, some techniques that are more appropriate and effective than others for specific groups of chemical constituents and equipment types in ground-water investigations.

Most larger equipment, such as a drilling rig or a support truck, is cleaned between sampling locations using a high-pressure, hot-water power wash system. This system frequently employs an initial soapy wash and scrubbing with brushes to remove larger soil particles and contaminants from surface areas. This is followed by a high-pressure, clean-water (potable water) wash to remove soap and contaminants. Equipment is usually allowed to air dry before being placed onto elevated racks or Visqueen for storage before use at the next location.

This type of power wash system should not be confused with or replaced by the type of power wash system found at the local car wash. Most car washes recirculate water and use additives such as glycerin to provide a shine to cleaned vehicles. Both of these practices make it impossible to have good quality control on decontamination efforts in a ground-water investigation.

Hot-water power wash systems are often incorrectly referred to as "steam cleaning," probably because the fine, mist-like hot-water spray that is emmitted from the high pressure spray nozzle resembles steam. Steam cleaning technically refers to the application of high pressure steam to remove contaminants and soil from larger pieces of field equipment, such as drilling rigs. The primary advantage of steam cleaning is that the volume of decontamination waste water is minimized. Two significant disadvantages, however, limit its use: (1) the lack of adequate pressure to effectively dislodge large particles such as clayey soils or residues such as oils; and (2) logistical difficulties in generating a sufficient source of steam under field conditions.

Most state and federal regulatory agencies will specify the use of "steam cleaning" or high pressure hot water power wash methods for drilling rigs, associated tools, and support vehicles. There are exceptions, however. Mickam et al. (1989) indicate that the Massachusetts Department of Environmental Quality Engineering, in its draft decontamination program, states that drilling tools, including augers and split-spoon samplers, must undergo the following decontamination procedures before steam-cleaning:

- immersion and scrubbing in a mixture of detergent (Alconox) and water;
- rinse with clean water;
- rinse with ispropyl alcohol, methanol, or acetone; and
- multiple rinses with distilled water.

This may be a manageable procedure for small equipment such as split-spoon samplers and wrenches, however, a significant exposure hazard can be associated with working with the large volumes of concentrated solvents necessary for decontaminating larger equipment such as augers, drill rod, and bits. In addition, large quantities of potentially hazardous waste would be generated by this cleaning procedure; these materials would be costly to manage and dispose of properly.

The variability in decontamination procedures employed for cleaning small equipment, such as sampling devices, is considerable. Some agencies have developed analyte-specific protocols, while others have developed procedures on the basis of equipment type and material(s) of construction of the equipment to be cleaned. Usually, one or a combination of several of five solutions are typically specified in field decontamination procedures: tap (potable) water, dilute acid, solvent, distilled or deionized water, and laboratory-grade phosphate-free detergent. The number and sequence of use of these solutions is usually the source of variation between decontamination protocols.

To generalize, for sampling devices used to sample for inorganics, such as metals, the most commonly used decontamination procedure is as follows:

- tap water and soapy water wash;
- rinse with dilute acid solution (10% nitric or hydrochloric); followed with
- rinses with distilled water or deionized water.

When sampling devices are to be used to sample for organics, the decontamination procedure typically includes:

- initial wash with water (tap, distilled, or deionized) and laboratory-grade detergent;

- solvent rinse (acetone, hexane, isopropyl alcohol, methanol alone, or in some combination); then
- distilled water or deionized water rinses.

Considering the number of subsurface investigations associated with underground petroleum product storage systems, pipelines, and terminals, it is unfortunate that decontamination procedures addressing petroleum hydrocarbons are commonly overlooked in decontamination procedure documents. Solvents such as acetone and hexane are commonly used to degrease equipment. However, this practice can cause interference in analyses for dissolved-phase hydrocarbons and product-specific additives. Other problems include generating hazardous wastes and the potential for the solvents to degrade parts (i.e., rubber or plastic) on some field equipment. Specially formulated laboratory-grade detergents, such as "Detergent 8," are more effective than most solvents at removing petroleum hydrocarbon residues from the surfaces of field equipment. Additionally, detergents eliminate personnel exposure hazards associated with solvents, and greatly reduce the cost associated with wastewater disposal.

For practical purposes, decontamination programs can be simplified to using an initial rinse with water (tap, distilled, or deionized), followed by a wash and scrubbing with laboratory-grade detergent (Alconox, Liquinox, or Detergent 8), followed by two to three rinses with distilled water. This method has been used in ground-water monitoring programs for inorganics, organics, pesticides, and petroleum hydrocarbons, and for concentrations of these compounds ranging from percent to low parts-per-billion levels. Field blanks should be used to determine the effectiveness of this and other decontamination practices. Table 15.3 presents a summary of decontamination procedures that have been implemented during a variety of field investigations.

Table 15.3.   Currently Available Decontamination Protocol Options.

Physical Decontamination
- air blasting
- wet blasting
- dry ice blasting
- high pressure Freon cleaning
- ultrasonic cleaning
- vacuum cleaning
- physical removal/scrubbing

Chemical Decontamination
- water wash
- tap water followed by deionized/distilled water rinses
- water, laboratory-grade detergent wash, distilled/dionized water rinses
- pesticide-grade solvent rinses in combination with distilled/deionized water rinses
- high pressure steam cleaning
- high pressure hot-water power wash
- hydrolazer
- acid rinses in combination with distilled/deionized water rinses

INHERENT PROBLEMS WITH DECONTAMINATION TECHNIQUES

A review of the various federal and state guidelines demonstrates that materials that are themselves hazardous by definition (i.e., acids and solvents), are commonly incorporated into many decontamination procedures. Other methods, such as steam cleaning or hot-water power washes, require the use of potable water supplies and a variety of support equipment. It is, therefore, apparent that the method of decontamination selected for a given project must be evaluated with respect to its potential to impact sample integrity by imparting potential contaminants to the equipment being cleaned. Table 15.4 presents a summary of some of the more common potential sources of contamination associated with decontamination procedures.

Table 15.4. Potential Sources of Contamination Associated with Decontamination Procedures.

- use of "contaminated" potable water supplies—(i.e., contaminated due to the presence of bacteria, organic compounds, metals, or other objectionable substances)

- use of contaminated supplies of commercially prepared distilled water (i.e., contaminated due to the presence of platicizers and organic compounds)

- contamination of samples with residues of fluids used for decontamination, such as dilute acids or solvents

- oil spray from unfiltered exhaust from generators used as power sources

- volatile organic contamination associated with equipment exhaust systems (steam cleaners, generators, support trucks, etc.)

- use of ethylene glycol (antifreeze) in decontamination equipment (e.g., hot-water power wash spray nozzles and hoses) to prevent freezing in extremely cold weather

- hydraulic fluids, oils, gasoline, or diesel fuel used to operate generators and other support equipment

With any decontamination program, solid and/or liquid wastes will be generated. The exact nature and volume of the waste generated is dependent on the equipment being decontaminated, the method of decontamination, QA/QC performance standards, the amount of decontamination required, and the contaminants of concern. Depending upon the decontamination method(s) used and the type of ground-water investigation being conducted, it is possible that decontamination wastes may be classified as hazardous by virtue of the decontamination fluids used (i.e., nitric acid, acetone, methanol, etc.) and/or contaminants encountered during the investigation. Generally, when working at hazardous waste sites, management of hazardous decontamination wastes is not a major issue, although it can result in substantial budget increases when specially designed decontamination facilities must be constructed and wastes must be containerized and sampled and the samples analyzed to provide for appropriate disposal. However, at nonhazardous sites where the ground-water quality monitoring program is designed to monitor mainly nonhazardous constituents (i.e., inorganic parameters such as chloride or nitrate), the potentially hazardous wastes generated by decontamination procedures can pose a major problem. Compliance with federal hazardous waste regulations becomes a major issue that must be resolved. This is typically very time-consuming and costly and, consequently, can have the undesired effect of encouraging improper management of hazardous materials and wastes. Because of these potentially significant problems, an attempt is usually made to avoid the use of solvents and acids in decontamination programs.

QUALITY ASSURANCE/QUALITY CONTROL
COMPONENTS OF DECONTAMINATION PROTOCOLS

As with all other components of a ground-water investigation, it is necessary to monitor the effectiveness of decontamination protocols. This is done to verify that the contaminants of concern are removed from all equipment being decontaminated so that any data generated from samples collected for chemical analysis during the investigation can be considered valid and uncompromised.

The QA/QC segment of the decontamination protocol must address several key issues including:

- location and construction of decontamination area;
- movement of clean and contaminated equipment in and out of the decontamination area;
- preliminary cleaning of all equipment to be used in a project at an offsite location prior to being permitted access to the site;
- segregation of clean and contaminated equipment;
- controls to ensure that cleaned equipment does not become contaminated prior to use (i.e., placing cleaned auger flights on elevated racks or plastic sheeting);
- controls to ensure that the equipment used for decontamination will not in itself act as a source of contamination (i.e., installing exhaust collectors on generators);
- chemical verification of the suitability of the potable water supply; and
- use of rinse (field) blanks and/or "wipe" samples in conjunction with trip blanks to verify the effectiveness of decontamination procedures.

Should any of the approved decontamination procedures be modified during the course of the ground-water investigation, it is critical to thoroughly document all changes, and provide justification for any changes. Under these circumstances, the use of rinse blanks or wipe samples can be even more critical.

SUMMARY

It is not possible to write a decontamination program that will be applicable to every field situation. Each ground-water investigation will be governed by site-specific physical and chemical variables that will direct the process of selection of the most effective decontamination method on a task-specific and project-specific basis. Once these variables have been defined, however, it is possible to develop a workable decontamination program incorporating procedures that will ensure that all data generated by the investigation are representative of site conditions and that the results of the study are not compromised.

Work is currently underway by ASTM to establish standardized decontamination procedures designed to provide continuity for decontamination programs conducted across the country. These standards will be flexible enough to permit addressing site-specific requirements. Until these standards are established, however, managers of ground-water investigations will need to draw upon available resources to develop effective decontamination programs.

REFERENCES

Fetter, C. W., and R. A. Griffin. "Field Verification of Noncontaminating Methodology for Installation of Monitoring Wells and Collection of Ground Water Samples," *Proceedings of the Second National Outdoor Action Conference on Aquifer Restoration, Ground Water Monitoring and Geophysical Methods.* Vol. 1. National Water Well Association, Dublin, Ohio, 1988.

Keely, J. F., and K. Boateng. "Monitoring Well Installation, Purging and Sampling Techniques Part 1: Conceptualizations," *Proceedings of the FOCUS Conference on Northwestern Ground Water Issues.* National Water Well Association, Dublin, Ohio, 1987.

Matteoli, R. J., and J. M. Noonan. "Decontamination of Rubber Hose and "Teflon" Tubing for Ground Water Sampling," *Proceedings of the First National Outdoor Action Conference on Aquifer Restoration, Ground Water Monitoring and Geophysical Methods.* National Water Well Association, Dublin, Ohio, 1987.

Mickam, J. T., R. Bellandi, and E. C. Tifft, Jr. "Equipment Decontamination Procedures for Ground Water and Vadose Zone Monitoring Programs: Status and Prospects," *Ground Water Monitoring Review* 9(2): (1989).

U.S. Environmental Protection Agency. Region IV Standard Operating Procedures and Quality Assurance Manual. Athens, Georgia (1986).

GLOSSARY

The terms and definitions in this glossary have been compiled in part from existing documents, and in part from contributions from authors of individual chapters of this book. The following documents were consulted in the assembly of the glossary:

ASTM Standard D653-87, *Terminology Relating to Soil, Rock and Contained Fluids.*

American Geological Institute, *Glossary of Geology,* 1987.

American Geological Institute, *Dictionary of Geological Terms,* 1974.

Fetter, C. W., Jr., *Applied Hydrogeology,* 1980.

Campbell, M. D., and J. H. Lehr, *Water Well Technology,* 1975.

Driscoll, F. G., *Groundwater and Wells,* 1986.

AAS: (*See* Atomic Absorption Spectroscopy).

Abandonment: The complete sealing of a well or borehole with neat cement or bentonite grout or other low hydraulic conductivity materials to restore the original hydrogeologic conditions and/or to prevent contamination of an aquifer. Also called decommissioning.

Absorption: The process by which one substance is taken into the body of another substance; the penetration of molecules or ions of one or more substances (gas, liquid, or solid) into the interior of another substance. For example, in hydrated bentonite, the planar water that is held between the mica-like layers is the result of absorption.

Accelerator: A substance used to reduce the time required for curing of neat cement; examples are calcium chloride, gypsum, and aluminum powder.

Accuracy: The closeness of the results of a measurement to the true value of the quantity measured.

ACGIH: American Conference of Governmental Industrial Hygienists.

Acid: A substance that releases hydrogen ions when dissolved in water; characterized by a pH less than 7. A strong acid will release a large proportion of hydrogen ions, whereas a weak acid will release a small proportion of hydrogen ions.

Acidification: To make into an acidic solution through the addition of an acid.

Acidity: The ability of a water solution to neutralize an alkali or base.

Acidization: The process of forcing acid through a well screen or into limestone, dolomite, or sandstone making up the wall of an open borehole. The general objective of acidization is to clean incrustations from the well screen or to increase hydraulic conductivity of the aquifer materials surrounding a well by dissolving and removing a part of the rock constituents.

Acrylonitrile butadiene styrene (ABS): A thermoplastic material produced by combining varying ratios of three different monomers to produce well casing with good heat resistance and impact strength.

ACS: American Chemical Society.

Activity coefficient: A fractional number that, when multiplied by the molar concentration of a substance in solution, yields the chemical activity.

Adapter: A device used to connect two different sizes or types of casing threads. Also known as a sub, connector, or coupling.

Adsorption: The process by which a gas, vapor, dissolved material, or very small particle adheres to the surface of a solid; the attraction and adhesion of ions from an aqueous solution to the solid soil or rock surfaces with which they are in contact.

Advection: The process by which solutes are transported with and at the same rate as moving ground water.

Aeration: The process by which air becomes dissolved in water.

Aerobic: An action or process conducted in the presence of oxygen.

Airborne geophysics: Geophysical surveys carried out using aircraft as a platform for the instruments.

Air lift pump: A device consisting of two pipes, with one inside the other, to withdraw water from a well. The lower ends of the pipes are submerged, and compressed air is delivered through the inner pipe to form a mixture of air and water. This mixture rises in the outer pipe to the surface because the specific gravity of this mixture is less than that of the water column.

Air rotary drilling: A drilling technique in which compressed air is circulated down the drill rods and up the open hole. The air simultaneously cools the bit and removes the cuttings from the borehole.

Air rotary with casing driver: A drilling technique that uses conventional air rotary drilling while simultaneously driving casing. The casing driver is installed in the mast of a top-head drive air rotary drilling rig.

Aliphatic hydrocarbons: A class of organic compounds characterized by a straight or branched chain arrangement of the constituent carbon atoms joined by single covalent bonds with all other bonds to hydrogen atoms. Examples include propane, ethylene, acetylene, and cyclohexane.

Aliquot: One of a number of equal-sized portions of a water sample that is being analyzed.

Alkalinity: The ability of the salts contained in water to neutralize acids. Materials that exhibit a pH of greater than 7 are alkaline.

Alluvium: A general term for clay, silt, sand, gravel, or similar unconsolidated material deposited during comparatively recent geologic time by a stream or other body of running water as a sorted or semisorted sediment in the bed of the stream or on its floodplain or delta, or as a cone or fan at the base of a mountain slope.

Aluminum powder: An additive to neat cement that produces a strong, quick-setting cement that expands upon curing.

Anaerobic: An action or process conducted in the absence of oxygen.

Analyte: A specific compound or element of interest undergoing chemical analysis.

Anion: An atom or radical carrying a negative charge.

Annular sealant: Material used to provide a seal between the borehole and the casing of a well. Annular sealants should have a hydraulic conductivity less than that of the surrounding geologic materials and be resistant to chemical or physical deterioration.

Annular space: The space between casing or well screen and the wall of the drilled hole, or between drill pipe and casing, or between two separate strings of casing. Also called annulus.

Annulus: (*See* annular space).

Antecedent moisture: The soil moisture present before a particular precipitation event.

API: American Petroleum Institute.

Appendix VIII constituents: A list of 297 toxic constituents (40 CFR Part 261) which, if present in a waste, may make the waste hazardous. The waste containing these constituents poses a substantial hazard to human health or the environment when improperly treated, stored, transported, or disposed.

Aquiclude: A saturated but poorly permeable formation, or group of formations, that impedes ground-water movement and does not yield water freely to a well or spring. An aquiclude may transmit appreciable water to or from adjacent aquifers and, where sufficiently thick, may constitute an important ground-water storage unit.

Aquifer: A geologic formation, group of formations, or part of a formation that is capable of yielding a significant amount of water to a well or spring.

Aquifer, confined: An aquifer that is overlain by a confining bed. The confining bed has a significantly lower hydraulic conductivity than the aquifer.

Aquifer, unconfined: An aquifer in which there are no confining beds between the zone of saturation and the ground surface. There will be a water table in an unconfined aquifer; water-table aquifer is a synonym.

Aquifer test: A test involving the withdrawal of measured quantities of water from, or the addition of water to, a well, and the measurement of resulting changes in head in the aquifer, both during and after the period of discharge or addition. Examples are pumping tests, slug tests, bail tests, and pressure tests.

Aquitard: A geologic formation, group of formations, or part of a formation of low hydraulic conductivity that is typically saturated but yields very limited quantities of water to wells.

Aromatic hydrocarbons: A class of organic compounds containing one or more benzene-type ring structures or cyclic groups with very stable bonds through the substitution of a hydrogen atom for an element or compound. Examples include benzene, toluene, naphthalene, and chlorobenzene.

Artesian: Ground water occurring under greater than atmospheric pressure.

Artesian (confined) aquifer: An aquifer bounded by aquicludes or confining beds, and containing water under artesian conditions.

Artesian well: A well deriving water from an artesian or confined aquifer in which the water level stands above the top of the aquifer.

Artificial recharge: The process by which water can be added to an aquifer by man. Dug basins, drilled wells, or the spreading of water across the land surface are all means of providing artificial recharge.

Assessment monitoring: An investigative monitoring program under Resource Conservation and Recovery Act requirements that is initiated after the presence of a contaminant in ground water has been detected. The objectives of this type of program are to determine the concentration of constituents that have contaminated the ground water and to quantify the rate and extent of migration of these constituents.

Assessment plan: The written detailed plan drawn up by the owner/operator of a waste management unit that describes and explains the procedures that the owner/operator intends to take to perform assessment monitoring.

ASTM: American Society for Testing and Materials. An organization concerned with the development of standards on characteristics and performance of materials, products, systems, and services, and the promotion of related knowledge.

Atmosphere: The pressure exerted by the air at sea level (14.696 psi), which will support a column of mercury 760 mm high (about 30 inches).

Atomic absorption spectroscopy (AAS): An analytical technique used for measurement of metal ion concentrations in solutions.

Attenuation: The reduction or removal of constituents in the ground water by the sum of all physical, chemical, and biological factors acting upon the ground water. Also called retardation.

Auger: Any rotary drilling device in which the cuttings are mechanically continuously removed from the borehole during the drilling operation without the use of fluids. Used in soils or unconsolidated geologic materials.

Auger flights: Winding metal strips welded to the auger sections that carry drill cuttings to the surface during drilling; may also refer to individual sections of augers used in drilling.

Background mean: The arithmetic average of a set of data, used as a control value in subsequent statistical tests.

Background monitoring: A schedule of sampling and analysis under Resource Conservation and Recovery Act requirements that is completed during the first year of monitoring. All wells in the monitoring system must be sampled on a quarterly basis to determine concentrations of drinking water, ground-water quality, and contamination indicator parameters. For each upgradient well, at least four replicate measurements must be made for the indicator parameters.

Background variance: A measure of how far an observation value departs from the mean. Background refers to the observations used for control in subsequent statistical tests.

Backwashing: A method of filter pack emplacement whereby the filter pack material is allowed to fall freely through the annulus while clean, fresh water is simultaneously pumped down the well, through the screen, and up the annulus to the surface.

Backwashing (well development): The surging effect or reversal of water flow in a well that removes fine-grained material from the formation surrounding the borehole and helps prevent bridging.

Bailer: A long, narrow tubular device with an open top and a check valve at the bottom that is used to remove water and/or drill cuttings from the borehole or well.

Bail test: A single-well test conducted to determine the in-situ hydraulic conductivity of low to moderate hydraulic conductivity formations by the instantaneous removal of a known quantity of water from a well and the subsequent measurement of the resulting well recovery.

Bar: A unit of pressure equal to one million dynes per square centimeter, or 0.98 atmospheres.

Barium sulfate: A natural additive used to increase the density of drilling fluids.

Barrier boundary: An aquifer-system boundary represented by a rock mass that is not a source of water. Also called a no-flow boundary.

Barrier well: A pumping well used to intercept a plume of contaminated ground water. Also, a recharge well that delivers water to, or in the vicinity of, a zone of contamination under sufficient head to prevent the further spreading of the contaminant.

Base: A substance that releases hydroxyl ions when dissolved in water, or that takes up hydrogen ions.

Base flow: The flow of streams composed solely of ground-water discharge.

Basement: The oldest rocks recognized in a given area; generally, a complex of metamorphic and igneous rocks that underlies all sedimentary (consolidated or unconsolidated) formations.

Basic: Having a pH greater than 7.0.

Bedrock: A general term for the rock that underlies unconsolidated material.

Bentonite: A hydrous aluminum silicate clay mineral available in powdered, granular, or pellet form and used to provide a seal between the well casing and borehole. Bentonite is also added to drilling fluid to impart specific characteristics (i.e., increased viscosity) to the fluid.

Bicarbonate: A salt containing a metal and the radical HCO_3; also refers to the radical alone.

Bioassay: A method used to determine the toxicity of specific contaminants. A number of individuals of a sensitive species are placed in water containing varying concentrations of the contaminant for a specified period of time and the loss in population is tabulated.

Biochemical oxygen demand (BOD): A measure of the dissolved oxygen consumed by microbial life while assimilating and oxidizing the organic matter present in water.

Biodegradation: The breakdown of chemical constituents through the biological processes of naturally occurring organisms.

Bit: The cutting tool attached to the bottom of the drill stem. Bit design varies according to the type of drilling equipment used. Bits are selected for their ability to drill in various types of formations. Examples include roller cone and drag-type bits.

Bladder pump: A pump consisting of a flexible bladder contained within a rigid cylindrical body. The lower end of the bladder is connected through a check valve to the intake port, while the upper end is connected to a sampling line that leads to ground surface. A second line, the gas line, leads from ground surface to the annular space between the bladder and the outer body of the pump. After filling, under hydrostatic pressure, application of gas pressure causes the bladder to collapse, closing the check valve and forcing the sample to ground surface through the sample line. Gas pressure is often provided by a compressed air tank, and commercial models generally include a control box that automatically switches the gas pressure off and on at appropriate intervals.

Blooey line: Air discharge line used during air rotary drilling operations.

Borehole: A hole drilled or bored into the earth, usually for exploratory or economic purposes; a hole into which casing, screen, and other materials may be installed to construct a well.

Borehole geophysics: Techniques that use a sensing device that is lowered into a borehole or well for the purpose of characterizing geologic formations and their associated fluids. The results can be interpreted to determine lithology, resistivity, bulk density, porosity, permeability, and moisture content and to define the source, movement, and physical/chemical characteristics of ground water.

Borehole log: The record of geologic units penetrated, drilling progress, depth, water level, sample recovery, volumes and types of materials used, and other significant facts regarding the drilling of an exploratory borehole or well.

Bore volume: The volume of water contained in a monitoring well and the adjacent filter pack material present within the borehole in which the well was installed.

Bridge: An obstruction within the borehole annulus that may prevent circulation (during drilling) or proper emplacement of filter pack or annular seal materials (during well construction).

Bridge seal: An artificial plug set to seal off specific zones in the abandonment of a well or borehole.

Bridge-slot screen: A well screen that is manufactured on a press from flat sheets that are perforated, rolled, and seam welded, and in which the slots are vertical and occur as two parallel openings longitudinally aligned to the well axis.

Bridging: The development of gaps caused by obstructions in either grout or filter pack materials during emplacement. Also refers to blockage of particles in natural formation materials or artificial filter pack materials that may occur during well development.

Brine: A concentrated saline solution, especially of chloride salts.

BTEX: Abbreviation for benzene, toluene, ethylbenzene, and xylenes. Aromatic compounds that are the most water soluble of the major gasoline components, and therefore common indicators of gasoline contamination.

Buffer: Any substance or mixture of compounds that, when added to a solution, is capable of neutralizing both acids and bases without appreciably changing the original acidity or alkalinity of the solution.

Bulk density, soil: The mass of dry soil per unit bulk volume which is determined before drying to a constant weight at 105°C.

Buried valley: A depression in a preexisting land surface or in bedrock now covered by younger, usually unconsolidated deposits, especially a preglacial valley filled with outwash, covered by glacial drift.

Cable tool drilling: A drilling technique in which a drill bit attached to the bottom of a weighted drill stem is raised and dropped to crush and grind formation materials. In unconsolidated formations, casing is usually driven as drilling proceeds to prevent collapse of noncohesive materials into the borehole.

Calcereous: Containing calcium carbonate, calcium, or lime.

Calcium hydroxide: A primary constituent of neat cement.

Calibration: The evaluation of the accuracy of an instrument. Calibration is accomplished by measuring acceptable standards and determining any difference between the standard known value and the reading of the instrument. Calibration standards should be traceable to the National Bureau of Standards whenever possible.

Caliper logging: A logging technique used to determine the diameter of a borehole or the internal diameter of casing through the use of a probe with one or more spring-loaded expanding prongs. Caliper logging indicates variations in the inside diameter of a vertical borehole or well casing.

Capillary forces: The forces acting on soil moisture in the unsaturated zone, attributable to molecular attraction between soil particles and water.

Capillary fringe: The zone immediately above the water table, in which water is drawn upward by capillary attraction, and the interstices (voids) are filled with water under pressure less than atmospheric.

Capillary potential: The amount of work that must be done per unit quantity of pure water in order to transport reversibly and isothermally an infinitesimal quantity of water (identical in composition to the soil water) from a pool at the elevation and at the external gas pressure of the point under consideration, to the soil water. Same as matric potential.

Capillary pressure: The difference in pressure across the interface between two immiscible fluid phases concurrently occupying the interstices of any porous material. This difference is due to the tension of the interfacial surface, the value of which depends upon the surface curvature.

Capillary water: Water held in tiny voids in soil or rock by capillary forces.

Carbonate: A sediment formed by the organic or inorganic precipitation from aqueous solution of carbonates of calcium, magnesium, or iron; a mineral compound characterized by a fundamental anionic structure of CO_3^{-2}.

Carbonate rocks: Rocks consisting chiefly of carbonate minerals, such as limestone ($CaCO_3$) and dolomite [$CaMg(CO_3)_2$].

Casing: An impervious, durable pipe placed in a borehole to prevent the walls of the borehole from caving, and to seal off surface drainage or undesirable water, gas, or other fluids and prevent their entrance into the well.

Casing driver: A device fitted to the top-head drive of a rotary rig that is used to advance casing into the subsurface.

Casing, flush-coupled: Casing joined with a coupling having the same outside diameter as the casing, but with two female threads. The inside diameter of the male-threaded coupling is approximately 3/16 inch smaller than that of the casing. Flush-coupled casing has thinner walls than flush-joint casing.

Casing, flush-joint: Casing with an outside diameter at the finished joint equal to the outside diameter of the remainder of the casing; has a male thread at one end and a female thread at the other. No coupling is used to join flush-joint casing.

Casing, protective: A section of large-diameter pipe that is emplaced over the upper end of a smaller diameter monitoring well casing to provide structural protection to the well and restrict unauthorized access into the well.

Casing, surface: Pipe used to stabilize a borehole near the surface during and following the drilling of the borehole.

Casing, temporary: A string of casing temporarily set in the borehole to stabilize a section of the formation and/or to prevent leakage into and out of the formation and to allow drilling to continue to a greater depth.

Cation: An atom or radical carrying a positive charge.

Cation exchange capacity (CEC): A measure of the availability of cations that can be displaced from sites on solid surfaces and that can be exchanged for other cations. For geologic materials, CEC is expressed as the number of milliequivalents of cations that can be exchanged per 100 grams of dry sample.

Caving: The inflow of unconsolidated material into a borehole that occurs when the borehole walls lose their cohesive strength. Also called sloughing.

Cavitation: Corrasive and corrosive effect of collapsing of bubbles produced by decrease of pressure due to increase of velocity in a water stream at the point where pressure is increased. Occurs in electric submersible pumps at the impellers.

Cement bond log: A downhole geophysical log used to determine the integrity of the cement bond to the casing.

Cement (neat cement): A mixture of calcium aluminates and silicates that is made by combining lime and clay while heating. It is mixed with water and emplaced in the annular space to form a seal between the casing and the borehole.

Cement, quick-setting: Cement of special composition and fineness of grind that sets much more quickly than ordinary cement. This cement is generally used for plugging cavities in boreholes, such as large fractures and limestone solution channels.

Cementing: The emplacement of a cement slurry by various methods so that it fills the space between the casing and the borehole wall in a predetermined interval in the well. This secures the casing in place and excludes water and other fluids from entering the borehole.

Center plug: A plug within the pilot assembly of a hollow-stem auger that is used to prevent formation materials from entering the stem of the lead auger during drilling.

Centralizers: Guides that are used to center the casing in the borehole to ensure effective placement of filter pack or annular seal materials.

Centrifugal pump: A pump that moves a liquid by accelerating it radially outward in an impeller to a surrounding spiral-shaped casing.

CERCLA: Comprehensive Environmental Response, Compensation, and Liability Act, passed by Congress in 1980 and administered by the U.S. Environmental Protection Agency.

Chain of custody: Method for documenting the history and possession of a sample from the time of its collection, through its analysis and data reporting, to its final disposition.

Check valve: Ball and spring valves on core barrels, bailers, and sampling devices that are used to allow water to flow in one direction only.

Chemical activity: The molal concentration of an ion multiplied by a factor known as the activity coefficient.

Chemical oxygen demand (COD): The amount of oxygen, expressed in parts per million, consumed under specified conditions in the oxidation of organic and oxidizable inorganic matter in wastewater, corrected for the influence of chlorides.

Chemical spike (spike): A sample that contains a measured amount of a known analyte, used for determining matrix interferences.

Chemical standards: Materials made from ultrapure compounds used to calibrate laboratory analytical equipment.

Chisel bit: A beveled, heavy steel bit used in cable tool drilling.

Chlorination: The process of introducing a chlorine solution into a well for the purpose of destroying nuisance bacteria.

Circulate: To cycle drilling fluid through the drill pipe and borehole while drilling operations are temporarily suspended, to condition the drilling fluid and the borehole before hoisting the drill pipe, and to obtain cuttings from the bottom of the well before drilling proceeds.

Circulation: The movement of drilling fluid from the suction pit through the pump, drill pipe, bit, and annular space in the borehole, and back again to the suction pit. The time involved is usually referred to as circulation time.

Circulation, loss of: The loss of drilling fluid into the formation through crevices or by infiltration into a porous medium.

Clay (grain size): A rock or mineral fragment or a detrital particle of any composition having a diameter smaller than 1/256 mm (0.002 mm).

Clay (mineral): A plastic, soft, variously colored earth, commonly a hydrous silicate of aluminum, formed by the decomposition of feldspar and other aluminum silicates.

Coefficient of variation: The standard deviation divided by the mean of a set of data. The coefficient of variation can be expressed as a percentage by multiplying the number obtained by 100.

Coliform: Group of several types of bacteria that are found in the alimentary tract of warm-blooded animals; often used as an indicator of animal and human fecal contamination of water.

Collapse strength: The capability of a casing or well screen to resist collapse by any or all external loads to which it is subjected during and after installation.

Common depth point: A seismic reflection survey method where subsurface reflectors are revealed by gathering and stacking seismic traces from the same point(s) on the reflector in order to attenuate noise and events not related to the main reflections. Also referred to as common midpoint.

Common ion effect: The decrease in the solubility of a salt dissolved in water already containing some of the ions of the salt.

Common offset: A seismic reflection survey method where subsurface reflectors are revealed by stacking traces from a single source/geophone combination. This common offset method attempts to find a window where other seismic waves do not interact with and/or obliterate the reflected signal.

Complexation: Electrostatic association of positively charged metal ions and negatively charged organic matter, usually with two or more points of attachment. Also called chelation.

Components of variability: The characteristics that vary between statistical populations, such as well logs and analytical lab errors.

Compound: A chemical combination of two or more elements in definite ratios by weight in which the set of characteristics of each element is lost.

Compressive strength: A measure of the stress that a substance can bear under compression without deformation.

Concentrated: Describes a solution that contains a relatively large quantity of solute.

Concentration profiles: Graphic representations of the horizontal and vertical locations of contaminant concentration levels on maps and cross-sections.

Conductance: The ability of a solution to carry an electrical charge; related to the total ion concentration.

Conductivity (electrical): A measure of the quantity of electricity transferred across a unit area, per unit potential gradient, per unit time. It is the reciprocal of resistivity.

Cone of depression: A depression in the water table or potentiometric surface that has the shape of an inverted cone and develops around a well from which water is being withdrawn.

Cone of impression: A conical mound on the water table that develops in response to well injection whose shape is identical to the cone of depression formed during pumping of the aquifer, but inverted.

Confined aquifer: (*See* Aquifer, confined)

Confining bed: A body of geologic material of low hydraulic conductivity that is stratigraphically adjacent to one or more aquifers. It may lie above or below the aquifer. Also called an aquiclude.

Connate water: Water that was deposited simultaneously with the geologic formation in which it is contained.

Conservative substance: A dissolved constituent that moves as fast as water in the ground-water system; a substance not subject to attenuation beyond that occurring by dispersion or diffusion.

Contaminant: Any physical, chemical, biological, or radiological substance or matter in air, soil, or water that has an adverse impact.

Contamination: The introduction into air, soil, or water of any chemical material, organic material, live organism, or radioactive material that will adversely affect the quality of the medium.

Continuous geophysical measurements: The practice of taking and recording geophysical measurements continuously along a traverse line. Commonly used in electromagnetic, magnetic, radar, and also downhole survey methods. *See* Station geophysical methods.

Continuous sampling tube system: Thin-wall sampling tube attached in advance of the cutting head of the hollow-stem auger that allows undisturbed samples to be taken continuously while the augers are rotated.

Consolidated formation: Naturally occurring geologic formation that has been lithified. The term is sometimes used interchangeably with the word "bedrock."

Continuous slot wire-wound screen: A well screen that is made by winding and welding triangular-shaped, cold-rolled wire around a cylindrical array of rods. The spacing of each successive turn of wire determines the slot size of the screen.

Coprecipitation: The carrying down by a chemical precipitate of substances normally soluble under the conditions of precipitation.

Core: A continuous columnar sample of the lithologic units extracted from a borehole. Such a sample preserves stratigraphic contacts and structural features.

Core barrel sampler: A reaming shell and length of tubing used during air or mud rotary drilling to collect formation samples in both consolidated and unconsolidated formations. Core barrels may be single- or double-walled, and of a swivel or rigid type.

Core lifter: A tapered split ring inside the core bit and surrounding the core. On lifting the rods, the taper causes the ring to contract in diameter, seizing and holding the core.

Corrosion: The adverse chemical alteration that converts elemental metals back to more stable mineral compounds, and that affects the physical and chemical properties of the metal.

Corrosive environments: Subsurface zones containing ground water or soil corrosive to metallic materials.

Cost-plus contract: Contracts that list specific costs associated with performing the work and include a percentage of those costs as an additional amount that will be paid to perform a job.

Coupling: A connector for drill rods, pipe, or casing with identical threads, male and/or female, at each end.

Cross-contamination: The movement of contaminants between aquifers or water-bearing zones through an unsealed or improperly sealed borehole.

Curie: The quantity of any radioactive material giving 3.7×10^{10} disintegrations per second. A picocurie is one trillionth of a curie, or a quantity of radioactive material giving 22.2 disintegrations per minute.

Cutter head: The auger bit located at the lead end of the auger column that breaks up formation materials during drilling.

Cuttings: Formation materials obtained from a borehole during the drilling process. Also called drill cuttings.

Darcy's law: An empirical law that states that the velocity of flow through a porous medium is directly proportional to the hydraulic gradient, assuming that the flow is laminar and inertia can be neglected.

Datum: An arbitrary surface (or plane) used in the measurement of heads (*See* NGVD).

Decant: To pour from one vessel into another; to draw off without disturbing sediment or lower liquid layers.

Decontamination: A variety of processes used to clean equipment that has contacted formation material or ground water that is known to be or suspected of being contaminated.

Dedicated (sampling equipment): Sampling equipment that is reserved for use in only one monitoring well.

Degas: To remove dissolved gases from a liquid.

Deionized water: Water from which all free ions have been removed.

Dennison sampler: A specialized sampler of a double-tube core design with a thin inner tube that permits penetration in extremely stiff or highly cemented unconsolidated deposits, while collecting a thin-wall sample.

Density: The mass or quantity of a substance per unit volume. Metric units are kilograms per cubic meter or grams per cubic centimeter. *See* Specific gravity.

Density log: A downhole geophysical tool which uses a gamma ray source and detector to measure backscattered gamma rays from the formation to obtain bulk formation density indirectly. *See* Gamma-gamma log.

Desorption: The process of removing a sorbed substance by the reverse of adsorption or absorption.

Detection limit: The lowest concentration of a chemical that can be reliably reported to be different from zero concentration.

Detection monitoring: A program of monitoring under Resource Conservation and Recovery Act requirements conducted for the express purpose of determining whether or not there has been a contaminant release to ground water.

Detergent: Any material with cleaning powers, including soaps, synthetic detergents, many alkaline materials, and solvents.

Development: The act of repairing damage to the formation caused during drilling procedures, and increasing the porosity and hydraulic conductivity of the materials surrounding the screened portion of the well.

Diagenesis: The chemical and physical changes occurring in sediments before consolidation or while in the environment of deposition.

Diatomaceous earth: A cement additive (composed of siliceous skeletons of diatoms) used to reduce slurry density and increase water demand and thickening time; reduces set strength of the cement.

Dielectric: Substance having a very low electrical conductivity.

Differential pressure: The difference in pressure between the hydrostatic head of the drilling fluid–filled or empty borehole, and the formation pressure at any given depth.

Diffusion (molecular): Process whereby ionic or molecular constituents move under the influence of their kinetic activity in the direction of their concentration gradient; the spreading of a dissolved constituent due to the random motion of the constituent molecules.

Digital computer model: A model of ground-water flow in which the aquifer is described by numerical equations with specified values for boundary conditions solved on a digital computer.

Dilute: Describes a solution that contains a relatively small quantity of solute.

Direct mud rotary drilling: A drilling technique in which a drilling fluid is pumped down the drill rod and through the bit and circulates back to the surface by moving up the annular space between the drill rods and the borehole.

Discharge area: An area in which there are upward components of hydraulic head in the aquifer. Ground water flows toward the surface in a discharge area and may escape as a spring, seep, or baseflow, or by evaporation and transpiration.

Discharge velocity: An apparent velocity, calculated from Darcy's law, which represents the flow rate at which water would move through an aquifer if the aquifer were an open conduit. Also called specific discharge.

Dispersion: A process of contaminant transport that occurs as a result of mechanical mixing and molecular diffusion; the extent to which a liquid substance introduced in a ground-water system spreads as it moves through the system.

Dispersion coefficient: A measure of the spreading of a flowing substance due to the nature of the porous medium, with its interconnected channels distributed at random in all directions.

Dispersivity: A geometric property of a porous medium that determines the dispersion characteristics of the medium by relating the components of pore velocity to the dispersion coefficient.

Disposal facility (hazardous waste): A facility as defined in 40 CFR 260.10 in which hazardous waste is intentionally placed into or on land or water, and at which waste will remain after closure of the facility.

Dissociation: The splitting up of a compound or element into two or more simple molecules, atoms, or ions. Usually applied to the effect of the action of heat or solvents upon dissolved substances. The reaction is reversible and not as permanent as decomposition; that is, when the solvent is removed, the ions recombine.

Dissolution: The process of dissolving.

Dissolved: Generally, the fine intermingling of a small amount of material (commonly a solid) in a large amount of material (commonly a liquid) to form a combination that will not settle upon standing (a solution). For environmental samples, the operational definition differs somewhat from the general definition; a dissolved constituent is one that can pass through a filter with pores that are 0.45 micrometer (um) in diameter.

Dissolved solids: The weight of matter in true solution in a stated volume of water, including both inorganic and organic matter.

Distilled water: Water that has been purified through a separation process in which the water is converted to vapor and the vapor is then condensed to a liquid.

Distribution coefficient: The quantity of a solute, chemical or radionuclide sorbed by a solid per unit weight of solid divided by the quantity of solute dissolved in water per unit volume of water.

DNAPL: Acronym for dense, nonaqueous phase liquid (*See* Sinker).

Dolomite: A carbonate sedimentary rock composed predominantly of $CaMg(CO_3)_2$.

Downgradient: In the direction of decreasing hydrostatic head.

Downgradient well: A well that has been installed hydraulically downgradient of a site and is capable of detecting the migration of contaminants from a site. RCRA regulations require the installation of three or more downgradient wells depending on the site-specific hydrogeological conditions and potential zones of contaminant migration.

Down-the-hole hammer: A pneumatic drill operated at the bottom of the drill pipe by air pressure provided from the surface.

Drainage well: A well that is installed for the purpose of draining swampy land or disposing of storm water, sewage, or other wastewater at or near the land surface.

Drawdown: The vertical distance that the potentiometric surface is lowered due to the removal of ground water by a pumping well; lowering of the water surface in a well and water-bearing zone resulting from the discharge of water from the well.

Drill collar: A length of heavy, thick-walled pipe used to stabilize the lower drill string, to minimize bending caused by the weight of the drill pipe and to add weight to the bit.

Drill cuttings: Fragments or particles of soil or rock, with or without free water, created by the drilling process. Also called cuttings.

Drilling fluid: A water- or air-based fluid used in the well drilling operation to remove cuttings from the borehole, to clean and cool the bit, to reduce friction between the drill string and the sides of the borehole and to hold the borehole open during the drilling operation. Also called mud or drilling mud.

Drill pipe: Special pipe used to transmit rotation from the rotating mechanism to the bit. The pipe also transmits weight to the bit and conveys air or fluid, which removes cuttings from the borehole and cools the bit.

Drill rod: Hollow flush-joined or coupled rods that are rotated in the borehole that are connected at the bottom to the drill bit and on the top to the rotating or driving mechanism of the drilling rig.

Drill string: The string of pipe that extends from the bit to the driving mechanism and that serves to rotate the bit.

Drive block: A heavy weight used to drive pipe or casing through unconsolidated material.

Drive couplings: Heavy-duty couplings used to join sections of heavy-wall drive casing that are specifically designed to withstand the forces during driving.

Drive head: A component fastened to the top of a driven casing to take the blow of the drive block.

Driven well: A well that is driven to the desired depth, either by hand or machine; may employ a well point or alternative technology.

Drive shoe: A forged steel collar with a cutting edge fastened onto the bottom of a driven casing to shear off irregularities in the hole as the casing advances; designed to withstand drive pressures to protect the lower edge of the casing as it is driven.

Drop hammer: A weighted device used to drive samplers (i.e., split-spoon samplers) during drilling and sampling.

Dry well: A borehole or well that does not extend into the zone of saturation. Also, a well completed within the saturated zone in formation materials that do not yield water to wells.

Dual-wall reverse circulation: A drilling technique in which the circulating fluid is pumped down between the outer casing and the inner drill pipe, through the drill bit and up the inside of the drill pipe.

Dynamic equilibrium: A condition in which the amount of recharge to an aquifer equals the amount of natural discharge.

Effective grain size: The particle grain size of the sample where 90% represents coarser-size grains and 10% represents finer-size grains, i.e., the coarsest diameter in the finest 10% of the sediment. Also called effective diameter or effective size.

Effective porosity: The amount of interconnected pore space through which fluids can pass, expressed as a percent of bulk volume. Part of the total porosity may be occupied by static fluid being held to the mineral surface by surface tension or in dead-end pores, so effective porosity may be less than total porosity.

Effluent: A waste liquid from a manufacturing or treatment process, in its natural state or partially or completely treated, that is discharged into the environment.

Effluent stream: A stream or reach of a stream in which the flow is being supplemented by, or consists solely of, the inflow of ground water. Also called gaining stream.

Eh: A measure of the electron balance in an environmental sample; the numerical indication of oxidation-reduction conditions, much as hydrogen-ion concentration or pH are measures of acid-base conditions.

Electrical conductance: A measure of the ease with which a conducting current can be caused to flow through a material under the influence of an applied electric field. It is the reciprocal of the electrical resistivity and is measured in micromhos per centimeter. Also called specific conductance.

Electrical earth resistivity: A surface geophysical method in which a known current is applied to a pair of electrodes in the ground and the resulting voltage is measured at a second set of electrodes. The apparent resistivity of earth materials between and below the electrodes is calculated knowing the current applied, voltage measured, and the geometry and spacing of the electrodes. Particularly useful at sites characterized by settings having minimal quantities of high resistance materials. Can be used to map both formation materials and certain types of water quality changes.

Electrical resistivity: A characteristic of earth materials and their contained fluids (independent of geometry). Resistivity is measured in ohm-feet or ohm-meters. The reciprocal of electrical conductivity.

Electric logging: Logging techniques used in fluid-filled boreholes to obtain information concerning the porosity, hydraulic conductivity, and fluid content of the formations drilled, based on the dielectric properties of the aquifer materials. Usually refers to a resistivity log but sometimes used to indicate resistance and spontaneous potential logs.

Electric submersible pump: A pump that consists of a rotor contained within a chamber and driven by an electric motor. The entire device is lowered into the well with the electrical cable and discharge tubing attached. A portable power source and control unit remains at the surface. Electric submersible pumps used for ground-water sampling are constructed of inert materials and secure seals to prevent sample contamination by lubricants.

Electrolyte: A chemical that dissociates into positive and negative ions when dissolved in water, thereby increasing the electrical conductivity.

Electromagnetic (EM) conductivity: A surface geophysical method in which induced currents are produced and measured in conductive formations from electromagnetic waves generated at the surface. Can be used to define shallow ground-water zones characterized by high dissolved solids content.

Element: Any substance that cannot be separated into different substances by ordinary chemical methods. Each element has its own set of characteristics, which differs from that of all the other elements.

Equilibrium: Condition that exists in a system when the phases of the system do not undergo any change of properties with the passage of time; the state in which the action of multiple forces produces a steady balance, resulting in no change over time.

Equilibrium constant: The number defining the conditions of equilibrium of a particular reversible chemical reaction.

Equipment blank: Chemically pure solvent (typically reagent-grade, distilled, or deionized water) that is passed through an item of field sampling equipment and returned to the laboratory for analysis, to determine the effectiveness of equipment decontamination procedures.

Equipotential: Equal pressure.

Equipotential line: A line in a two-dimensional ground-water flow field along which the total hydraulic head is the same at all points; a line drawn between points of equal pressure.

Equipotential surface: A surface in a three-dimensional ground-water flow field such that the total hydraulic head is the same everywhere on the surface.

Equivalent weight: The concentration in parts per million of a solute multiplied by the valence charge and then divided by its formula weight in grams.

Established ground surface: The permanent elevation of the ground surface at the site of a well upon completion.

Esters: A class of organic compounds derived by the reaction of an organic acid with an alcohol.

Evaporation: The process by which water passes from the liquid to the vapor state.

Evapotranspiration: The sum of evaporation plus transpiration.

Exfiltration: The leakage of effluent or water from a pipe (i.e., a sewer pipe) or other structure into the subsurface.

Extractable organic compounds: Organic compounds dissolved in a water sample that transfer almost entirely to an organic solvent such as hexane or methylene chloride when the solvent and water sample come into contact with each other.

False negative: Contamination is present, but the results of sample analyses fail to indicate its presence.

False positive: No contamination is present, but the results of sample analyses indicate its presence.

Fault: A fracture or a zone of fractures along which there has been displacement of the sides relative to one another, parallel to the fracture.

Field blank: A laboratory-prepared sample of reagent-grade water or pure solvent that is transported to the sampling site for use in evaluation of field sampling procedures. *See* Equipment blank and Trip blank.

Field capacity: The maximum amount of water that a soil can hold against the pull of gravity. Also called specific retention.

Filter cake: The suspended solids from a drilling fluid that are deposited on the walls of a borehole in a porous medium during the process of drilling. Also called mudcake.

Filter pack: Sand or gravel that is generally uniform, clean, and well-rounded and that is placed in the annulus of the well between the borehole wall and the well screen to prevent formation material from entering through the well screen, and to stabilize the adjacent formation.

Filter pack ratio: A ratio used to express size differential between the formation materials and the filter pack that typically refers to either the average grain size (D50) or the 70% (D70) retained size of the formation material.

Filtrate invasion: The movement of drilling fluid into the formation adjacent to the borehole that occurs when the weight of the drilling fluid substantially exceeds the natural hydrostatic pressure of the formation.

Filtration: The operation of separating suspended solids from a liquid (i.e., a water sample) by forcing the mixture through a porous barrier (i.e., filter medium; a 0.45-micron filter is most commonly used).

Fine-grained soil (or geologic material): Soil or material consisting of mostly clay and silt, more than 50% by weight, smaller than 0.074 mm in diameter.

Fixed-price contracts: Contracts that list the manpower, materials, and additional costs needed to perform the work specified as a fixed cost payable upon completion.

Floaters: Lighter-than-water fluids, generally petroleum hydrocarbons or other organic liquids, capable of forming an immiscible layer that can float on the water table.

Float shoe: A drillable valve attached to the bottom of a drive casing.

Flocculation: The agglomeration of finely divided suspended solids into larger, usually gelatinous particles through electrical charge alignment of particles.

Flow, steady: The flow that occurs when, at any point in the flow field, the magnitude and direction of the specific discharge are constant in time.

Flow, unsteady: The flow that occurs when, at any point in the flow field, the magnitude or direction of the specific discharge changes with time. Also called transient flow or nonsteady flow.

Flow lines: Lines indicating the direction followed by ground water toward points of discharge. Flow lines are perpendicular to equipotential lines.

Flow meter: A tool used to monitor fluid flow rates in cased or uncased boreholes using low-inertia impellers or through changes in thermal conductance as liquids pass through the tool.

Flow net: A set of intersecting equipotential lines and flow lines representing a two-dimensional steady flow field in porous media.

Flow path: The subsurface course that a water molecule or solute would follow in a given ground-water velocity field; the direction of movement of ground water and any contaminants that may be contained therein, as governed principally by the hydraulic gradient.

Flow-through well: A small-diameter monitoring well that penetrates all or a significant portion of the aquifer and is designed to create minimal distortion of the flow field in the aquifer.

Fluid conductivity log: Borehole geophysical log made by a tool that measures the specific conductance of borehole fluids. Units are usually in micromhos/cm.

Fluid loss: Measure of the relative amount of fluid (filtrate) lost through permeable formations when the drilling fluid is subjected to a pressure differential.

Fluid potential: The mechanical energy per unit mass of fluid at any given point in space and time with regard to an arbitrary state and datum.

Fluoropolymers: Man-made materials consisting of different formulations of monomers molded by powder metallurgy techniques that exhibit anti-stick properties and resistance to chemical and biological attack. Polytetrafluoroethylene is one type of fluoropolymer.

Flush-coupled casing: (*See* Casing, flush-coupled).

Flushing: The act of causing rapid removal of water or other fluids from a bore-hole or a well.

Flush-joint casing: (*See* Casing, flush-joint).

Flux: The rate of transfer of a quantity (water, heat, etc.) across a surface.

Fly ash: An additive to cement that increases sulfate resistance and early compressive strength.

Formation: A mappable unit of consolidated or unconsolidated geologic material characterized by a degree of lithologic homogeneity.

Formation fluid: The natural fluid present in a formation.

Fracture: A break in a rock formation due to structural stresses. Faults, shears, joints, and planes of fracture cleavage are all types of fractures.

Frequency domain: An instrument system that functions by taking measurements as a function of frequency or at a given frequency. *See* Time domain.

Friable: Easily crumbled. For example, a sandstone that is poorly cemented.

Fully penetrating well: A well installed in a borehole drilled to the bottom of an aquifer, and constructed in such a way that it is screened through the entire thickness of the aquifer.

Gaining stream: (*See* Effluent stream).

Gamma-gamma log: A borehole geophysical log created by a tool consisting of a source of gamma radiation as well as a detector; measures bulk density of the formation and contained fluids indirectly through the observation of back-scattered gamma rays from the formation.

Gas chromatography (GC): An instrumental method for separating and identifying organic compounds, and measuring their concentrations. The various compounds pass through the chromatographic column at different rates; this time of travel through the column (called retention time) forms the basis for compound identification. Detection limits are commonly in the range of 1 to 10 micrograms per liter.

Gas chromatography/mass spectroscopy (GC/MS): A tandem instrumental method for separating, identifying, and quantifying organic compounds. The GC separates the compounds. Compound identification is based on the compound retention time in the GC and on the mass spectral pattern. Compound quantification is normally done by measuring peak heights in the mass spectra. Detection limits are commonly 5 to 10 micrograms per liter.

Gas-drive piston pumps: Gas-drive piston pumps are similar in principle and operation to bladder pumps. The bladder is replaced by a piston (two pistons in some cases) and an arrangement of check valves. They are constructed of inert materials and are of a diameter suitable for use in 2-inch (5-cm) diameter sampling wells.

Gas-drive pump: A pump that utilizes the mechanical process of lifting a column of water from within a pump body and attached discharge tubing, in which a pressurized gas is used as the lifting agent.

GC: (*See* Gas chromatography).

GC/MS: (*See* Gas chromatography/mass spectroscopy).

Gel strength: A measure of the capability of a drilling fluid to maintain suspension of particulate matter in the mud column when the mud pump is off.

Geophysics: In the broad sense, this term implies physics of the earth. In a more contemporary sense, the term is used to refer to the field and methods of measurements used to assess relatively shallow subsurface conditions by nondestructive means.

Glacial drift: A general term for unconsolidated sediment transported by glaciers and deposited directly on land or in the sea.

Goodness of fit: A statistical test to determine the likelihood that sample data have been generated from the population that conforms to a specified type of probability distribution.

Graded: An engineering term pertaining to the variation of sizes in a soil or an unconsolidated sediment; a soil consisting of particles of several or many sizes or having a uniform or equable distribution of particles from coarse to fine. Well-graded materials have many sizes, whereas poorly graded materials are more uniform in size.

Gradiometer: A device for measuring the vertical gradient of the earth's magnetic field using an arrangement of two magnetometer sensors, one above the other.

Grain size: The general dimensions of the particles in a sediment or rock, or of the grains of a particular mineral that make up a sediment or rock. It is common for these dimensions to be referred to with broad terms, such as fine, medium, coarse. One widely used grain size classification is the Udden-Wentworth grade scale.

Gravel pack: An outdated term used to describe gravel or other permeable filter material placed in the annular space around a well screen to prevent the movement of fine material into the well casing, to stabilize the formation, and to increase the ability of the well to yield water. Preferred term is filter pack.

Gravimetric method: Procedure for determining the soil moisture content.

Ground-penetrating radar: A geophysical method used to identify subsurface formations that will reflect electromagnetic radiation; useful for defining the boundaries of buried trenches and other subsurface features on the basis of time-domain reflectometry.

Ground water: Water beneath the land surface contained in interconnected pores in the saturated zone that is under hydrostatic pressure. The water that enters wells and issues from springs.

Ground water, confined: The water contained in a confined aquifer. Pore-water pressure is greater than atmospheric at the top of a confined aquifer.

Ground water, perched: The water in an isolated saturated zone located within the vadose zone. It is the result of the presence of a layer of material of low hydraulic conductivity. Perched ground water will have a perched water table.

Ground water, unconfined: The water in an aquifer in which there is a water table.

Ground-water basin: A rather vague designation pertaining to a ground-water reservoir that is more or less separate from neighboring ground-water reservoirs. A ground-water basin could be separated from adjacent basins by geologic or hydrologic boundaries.

Ground-water flow: The movement of water through openings in sediment and rock that occurs in the zone of saturation.

Ground-water reservoir: The earth materials and the intervening open spaces that contain ground water.

Grout: A fluid mixture of cement and water (neat cement) with various additives, or a bentonite slurry of a consistency that can be forced through a pipe and emplaced in the annular space between the borehole and the casing to form a low hydraulic conductivity seal.

Grouting: The operation by which grout is placed between the casing and the wall of the borehole to secure the casing in place and to exclude water and other fluids from moving into and through the borehole.

Grout shoe: A "plug" fabricated of relatively inert materials that is positioned within the lowermost section of a permanent casing and fitted with a passageway, often with a flow check device, through which grout is injected under pressure to fill the annular space. After the grout has set, the grout shoe is usually drilled out.

Gypsum: Calcium sulfate; an additive to cement slurries that produces a quick-setting, hard cement that expands upon curing.

Half-life: For radioactive substances, the half-life is the length of time required for half of a given amount of material to disintegrate through radiation. In a chemical sense, the half-life is the time required for one-half of a given material to undergo a chemical reaction.

Halogenated hydrocarbons: Organic compounds containing one or more halogens (i.e., fluorine, chlorine, bromine, and iodine).

Hand auger: Any of a variety of hand-operated devices for installing shallow holes into the ground.

Hardness: A measure of the amount of alkali metal ions, principally calcium, magnesium, and iron, dissolved in water.

Hazardous waste: Any waste or combination of wastes that pose a substantial present or potential hazard to human health or living organisms.

Hazardous waste constituent: A constituent that causes a waste to be classified as hazardous, based upon the criteria cited in 40 CFR 261.2 and 261.3.

Hazardous waste management: The collection, source separation, storage, transportation, processing, treatment, recovery, and disposal of hazardous waste.

Hazardous waste management area: The area within a facility's property boundary that encompasses one or more hazardous waste management units or cells.

Head, static: The height above a standard datum of the surface of a column of water (or other liquid) that can be supported by the static pressure at a given point. The static head is the sum of the elevation head and the pressure head. *See* Head, total.

Head, total: The sum of three components at a point: (1) elevation head, h_e, which is equal to the elevation of the point above a datum; (2) pressure head, h_p, which is the height of a column of static water that can be supported by the static pressure at the point; and (3) velocity head, h_v, which is the height the kinetic energy of the liquid is capable of lifting the liquid.

Head loss: That part of potential energy that is lost because of friction, as water flows through a porous medium.

Headspace: The empty volume in a sample container between the water level and the cap.

Heat of hydration: Exothermic or heat-producing reaction that occurs during the curing of cement.

Heaving sand: Saturated sands encountered during drilling in which the hydrostatic pressure of the formation is greater than the pressure in the borehole, causing the sand to move up into the borehole.

Heavy metals: Metallic elements, including the transition series, which include many elements required for plant and animal nutrition in trace concentrations, but which become toxic at higher concentrations. Examples are mercury, chromium, cadmium, and lead.

Henry's law: An empirical law that states that the amount of gas or vapor dissolved in water is proportional to the pressure of the gas in contact with the water at a specified temperature; that is, the higher the gas pressure, the higher the gas concentration in the water.

Heterogeneous: Pertaining to a substance having different characteristics in different locations. A synonym is nonuniform.

High-yield drilling clay: A classification given to a group of commercial drilling-clay preparations having a yield of 35 to 50 bbl/ton and intermediate between bentonite and low-yield clays. High-yield drilling clays are usually prepared by peptizing low-yield calcium montmorillonite clays or, in a few cases, by blending some bentonite with the peptized low-yield clay.

High-yield monitoring well: A relative term referring to a well capable of quick recovery after it has been purged of stagnant water standing in the well casing; a well in which samples can be collected immediately after purging.

Hollow-stem auger drilling: A drilling technique in which hollow, interconnected flight augers, with a cutting head, are pressed downward as the auger is rotated.

Homogeneous: Pertaining to a substance having uniform characteristics throughout. A synonym is uniform.

Hydraulic conductivity: A coefficient of proportionality that describes the rate at which a fluid can move through a permeable medium. It is a function of both the medium and of the fluid flowing through it; also defined as the quantity of water that will flow through a unit cross-sectional area of a porous material per unit of time under a hydraulic gradient of 1.00 (measured at right angles to the direction of flow) at a specified temperature.

Hydraulic connection: The hydraulic relationship between two different lithologic layers.

Hydraulic diffusivity: A property of an aquifer or confining bed, defined as a ratio of the transmissivity to the storativity.

Hydraulic fracturing: The fracturing of a rock by pumping fluid under high pressure into a well for the purpose of increasing hydraulic conductivity.

Hydraulic gradient: The change in total head with a change in distance in a given direction. The direction is that which yields a maximum rate of decrease in head.

Hydraulic jetting: When jetting is used for well development, a jetting tool with nozzles and a high pressure pump is used to force water outwardly through the screen, the filter pack, and sometimes into the adjacent geologic unit.

Hydrocarbon: A compound composed of the elements carbon and hydrogen.

Hydrodynamic dispersion: The process by which ground water containing a solute is diluted with uncontaminated ground water as it moves through an aquifer.

Hydrograph: A graph that shows some property of ground water or surface water as a function of time.

Hydrologic equation: An expression of the law of mass conservation for purposes of water budgets. It may be stated as inflow equals outflow, plus or minus changes in storage. Also called the continuity equation.

Hydrologic unit: Geologic strata that can be distinguished on the basis of capacity to yield and transmit fluids. Aquifers and confining units are types of hydrologic units. Boundaries of a hydrologic unit may not necessarily correspond either laterally or vertically to geologic formations.

Hydrolysis: The chemical decomposition of a compound or splitting of a compound by reaction with water; reaction of water molecules with dissolved organic or inorganic constituents to form new compounds.

Hydrophilic: Attracted to or preferentially dissolved in water. The opposite of hydrophobic.

Hydrophobic: Tending to avoid being dissolved in water. Attracted more to nonpolar liquids (such as oils) or solids. The opposite of hydrophilic.

Hydrostatic head: The pressure exerted by a column of fluid, usually expressed in pounds per square inch (psi). To determine the hydrostatic head at a given depth in psi, multiply the depth in feet, by the density in pounds per gallon, by 0.052.

Hydrostratigraphic unit: A formation, part of a formation, or a group of formations in which there are similar hydrologic characteristics, allowing for grouping into aquifers or confining layers.

Hygroscopic water: Water that clings to the surfaces of soil particles in the vadose zone.

Hysteresis: A change in the shape of a soil water characteristic curve, which differs depending upon whether soil sorption or desorption occurs.

ICAP: (*See* Inductively coupled argon plasma).

Ideal gas: A gas having a volume that varies inversely with pressure at a constant temperature, and that also expands by 1/273 of its volume at 0°C for each degree rise in temperature at constant pressure.

Igneous rock: A rock that solidified from molten matter that originated within the earth.

Imaging methods: A remote sensing method in which data are displayed in the form of an image, as in a photograph along a flight path. This form of display is often used when measuring natural radiation or radiation from a source that has been reflected by the earth, usually electromagnetic waves (ultraviolet, visible, infrared, microwave, or radar). *See* Nonimaging methods.

Immiscible: Fluids that are not significantly soluble in water.

Incrustation (encrustation): The process by which a crust or coating is formed on the well screen and/or casing; typically, through chemical or biological reactions.

Indicator parameters: Chemical parameters specified for analysis under the Resource Conservation and Recovery Act requirements as indicators of ground-water contamination, including pH, specific conductance, total organic carbon (TOC), and total organic halogens (TOX).

Induction log: A borehole geophysical log created by a logging tool used to measure pore fluid conductivity.

Inductively coupled (argon) plasma (ICP or ICAP): An instrumental method to measure metal ion concentrations in solution. Several ions can be quantified simultaneously, resulting in lower cost than atomic absorption spectroscopy, especially when analyses for several metals are needed.

Inert: Immune to chemical or biological action.

Infiltration: The flow of water downward from the land surface into and through the interstices (pores) in soil and rock.

Infiltration capacity: The maximum rate at which infiltration can occur under specific conditions of soil moisture. For a given soil, the infiltration capacity is a function of the water content.

Influent stream: A stream or reach of a stream in which water flows from the streambed into the ground. Also called losing stream.

Inhibitor (mud): Substances generally regarded as drilling mud contaminants, such as salt and calcium sulfate, are called inhibitors when purposely added to mud so that the filtrate from the drilling fluid will prevent or retard the hydration of formation clays and shale.

Injection well: A well used for injecting fluids into an underground stratum.

Inorganic constituent/compound: Any substance that is not a compound of carbon, with the exception of carbon oxides and carbon disulfide.

Inorganics: Chemicals without organic carbon, including metals and other ions such as chloride, sulfate, and nitrate.

In situ: In the natural or original position; in place.

In situ measurements: Measurements made to describe the properties and conditions of a soil, rock, or pore fluids in their natural or original position (in place) without removing or disturbing the materials.

Interference: The condition occurring when the area of influence of a pumping well comes into contact with or overlaps that of a neighboring well, as when two wells are pumping from the same aquifer or are located near each other.

Interflow: The lateral movement of water in the vadose zone during and immediately after a precipitation event. The water moving as interflow discharges directly into a stream or lake.

Intermediate zone: That part of the unsaturated zone below the root zone and above the capillary fringe.

Intermittent stream: A stream that flows only part of the time.

Interstice: A void in a body of otherwise solid material.

Intrinsic permeability: A term describing the relative ease with which a porous medium can transmit a liquid under a hydraulic or potential gradient. It is distinguished from hydraulic conductivity in that it is a property of the porous medium alone and is independent of the nature of the liquid or the potential field.

Ion: An element or compound that has gained or lost an electron, so that it carries a charge.

Ion balance: A calculation, done in terms of equivalents or milliequivalents, to show the relative amounts of positive and negative charges reported in laboratory results for a solution. All solutions are neutral, so an imbalance in charge results from faulty analysis or failure to analyze for an important ion. Also called anion-cation balance or cation-anion balance.

Ion exchange: A process in which an ion in a mineral lattice is replaced by another ion that was present in a aqueous solution.

Iron bacteria: Bacteria that can oxidize or reduce iron as part of their metabolic process.

Isocon: A line drawn on a map to indicate equal concentrations of a solute in ground water.

Isohyetal line: A line drawn on a map of the ground surface, along which all points receive equal amounts of precipitation.

Isotropic: Having identical properties in all directions.

Jet percussion: A drilling process which uses a wedge-shaped drill bit that discharges water under pressure while being raised and lowered to loosen or break up materials in the borehole.

Jetting: When applied as a drilling method, water is forced down through the drill rods or casing and out through the end aperture. The jetting water then transports the generated cuttings to the ground surface in the annulus of the drill rods or casing and the borehole. The term jetting may also refer to a development technique (*See* Hydraulic jetting).

Karst topography (karst): A topographic area that has been created by the dissolution of a carbonate rock terrain; characterized by sinkholes, caverns, and lack of surface streams.

Kelly: A hollow steel bar that is in the main section of a rotary drill string, to which power is directly transmitted from the rotary table to rotate the drill pipe and bit.

Ketones: A class of organic compounds in which a carbonyl group is bonded to two alkyl groups.

KHz: A unit of frequency equal to 1,000 cycles per second.

Kinematic viscosity: The ratio of dynamic viscosity to mass density. It is obtained by dividing dynamic viscosity by the fluid density. Units of kinematic viscosity are square meters per second.

Knock-out plate: A nonretrievable plate wedged within the auger head that replaces the traditional pilot assembly and center rod that is used to prevent formation materials from entering a hollow-stem auger.

Lagoon: A shallow pond in which wastewater is stored or partially treated by exposure to sunlight, oxygen, and microorganisms. *See* Surface impoundment.

Laminar flow: Flow in which the fluid particles follow paths that are smooth, straight, and parallel to the channel wall. In laminar flow, the viscosity of the fluid damps out turbulent motion. Compare with Turbulent flow.

Landfill: A disposal facility or part of a facility in which waste is placed in or on the land, and which is not a land treatment facility, a surface impoundment, or an injection well.

Law of mass action: The physical law stating that for a reversible chemical reaction the rate of reaction is proportional to the concentrations of the reactant.

Leach: To wash or drain by percolation.

Leachate: A solution produced by the movement or percolation of liquid through soil or solid waste, and the subsequent dissolution of certain constituents in the water.

Leachate management system: In a landfill, a method of collecting leachate and directing it to a treatment or disposal area.

Leakage: The flow of water from one hydrogeologic unit to another. The leakage may be natural, as through a semiimpervious confining layer, or man-made, as through an uncased well.

Leaky confining layer: A low hydraulic conductivity layer that can transmit water at sufficient rates to furnish some recharge to a well pumping from an underlying aquifer. Also called an aquitard.

Less than detection limits: A phrase which indicates that a chemical constituent was either not identified or not quantified at the lowest level of sensitivity of the analytical method being employed by the laboratory. Therefore, the chemical constituent either is not present in the sample, or it is present in such a small concentration that it cannot be measured by the analytical procedure.

Lift: In a pumping scenario, the vertical distance from the pumping level to the point of discharge of the water, plus the friction loss in the discharge pipe.

Limestone: Sedimentary rock made primarily of calcium carbonate.

Lineament: A natural linear surface feature longer than 1500 meters.

Liner: A continuous layer of natural or man-made materials lining the bottom and/or sides of a surface impoundment, landfill, or landfill cell that restricts the downward or lateral escape of hazardous waste, hazardous waste constituents, or leachate.

Lithology: The systematic description of rocks, in terms of mineral composition and texture.

Loading rate: The rate of application of a material to the land surface.

Loess: A homogeneous, nonstratified, unconsolidated deposit consisting predominantly of silt-size particles, with subordinate amounts of very fine sand and/or clay, deposited primarily by the wind.

Losing stream: (*See* Influent stream).

Lost circulation: The result of drilling fluid escaping from the borehole into the formation by way of crevices or porous media.

Louvered screen: A well screen with openings that are manufactured in solid-wall metal tubing by stamping outward with a punch against dies that control the size of the openings. Also called a shutter screen.

Low-solids muds: A designation given to any type of mud in which high-performing additives have been partially or wholly substituted for commercial or natural clays.

Low-yield monitoring wells: A relative term referring to a well that cannot recover in sufficient time after well evacuation to permit the immediate collection of water samples.

Lysimeter: Vadose zone sampling device used to collect soil pore water via suction in the unsaturated zone. Lysimeters are capable of retaining the accumulated water within the sampling vessel.

Machine-slotted casing: Well screens fabricated from standard casing in which slots of a predetermined width are cut into the casing at regular intervals using machining tools.

Magnetometer: A surface geophysical method for measuring the strength of the earth's magnetic field. Variations in the earth's magnetic field may be caused by changes in magnetic properties of soil and rock conditions or by the presence of ferrous metals.

> • *Total Field Magnetometer:* One which usually measures the total field intensity as opposed to the vertical component, a horizontal component, or the gradient of the field.

- *Gradient Magnetometer:* One which measures the vertical (or horizontal) gradient of the magnetic field.
- *Fluxgate Magnetometer:* A type of magnetometer capable of making continuous measurements of the magnetic field component along the axis of its ferrous core.
- *Proton Procession Magnetometer:* A type of magnetometer that measures the magnetic field by proton procession within the earth's field. Because of its design, must be cycled on and off between measurements and, therefore, cannot be used to obtain continuous measurements.

Manifest: A shipping document required for all hazardous waste shipments, naming the material, quantity, point of origin, shipping route, and destination.

Manometer: A device used to measure fluid or vapor pressures; consists of a tube filled with a liquid so that the level of the liquid is determined by the fluid pressure, and the height of the liquid may be read from a scale.

Marsh funnel: A device used to measure drilling fluid viscosity in which the time required for a known volume of drilling fluid to drain through an orifice is measured and calibrated against a time for draining of an equal volume of water.

Marsh funnel viscosity: A measure of the number of seconds required for 1 quart (946 mL) of a given fluid to flow through the Marsh funnel.

Matrix: The solid framework of a porous medium.

Mature karst: Karst environment in which the physical features (e.g., sinkholes, caves) are well defined. *See* Karst topography.

Maximum contaminant level: The highest concentration of a solute permissible in a public water supply, as specified in the national Primary Drinking Water Standards established under the Safe Drinking Water Act (SDWA) by the United States Environmental Protection Agency.

Maximum value: In a set of data, the measurement having the highest numerical value.

Mean: The sum of all measurements collected over a statistically significant period of time (e.g., one year), divided by the number of measurements. Commonly called the average.

Mechanical joining: The use of threaded connections to join two sections of casing.

Median: The middle point in a set of measurements ranked by numerical value. If there are an even number of measurements, the median is the mean of the two central measurements.

Mesh: A measure of fineness of a woven material, screen, or sieve; e.g., a 200-mesh sieve screen with a wire diameter of 0.0021 inch (0.0533 mm) has an opening of 0.074 mm, or will pass a particle of 74 microns.

Metal: A chemical element, usually characterized by lustrous appearance, malleability, and the ability to conduct electricity; tends to donate electrons and thereby become positively charged. Over three-quarters of all elements are metals.

Metal detector: A surface geophysical survey method in which induced currents are produced within ferrous and nonferrous metals by a transmitter coil. The resulting magnitudes of these induced currents are a measure of the presence of metals.

Metamorphic rock: A rock that has been derived from preexisting rocks by mineralogicol, chemical, and structural changes, essentially in the solid state, in response to marked changes in temperature, pressure, shearing stress, and chemical environment at depth in the earth's crust.

MHz: A unit of frequency equal to 1,000,000 cycles per second.

Micro gravity: A surface geophysical survey method in which measurements of the earth's gravitational field are made at various locations over an area of interest in order to associate variations in the field with differences in subsurface density distributions associated with changes in soil, rock, and cultural factors. Usually done on small spatial scale, measurements are usually made with small distance between stations and measuring very small variations in the gravitational field.

Mineralization: Increases in concentration of one or more inorganic constituents as the natural result of contact of ground water with geologic formations.

Mineralogy: The study of minerals, including their formation, occurrence, properties, composition, and classification.

Minimum value: In a set of data, the measurement having the lowest numerical value.

Miscible: Soluble in water to a significant degree; two or more fluids that are mutually soluble.

Mobilize: To accelerate the movement of a contaminant in the ground-water system by changing the prevailing chemical conditions. Examples: addition of acid will mobilize metals; addition of an organic solvent will mobilize oils or PCBs.

Model: A conceptual, mathematical, or physical system obeying certain specified conditions, whose behavior is used to understand the physical system to which it is analogous in some way.

Molality: A measure of chemical concentration. A one-molal solution has one mole of solute dissolved in 1000 grams of water. One mole of a compound is its formula weight in grams.

Molecular diffusion, coefficient of: The component of mass transport flux of solutes (at the microscopic level) due to variations in solute concentrations within the fluid phases. Also called diffusion coefficient.

Monitoring well: A well that is constructed by one of a variety of techniques for the purpose of extracting ground water for physical, chemical, or biological testing, or for measuring water levels.

Montmorillonite: A group of clay minerals that swell upon wetting and shrink upon drying; a clay mineral commonly used as an additive to drilling muds. The main constituent in bentonite.

Mounding: A phenomenon usually created by the recharge of ground water from a man-made structure into a permeable geologic material. Associated ground-water flow will be away from the man-made structure in all directions.

Mud: (*See* Drilling fluid).

Mud additive: Any material added to a drilling fluid to alter its chemical or physical properties.

Mud balance: A balance used to determine the density of a drilling mud.

Mud pit: Usually a shallow, rectangular, open, portable container with baffles into which drilling fluid and cuttings are discharged from a borehole, and that serves as a reservoir and settling tank during recirculation of the drilling fluids. Under some circumstances, an excavated pit with a lining material may be used.

Multiple level monitoring well: A single-hole monitoring well in which devices are installed that are capable of sampling at multiple levels within a formation or a series of formations.

Nanosecond: One billionth of a second (10^{-9}). This is the unit of travel time used in ground-penetrating radar to measure the travel time between the surface and the reflector of the EM wave and back to the surface.

National priority list (NPL): The U.S. EPA list of uncontrolled hazardous waste sites eligible for cleanup under CERCLA (Superfund).

Natural clays: Clays that are encountered when drilling various formations, whose yield may vary greatly, and that may or may not be purposely incorporated into the drilling mud.

Natural gamma log: A borehole geophysical log produced by a tool that measures the natural gamma radiation emitted by geologic formations. It can be used to delineate subsurface rock types.

Naturally developed well: A well in which the natural formation materials are allowed to collapse around the well screen, and fine formation materials are removed using standard development techniques.

Neat cement: A mixture of portland cement (ASTM C-150) and water in the proportion of 5 to 6 gallons of clean water per bag (94 pounds or 1 cubic foot) of cement.

Nested wells: A series of single-cased monitoring wells that are closely spaced, but with screens at different depths. Also used to refer to a set of multiple level wells constructed in a single borehole. Also called a well nest.

Neutralization: The inorganic reaction of an acid and a base to create a salt and water.

Neutron log: A borehole geophysical log obtained by lowering a tool containing a radioactive element, which is a source of neutrons, and a neutron detector into the well. The neutron log measures the amount of water present and, hence, the porosity of the formation. In the vadose zone, the neutron log measures moisture content. Also called a porosity log.

NGVD: National Geodetic Vertical Datum, commonly referred to as mean sea level.

Nominal diameter: A term used to describe standard sizes for pipe or casing from 1/8 in. to 12 in. (3.2 mm to 305 mm) in diameter, specified on the basis of the inside diameter. Depending on the wall thickness, the inside diameter may be less than or greater than the number indicated. As an example, 2-inch nominal Schedule 40 PVC casing has a standard outside diameter of 2.375 inches, and an inside diameter of 2.067 inches; 2-inch nominal Schedule 80 PVC casing has the same outside diameter, but has an inside diameter of only 1.939 inches.

Nondedicated sampling equipment: Equipment used to sample more than a single sampling point.

Nonequilibrium type curve: A plot on logarithmic paper of the well function $W(\mu)$ as a function of μ.

Nongraded: An engineering term pertaining to a solid or an unconsolidated sediment consisting of particles of essentially the same size.

Nonimaging methods: Remote sensing methods in which data are displayed in the form of a profile of variations along a flight path; for example, the variation of the earth's magnetic field. *See* Imaging methods.

Nonpoint source: A source of contamination in which the contaminant enters the receiving water in an intermittent and/or diffuse manner.

Nonpolar: A molecule possessing a uniform charge distribution, with no positive or negative end; opposite of polar.

Normal distribution: The character of data that follows the Gaussian distribution (bell) curve.

Nuclear logs: Downhole geophysical logs using instruments that measure either in situ natural radiation or the response of introduced radiation. For example, natural gamma, radiation, gamma-gamma (density), and neutron (porosity) logs.

Number of less-than-detection-limit values: The number of times a chemical parameter was not detected by a given analytical procedure over a statistically significant period of time (e.g., one year).

NWWA: National Water Well Association.

Observation well: A nonpumping well used to observe the elevation of the water table or the potentiometric surface. An observation well is generally of larger diameter than a piezometer and may be screened throughout the thickness of the aquifer.

Octanol-water partition coefficient: A coefficient representing the ratio of solubility of a compound in a nonpolar substance (octanol) to its solubility in a relatively polar substance (water). As the octanol-water partition coefficient increases, water solubility decreases, as does mobility in a ground-water system.

Oil air filter: A filter or series of filters placed in the air flow line from an air compressor to reduce the oil content of the air.

Oil trap: A device used to remove oil from the compressed air discharged from an air compressor.

Olefins: Organic compounds with one or more carbon-carbon double bonds in a chain or branched-chain structure. Examples include ethylene (ethene) and butadiene.

Open hole: The uncased portion of a well completed in bedrock that is open to the formation.

Organic: Being, containing, or relating to carbon compounds, especially in which hydrogen is attached to carbon, whether derived from living organisms or not; usually distinguished from inorganic or mineral.

Organic compound: Chemicals containing carbon, with the exception of carbon dioxide (CO_2) and carbonates (such as calcium carbonate, $CaCO_3$).

Organic polymers: Drilling fluid additives comprised of long-chained, heavy organic molecules used to alter the physical and chemical characteristics of the fluid.

Organic vapor analyzer: A field monitoring device used to determine the concentration of organic compounds in air using a flame ionization or photoionization detection system.

O-Ring seal: A rubber seal emplaced between the threaded connections of casing or hollow-stem auger sections to prevent leakage and infiltration of fluids.

Outcrop: An exposure of bedrock or strata projecting through overlying unconsolidated sediments or soils.

Outwash: Glacial drift deposited by meltwater streams beyond an active glacier.

Overburden: All geologic material (loose soil, sand, gravel, etc.) that lies above bedrock.

Overland flow: The flow of water over a land surface due to direct precipitation. Overland flow generally occurs when the precipitation rate exceeds the infiltration capacity of the soil, and depression storage is full.

Overpumping: A well development technique that alternately starts and stops a pump to raise and drop the column of water in the borehole in a surging action.

Oxidation: A chemical reaction in which there is an increase in valence resulting from a loss of electrons.

Oxidizing acid: An acid (e.g., HNO_3) that tends to lose electrons in a reaction.

Packer: A transient or dedicated device placed in a well or borehole that isolates or seals a portion of the well, well annulus, or borehole at a specific level.

Partial penetration: Describes the situation in which the screened portion of the well is less than the full thickness of the aquifer.

Partially penetrating well: A well constructed in such a way that it draws water directly from a fractional part of the total thickness of the aquifer. The fractional part may be located at the top or the bottom of the aquifer or anywhere in between.

Parts per billion (ppb): Unit weight of solute per billion unit weights of solution (solute plus solvent), corresponding to a weight-percent. Metric equivalent is micrograms per liter (liquid) or micrograms per kilogram (solid).

Parts per million (ppm): Unit weight of solute per million unit weights of solution (solute plus solvent), corresponding to a weight-percent. Metric equivalent is milligrams per liter (liquid) or milligrams per kilogram (solid).

PCBs: (*See* Polychlorinated biphenyls).

Perched ground water: Unconfined ground water separated from an underlying main body of ground water by unsaturated material.

Percolate: The water moving by gravity or hydrostatic pressure through interstices of unsaturated rock or soil.

Percolation: Downward movement of percolate under gravity or hydrostatic pressure.

Perennial stream: One which flows continuously. Perennial streams are generally fed in part by ground water.

Peristaltic pump: A low-volume suction pump. The compression of a flexible tube by a rotor results in the development of suction.

Permafrost: Perennially frozen ground, occurring wherever the temperature remains at or below 0°C for two or more years in a row.

Permeability: (*See* Intrinsic permeability).

Permeameter: A laboratory device used to measure the intrinsic permeability and hydraulic conductivity of a solid or rock sample.

pH: A measure of the acidity or alkalinity of a solution, numerically equal to 7 for neutral solutions, increasing with increasing alkalinity and decreasing with increasing acidity. Originally stood for the words potential of hydrogen.

Phreatic water: Water in the zone of saturation.

Piezometer: An instrument used to measure head at a point in the subsurface; a nonpumping well, generally of small diameter, that is used to measure the elevation of the water table or potentiometric surface. A piezometer generally has a short screen through which water can enter.

Piezometer nest: A set of two or more piezometers installed close to each other horizontally but screened to different depths; used to define vertical gradients.

Piezometric surface: The surface defined by the levels to which ground water will rise in tightly cased wells that tap an aquifer. *See* Potentiometric surface.

Pilot assembly: The assembly placed at the lead end of the auger, consisting of a solid center plug and a pilot bit.

Piston pump: A pump that consists of a piston rod, cylinder, and check valve. Piston pumps force water to the surface through positive displacement.

Plug, casing: A plug of drillable material made to correspond to the inside diameter of the casing. During cementing, plugs are pumped to the bottom of the casing to force all cement outside of the casing.

Plugging: The complete filling of a borehole or well with a low hydraulic conductivity material to prevent flow into and through the borehole.

Plumb: Vertical.

Plume: A body of contaminated ground water originating from a specific source and influenced by such factors as the local ground-water flow pattern, density of the contaminant, and character of the aquifer.

Plume characterization: A study that provides information on concentration profiles and rates of migration of a contaminant plume.

Point source: Any discernible, confined, and discrete source of contamination.

Polar: A molecule possessing an uneven distribution of charge, so that one end is predominantly positive and another end is predominantly negative. Water is polar; many oils are not.

Pollutant: Any substance that renders water unfit for a given use.

Polychlorinated biphenyls (PCBs): Fire-resistant organic fluids used in making plastics and as insulation in heavy-duty electrical equipment.

Polyethylene: A plastic composed of synthetic crystalline polymer of ethylene ($H_2C:CH_2$). Polymer may be low density (branched) or high density (linear).

Polymeric additives: The natural organic colloids developed from the guar plant that are used for viscosity control during drilling.

Polynuclear aromatic hydrocarbons (PNAs or PAHs): A large group of multi-ring organic compounds, some of which are Priority pollutants. Found naturally in heavy petroleum residues such as tars.

Polypropylene: A plastic composed of synthetic crystalline polymer of propylene (C_3H_5)n.

Polyvinyl chloride (PVC): A thermoplastic produced by combining PVC resin with various types of stabilizers, lubricants, pigments, filler, and processing aids often formulated to produce rigid well casing.

Pore: An opening, void, or interstice in a mass of soil or rock.

Pore water: Water that occupies an open space between solid soil particles. Also called interstitial water.

Porosity: The ratio of volume of void spaces in a rock or sediment to the total volume of the rock or sediment.

Porosity, primary: The void spaces that were created at the time a soil or rock unit was deposited.

Porosity, secondary: The void spaces (typically fractures or solution channels) that were created after the time a soil or rock unit was deposited.

Porosity log: (*See* Neutron log).

Portland cement: Cement specified under ASTM Standard C-150.

Potable water: Water that is suitable for human or animal consumption.

Potentiometric data: Ground-water surface elevations obtained at wells and piezometers that penetrate a water-bearing formation.

Potentiometric surface: An imaginary surface representing the static head of ground water. Where the head varies with depth in the aquifer, a potentiometric surface is meaningful only if it describes the static head along a particular specified surface of stratum in that aquifer. More than one potentiometric surface is required to describe the vertical distribution of head. The water table is a particular potentiometric surface. Used synonymously with piezometric surface.

Precipitate: Material that will separate out of solution as a solid under changing chemical and/or physical conditions.

Precipitation: The formation of solids out of constituents that were once dissolved. Precipitation is caused by a change in conditions, such as temperature, chemical concentration, or the presence of seed particles to begin the process.

Precision: The reproducibility or repeatability of a measurement. The closeness of each number of similar measurements to their arithmetic mean.

Preservation: Storage conditions or addition of a reagent that will minimize the change in concentration of a constituent of interest until analyses can be performed.

Preservative: An additive used to protect a sample against decay or spoilage, or to extend the holding time for a sample.

Pressure, hydrostatic: The pressure exerted by the weight of water at any given point in a body of water at rest.

Pressure grouting: A process by which a grout is confined within the borehole or casing by the use of retaining plugs or packers, and by which sufficient pressure is applied to drive the grout into the annular space or other zone to be grouted.

Priority pollutants: A group of approximately 130 chemicals (about 110 are organics) that appear on a U.S. EPA list because they are toxic and relatively common in industrial discharges.

Profiling: A surface geophysical survey technique in which the measuring system is moved over an area of measurement (usually along a line or grid) with the objective of determining how measurements vary with location. Applications include resistivity and electromagnetics. *See* Sounding.

Psychrometer: An instrument used to measure soil humidity by means of a thermocouple which is cooled below the dew point by means of the Peltier effect. The water on the thermocouple evaporates, causing the junction temperature to be depressed below the ambient temperature. The wet bulb temperature depression persists until all the water has evaporated; the thermocouple then returns to the ambient temperature.

PTFE: Polytetrafluoroethylene, one of several fluoropolymers.

Public water supply: A water supply system in which there is a vendor and customers; the vendor may be a private company, a municipality, or other governmental agency.

Puddled clay: A mixture of bentonite, other expansive clays, fine-grained material, and water, in a ratio of not less than 7 pounds of bentonite or expansive clay per gallon of water. It may be composed of not less than 50% expansive clay, with the maximum size of the remaining portion not exceeding that of coarse sand.

Pulling casing: To remove the casing from a well.

Pumping test: An aquifer test conducted by pumping a well for a period of time and observing the change in hydraulic head in observation wells. A pumping test may be used to determine the capacity of the well and the hydraulic characteristics of the aquifer.

Pumping water level: The elevation of the surface of the water in a well, or the water pressure at the top of a flowing artesian well, after a period of pumping at a specified rate.

Purge water: Water extracted from wells undergoing well evacuation.

Purging: (*See* Well evacuation).

PVC: (*See* Polyvinyl chloride).

QA/QC: Quality Assurance/Quality Control.

Qualified professional in geology: A professional, by degree, experience, or certification, specializing in the study of earth science or geology.

Quality assurance: A management function based on the establishment of quality control protocols, and on the evaluation and documentation of their outcomes. For example, an evaluation of the precision of a laboratory instrument based on the results of a series of duplicate sample analyses.

Quality control: Technical and operational procedures that investigate and confirm the proper conduct of all those field, sample transportation, and laboratory activities necessary to assure accuracy and precision in the data. For example, analysis of field blanks, duplicate samples, and laboratory spikes.

Radial flow: The flow of water in an aquifer toward a vertically oriented well.

Radioactivity: A characteristic of certain chemicals which emit particles (alpha or beta) or rays (gamma) at a rate specific to the particular substance.

Radioisotope: An unstable isotope of an element that decays or disintegrates spontaneously, emitting radiation. Also called radionuclide.

Radius of influence: The radial distance from the center of a well under pumping conditions to the point at which there is no lowering of the water table or potentiometric surface.

Rate of migration: The time ground water or a solute in ground water takes to travel from one stationary point to another. Generally expressed in units of time/distance.

RCRA: Resource Conservation and Recovery Act, passed by Congress in 1976 and administered by the U.S. EPA.

Reamer: A bit-like tool, generally run directly above the bit and used to enlarge and maintain a straight borehole.

Reaming: A drilling operation used to enlarge a borehole.

Recharge: The addition of water to the ground-water system by natural or artificial processes.

Recharge area: An area in which there are downward components of hydraulic head in the aquifer. Infiltration moves downward into the deeper parts of an aquifer in a recharge area.

Recharge basin: A basin or pit excavated to provide a means of allowing water to infiltrate into the ground at rates exceeding those that would occur naturally.

Recharge boundary: An aquifer system boundary that adds water to the aquifer. Streams and lakes are typical recharge boundaries.

Recovery: The rise in the water level in a pumping well and nearby observation wells after ground-water pumpage has ceased.

Redox: Abbreviation for an oxidation-reduction reaction in which an atom or molecule loses electrons to another atom or molecule. Oxidation is the loss of electrons; reduction is the gain in electrons. Also called oxidation-reduction.

Reduction: A chemical reaction in which there is a decrease in valence as a result of gaining electrons.

Regolith: The upper part of the earth's surface that has been altered by weathering processes. It includes both soil and weathered bedrock.

Regulated unit: A hazardous waste management unit. The boundaries of regulated units will define the extent of the hazardous waste management area.

Rehabilitation: The restoration of a well to its most efficient condition, using one or a combination of a variety of chemical and mechanical techniques.

Remote sensing: The process of obtaining information about the surface of the earth from aircraft or satellites. These are usually measurements of either natural radiation or radiation from a source that has been released from the earth. Also includes nonradiation measurements, such as those of magnetic field.

Representative sample: A sample that retains the chemical and physical character of the in-place ground water.

Resistance log: A borehole geophysical log made by lowering a single electrode into the well with another electrode at the ground surface; measures the overall electrical resistance of the formation and drilling fluid between the surface and the probe.

Resistivity: The electrical resistance offered to the passage of a current, expressed in ohm-meters; the reciprocal of conductivity.

Resistivity log: A borehole geophysical log made by lowering a tool with two current electrodes into the borehole and measuring the resistivity between two potential electrodes. It measures the electrical resistivity of the formation and contained fluids near the probe.

Resolution: The value of the smallest unit of measurement that may be distinguished using a particular instrument or method.

Retardation: Preferential retention of contaminants in the subsurface by one or more physical, chemical, or biological factors. Also called attenuation.

Rig: The machinery used in the construction or repair of wells and boreholes.

Riser: A casing (usually steel or PVC) that extends from the well screen or open section of the well to above the ground surface.

Root zone: The zone from the land surface to the depth penetrated by plant roots. The root zone may contain part or all of the vadose zone, depending upon the depth of the roots and the thickness of the vadose zone.

Rotary drilling: A common drilling method that is a hydraulic process using a rotating drill pipe, at the bottom of which is attached a hard-toothed drill bit. The drill cuttings at the bottom of the hole are carried up by the circulation of a fluid (drilling mud), down through the drill pipe, and up the annulus between the drill pipe and the borehole.

Rotary table drive: Hydraulic or mechanical drive on a rotary rig, used to rotate the drill stem and bit.

Runoff: That part of precipitation that flows to surface streams. Direct or overland runoff is that portion of rainfall which is not absorbed by soil, evaporated, or transpired by plants, but finds its way into streams as surface flow. That portion which is absorbed by soil and later discharged to surface streams is ground-water runoff.

RVCM: Residual vinyl chloride monomer.

Safe yield: The amount of naturally occurring ground water that can be economically and legally withdrawn from an aquifer on a sustained basis without impairing the native ground-water quality or creating an undesirable effect such as environmental damage. It cannot exceed the increase in recharge or leakage from adjacent strata plus the reduction in discharge, which is due to the decline in head caused by pumping.

Saline: Containing relatively high concentrations of salts, usually predominantly sodium chloride.

Saline-water (salt-water) encroachment: The movement, as a result of human activity, of saline ground water into an aquifer formerly occupied by fresh water. Passive saline-water encroachment occurs at a slow rate, due to a general lowering of the freshwater potentiometric surface. Active saline-water encroachment proceeds at a more rapid rate, due to the lowering of the freshwater potentiometric surface below sea level.

Salt: For inorganic substances, the combination of a cation and an anion. Can also be defined as the nonwater product of the reaction between an acid and a base.

Sample density: The number of samples or measurements made per unit area. Examples are the number of borings per acre, the number of geophysical measurements per acre, or the spacing between geophysical traverse lines.

Samples: Solid and/or liquid materials obtained from the borehole during the drilling and/or formation sampling process that provide geological information. Also, liquid materials obtained from a well and used to establish water quality of the zone in which the well screen is set.

Sampling and analysis plan: A detailed document describing the procedures used to collect, handle, and analyze ground-water samples. The plan should detail all quality control measures that will be implemented to ensure that sample collection, analysis, and data presentation activities meet any prescribed regulatory requirements.

Sandstone: A sedimentary rock composed of abundant rounded or angular fragments of sand, and more or less firmly consolidated by a cementing material.

Sanitary landfill: A solid waste disposal facility in which the disposal of solid and, in some instances, semisolid and liquid wastes is conducted by burying the material in trenches excavated to shallow depths, usually in unconsolidated materials.

Saturated zone: The zone in which the voids in the rock or soil are filled with water at a pressure greater than atmospheric. The water table is the top of the saturated zone in an unconfined aquifer.

Schedule: Refers to pipe and casing wall thickness; wall thickness increases as the schedule number increases.

Schlumberger array: A particular arrangement of electrodes used to measure surface electrical resistivity, in which electrodes are spaced in an exponential progression from the center point.

Screen: (*See* Well screen).

SDWA: Safe Drinking Water Act, passed by Congress in 1974 and administered by the U.S. EPA.

Seal: The low hydraulic conductivity material, such as cement grout, bentonite, or puddled clay, placed in the annular space between the borehole wall and the permanent casing to prevent the downhole movement of surface water or the vertical mixing of water from different water-bearing zones.

Secure landfill: An engineered and geologically suitable waste disposal facility, protected by liners and a leachate collection system.

Sedimentary rocks: Rocks resulting from the consolidation of loose sediment (clay, silt, sand, and gravel) that has accumulated in layers.

Sedimentation: The process of deposition of the solid suspended particles out of water, usually when the water has little or no movement.

Sediment sump: A blank extension of casing beneath the well screen used to collect fine-grained material from the filter pack and adjacent strata. Also called rat hole or tail pipe.

Seepage velocity: The actual rate of movement of fluid through porous media. This is the discharge velocity divided by the effective porosity.

Segregation: The differential settling of filter pack materials that occurs in the annular space surrounding the well screen during placement by gravity (free fall).

Seismic prospecting: Any of the various geophysical methods for characterizing subsurface properties based on the analysis of elastic waves artificially generated at the surface. *See* Seismic reflection and Seismic refraction.

Seismic reflection: A surface geophysical method used to measure the depth and thickness of soil and rock layers by recording the seismic waves that have been reflected from geologic strata.

Seismic refraction: A method of determining subsurface geophysical properties by measuring the length of time it takes for artificially generated seismic compressional waves to pass through the ground.

Sensitivity: Referring to a laboratory method, the smallest increment of concentration that can be distinguished. For example, being able to distinguish 1.1 ppm from 1.0 ppm implies greater sensitivity than being able to distinguish 1.5 ppm from 1.0 ppm.

Set casing: To install casing in a borehole.

Shale: A fine-grained sedimentary rock formed by the consolidation of clay, silt, or mud. It is characterized by finely laminated structure and is sufficiently indurated so that it will not fall apart on wetting.

Shear strength: A measure of the shear or gel properties of a drilling fluid or grout; also, the maximum resistance of a soil or rock to shearing stresses.

Shelby tube: A thin-walled tubular device pressed into an open borehole to obtain an undisturbed core sample of unconsolidated strata.

Sieve analysis: Determination of the particle-size distribution of soil, sediment, or rock by measuring the percentage of the particles that will pass through standard sieves of various sizes.

Significant digits: The number of digits reported as the result of a calculation or measurement (exclusive of following zeros).

Significant figures: The number of meaningful digits reported as the result of a calculation or measurement (exclusive of zeros used for decimal placement). For example, the number 0.020 has two significant figures, 20. The first two zeros are used for decimal placement.

Silt: A rock fragment or detrital particle smaller than a very fine sand grain and larger than a coarse clay, having a diameter in the range of 1/256 to 1/16 mm (0.00016 to 0.0025 in.).

Single-cased well: An individual monitoring well installed with a limited-length well screen that is used to monitor a specific zone of a formation.

Single-point resistance log: A borehole log made by lowering a tool with a single electrode into the well with a second electrode at the ground surface. It measures the overall electrical resistivity of the formation and drilling fluid between the surface and the probe. *See* Resistance log.

Sinkers: Dense, free-phase organic liquids that coalesce in an immiscible layer at the bottom of a saturated geologic unit.

Sink hole: A depression caused by the slumping or collapse of rock and soil due to a solution cavity. A typical feature of limestone terrain.

Slip-fit box and pin connections: A type of coupling used to join two hollow-stem auger sections.

Slotted coupling: A device attached to the knock-out plate at the base of the lead auger that allows water to pass into the center of the auger during drilling while preventing the entrance of sediment or sand into the hollow stem.

Slotted well casing: Well screens that are fabricated by cutting slots of predetermined width at regular intervals by machining tools.

Sludge: The solid residue resulting from a process or wastewater treatment operation which also produces a liquid waste stream (effluent).

Slug test: A single well test conducted to determine the in situ hydraulic conductivity of low to moderate hydraulic conductivity formations by the instantaneous addition of a known quantity of water or a solid, cylindrical object of known displacement into a well, and the subsequent measurement of the resulting well recovery.

Slurry: A thin mixture of liquid, especially water, and any of several finely divided substances, such as cement or clay particles.

Smectite: A commonly used name for clay minerals that exhibit high swelling properties and a high cation exchange capacity.

Sodium bentonite: A type of montmorillonite clay added to drilling fluids to increase viscosity. *See* Bentonite.

Soil bulk density (dry): The ratio of the mass of dried soil to its total volume.

Soil bulk density (wet): An expression of the total mass of moist soil per unit volume.

Soil moisture content: The amount of water in the vadose zone expressed as a fraction of the total porous medium volume occupied by water. It is less than, or equal to, the porosity.

Soil moisture potential (total potential): The amount of work that must be done per unit quantity of pure water in order to transport reversibly and isothermally an infinitesimal quantity of water from a pool of pure water, at a specified elevation and at atmospheric pressure, to a pool of water identical in composition to the equilibrium soil solution (at the point under consideration), but in all other respects, being identical to the reference pool.

Soil moisture tension: The equivalent negative pressure in the soil water. It is equal to the equivalent pressure that must be applied to the soil water to bring it to hydraulic equilibrium through a porous permeable wall or membrane with a pool of water of the same composition.

Soil salinity: The amount of soluble salts in a soil. The conventional measure of soil salinity is the electrical conductivity of a saturation extract.

Soil texture: Relative proportions of the various soil separates in a soil as described by the classes of soil texture. The textural classes may be modified by the addition of suitable adjectives when coarse fragments are present in substantial amounts; for example, "stony silt loam" or "silt loam, stony phase." The sand, loamy sand, and sandy loam are further subdivided on the basis of the proportions of the various sand separates present.

Soil water characteristic curve: An experimentally derived graph showing the soil moisture percentage (by weight or volume) versus applied tension (or pressure). Points on the graph are usually obtained by increasing (or decreasing) the applied tension or pressure over a specified range. Also called soil moisture retention curve.

Soil-water pressure: The pressure (positive or negative), in relation to the external gas pressure on the soil water, to which a solution identical in composition with the soil water must be subjected in order to be in equilibrium through a porous permeable wall with the soil water.

Solids concentration or content: The total amount of solids in a drilling fluid, as determined by distillation, that includes both the dissolved and the suspended or undissolved solids. The suspended solids content may be a combination of high and low specific gravity solids and native or commercial solids. Examples of dissolved solids are the soluble salts of sodium, calcium, and magnesium. Suspended solids make up the mudcake; dissolved solids remain in the filtrate. The total suspended and dissolved solids contents are commonly expressed as percent by weight.

Solid-stem auger: An auger with a solid center, a cutting head and continuous flighting, that is rotated by a rotary drive head at the surface and forced downward by a hydraulic pulldown or feed device.

Solubility: The total amount of solute species that will remain indefinitely in a solution maintained at constant temperature and pressure in contact with the solid crystals from which the solutes were derived.

Solubility product: The equilibrium constant that describes a solution of a slightly soluble salt in water.

Solute: A substance dissolved in a solution; the substance present in a solution in the smaller amount. For convenience, water is generally considered the solvent, even in "concentrated" solutions with water molecules in the minority.

Solute transport: The net flux of solute through a hydrogeologic unit controlled by the flow of subsurface water and transport mechanisms.

Solution: A homogeneous mixture of two or more components. In ideal solutions, the movement of molecules in charged species is independent of each other; in aqueous solutions, charged species interact even at very low concentrations, decreasing the activity of the solutes.

Solution channel: A tubular or planar channel formed by solution in carbonate-rock (karst) terrains.

Solvation: The degradation of plastic well casing in the presence of very high concentrations of specific organic solvents.

Solvent: Any substance that can dissolve another substance; the major component in a solution. In ground-water applications, water is usually the solvent, except where there is a layer of another liquid chemical which itself may be a solvent for certain constituents.

Solvent cementing: A method of joining two sections of casing in which solvent is applied to penetrate and soften the casing pieces and fuse the casing together as the solvent cement cures. Also called solvent welding.

Sorption: The combined effect of adsorption and absorption.

Sorting: The degree of similarity, with respect to some particular characteristic (typically grain size) of the component parts in a mass of material.

Sounding: A surface geophysical survey technique in which a series of measurements are made at one location, with the objective of determining how measurements vary with depth. Applications include resistivity and electromagnetics. *See* Profiling.

Specific capacity: An expression of the productivity of a well, obtained by dividing the rate of discharge of water from the well by the drawdown of the water level in the well. Specific capacity should be described on the basis of the number of hours of pumping prior to the time the drawdown measurement is made. It will generally decrease with time as the drawdown increases.

Specific conductance: The ability of a cubic centimeter of water to conduct electricity; varies directly with the amount of ionized minerals in the water. Measured in micromhos per centimeter.

Specific gravity: The weight of a particular volume of any substance compared to the weight of an equal volume of water at a reference temperature (usually 4°C). *See* Density.

Specific retention: The ratio of the volume of water that a rock or sediment will retain against the pull of gravity to the total volume of the rock or sediment.

Specific weight: The weight of a substance per unit volume. The units are newtons per cubic meter.

Specific yield: The ratio of the volume of water that a rock or soil, after being saturated, will yield by gravity drainage to the volume of the rock or soil.

Spectrophotometer: Optical instrument used to compare the intensities of the corresponding colors of two spectra.

Spiked sample: An environmental sample (usually soil or water) to which a known amount of a specific constituent has been added. Generally used as a means of checking laboratory performance.

Split-spoon sampler: A hollow, tubular sampling device driven by a 140-pound weight below the depth of drilling to retrieve representative samples of the formation.

Spontaneous potential log: A borehole geophysical log made by a tool that measures the natural electrical potential that develops between the formation and the borehole fluids.

Spring: A discrete place where ground water flows naturally from a formation onto the land surface or into a body of surface water.

Stagnant water: Water that has remained standing in the well between sampling rounds, and which is typically purged to obtain a sample of ground water.

Stagnation point: A place in a ground-water flow field at which the ground water is not moving. The magnitude of vectors of hydraulic head at the point are equal but opposite in direction.

Standard deviation: The positive square root of the variance. The variance is the average of the squares of the differences between the actual measurements and the mean.

Standard dimension ratio: A ratio expressed as the outside diameter of casing divided by the wall thickness.

Static water level: The elevation of the top of a column of water in a monitoring well or piezometer that is not influenced by pumping or conditions related to well installation, hydrologic testing, or nearby pumpage.

Station geophysical measurements: The practice of measuring and recording geophysical data at a designated station located over the area of interest, usually along a traverse line over a grid. *See* Continuous geophysical measurements.

Storage (aquifer): The volume of water held in the interstices of a water-bearing unit.

Storage, specific: The amount of water released from or taken into storage per unit volume of a porous medium, per unit change in head.

Storativity: The volume of water an aquifer releases from or takes into storage per unit surface area of the aquifer, per unit change in head. It is equal to the product of specific storage and aquifer thickness. In an unconfined aquifer, the storativity is equivalent to the specific yield. Also called storage coefficient.

Strata: Beds, layers, or zones of rock.

Stratification: The layer structure of sedimentary rocks.

Stratigraphy: The science (study) of original succession and age of rock strata, also dealing with their form, distribution, lithologic composition, fossil content, and geophysical and geochemical properties. Stratigraphy also encompasses unconsolidated materials.

Structural anomaly: A geologic feature, especially in the subsurface, distinguished by geophysical, geological, or geochemical means, that is different from the general surroundings.

Subsidence: Surface caving or distortion brought about by collapse of deep mine workings or cavernous carbonate formations, or from overpumping of certain types of aquifers.

Superfund: *See* CERCLA.

Surface geophysics: Those geophysical methods that involve making measurements of subsurface conditions using instruments located on the ground surface. These include measurements such as radar, electromagnetics, resistivity, magnetics, gravity, and seismic methods.

Surface impoundment: A facility or part of a facility that is a natural topographic depression, man-made excavation, or diked area made primarily of earthen materials (although it may be lined with man-made materials), which is designed to hold an accumulation of liquid wastes or wastes containing free liquids. Examples of surface impoundments are holding, storage, settling, and aeration pits, ponds, and lagoons.

Surface runoff: Water flowing across the soil surface into a channel.

Surface seal: The seal at the surface of the ground that prevents the intrusion of contaminants from the surface into a well or borehole.

Surface tension: Free energy in a liquid surface produced by the unbalanced inward pull exerted by the underlying molecules upon the layer of molecules at the surface.

Surface water: That portion of water that appears on the land surface, i.e., oceans, lakes, rivers.

Surfactant: A substance capable of reducing the surface tension of a liquid in which it is dissolved. Used in air-based drilling fluids to produce foam, and during well development to disaggregate clays.

Surge block: A plunger-like tool consisting of leather or rubber discs sandwiched between steel or wooden discs that may be solid or valved, and that is used in well development. *See* Surging.

Surging: A well development technique in which a surge block is alternately lifted and dropped within the borehole adjacent to the screen to create a strong inward and outward movement of water through the well screen.

Swivel: A hose coupling that forms a connection between the mud pump and the drill string, and permits rotation of the drill string.

Tamping rod: A heavy metal rod used to tamp annular sealants or filter pack materials into place and prevent bridging.

Target monitoring zone: In detection monitoring programs, the ground-water flow path from a particular area or facility in which monitoring wells will be screened. The target monitoring zone should be a stratum (strata) in which there is a reasonable expectation that a vertically placed well will intercept migrating contaminants.

Teflon®: Registered trademark of E. I. Du Pont de Nemours and Company, Inc., Wilmington, Delaware, for polytetrafluoroethylene.

Telescoping: A method of fitting or placing one casing inside another, or of introducing a screen through a casing having a diameter larger than the diameter of the screen.

Temperature survey: A method used to determine temperatures at various depths in the wellbore, typically used to ensure the proper cementing of the casing or to find the location of inflow of water into the borehole.

Tensile strength: The greatest longitudinal stress that a substance can bear without pulling the material apart.

Tensiometer: A device used to measure the in situ soil water matric potential (or tension); a porous, permeable cup connected to a rigid tube which is attached to a manometer, vacuum gage, pressure transducer, or other pressure measuring device.

Tension: The condition under which pore water exists at a pressure less than atmospheric.

Test hole: A borehole designed to obtain information on geological and/or hydrological conditions.

Test pit: A shallow excavation made by a backhoe to characterize the subsurface.

Texture: The interrelationship between the size, shape, and arrangement of minerals or particles in a rock.

Thermoplastic materials: Man-made materials often used for well casing that are composed of different formulations of plastics such as PVC, that are softened by heating and hardened by cooling, and can be easily molded and extruded.

Thin-wall sampler: A hollow tubular sampling device that is pressed into the formation below the drill stem to retrieve an undisturbed sample. *See* Shelby tube.

Thread protectors: A steel box and pin used to plug each end of a drill pipe when it is pulled from the borehole to prevent foreign matter or abrasives from collecting on the threads, and to protect threads from corrosion or damage while transporting or in storage.

Till: Predominantly unsorted and unstratified glacial drift, deposited directly by and underneath a glacier without subsequent reworking by meltwater, and consisting of a heterogeneous mixture of clay, silt, sand, gravel, and boulders ranging widely in size and shape.

Time domain: An instrumentation system that functions by taking measurements as a function of time.

TOC: (*See* Total organic carbon).

Top-head drive: A drive for the rotary drill string in which the bottom sub of the hydraulic drive motor is connected directly to the drill rod.

Tortuosity: Sinuosity of the actual flow path in a porous medium; it is the ratio of the length of the flow path divided by the length of the sample.

Total dissolved solids: The total concentration of dissolved constituents in solution, usually expressed in milligrams per liter.

Total number of values: The number of measurements (including less than detection values) made for a chemical parameter over a statistically significant period of time (e.g., one year).

Total organic carbon: A measure of the carbon present in a sample as part of organic compounds.

Total organic halogen: A measure of the concentration of organic compounds that have one or more halogen atoms.

TOX: (*See* Total organic halogen).

Toxicity: The ability of a material to produce injury or disease upon exposure, ingestion, inhalation, or assimilation by a living organism.

Trace element: Any element present in minute quantities in soil or water.

Tracer: A substance used to determine the flow rate and direction of water movement in a stream or aquifer.

Transducer: A device that is actuated by power from one system and retransmits it, often in a different form, to a second system.

Transformation: Process of establishing correspondence between elements in one set of data to elements in another set of data, so that each element in the first set corresponds to a unique element in the second set.

Transmissivity: The rate at which water of a prevailing density and viscosity is transmitted through a unit width of an aquifer or confining bed under a unit hydraulic gradient. It is a function of properties of the liquid, the porous medium, and the thickness of the porous medium. Also called coefficient of transmissibility.

Transpiration: The process by which plants give off water vapor through their leaves.

Transport: Conveyance of solutes and particulates in flow systems.

Tremie method: Method in which filter pack materials or bentonite/cement slurries are pumped uniformly into the annular space of the borehole through the use of a tremie pipe.

Tremie pipe: A pipe or tube that is used to transport filter pack materials and/or annular sealant materials from the ground surface into the borehole/casing annulus of a monitoring well.

Trip blank: A sample container filled in the laboratory with reagent-grade, distilled, or deionized water that is transported to the sampling site, handled the same as other samples, then returned to the laboratory for analysis as a quality control measure to check sample handling procedures.

T-Test: A statistical method used to determine the significance of difference or change between sets of initial background and subsequent parameter values. Also known as the Student's T-test.

Turbidity: Cloudiness in water due to suspended and colloidal organic and inorganic material.

Turbulent flow: That type of flow in which the fluid particles move along very irregular paths. Momentum can be exchanged between one portion of the fluid and another. Compare with laminar flow.

Unconfined: A condition in which the upper surface of the zone of saturation forms a water table under atmospheric pressure.

Unconfined aquifer: An aquifer that has a water table, and that is not bounded above by a bed of distinctly lower permeability than that of the aquifer.

Unconsolidated: Naturally occurring geologic materials that have not been lithified.

Underreamer: A bit-like tool with expanding and retracting cutters for enlarging a drill hole below the casing.

Undulation: A periodic rise and fall of a surface.

Unified soil classification system: A standardized classification system for the description of soils that is based on particle size and moisture content.

Uniformity coefficient: The ratio of the 60% finer (D 60) grain size to the 10% finer (D 10) grain size (effective size) of a sample of granular material (refer to ASTM Standard Test Method D 2487).

Uniformly graded: A quantitative definition of a particle size distribution of a soil that consists of the majority of particles being of the same approximate diameter. A granular material is considered uniformly graded when the uniformity coefficient is less than about 5 (refer to ASTM Standard Test Method D 2487). Synonymous with the geologic term well sorted.

Unit-price contracts: Contracts that establish a fixed price for materials and manpower for each unit of work performed.

Unsaturated zone: The zone between the land surface and the water table. It includes the root zone, intermediate zone, and capillary fringe. The pore spaces contain water at less than atmospheric pressure, as well as air and other gases. Saturated bodies, such as perched ground water, may exist in the unsaturated zone. Also called vadose zone and zone of aeration.

Upconing: The upward migration of ground water from underlying strata into an aquifer caused by reducing hydrostatic pressure in the aquifer as a result of pumping.

Upgradient: In the direction of increasing static head.

Upgradient well: One or more wells placed hydraulically upgradient of a site, that are capable of yielding ground-water samples representative of regional conditions, and that are not affected by activities at the site.

Uppermost aquifer: The geologic formation, group of formations, or part of a formation that contains the uppermost saturated zone capable of yielding a significant amount of ground water to wells or springs.

U.S. EPA: United States Environmental Protection Agency.

U.S.G.S.: United States Geological Survey, a part of the Department of the Interior.

Vadose zone: The zone containing water under pressure less than that of the atmosphere, including soil, water, intermediate vadose water, and capillary water. This zone is limited above by the land surface and below by the zone of saturation. Also called the unsaturated zone and zone of aeration.

Vented cap: A cap with a small hole that is installed on top of the well casing.

Vicksburg sampler: A structurally strong thin-walled sampler for use in stiff and highly cemented unconsolidated deposits.

Viscosity: The property of a fluid describing its resistance to flow. Units of viscosity are newton-seconds per meter squared, or pascal-seconds. Also known as dynamic viscosity.

Void: Space in a soil or rock mass not occupied by solid matter.

Volatile constituents: Solid or liquid compounds that are relatively unstable at standard temperature and pressure, and undergo spontaneous phase change to the gaseous state.

Volatile organics: Liquid or solid organic compounds that exhibit a tendency to pass into the vapor state.

Walking beam: The beam of a cable tool rig that pivots at one end while the other end connected to the drill line is moved up and down, imparting the "spudding" action of the rig. Also called a spudding beam.

Washout nozzle: A tubular extension with a check valve utilized at the end of a string of casing through which water can be injected to displace drilling fluids and cuttings from the annular space of a borehole.

Wastewater treatment system: A collection of treatment processes designed and built to reduce the amount of suspended solids, bacteria, oxygen-demanding materials, and chemical constituents in wastewater.

Water budget: An evaluation of all the sources of supply and the corresponding discharges with respect to an aquifer or a drainage basin.

Water-cement ratio: The amount of mixing water in gallons used per sack of cement.

Water content: The ratio of the volume of soil moisture to the total volume of the soil. This is the volumetric water content, also called volume wetness.

Water table: The surface in a ground-water body at which the pore water pressure is atmospheric. It can be measured by installing shallow wells extending a few feet into the zone of saturation and then measuring the water level in those wells.

Water-table aquifer: An aquifer containing water under atmospheric conditions.

Weep hole: A small diameter hole (usually 1/4 in.) drilled into the protective casing above the ground surface that serves as a drain hole for water that may enter the protective casing annulus.

Weight (of drilling fluid): A reference to the density of a drilling fluid. This is normally expressed in either lb/gal, lb/ft^3, or psi hydrostatic pressure per 1000 ft of depth.

Well: Any excavation that is drilled, cored, bored, washed, fractured, driven, dug, jetted, or otherwise constructed when the intended use of such evacuation is for the location, monitoring, dewatering, observation, diversion, artificial recharge, or acquisition of ground water, or for conducting pumping or aquifer tests.

Well cap: A removable, watertight apparatus or device used to cover a well casing.

Well capacity: The rate at which a well will yield water.

Well cluster: Two or more wells completed (screened) at different depths in a single borehole or a series of boreholes in close proximity to each other. From these wells, water samples that are representative of different horizons within one or more aquifers can be collected.

Well completion diagram: A record that illustrates the details of a well installation.

Well contractor: Any person, firm, or corporation engaged in the business of constructing, testing, developing, or repairing a well or borehole.

Well development: The act of repairing damage to the borehole caused by the drilling process and removing fine materials (silts, clays) from formation materials so that natural hydraulic conditions are restored and yields are enhanced.

Well evacuation: Process of removing stagnant water from a well prior to sampling.

Well intake: (*See* Well screen).

Well interference: The result of two or more pumping wells, the drawdown cones of which intercept. At a given location, the total well interference is the sum of the drawdowns due to each individual well.

Well log: A record that includes descriptions of geologic formations, and well testing or development techniques used in well construction.

Well nest: (*See* Nested wells).

Well point: A sturdy, reinforced well screen and casing structure that can be installed by driving it into the ground.

Well screen: A filtering device that allows ground water to flow freely into a well from the adjacent formation, while minimizing or eliminating the entrance of sediment into the well.

Well seal: A device used to cover a well, or to establish or maintain a junction between the casing or curbing of a well and the piping or equipment installed therein, to prevent contaminated water or other material from entering the well at the surface.

Well vent: An outlet at the upper end of the well casing to allow equalization of air pressure in the well.

Well yield: The volume of water discharged from a well; measured in units of gallons per minute or cubic meters per day.

Wenner array: A particular arrangement of surface electrodes used to measure subsurface electrical resistivity of soil and rock.

Wilting point: The soil-moisture content below which plants are unable to withdraw soil moisture.

Withdrawal: The pumping of water from a well or wells.

X-ray diffraction: An analytical technique used for mineralogical characterization. A sample is exposed to a filtered and monochromatic beam of X-rays and the reflected energy is measured and used to identify soil colloid types, degree of interleaving, or interstratification, and variations in interplatelet spacings.

Yield: The quantity of water per unit of time that may flow, or be pumped, from a well under specified conditions.

Yield point: A measure of the amount of pressure, after the shutdown of drilling fluid circulation, that must be exerted by the pump upon restarting of the drilling fluid circulation, to start flow.

Zone of aeration: The zone above the water table in which the interstices are partly filled with air, including the capillary fringe. Also called vadose zone or unsaturated zone.

Zone of potential contaminant migration: Any subsurface formation or layer that is permeable and would preferentially channel the flow of contaminants away from a site.

Zone of saturation: A hydrologic zone in which all the interstices between particles of geologic material or all of the joints, fractures, or solution channels in a consolidated rock unit are filled with water under pressure greater than atmospheric.

LIST OF CONTRIBUTORS

Ashley, James W. Department of Environmental Conservation, 103 S. Main Street, Waterbury, VT 05676

Ballestero, Thomas P., Water Resources Research Center, University of New Hampshire, Durham, NH 03824-3525

Benson, Richard C., Technos, Inc., 3333 N.W. 21st Street, Miami, FL 33142

Braids, Olin, Blasland, Bouck & Lee, 14216 Banbury Way, Tampa, FL 33624

Bradbury, Ken, Wisconsin Geological Survey, 3817 Mineral Point Road, Madison, WI 53705

Dalton, Matthew G., Dalton, Olmsted & Fuglevand, Inc., 22125 17th Avenue S.E., Suite 110, Bothell, WA 98021

Davis, H. E. "Hank," Mobile Drilling Company, P.O. Box 610, Aptos, CA 95001

Evans, O. D., Tracer Research Corporation, 3855 N. Business Center, Tucson, AZ 85705

Gibbons, Robert, Waste Management, Inc., 3003 Butterfield Road, Oakbrook, IL 60521

Herzog, Beverly L., Groundwater Resources Section, Illinois State Geological Survey, 615 East Peabody Drive, Champaign, IL 61820

Huntsman, Brent E., Terran Corporation, P.O. Box 1410, Fairborn, OH 45324

Jehn, James L., Jehn and Wood, Inc., 725 South Broadway, Denver, CO 80209

Kraemer, Curtis A., 111 Van Cedarfield Road; Colchester, CT 06415

Makeig, Kathryn S., Radian Corporation, 2455 Horsepen Lane, Suite 250, Herndon, VA 22071

Maslansky, Carol J., Geo-Environmental Consultants, 122 Saxon Woods Road, White Plains, NY 10605

Maslansky, Steven P., Geo-Environmental Consultants, 122 Saxon Woods Road, White Plains, NY 10605

Nielsen, David M., Nielsen Ground-Water Science, Inc., 4686 State Route 605 South, Galena, OH 43021

Nielsen, Gillian L., Nielsen Ground-Water Science, Inc., 4686 State Route 605 South, Galena, OH 43021

Pennino, James D., Minnesota Pollution Control Agency, Division of Ground Water, 520 Lafayette Road, St. Paul, MN 55155

Preslo, Lynne M., Earth Science Practice, ICF Kaiser Engineers, 160 Spear Street, San Francison, CA 94105

Sara, Martin N., Waste Management, Inc., 3003 Butterfield Road, Oakbrook, IL 60521

Schalla, Ronald, Site Characterization and Assessment Section, Environmental Assessment Department, Battelle Pacific Northwest Laboratories, P.O. Box 999, Richland, WA 99352

Schuller, Rudolph, ERM, Inc., 855 Springdale Drive, Exton, PA 19341

Sevee, John, Sevee & Maher Engineers, 4 Blanchard Road, Cumberland Center, ME 04021

Shultz, James A., EA Science & Technology, RD #2, Box 91, Goshen Turnpike, Middletown, NY 10940

Smith, Stephen, Geo West Group, 8669 East San Alberto Street, Suite 101, Scottsdale, AZ 85258

Stoner, David W., Stears and Wheler, 10 Albany Street, Casenovia, NY 13035

Thompson, Glenn, Tracer Research Corporation, 3855 N. Business Center, Tucson, AZ 85705

Vitale, Rock J., Environmental Standards, Inc., 1220 Valley Forge Road, P.O. Box 911, Valley Forge, PA 19481

INDEX